The Old English Complex Plant Names: A Linguistic Survey and a Catalogue

MÜNCHENER UNIVERSITÄTSSCHRIFTEN

Texte und Untersuchungen zur Englischen Philologie
herausgegeben von
Christoph Bode, Helmut Gneuss und Hans Sauer

Band 39

PETER LANG
EDITION

Ulrike Krischke

THE OLD ENGLISH COMPLEX PLANT NAMES: A LINGUISTIC SURVEY AND A CATALOGUE

PETER LANG
EDITION

Bibliographic Information published by the Deutsche Nationalbibliothek
The Deutsche Nationalbibliothek lists this publication in the Deutsche Nationalbibliografie; detailed bibliographic data is available in the internet at http://dnb.d-nb.de.

Zugl.: München, Univ., Diss., 2010

Library of Congress Cataloging-in-Publication Data

Krischke, Ulrike.
The Old English complex plant names : a linguistic survey and a catalogue / Ulrike Krischke. — Peter Lang Edition.
pages cm — (Münchener Universitätsschriften; Band 39)
Includes bibliographical references and indexes.
ISBN 978-3-631-64269-6 — ISBN 978-3-653-03045-7 (E-Book)
1. Plant names, Popular—English (Old) 2. Plants—Nomenclature. 3. English language—Old English, ca. 450-1100—Glossaries, vocabularies, etc. 4. English language—Old English, ca. 450-1100—Etymology. I. Title.
PE685.P5K75 2013
427'.02—dc23

2013021614

D 19
ISSN 0178-1383
ISBN 978-3-631-64269-6 (Print)
E-ISBN 978-3-653-03045-7 (E-Book)
DOI 10.3726/978-3-653-03045-7

Peter Lang Edition is an Imprint of Peter Lang GmbH.

Peter Lang – Frankfurt am Main · Bern · Bruxelles · New York · Oxford · Warszawa · Wien

www.peterlang.de

For Maja, Sarah and Stephan

ACKNOWLEDGEMENTS

This book is a thoroughly revised version of my doctoral dissertation, which was accepted by the Ludwig-Maximilians-Universität, München (LMU) in 2010. Though only my name appears on the cover of this book, it would not have been possible withouth a great many people that have lent their support and time. They deserve my special thanks.

I am indebted to Prof. Dr. Hans Sauer and to Prof. Dr. Ursula Lenker, my first and second supervisors, for their constructive criticism and practical advice, for commenting on my views and for helping me develop my ideas. Prof. Dr. Helmut Gneuss supported me with many insightful reflections, in particular with regard to the philological aspects in my study, and assistance with the textual tradition. My colleagues from the English and American department at the University of Munich (LMU), Renate Bauer, Susanne Gärtner, Susanne Handl and Maria Sutor always listened to my considerations and helped me focus my ideas.

My special thanks go to Kathleen Rabl, who assisted in polishing my sentences and looked closely at the final version of this study. I am also especially grateful to my parents for carefully reading the manuscript and detecting inconsistencies in chapter titles, footnotes, etc. In addition, I am truly indebted to my husband, who maintained the various private computers and notebooks I was working with so efficiently that I never had to worry about viruses, losing files, creating backups or installing software.

Many friends have helped me facing the various challenges in my research work. Their support and concern enabled me to overcome setbacks and stay focused on my doctoral studies. I greatly value their friendship and I deeply appreciate their belief in me. Most importantly, none of this would have been possible without the love and patience of my family. This dissertation is dedicated to them.

Ulrike Krischke

Munich, 20.04.2013

CONTENTS

LIST OF ABBREVIATIONS AND SYMBOLS

Abbreviations as they occur in quotations are not listed here. For the abbreviations for titles of medieval texts or reference works, see 'List of abbreviated references' on pp. 15-7.

LANGUAGES

Agn	Anglo-Norman French
Alem	Alemannic
Arm	Armenian
C	Celtic
CL	Classical Latin
Dan	Danish
Du	Dutch
Flem	Flemish
Fr	French
Fris	Frisian
G	German
Gk	Greek
Gmc	Germanic
Goth	Gothic
Icel	Icelandic
IE	Indo-European
It	Italian
L	Latin
LG	Low German
Lith	Lithuanian
MDu	Middle Dutch
ME	Middle English
MedL	Medieval Latin
MHG	Middle High German
MLG	Middle Low German
ModE	Modern English
Norw	Norwegian
nWS	non-West Saxon
OE	Old English
OFr	Old French
OFris	Old Frisian
OHG	Old High German
OIcel	Old Icelandic
OInd	Old Indic
OIr	Old Irish
OLG	Old Low German
ON	Old Norse
OS	Old Saxon
OSlav	Old Slavonic
OWN	Old West Norse
preGmc	pre Germanic
Prov	Provençal
Rum	Rumanian
Skr	Sanskrit
Sw	Swedish
VL	Vulgar Latin
WGmc	West Germanic
WS	West Saxon

GRAMMAR AND TEXT

acc.	accusative
Adj	adjective
art.	article
BL	British Library
c.	century, circa
cf.	confer
cm	centimetre(s)
deadj	deadjectival
decl.	declension
fem.	feminine
fig., figs.	figure, figures
fn.	footnote

fol., fols.	folio, folios
ft.	feet
gen.	genitive
lit.	literally
ll.	lines
m	meter(s)
masc.	masculine
ms.	manuscript
N	noun
n.	neuter
no.	number
nom.	nominative
Num	numeral
pl.	plural
pp.	pages
Pron	pronoun
r.	recto
RDP	recoverably deletable predicate
s.v., s.vv.	sub voce, sub vocibus
sg.	singular
str.	strong
subsp.	subspecies
V	verb
v.	verso
var.	varietas
vs.	versus
wk.	weak

SYMBOLS

>	develop(s) into
<	developed from
→	see also the entry (or entries)
←	translates, renders
*	reconstructed item; date of birth
^{e}	'early', i.e. attested in *Épinal-Erfurt Glossary*, in *Corpus Glossary* or in both
/ /	phoneme
†	obsolete, date of death
⊂	is a subset of
⊄	is not a subset of
=	is equal to
CONCEPTS	lexemes in small capitals refer to cognitive concepts and categories
[FRAME]	lexemes in small capitals put in square brackets refer to frames

LIST OF FIGURES

LIST OF ABBREVIATED REFERENCES

Ælfric's *Homilies*	*Ælfric's Catholic Homilies*, ed. Malcolm Godden (1979).
Ælfric's *Lives of Saints*	*Ælfric's Lives of Saints*, ed. Walter W. Skeat (1881-1900).
American Heritage Dictionary	*The American Heritage Dictionary of the English Language*, ed. Anne H. Soukhanov et al. (1992).
Anglo-Saxon Minor Poems	*The Anglo-Saxon Minor Poems*, ed. Elliott van Kirk Dobbie (1942).
Antwerp-London Glossaries	"The Latin-Old English Glossaries in Plantin-Moretus MS 32 and British Museum MS Additional 32, 246", ed. Lowell Kindschi (1955).
Arundel-Psalter	*Der altenglische Arundel-Psalter*, ed. Guido Oess (1910).
Bierbaumer 1	*Der botanische Wortschatz des Altenglischen, 1. Teil: Das Læcebōc*, ed. Peter Bierbaumer (1975).
Bierbaumer 2	*Der botanische Wortschatz des Altenglischen, 2. Teil: Lācnunga, Herbarium Apuleii, Peri Didaxeon*, ed. Peter Bierbaumer (1976).
Bierbaumer 3	*Der botanische Wortschatz des Altenglischen, 3. Teil: Der botanische Wortschatz in altenglischen Glossen*, ed. Peter Bierbaumer (1979).
Bodley 730	"Die unveröffentlichten Glossare der Hs Oxford, Bodley 730", ed. Waltraut Ziegler (1981).
Botanica	*Botanica*, ed. Gordon Cheers (1999).
Brockhaus	*Brockhaus – Die Enzyklopädie* (1996).
Brussels Glossary	*Anglo-Saxon and Old English Vocabularies*, eds. Thomas Wright and Richard Paul Wülcker (1884), vol. 1, no. 9.
BT	*An Anglo-Saxon Dictionary*, ed. Joseph Bosworth and Thomas N. Toller (1882-1989).
Charters (Birch)	*Cartularium Saxonicum*, ed. Walter de Gray Birch (1885-1899).
Charters (Earle)	*A Hand·Book to the Land·Charters, and other Saxonic Documents*, ed. John Earle (1888).
Chronicle (Bately)	*The Anglo-Saxon Chronicle*, ed. Janet M. Bately (1986).
Cleopatra Glossary (Quinn)	"The Minor Latin-Old English Glossaries in MS. Cotton Cleopatra A.III", ed. John J. Quinn (1956).
Cleopatra Glossary (Stryker)	"The Latin-Old English Glossary in Ms Cotton Cleopatra A iii", ed. William G. Stryker (1952).
ClH	*A Concise Anglo-Saxon* Dictionary, ed. J.R. Clark Hall (1960).
Codex (Kemble)	*Codex Diplomaticus Aevi Anglo-Saxonici*, ed. John M. Kemble (1839-1845).

CONE	*A Corpus of Narrative Etymologies from Primitive Old English to Early Middle English and A Corpus of Changes*, eds. Roger Lass et al (forthc.).
Corpus Glossary	*A Second Anglo-Saxon Reader, Archaic and Dialectal*, ed. Henry Sweet and T.F. Hoad (1978), 1-101.
DMLBS	*Dictionary of Medieval Latin from British Sources: A-Sal*, eds. Ronald E. Latham, David R. Howlett and Richard K. Ashdowne (1975-).
DOE	*Dictionary of Old English: A-G Online*, eds. Angus Cameron and Antonette diPaolo Healey et al. (2007).
DOEC	*Dictionary of Old English Web Corpus*, eds. Antonette diPaolo Healey, John Price Wilkin, Xin Xiang (2007).
DOEPN	*The Dictionary of Old English Plant Names*, eds. Peter Bierbaumer et al. (2007-).
Durham Plant Name Glossary:	*Das Durhamer Pflanzenglossar: Lateinisch und Altenglisch*, ed. Bogislav von Lindheim (1941).
Épinal-Erfurt Glossary	*Old English Glosses in the Épinal-Erfurt Glossary*, ed. Jason D. Pheifer (1974).
Etymologiae	*Isidorus Hispalensis*: *Etymologiae XVII*, ed. Jaques André (1981).
Exeter Book	*The Exter Book*, eds. George P. Krapp and Elliott van Kirk Dobbie (1936).
Genaust	*Etymologisches Wörterbuch der botanischen Pflanzennamen*, ed. Helmut Genaust (2005).
Gneuss *Handlist*	*Handlist of Anglo-Saxon Manuscripts*, Helmut Gneuss (2001).
Gneuss *Handlist*, add.	"Addenda and Corrigenda to the *Handlist of Anglo-Saxon Manuscripts*", Helmut Gneuss (2003).
Gneuss *Handlist*, second add.	"Second Addenda and Corrigenda to the *Handlist of Anglo-Saxon Manuscripts*", Helmut Gneuss (2011).
Gospels	*The Four Gospels in Anglo-Saxon, Northhumbrian, and Old Mercian Versions*, ed. Walter W. Skeat (1871-1887).
Harley Glossary	*The Harley Latin-Old English Glossary*, ed. Robert Oliphant (1966).
Heizmann	*Wörterbuch der Pflanzennamen im Altwestnordischen*, ed. Wilhelm Heizmann (1993).
Herbarium	*The Old English Herbarium and Medicina de Quadrupedibus*, ed. Hubert Jan de Vriend (1984).
Hermeneumata	*Corpus Glossariorum Latinorum a Gustavo Loewe Inchoatum*, Georg Goetz (1888-1923).
Holthausen	*Altenglisches Etymologisches Wörterbuch*, ed. Ferdinand Holthausen (1934).
Hortulus	"De cultura hortorum", ed. Eckehart Stöve [no date].

ICBN	*International Code of Botanical Nomenclature*, or *Vienna Code* (2005).
Ker *Catalogue*	*Catalogue of Manuscripts Containing Anglo-Saxon*, Neil R. Ker (1957).
Kluge	*Etymologisches Wörterbuch der deutschen Sprache.* Walter Kluge and Elmar Seebold (2002).
Lācnunga	*Anglo-Saxon Remedies, Charms, and Prayers from British Library MS Harley 585: The* Lacnunga, ed. Edward Pettit (2001).
Lǣcebōc	*Kleinere angelsächsische Denkmäler. 1. Das Lǣcebōc*, ed. Günter Leonhardi (1905), 1-120.
Laud Herbal Glossary	*The Laud Herbal Glossary*, ed. J. Richard Stracke (1974).
Leechdoms	*Leechdoms, Wortcunning and Starcraft of Early England*, ed. Oswald Cockayne (1864-1866).
Lewis and Short	*A Latin Dictionary*, Charlton T. Lewis and Charles Short (1879).
Marzell	*Wörterbuch der deutschen Pflanzennamen*, ed. Heinrich Marzell (1943-1979).
MED	*Middle English Dictionary*, ed. Hans Kurath et al. (1952-2001).
Meritt (1945)	*Old English Glosses: A Collection*, ed. Herbert D. Meritt (1945).
Meritt (1961)	"Old English Glosses, Mostly Dry Point", ed. Herbert D. Meritt (1961).
Napier (1900)	*Old English Glosses*, ed. Arthur S. Napier (1900).
Naturalis historia	*Naturalis historia*, ed. Karl F. Th. Mayhoff (1906).
ODEE	*Oxford Dictionary of English Etymology*, Charles T. Onions (1966).
OED	*The Oxford English Dictionary*, eds. James Murray et al. (1884-1928).
Old English *Heptateuch*	*The Old English Version of the Heptateuch*, ed. Samuel J. Crawford (1922).
Omont Fragment	"A Newly-Found Leaf of Old English from Louvain", eds. Bella Schauman, Angus Cameron (1977).
Peri Didaxeon	*Peri Didaxeon*, ed. Max Löweneck (1896).
Pokorny	*Indogermanisches Etymologisches Wörterbuch*, ed. Julius Pokorny (1959).
Stowe Psalter	*The Stowe Psalter*, ed. A.C. Kimmens (1979).
Tacitus	*De Origine et Situ Germanorum Liber*, ed. Henry Furneaux (1990).
TOE	*A Thesaurus of Old English*, Jane Roberts, Christian Kay and Lynne Grundy (1995).
VEPN	*The Vocabulary of English Place-Names* (1997-).
Zupitza (1880)	*Ælfrics Grammatik und Glossa*, ed. Julius Zupitza (1880).

PART A

A LINGUISTIC SURVEY

1. INTRODUCTION

For out of olde feldes, as men seyth,
Cometh al this newe corn from yer to yere,
And out of olde bokes, in good feyth,
Cometh al this newe science that men lere.
(Geoffrey Chaucer, *The Parliament of Fowls*, ll. 22-5)

The extant Anglo-Saxon texts testify to the existence of c. nine hundred Old English names for plants, including flowers, trees, shrubs and herbs; that is, nine hundred words relating exclusively to botanical species and life forms. This is an impressive number in view of the fact that the entire vocabulary that survived from the Anglo-Saxon period amounts to approximately 35,000 Old English lexemes, according to rough counts,[1] and this number can easily be augmented when all the terms denoting parts of plants, fruits, flowers, and collectivities of plants are also taken into account.[2]

The corpus of Old English plant names flaunts a rich variety of morphological and semantic patterns. Especially the morphologically complex plant names,[3] which make up about one half of the corpus, are an interesting field of research. Their particular interest results from the fact that they include a wealth of information on plants used in medieval times and their perception by the people, as well as on psychological processes and linguistic strategies that were used in the attempt to put 'nature' into words. Their analysis requires the interplay of the linguistic disciplines of lexical morphology, structural and cognitive semantics, of contact linguistics and, as they enjoy a special status as *natural kind terms*,[4] even touches upon questions of botany and socio-cultural history.

The present investigation is a qualitative analysis of the etymological, morphological and semantic history of the Old English complex plant names[5] as a lexical field and an in-depth study of the plant names as individual entities – an

1 For the estimate of c. 24,000 words see Scheler (1977: 74); more recently, the compilers of the *DOEC* and the *DOE* put the number of words at 35,000 (Healey 1997: 47, fn. 4).

2 The section 'A Plant' in the *TOE* lists c. 1264 terms, of which c. 150 are adjectives related to the plant world or derived from plant terms (e.g. *ānstelede* 'one-stalked', *æppelberende* 'carrying apples') and c. 35 are verbs related to the plant world or derived from plant terms (e.g. *spryttan* '(to) sprout'). Some of the terms appear twice or even more often, as they are polysemous and can be listed in several positions in the thesaurus.

3 Complex lexemes, i.e. lexemes that are made up of more than one lexical morpheme (Bauer 1983: 28-30) are distinguished from monomorphemic lexemes (or simplexes, monemes).

4 For the evaluation of the Old English plant names as natural kind terms, see chapter 2.

5 By 'complex' I understand, here, all compounds and syntactic groups, excluding derivatives, i.e. prefixations and suffixations or zero-derivations (cf. 1.3).

attempt at explaining their original *motivations* (or etymological meanings), their morphological shapes and semantic structures. Among the major research questions addressed are, thus, the following: What were the processes that governed the 'naming of names'[6]? What can be said about the extra-linguistic motives that triggered the creation of the plant names and in which way did psychological processes interact with linguistic processes? How can we describe the names' morphological structures for the individual plant names and for the lexical field as a whole, and to what extent can language contact account for the enrichment of this part of the Old English lexicon?

A (very) brief history of the naming of plants

The naming of plants started a long time before the Anglo-Saxons conquered the British Isles:[7] The first plant names developed when the first humans began to communicate about plants, accounting for the fact that all basic human necessities and activities such as cooking, house-building, or healing have relied ever since on plants and the plant world, on human knowledge of the plants and their properties. Evidence of the early human preoccupation with plants can be found not only in the earliest written documents, but also in some of the earliest pieces of art. Among the first plant names documented in writing are 202 Egyptian hieroglyphs dating to the times of the Pharaohs (Pavord 2005: 57); Mesopotamian paintings, around 3,000 BC, already document men's cultivation of date palms, vines, cereals and the like (Pavord 2005: 35).

The early botanical lexicons, rooted in the vocabulary of illiterate cultures, were enriched by scholars of various disciplines. In European medico-botanical history, the continuously growing number of herbals and medical treatises reflects the increasing interest in plants and their uses throughout the centuries before and after Anno Domini. The Greek scholar Theophrastus (*c. 370-†290 BC), one of the first botanists in Europe, and often dubbed the 'father of botany', described c. 500 plants in his *Historia plantarum*[8] (Pavord 2005: 21), and his compatriot, the Greek medic Pedanius Dioscorides (*40 AD-?) recorded in *De materia medica*[9] about three centuries later the morphological peculiarities and healing virtues of already as many as 600 plants. The scientific treatise *Naturalis historia*[10] of Dioscorides' Roman contemporary, the army commander, lawyer and naturalist Pliny (*23 AD-†79 AD), deals with the healing qualities of no less than 800 plants (Pavord 2005: 62).

The vigorous scientific activity in the field of botany was prompted by different motives. Theophrastus, a renowned scholar[11] and devoted botanist was searching for the order inherent in the world of plants, focusing on the descriptions and

6 In reference to Anna Pavord's (2005) work on the history of the naming of plants *The Naming of Names. The Search for Order in the World of Plants*.

7 For an interesting introduction into the history of botany and the naming of plants, see Pavord (2005); for a brief, but very informed overview, see Birkhan (2012).

8 Ed. by F. Wimmer (1842) as *Theophrastus Eresii Historia Plantarum*.

9 Transl. into English by Beck (2005).

10 Ed. by Karl F. Th. Mayhoff (1906).

11 Theophrastus was Aristotle's pupil and, on his death, his successor at the Peripatetic school in Athens (Pavord 2005: 49).

classifications of plants within the plant kingdom. By contrast, Dioscorides and Pliny, and on a smaller scale also the famous Greek physician and promoter of the doctrine of the four humours, Galen of Pergamon (*c. 129-†c. 200 AD), were interested in rather practical aspects of the plant world since they concentrated on the healing qualities of the medicinal plants. It is reported that especially Dioscorides, who worked as *medicus legionis* for presumably about 6,500-7,000 soldiers, had noticed a widespread ignorance of the therapeutical values of plants among his colleagues and therefore attempted a systematic description of medical plants in a way that guaranteed their quick identification and a quick preparation of remedies (Pavord 2005: 73). From Dioscorides and Pliny onwards, plants became an important interest of the intellectual elite, not necessarily as an object of interest in their own right, but rather with regard to their potential in curing sick people and in alleviating ailments.

The herbal and medical texts of the early Middle Ages clearly follow the antique tradition, albeit with alternations and additions, allowing at times for a different flora due to different climatic conditions as well as for cultural differences between the Graeco-Roman world and the peoples receiving these texts. The antique *materia medica* was retained, but modified by the needs of the early medieval world, and tended more towards providing practical advices in the form of prescriptions and charms than towards retaining the more "textbook approach to medical botany" (van Arsdall 2002: 71) that the antique authors had used.

The early medieval medical texts in Latin were written, copied, or compiled in learned, often monastic environments to the end of serving as handbooks for physicians. Characteristically, the entries in these texts are arranged according to plants, give synonyms and provide descriptions of the plants' preferred habitats and their outward appearances as well as accounts of their therapeutic values. Often, they are accompanied by stylised pictures. The medico-botanical texts of the Anglo-Saxons written in the vernacular,[12] such as the Old English *Herbarium*, the *Lācnunga* or the *Lǣcebōc*, fit quite neatly into the early Continental scheme of (Latin) medico-botanical texts (cf. chapter 4.). They are integral parts of the pan European medical tradition, which was equipped with a common corpus of texts and promoted an equally common corpus of medical lore:

> Small parts of them could be unique to Anglo-Saxon England, or to another locale, and there might be good reasons for later scholars to try to isolate what may be uniquely Anglo-Saxon, but the isolated pieces become fully comprehensible only when considered as parts of a larger medical continuum – essential pieces in a large and complex puzzle (van Arsdall 2002: 75).

---

12 At least two compilations containing medical texts in Latin can be assigned clearly to the Anglo-Saxons: those in the *Canterbury Classbook* (in manuscript Cambridge, University Library, Gg. 5. 53, middle of the 11th century; origin and provenance Canterbury; Gneuss *Handlist*, no. 12), and in the *Ramsey Scientific Compendium* (manuscript Oxford, St John's College, 17 and British Library, Cotton Nero C. vii, fols. 80-8; Ker *Catalogue*, no. 360); for an overview, see M.L. Cameron (1983a, 1983b, 1993) and chapter 4.5.

The tenor these Anglo-Saxon texts share is thus certainly the promotion of medical knowledge, yet not on principle the promotion of names of plants – the large number of botanical terms these texts have passed down to us is rather a more than welcome side-effect, prompted by the fact that early medieval medicine relied heavily on herbal ingredients in the preparation of drugs and potions (M.L. Cameron 1993: 101).

What is so special about the Anglo-Saxon medico-botanical texts is their use of the vernacular in a time when the Continent still wrote in Latin, the medieval *lingua franca* of learning and education: Bald's *Leechbook* (the first two parts of the *Lǣcebōc*; see 4.3.1.), for instance, "is (together with the accompanying third book of recipes) the oldest English medical work to survive in anything like a complete form, and it is also the oldest to survive in a European language other than Greek or Latin" (Cameron 1983a: 153), and the Old English *Herbarium* is the first translation of a medico-botanical text into a European vernacular (d'Aronco 2008: 113).[13] Thus, very early, the texts produced on the British Isles recorded ample medico-botanical terminology in the vernacular language of the Anglo-Saxons, Old English, and, simultaneously, testify to its importance:

> Surviving manuscripts provide firm evidence for the role of the vernacular. Anglo-Saxon culture was a bilingual one [...]. The gradual dismissing of Latin in favour of the vernacular or the parallel use of both languages is evident in many 'instructional' fields, such as computus and medicine, including herbaries (Lendinara 2007: 69-70).

The results of the Anglo-Saxons' use of their vernacular Old English in the field of botany can be observed not only in the number of manuscripts that have survived and in the number of those whose existence can be assumed on fairly reasonable grounds (cf. 4.3.), but also in the large number of botanical terms recorded in these documents. In its European context, the Old English botanical lexicon is exceptional in being the earliest one, and with c. 900 plant names (cf. Bierbaumer 1975-1979), also comparatively large. From Old West Norse for instance, the language spoken in Scandinavia, Iceland and on the Faroe Islands between c. 1200 and 1540, i.e. significantly later than Old English, only 336 plant names survive (Heizmann 1993: xvii).[14]

---

13 The first written records of Old English are dated to the 7th century, e.g. the runic inscriptions on the *Frank's Casket* and the *Épinal Glossary*. Among the first written documents of Old French and Old High German languages are the *Strassburg Oaths* (*Les Serments de Strasbourg*; ed. by Koschwitz 1964: 2-3) c. 842, recording Old French and Old High German, the Old French hagiographic text *Sequence of Saint Eulalia* (*La Séquence de Sainte Eulalie*), around 880 (ed. by Koschwitz 1964: 4-5), the 8th century Old High German *Hammelburger Markbeschreibung* and the *Würzburger Markbeschreibung*, ed. by Steinmeyer (1916; cf. Hutterer 1975: 306). The textual recording of Old Low German starts in the 9th century; the best-known texts are the Old Saxon *Heliand* and the Old Saxon *Genesis*, both from the ninth century (Gallée 1993: 3ff.).

14 The number 336 relates to the native botanical vocabulary only. In his *Wörterbuch der Pflanzennamen im Altwestnordischen*, Heizmann records in total c. 550 plant names, 214 of which are foreign words (i.e. mainly Latin words not assimilated into the vernacular; cf. Heizmann 1993: xvii).

1.1. AIMS AND METHODS

Although the botanical lexicon of the Anglo-Saxons has been the focus of linguistic research before, it has never been analysed in its entirety.[15] Valuable research was done either on individual linguistic aspects such as, for instance, on the extent of loan influence in the Old English *Herbarium* (d'Aronco 1988) or on various linguistic facets when taking into account only smaller text corpora (Sauer 1992b, 1995, 2000, etc.). However, a linguistic analysis of the entire corpus of Old English plant names, including morphological and semantic features, and also taking into account their diachronic stratification, has remained a research lacuna that the present survey aims to address. The possible outcome is, indeed, intriguing, since the systematic study of the information encoded in the Old English plant names may provide insights into important aspects of Anglo-Saxon culture and civilisation, as well as this language community's linguistic capabilities, while contributing also to the understanding of the texts in which this vocabulary was used.

> It is the basic function of lexemes to serve as labels for segments of extralinguistic reality that for some reason or another a speech community finds nameworthy. [...] Looked at from this point of view, the vocabulary of a language is as much a reflection of deep-seated cultural, intellectual and emotional interests, perhaps even of the whole *Weltbild* of a speech community as the texts that have been produced by its members (Kastovsky 1992: 291).

Cultural, intellectual and emotional interests, as well as *foci* of human observation can be disclosed in semantic-cognitive analyses of the plant names (see chapter 6.). Recurring attributes encoded in plant names are, e.g. specific colours or shapes observable in plants: compare for instance, the use of the colour-relating adjectives in the plant names *brūn(e)wyrt* 'brownwort, water-betony, etc.' lit. 'brown wort/plant', *rēade wudufille* 'asparagus' lit. 'red wood-chervil' and *hollēac* '? shallot' lit. 'hollow leek'.[16]

---

15 Chapter 3 provides an overview of previous scholarship on the botanical lexicon and the botanical texts.

16 The Old English complex plant names are all listed in the Catalogue (part B), where they appear with a literal translation, their scientific botanical identification(s), their Modern English equivalents and their names in German. Troughout this volume, that is, in the main body of text, the Old English plant names are only accompanied by their Modern English equivalents and a literal translation of the plant name where it is necessary for the understanding. When an Old English plant name is polysemous, that is, when it denotes more than one plant, the first plant in the Catalogue is listed and the fact that it is polysemous is marked by 'etc.'. For instance, OE *ācmistel* denotes (i) 'Loranthus europaeus L. / oak mistletoe / Europäische Riemenblume' or (ii) 'Viscum album L. / mistletoe / Weiße Mistel'; in the main body of text, *ācmistel* is accompanied by the information 'oak mistletoe, etc.' lit. 'oak-mistletoe'. When the reader is advised to consult the Catalogue for further details on an Old English plant name, it is marked as follows: see Catalogue, s.v. *ācmistel*.

These attributes are not always encoded directly as in the examples given above. As *hræfnes fōt* 'crowfoot, buttercups' lit. 'raven's foot' demonstrates, metaphorical allusions – here: to the shape of the leaves, which are compared to raven's feet (see Catalogue, s.v. *hræfnes fōt*) – are often employed in the naming process. In-depth insights into the extensive knowledge of medical and botanical issues of the Anglo-Saxons allow plant names such as *bānwyrt* 'comfrey' lit. 'bone-wort', and *āt(t)orlāþe* 'betony, etc.' lit. 'poison-harm', which encode the medical benefits of comfrey, respectively of betony. For worn-out shanks, for instance, it is recommended to apply a mixture of *bānwyrt* and the white of egg: "Gif scancan synd forode nim banwyrt, gecnuwa, geot æges þæt hwite, meng tosomne scancforedum men" (*Lǣcebōc*, 21/16-7) ['When the shanks are worn out, take *bānwyrt*, pound it, take the white of egg, mix [them] together [and apply it] on the man with the worn-out shanks']. OE *āt(t)orlāþe* is thought to be an ailment against poison-induced sicknesses, as the plant name, lit. 'poison-harm', implies (see Catalogue, s.v. *āt(t)orlāþe*).

In the corpus of c. 900 Old English botanical terms, morphologically simple and morphologically complex plant names balance each other with 467 complex and c. 430 simple plant names (i.e. which are, at least, synchronically simple). Of these two groups, complex plant names in particular feature a wealth of the aforementioned information: from the initial perception of the plants to the coding of specific features, from psychological processes to morphological and semantic strategies wrought in the plants' names. The focus of the present study is, therefore, centered on plant names that are compounds or syntactic groups (excluding simplexes and derivatives).[17]

The investigation of complex lexemes offers insights into the linguistic creativity of the early English language and provides data for a study of Old English word-formation, a desideratum often called for (e.g. Gneuss 1993: 139; Johnston 2011: 67). Clothing extra-linguistic reality in complex words is not only an extremely economic linguistic procedure, but also a most creative process, and often, morphological processes, figurative language use and loan influences combine in one plant name. The metaphorical plant name *dæges ēage* 'daisy', for instance, links processes of word-formation with figurative language use and reveals, simultaneously, the impact the medieval *lingua franca* Latin had on the Old English botanical lexicon: *dæges ēage* is a combination of two nouns, the first of which is in the genitive case, {*dæg*}+{*es*} 'day's' + {*ēage*} 'eye'. The plant name is based on a relation of metaphorical similarity between the plant's ray of petals opening at daybreak and a person's eye opening in the morning. That is to say, the botanical information on the most salient parts of the plant, namely, the plant's habit of opening at sunrise and closing at sunset is encoded metaphorically, via the comparison with people opening their eyes at daybreak, after sleeping, and closing them in the evening. Beyond that, as the Latin plant names *sōlis oculus* and *oculus diēī* suggest, *dæges ēage* was probably created on these Latin lemmas (although it never glosses one of them in the extant Old English texts; see Catalogue, s.v. *dæges ēage*). Thus, complex Old English plant names such as OE *dæges ēage* prove to be not only veritable mines of semantic and morphological

17 On the corpus of plant names investigated in the present survey, see further 1.3.

information, but also offer themselves to be examined for the influences on Old English exerted by other languages such as Latin.

A synchronic-diachronic approach

The chief purpose of the present survey is two-fold: First, to describe the synchronically transparent and morphologically analysable patterns of the Old English complex plant names and to trace their etymological histories. Second, to relate the insights gained in this process to one another and to identify major morphological and semantic patterns that show up in the lexical field of Old English complex plant names as a whole. These aims require a synchronic-diachronic approach.

The corpus of complex plant names is selected synchronically from within the lexical corpus of Old English and the morphological analysis is carried out within the system of Old English, that is, synchronically, too. The question of whether it is justified to speak of 'synchrony' in view of about 600 years of language use and, of course, of language development is a matter of dispute; yet, this concession must be made in view of the textual situation of the Anglo-Saxon period, which is characterised by the fact that the majority of manuscripts recording Old English prose and poetry that survived were written in the late tenth and in the eleventh century in the variety termed (late) West Saxon, while at least some of the texts are known to have been composed earlier or in a different dialect area.[18] The diachronic dimension is, nevertheless, touched upon in view of morphological particularities such as morphological variation, processes of attrition and doublets, in addition to the discussion of the diachronic stratification of the complex botanical terms from Indo-European to Old English (chapters 7.).

The semantic analysis is partly diachronic, partly synchronic. The synchronic part attempts to discover the meaning of the component parts of the plant names in Old English, while the diachronic part investigates the etymological meanings or *motivations* of the plant names. As plant names are usually not arbitrary in origin but refer to some attributes of the plants so named, it may be asked, on the one hand, which features of the plant have contributed to the plant's name, and, on the other, whether the motives behind a particular plant name were transparent or not to the speakers of Old English. This requires not only some knowledge about the historical or botanical facts under which the names were given, such as language contact, the educational background of the compilers or scribes and the morphology of the plants, but also depends on diachronic linguistic processes.

The diachronic angle is taken in all cases where processes of attrition have impaired the morphological and/or the semantic make-up of the (original) plant names, or where processes of folk etymology have changed the original form and/or meaning, adding further dimensions, such as a secondary, phonological or semantic motivation. It must be noted, however, that even when plant names have become synchronically completely opaque with regard to their motivation, as in the case of *wermōd* (see

18 This does not mean that these texts were received only in Wessex: "texts in this late West Saxon were written and read in other parts of the country too, in Kent (Canterbury), in Mercia (Worcester) and indeed even in Northumbria (York)" (Gneuss 1972: 63); on the origins of 'Standard Old English', see Gneuss (1972).

Catalogue, s.v. *wermōd*, and 7.5.), their denotation – in this case 'wormwood' – still can be perfectly discernible to the speakers. In most cases, however, the question of whether the reasons lying behind a particular name were semantically transparent or opaque (despite morphological analysability) to a majority of Anglo-Saxon speakers can be answered only tentatively.[19]

Semasiology & Onomasiology

As one of the central issues of the present survey is also to trace the etymological history of the plant names, i.e. to find the non-linguistic attributes of the plants encoded in the plants' names (their *motivation*), the analysis oscillates, at times, between a semasiological and an onomasiological perspective. While the semasiological approach starts from the Old English complex plant names and proceeds to the morphological and semantic analysis of the individual items, the onomasiological approach is concerned with botanical issues first.

The onomasiological approach asks, for instance, how identical or similar features of plants can be encoded differently in the names (cf. chapter 6.5.). To give an example: The plant prickly cnicus is denoted by *smælþistel* lit. 'small thistle' and by *unbrāda þistel* lit. 'not broad thistle'; the plant pennyroyal is called *dweorged(w)os(t)(l)e* lit. 'dwarf's *dost*'. Interestingly, while all three plant names verbalise the motive 'narrowness, smallness' in the first element, they do it quite differently. OE *smælþistel* expresses the motive directly, via the adjective *smæl* 'small', thus implying a contrast between the genus *þistel* 'thistle' and the species *smælþistel*. From this follows yet another implication, as, simultaneously, the (vertical) polysemy of *þistel* becomes obvious: On the one hand, *þistel* denotes the super-ordinated genus and, on the other, a sub-ordinated taxonym of the genus, the species '*þistel* with broad leaves', a co-hyponym of *smælþistel*. The synonym *unbrāda þistel* conceptualises the negation of the antonym of *smæl*, namely, *unbrād* 'not broad' as modifying *differentia specifica*. Both synonyms, though similar in motive ('narrow leaves' + 'a kind of thistle') and morphology (Adj+N), thus, make use of different cognitive procedures. Yet another different approach is exemplified in the verb-alisation of 'smallness' in *dweorged(w)os(t)(l)e*: here, the smallness of the plant is echoed in the metaphorical association of the plant with a *dweorg* 'dwarf' (see Catalogue, s.vv. *smælþistel*, *unbrāda þistel*, *dweorged(w)os(t)(l)e*).

Different paths of Old English plant name genesis can be pursued when language contact comes into play. The foreign plant name is borrowed into Old English along with its denotatum (e.g. *pīn* 'pine' < L *pīnus* 'pine'), or is decomposed along the language rules of the source language and recomposed with native language means, thus producing loan formations (or calques; 7.4.3.). In the first case, the borrowed word may be perceived as alien and efforts are made to integrate the loan into the receiving language system. This can even lead to phonological and semantic re-motivation, as happened to, e.g., L *levisticum*, which is not a transparent lexeme in the system of Old English and was, therefore changed into OE *lufestice* lit. 'love-stich'

---

19 For some information on the people involved in the production and use of the Anglo-Saxon medico-botanical texts, see chapter 2.2.2.

'lovage' (see Catalogue, s.vv. *pīn*, *lufestice*; chapter 7.5.). Loan formations, on the other hand, can display various degrees of morphological closeness to the terms they render.

A bipartite structure

To achieve the aims presented above, the plant names have to be 'attacked' from different linguistic and non-linguistic perspectives.[20] The *motivations* underlying the plant names, for instance, are identified on the basis of natural scientific and botanical knowledge, while the *associative relations* pertaining between the motivations and denotata of the elements of the plant names are analysed on the basis of findings in the discipline of *gestalt* psychology.[21] Again, morphological and etymological issues are observed from a linguistic perspective.

Therefore, the survey is bipartite: the first part, part A, is concerned with the theoretical outlines, while the second part, part B, consists in a detailed investigation of all complex plant names in the corpus.

The summary account of the previous research on Old English plant names (chapter 3), including a brief introduction into the nature of the Anglo-Saxon botanical texts and the various circumstances in which the Anglo-Saxons came into contact with plants (and, hence, with their names), is followed by the theoretical part, which falls into the categories 'morphology and word-formation' (chapter 5), 'semantics' (chapter 6) and 'diachronic stratification & the effects of language contact' (chapter 7). Different methodological approaches to the subject are presented and the methods of analysis are discussed in view of their applicability to the corpus of Old English complex plant names. The theoretical part not only forms the background for the detailed analysis of the Old English plant names but also draws on findings gained there and discusses them in evaluative synopses within each chapter, but also summarily in chapters 8 (with regard to the synonyms *bēam* and *trēow*) and 9.

The second part of the survey, the Catalogue (part B), provides the individual analysis of all the plant names in the corpus. Following the theoretical framework established in the first part, the plant names are accompanied by the following information (see fig. 1): a literal translation of the Old English name; the denotatum according to the modern scientific nomenclature and the Modern English and Modern German plant names; the number of occurrences and, if necessary, comments on the identification; illustrations in Anglo-Saxon manuscripts; Latin equivalents and Old English synonyms in the Anglo-Saxon manuscripts; the etymology; information on the morphological shape and structure; the motivation; the associative relations existing between the plant concepts and the elements of the names and, finally, cross-references.[22]

20 In allusion to Firbas' statement that "Language is a formidable fortress that must be attacked from all sides" (Firbas 1992: xii).

21 Gestalt psychology is the branch of psychology that investigates human categorisation systems and aims at defining general principles of human perception (*Brockhaus*, s.v. *Gestaltpsychologie*).

22 For more details, see the introduction to the Catalogue (part B).

bān-wyrt, lit. 'bone-wort'
(i) 'Bellis perennis L. / daisy / Gemeines Gänseblümchen'; (ii) 'Centaurium umbellatum Gilibert / common centaury / Tausendgüldenkraut'; (iii) 'Symphytum officinale L. / comfrey / Große Beinwurz, Beinwell'; (iv) 'Viola L. / violet / Veilchen'
DOE (s.v. *bān-wyrt*): 23 occ. ("in medical recipes and glossaries").
ILLUSTRATION in manuscript London, British Library, Cotton Vitellius C. iii, fol. 62r (Ohlgren 168(84)).
LATIN EQUIV. and OE synonyms in the manuscripts: *centauris* (with *minor*), *consolida*, *viola* (1x with *herba*), *viola aurosa*, *viola purpurea*; *clǣfre*, *hōfe*.
ETYMOLOGY: WGmc compound (Carr 1939: 96); OHG *beinwurz* (cf. Björkman 1901: 207). MORPHOLOGY: N+N; endocentric, rectional. MOTIVATION: illness & botanical information; Symphytum officinale L. is said to heal broken bones; cf. also OHG *beinuuella*, *beinwalla*, G *Beinheil*, *Beinbruchwurzel*, etc.; L *consolida minor* (Marzell, s.v. *Symphytum officinale* L.); see also →*brȳsewyrt*.
ASSOCIATIVE RELATIONS: conceptual contiguity & taxonomic sub-ordination.
▫ →*bān*, →*wyrt*
▫ →*grēate bānwyrt*

Fig. 1: Example of an entry in the Catalogue: *bānwyrt*.

The two parts of the survey relate to each other and are conceived as two complementary approaches to the same corpus: the theoretical deliberations and the ensuing analysis of the plant name corpus as a whole draw on the results of the detailed analysis of the individual plant names in the second part of the survey. The survey is concluded with three indexes: an *Index of word forms*, an *Index of the botanical names* with their Old English counterparts and an *Index of subjects*.

1.2. THE ANALYTICAL FRAMEWORK

Lexicogenesis – the genesis of (a part of) a lexicon – includes, according to Geeraerts (2002: 4), all

> […] mechanisms for introducing new pairs of word forms and word meanings – all the traditional mechanisms, in other words, like word formation, word creation (the creation of entirely new roots), borrowing, blending, truncation, ellipsis, folk etymology and others, that introduce new items into the onomasiological inventory of a language.[23]

Of the various attempts at classifying these mechanisms into one coherent system, none is universally accepted and, in general, they are treated separately. In semasiological approaches to meaning, semantic change in individual lexemes or in lexical fields are discussed under the heading *semantics*, while the combination of morphemes is confined to the sub-discipline *lexical morphology* and the enlargement of the lexicon resulting from language contact (e.g. through borrowing or loan translation) to the area of *etymology* (in a stricter sense).[24] However, this separation is often foiled by the very fact that one linguistic discipline is forced to use (parts of) the other when it comes to explaining certain characteristics, as, for instance, the sub-classification of rectional, non-verbal nexus compounds draws on syntax and semantics, when Marchand (1969) is followed (chapter 5.3.2.2.1.2.).

Blank's (1997; 2001a) approach to the lexicon from a semantic-cognitive point of view is integrative. His semantic typology of lexicogenetic mechanisms is based on the concept of *associative relations* existing between a source and a target concept. In his understanding, all lexicogenetic mechanisms are linked by a shared set of associative relations:

> [...] we have to develop a semantic typology for all types of lexical innovation, such as semantic change, derivation, compounding, gender change, conversion, the different types of idioms, but also for blending, agglutination, deglutination, onomatopoeia as well as loan words and their subsequent semantic changes. Recent studies [...] have shown that, although these types of lexical innovation are formally completely different, they rely semantically on a small set of associative relations between source and target concepts (Blank 2001b: 6).

---

23 For definitions of lexicogenesis, see also, e.g., Kastovksy (1982: 151), Schmid (2005: 71-3), and Tournier (1988: 47-8).

24 The field of 'etymology' can be understood as quite all-embracing when it is defined as "the application, at the level of the individual word, of methods and insights drawn from many different areas of historical linguistics, in order to produce a whole account of that word's history" (Durkin 2009: 2), but often, the term is narrowed down to the outcome of language contact and the diachronic stratification of lexical fields.

The associative relations Blank refers to go back to three principles Aristotle postulated as being at the basis of the process of remembering, namely, *similarity*, *contiguity* and *contrast* (see chapter 6.3.). According to Blank, formally completely different processes such as word-formation, semantic change, and loans and their subsequent changes are based on the same set of associative relations: on different forms of similarity, contiguity and contrast.

Blank's semantic typology will serve as a theoretical framework in the present survey of the Old English complex plant names. Starting from more traditional (represented by Bauer 1983) and structuralist (represented by Marchand 1969) approaches to the morphology of complex lexemes and their morpho-semantics, the present survey identifies theoretical inconsistencies, where, it is argued, alterations have to be made insofar as the concepts of syntactic-semantic deep structure and of types of semantic contents (cf. chapter 5.3.2.2.1.2.) have to be rejected and the analysis has to be completed on the level of cognitive-associative processes, i.e. using the notion of 'associative relations' (cf. chapters 6.3.). In a similar way, processes of folk etymology (cf. chapter 7.5.) can be described on the basis of the same cognitive processes.

An example may illustrate this: The semantics of the complex plant names *fīflēafe* lit. 'five leaves' in contrast to *fīffingre* lit. 'five fingers', both denoting the plant 'Potentilla reptans L. / fiveleaf, cinquefoil', cannot be explained comprehensively on the basis of a syntactic paraphrase. Following Marchand (1969), the relation between the first and the second element of *fīflēafe* is one of possession, since the plant name can be paraphrased as 'The plant has five leaves', while the relation between the elements of *fīffingre* is one of resemblance, since this plant name can be paraphrased as 'The plant's leaves resemble (a hand with) five fingers'. This, however, in my view, fails to cover the complexity of both plant names. As is explained in more detail in chapter (5.3.2.2.1.2 and 5.3.2.2.2.), syntactic paraphrases are highly idiosyncratic and remain on the level of language: *fīflēafe* could equally well be paraphrased as 'Five leaves grow out of the stem of this plant', or as 'The leaves are divided into five leaflets'. The explanation within the analytical framework of Blank's semantic typology, in contrast, addresses the relation between the denotata of the two linguistic signs composing the plant name – which is an extra-linguistic one: The plant 'fiveleaf' and its leaf (denoted by the second element in *fīflēafe*) are related via conceptual contiguity, or, more precisely, of meronymy. The information that the leaves of *fīflēafe* are divided into five parts can be gained only on botanical grounds; it is alluded to differently in the two plant names: *fīffingre* alludes to it metaphorically, as a comparison to a hand consisting of five 'fingers' is drawn (Catalogue, s.v. *fīffingre*), while *fīflēafe* refers to it in a straightforward description when it makes use of the element 'leaf' (Catalogue, s.v. *fīflēafe*).

A major focus of this survey is, therefore, to pool all linguistic and non-linguistic information on the plants and their names and to examine where and how these different kinds of information meet to form linguistic signs. It will also be a focus of the present survey to find patterns of a more general kind in the lexicon of complex plant names as a whole. To these ends, the formal make-up of the complex plant names, i.e. their morphological shapes and structures, will provide the basis for the ensuing semantic analysis as, here, the component elements of the complex plant names can be related to their denotata, which, in turn, are related to the extra-linguistic motives, i.e., to the salient attributes of the plants verbalised in their names, and to the concept of the plant itself. Eventually, the etymological histories of the plant names will allocate the plant names to cultural and historical events which can reach as far back as to Indo-European times but can also reflect more recent developments such as the language contact between Old English and Latin, e.g. in the wake of the Benedictine Reform.

Morphology

The theoretical first part of this survey, thus, starts with an introduction into the morphology of the Old English complex plant names, putting a special focus on the aspects of word-formation (chapter 5). The limits of the morphological approach are investigated especially with regard to the classification of the composite lexemes according to their morphological structure, as aforementioned. The diachronic view has to be taken, for instance, in the analysis of plant names affected by processes of attrition, such as, for example, the question of analysis of intermediate *-e-*, e.g., in *crāwelēac* 'crow garlic, etc.', or in the evaluation of how to interpret compounds with different declensions from that of the simplexes; cf., for instance, the discussion of final *-e* in *fīflēafe* 'fiveleaf' (chapter 5.3.3.4.; see also Carr 1939: 237-50).

It is argued that an approach to the structure of complex lexemes that does not go beyond the level of intra-linguistic knowledge, i.e. which operates within the boundaries of the language, cannot be satisfying. Instead, an analysis of the elements of complex lexemes must draw on the underlying, extra-linguistic concepts denoted by these elements. Here, the first interrelation between morphology and semantics is exposed, and the current definition of exocentric compounds is revised (chapter 5.3.2.2.2.).

Semantics

One of the two central issues of the semantic analysis (chapter 6) will be to discover the salient features of the plants encoded in the plant names: their *motivation*. Since these features can be encoded differently – literally or figuratively –, the manner in which they find their way into the plant names will be carefully examined. The *associative relations* between the extra-linguistic motivating source concept and the extra-linguistic target concepts of the plant names link the morphological structure of the plant names to their semantic content. To the end of identifying the motives and the associative relations, linguistic and botanical information have to be checked against each other.

Diachronic stratification & language contact

The focus of chapter 7 will be partly on the chronological layers present in the corpus of the Old English complex plant names, which range from the Indo-European stratum (present in the component parts of the plant names) via Germanic and West Germanic to the Old English stratum. The other part is concerned with the fact that the chronological layers contain interference features from other languages. The loan influences are investigated in detail with regard to their integration into the system of Old English, but also with regard to their 'typology', namely, as forming hybrid plant names (chapter 7.4.2.1.) or as generating the formation of words out of the stock of native (Germanic) material modelled on a foreign name (different types of loan formations; chapter 7.4.3.).

1.3. CORPUS

The present survey includes all complex plant names recorded in Anglo-Saxon texts, i.e. in total 467 terms. The term *complex* serves here to refer to all compounds and syntactic groups that denote plants, but it excludes all derivatives such as prefixations and suffixations.

The corpus of plant names is taken mainly from Bierbaumer's three volumes of *Der botanische Wortschatz des Altenglischen* (1975-1979), which represent, with about 900 entries, the most comprehensive inventory of Old English plant names. Prior to Bierbaumer, parts of the Old English plant names had been collected in editions of Anglo-Saxon medico-botanical texts, respectively, in studies of these texts (cf., for example, Cockayne 1864-1866, Grattan and Singer 1952, d'Aronco 1988, Sauer 1992b, 1996, 2000, etc., and d'Aronco and Cameron 1998). Bierbaumer's work, though extensive, is likely to lack some plant names, since the charters, the poetical texts, and the Anglo-Saxon toponyms were not included in his corpus of texts.[25] An analysis of these texts and the linguistic analysis of the toponyms are very likely to produce lexemes not yet considered. Since such an analysis would go beyond the scope of this survey, and as Bierbaumer still covers the major potential textual sources for plant names[26] and since, even if a list of plant names not appearing in Bierbaumer were available, the correct identification of these plants would still be lacking, it seems justified to concentrate in the present work on the plant name lexicon recorded in Bierbaumer's three volumes.

For some of the plant names not recorded in Bierbaumer's *Wortschatz*, however, the *DOE* and others provide the denotata (among other information); in which cases, i.e. when the information given was sufficient for a thorough analysis, I have included the plant names in my corpus. The c. 21 plant names thus presented in the Catalogue are marked as 'not in Bierbaumer (1, 2, 3)' and the source for the identification is given. To give two examples, the first of which I included in the Catalogue, because the identification was provided by Hooke (2003), while the second was not included for lack of proper identification of the plant: First, *æscapuldre* lit. 'ash apple tree', identified as 'service-tree' by Hooke (2003: 29) is attested only once in the Old English *Charters* ((Birch), 2: 162, no. 542), and is, according to Hooke (2003: 29),

---

25 His corpus includes all Anglo-Saxon medico-botanical texts and the glosses and glossaries (cf. Bierbaumer 1: vi; 2: v-xii, 3: vii-xli).

26 In poetic texts, for instance, plant names seldom appear, and when they do, it is not their identity or characteristics, which is important, buth rather what they reflect: "through the representation of the natural world, Old English poets reflected on their powers and limitations, their fears and hopes, and their relationship to both the world and its creators" (Neville 2003: 291).

associated with a wood in Pendock, Worcestershire. Second, *gātepyrne* '? goat's thorn' (*DOE*, s.v. *gātepyrne*), a hapax legomenon, too, appears in the following entry of the *Charters* ((Birch), 3: 240, no. 1036): "⁊ lang dices on east hæma gemære on þone bige to þan heafdan on gateþyrnan" ['along the dike on the boundary of the eastern dwelling in the bending to the head at the goat's thorn'];[27] Carr (1939: 317) includes *gātepyrne* in his group of secondary compounds.

The scope of this study is further defined by morphological, semantic and etymological criteria. Morphologically speaking, it includes all complex Old English plant names that are made up of at least two free lexical morphemes, i.e. synchronically analysable plant names. Etymologically speaking, it includes all loan words that have changed their form due to folk etymological processes and can now be analysed within the system of Old English, such as, for instance, *wulfesteors* 'wolf's teasel' lit. 'wolf's penis' (probably originally from OE *wulfestǣsel* lit. 'wolf's teasel'; see Catalogue, s.v. *wulfestǣsel*), or *bisceopwyrt* 'marsh mallow' (with *bisceop-* from < L *hibiscum*; see Catalogue, s.v. *bisceopwyrt*).

Loan formations and hybrid compounds are included as well: examples are *brōcminte* 'water-mint' (OE *minte* < L *ment(h)a*; see Catalogue, s.vv. *brōcminte*, *minte*), and *hundestunge* 'dog's tongue' lit. 'dog's tongue', a loan translation based on L *lingua canis* lit. 'dog's tongue' (itself a loan translation of Gk *κυνόγλοσσον*; see Catalogue, s.v. *hundestunge*). Doubtful cases are included rather than excluded, but loans from Latin or Greek that have not been assimilated to match the Old English language system (i.e. *foreign words*; cf. chapter 7.5.) are excluded.

Semantically speaking, all compounds and syntactic groups that can be subsumed under one of the following categories are included: All Old English complex plant names that certainly or probably denote specific botanical entities, i.e. that are *natural kind terms* (see chapter 2.), are included, such as *bisceopwyrt* 'marsh mallow', *candelwyrt* 'great mullein', *pīntrēow* 'pine-tree' or *wudusūræppel* 'crab-apple'. When the plant name itself (i.e. the natural kind term) is not attested independently, the complex lexemes denoting parts or natural products of these plants such as, for example, *fingeræppel* 'date' and *dracanblōd* 'resin of dragon's blood tree' are considered. The hybrid loan translation *dracanblōd*, for example, which is made up of the loan word *draca* 'dragon, devil' (L *draco*, *dracōnis* < Gk *δράκων*) and OE *blōd* 'blood', denotes the resin of the dragon's blood tree and is included, since, although the tree is, apparently, part of the plant world of the Anglo-Saxons, none of the attested lexemes from Old English refer to the tree itself (see Catalogue, s.v. *dracanblōd*).[28] In a similar way, the tree bearing pomegranates is not represented in the Old English

27 Due to the uncertain identification, I have not included *gātepyrne* in my corpus of complex plant names; however, *æscapuldre* was included.

28 The resin was a good already traded in Dioscorides' times (Gk *αἷμα δρακόντιον*, lit. 'dragon's blood'; cf. Marzell, s.v. *Dracaena* L.); cf. OHG and OLG *drachenbluot* (Carr 1939: 313). Another example is OE *ele* whose primary meanings are 'oil' and 'olive', but which, obviously, could also refer to the olive-tree – as is implied by its use in the compound *elebearu* 'grove of olive trees'. The meaning 'olive-tree', however, is not attested independently in the surviving Old English texts (*DOE*, s.v. *ele*). Obviously, *ele* is not a complex formation and therefore not included as a separate entry in my corpus, but it occurs, for example, in *elebēam* and *eletrēow*.

lexicon by a lexeme on its own, while its fruit – *affricanisc æppel* lit. 'African apple' – is. In contrast, complex lexemes such as *æscrind* 'rind of the ash', *æsprind* 'rind of the asp' and *alorrind* 'rind of the elder', where the plant names *æsc*, *æspe* and *alor* are attested independently, are not included, as the aim of the present study is to unravel the ways in which the vernacular names of botanical taxa, i.e. natural kind terms, were formed, excluding the names of parts of plants by definition.

Not included in the Catalogue are, further, compounds in which one of the elements is an otherwise attested plant term but which, as a whole, do not refer to plants, such as e.g., *lēactūn* '(herb-)garden', *ācbearo* 'oak grove', or *mǣrþorn* 'boundary-hawthorn'. Also excluded are all adjectives and verbs derived from plant names, parts of plants or life forms, and adjectives that refer to different states of ripeness, such as for example *grēne* 'green', *cnēowiht* 'cloved', *æppelberende* 'bearing apples' or *gepiperian* 'to pepper'. These lexemes can appear, however, as parts of complex plant names, e.g., *cnēowiht* 'cloved' in *cnēowihte wenwyrt* 'figwort' lit. 'knotty wenwort', or *rēad* 'red' in *rēad(e) clǣfre* lit. 'red clover' 'bloody dock'. Compounds with *cyn* 'kind (of)' as a second element, i.e. compounds not denoting specific plants (e.g., *hwǣtecynn* 'kind of wheat'), are not included, either.

It is sometimes difficult to decide whether a lexeme can be counted as part of the Old English botanical lexicon or not. OE *eolhsand* 'amber', for instance, is regarded as a mineral, even though it is made of the solidified resin of prehistoric trees, i.e. is of plant origin. With regard to OE *eolhsand* glossing L *electrum*, the *DOE* (s.v. *eolhsand*) remarks that it appears "in context of source for all glosses, 'alloy of gold and silver' or perhaps 'brass'". Similarly problematic is frankincense, which can be referred to by a variety of Old English lexemes – *stōr*, *anstōr*, *beorning*, *hwīt stōr* and *recels*. Frankincense is a product of the African shrub Boswellia carteri L., the 'luban-tree' or, more precisely, the air-dried resin of this shrub (*Brockhaus*, s.v. *Weihrauch*). Although its production does not involve too complex a process or the use of different ingredients, as the resin must only be dried to be ready for use, its use in church before and during mass places frankincense rather into the religious frame of knowledge than into a botanical context.[29]

1.4. SUMMARY

In the opening chapter, the general outlines of the present study were presented. After a brief introduction into the texts recording the c. 900 Old English plant names, the process of naming the plants was embedded in the matrix of botanical and medical writing from antiquity to the early Middle Ages. The outstanding nature of the corpus

---

29 A borderline case is also the hapax legomenon OE *ambrosie* 'perfume or name of a plant'. According to Feulner (2000: 70), the botanical meaning is not identifiable and the lexeme refers rather to some kind of heavenly scent than to a concretely identifiable plant. The fact, however, that it is referred to as "þære wyrte", i.e. as a plant, in the translation of *Vita S. Guthlaci*, and that the Latin lemma *ambrosia* is glossed several times by the Old English plant name *hindheolaþ* 'water-hemp' can justify its treatment as a plant name.

of Old English plant names as the oldest and largest corpus of this kind in a European vernacular in Europe was pointed out.

The major aim of the present study was then presented. This is, on the one hand, to trace the individual 'histories' of the complex Old English plant names, including their morphological make-up, semantic aspects and their etymological histories, with a special focus on loan influence. On a larger scale, it is attempted to identify structures within the botanical lexicon of Old English such as, for instance, correlation of specific morphological patterns with semantic patterns.

Methodologically, traditional and structuralist morphological and semantic approaches are complemented with cognitive semantics and *gestalt* psychology, as the study intends to explain the processes that may have been at work in the formation of the complex plant names – from the recognition of specific features in the morphology of the plant, for instance, via the associations that are elicited by these features and the morphological processes used to transform this semantic information into lexical material. Further, the approach to the corpus has to be a synchronic-diachronic one, as different diachronic strata have become evident, and as diachronic processes, such as folk etymology and attrition, have to be observed and commented upon in order to arrive at a comprehensive understanding of the plant names.

The survey is meant to provide insights into the Anglo-Saxons' linguistic abilities, including understandings of cultural, especially medico-botanical issues and plant taxonomy, but it also may serve as a starting point towards a large-scale investigation of Old English word-formation.

2. TAXON, NATURAL KIND TERM, LEXEME?, OR: WHAT IS (IN) A PLANT NAME?

Research on a corpus of vernacular plant names quickly and inevitably throws the researcher amidst linguistic, botanical and philosophical interpretations of the subject, as the analysis is essentially situated on the interface between these three disciplines. The relation between *botanical taxon* and *vernacular plant name* is obvious, since vernacular plant names denote botanical taxa, i.e., plants. Especially where two or more vernacular plant names denote one and the same botanical taxon, the analysis must draw on the discipline of botany as the science which provides an exact definition of the denotatum, an unequivocal scientific name and, consequently, the possibility of identifying the denotatum with all its characteristics. With respect to this relation, the analysis of vernacular plant names must be interdisciplinary, broaching the subject from both a *linguistic* and a *botanical* angle.[30]

From a *philosophical* point of view, plant names are of special interest because they "are object-groupings that reflect real distinctions in nature – in particular, distinctions that play an explanatory role in (accurate) scientific theories" (Gross 2005: 1). In contrast to other lexical items, plant names are "not *arbitrary*" (Koslicki 2008: 789). While, for instance, 'keys' or 'tables' are "artifactual kinds – objects grouped together in virtue of our intending them to fulfill certain functions" (Gross 2005: 1), plants are perceived as *natural kinds*, the plant names being, accordingly, *natural kind terms*. From this it follows that the representation and classification of plant names in human (mental) categorisation systems reflects, to a certain extent, real distinctions in nature. The philosophical approach is thus connected with the extra-linguistic, botanical and the linguistic ones: the linguistic make-up of complex plant names reflects pathways of human categorisation strategies, based on extra-linguistic, botanical 'facts'.

In their capacity of natural kind terms, plant names resemble anthroponyms, toponyms, hydronyms, and the like, yet without being considered 'names' or 'proper nouns' (in contrast to 'common nouns') in research into onomastics.[31] With regard to the etymological history of names, Durkin (2009) remarks that "the study of name etymologies depends to an unusual degree on close consideration of extra-linguistic factors" (Durkin 2009: 266) and that "the vast majority of names are not arbitrary in origin" (Durkin 2009: 268). Research has revealed that the semantic origins of, e.g., personal names may be as varied as ranging from "some attribute of a person so

30 In 2001, a cross-linguistic study of a sample of Modern English, German and French plant names appeared, which starts from similar preliminaries as concerns the relation between vernacular and scientific terminology (Sauerhoff 2001). As learned about this study only after the submission of my thesis it was not possible to include it.

31 For an important recent study on names, see Anderson (2007).

named [...]" to "[...] qualities which are particularly esteemed, or simply [to the fact that] they are deemed to have an impressive or pleasant 'ring' to them" (Durkin 2009: 268). In a similar way, as the present survey will show, plants are named after certain characteristics that are particularly valuable or eye-catching or otherwise conspicuous, in addition to the fact that their morpho-semantics reflect real distinctions in nature. As such, their description both in terms of attributes or features and of hyp(er)onymy (see chapter 6.) relies on encyclopaedic, botanical knowledge. In contrast to *names* (Anderson 2002: 2f.), however, plant names do not *refer* to the particular specimens of the plants they name (in the sense of Lyon's 1977 use of *reference*) as anthroponyms and toponyms do. While terms such as *Mary* or *London* refer to individual people or places in a one-to-one relationship (Durkin 2009: 267), plant names are *denotations* (Lyons 1977: 1, 207), i.e. they denote any of a whole class of plants that exhibit certain characteristics.[32]

In their function as natural kind terms, it was stated above, plant names reflect real distinctions and relationships of extra-linguistic reality. These distinctions are expressed morphologically and semantically – in fashion similar in many ways to that of scientific taxonomy, although folk taxonomies are not necessarily congruent with scientific botanical categorisation (see chapter 2.1.1.).[33] There are, indeed, a number of cases where scientific botanical facts are 'mis'-represented: cases like the vertical polysemy in OE *more* 'wild carrot', for instance – a synonym and, at the same time, also a hyperonym of *wealhmore* (i) 'common parsnip', (ii) 'wild carrot' – prove that humans are perfectly able to cope with 'imperfect' non-scientific categorisation systems.

The vernacular plant names' relations to botanical taxa are first and foremost relations of denotation: vernacular plant names denote particular sets of plants in extra-linguistic reality, as, for instance, OE *dæges ēage* 'daisy' demonstrates: The denotations of the component parts of the plant name, their relation to each other and to the plant itself – in our example lit. 'day's eye' – point to specific, nameworthy habits or features of the plant, called in the present survey its *motivation* (chapter 6.2.). In order to explain this *motivation* – the *motives* underlying the component elements of the plant names – we have to leave linguistics and enter the discipline of botany. It is botany that provides us with information on the morphology, the growing habits or other characteristics of plants – the motives that we find encoded in the plant names.

It should, however, be kept in mind that what appears to have been a botanical 'fact' to the Anglo-Saxons is not necessary one in modern botany. The plant house-leek, for instance, was ascribed the power to ward off thunder and lightening, as its

32 For the terminology, compare, further Lipka (2002: 62): "What is important at the moment is that both *reference* and *denotation*, as used by Lyons, fall together in Coseriu's *designation*, but are strictly separated by him from *signification*. Coseriu's concept of *meaning* (*signification*, *Bedeutung*) excludes extra-linguistic objects and relations, and is therefore restricted to language itself".

33 An important study into folk taxonomic structures is the investigation into the botanical terminology of the indigeneous people of Mexico, the Tzeltal (Berlin, Breedlove and Raven 1973; 1974); for an investigation into Anglo-Saxon botanical taxonomy based on Ælfric's *Glossary*, see Krischke (2011).

Old English name, *þunorwyrt* lit. 'thunder-wort', implies. According to modern scientific insights, this plant does not possess this quality, neither does the plant named *þunorclǣfre* lit. 'thunder-clover', which denotes the house-leek as well, and encodes the same motive as *þunorwyrt* (see Catalogue, s.vv. *þunorclǣfre*, *þunorwyrt*). In addition, notwithstanding the fact that both vernacular plant names and scientific plant names refer to the same set of botanical taxa, the scientific and the vernacular nomenclatures are not congruent – not from the point of view of naming system, nor from the point of view of clarity and un-equivocality. As mentioned before, scientific nomenclature is unambiguous and the rank of botanical taxa is reflected in the nomenclatural system while this is not always the case in folk taxonomies like the Old English one (see 2.1.1.).

2.1. ISSUES OF TAXONOMY

2.1.1. FOLK TAXONOMY & SCIENTIFIC TAXONOMY

The editorial committee of the Seventeenth International Botanical Congress in 2005 in Vienna defined names of botanical taxa as "unambiguous names for organisms; [...] essential for effective scientific communication" (*ICBN*, preface) and specified 'botanical terminology' to include

> [...] on the one hand [...] terms that denote the ranks of taxonomic groups or units, and on the other hand [...] the scientific names that are applied to the individual taxonomic groups of plants. The purpose of giving a name to a taxonomic group is not to indicate its characters or history, but to supply a means of referring to it and to indicate its taxonomic rank (*ICBN*, preamble).

Following the definition as specified by the *ICBN*, scientific botanical taxa fulfill the double function of denoting taxonomic groups or units on the one hand and of indicating taxonomic ranks of taxonomic groups and individuals on the other hand. Thus, the plant name Prunus domestica L., for instance, denotes the taxonomic group commonly known as 'plum' in English speaking communities. Further, it indicates that the plant denoted by Prunus domestica is of the genus Prunus, which belongs to the subfamily of Prunoidae, itself a part of the family of Rosaceae (cf. Groß 2002: 17).

Plant name systems in vernacular languages do not meet the rigid criteria of scientific taxonomy and terminology, despite the vernacular plant names' capacity to 'name' or 'denote' the same biodiversity and to indicate taxonomic ranking. As is shown in Krischke (2011), an analysis of the corpus of complex plant names in Ælfric's *Glossary*, many endocentric formations in the vernacular, such as, for instance, *æppeltrēow* 'apple-tree', do provide taxonomic ranking in a manner analogous to that of scientific botanical terms, but exocentric formations such as, for instance, *dæges ēage* 'daisy', are not within this classification system: Similar to the binary structure of modern scientific taxonomy (cf. Krischke 2011: 83-4), one element of *æppeltrēow* indicates taxonomic rank, namely, *trēow* 'tree', while the other, here the first element, *æppel* 'apple', specifies that the denotatum is an 'apple' tree, in contrast to, for instance, the 'plum' tree or the 'oak' tree. This information structure is, in contrast, not true for the exocentric plant name *dæges ēage* lit. 'day's eye': neither the first (*dæg* 'day'), nor the second element, *ēage* 'eye', suggest a plant name or a botanical taxonomic rank; people encountering this lexeme for the first time cannot detect its meaning 'daisy' solely on the basis of the two elements.

2.1.2. POLYSEMY & SYNONYMY

Polysemy presents the researcher with another major challenge: folk classifications are replete with polysemous plant names, thus defying the claim for unambiguousness of scientific taxonomy. As Sauer (1995: 301) pointed out in his analysis of Old and Middle English plant names, the variation

> can be due to a number of reasons, for example: to imprecise botanical knowledge; to different characteristics or properties after which the plant could be named; to regional or dialectal differences; to diachronic developments [...].

Thus, for instance, OE *bānwyrt* can denote

(i) 'Bellis perennis L. / daisy / Gemeines Gänseblümchen
(ii) 'Centaurium umbellatum Gilibert / common centaury / Tausendgüldenkraut'
(iii) 'Symphytum officinale L. / comfrey / Große Beinwurz, Beinwell'
(iv) 'Viola L. / violet / Veilchen'

while Bellis perennis L., for instance, in turn, can be rendered by

(i) OE *bānwyrt*
(ii) OE *brȳsewyrt*
(iii) OE *dæges ēage*.

In the corpus of Old English complex plant names, this phenomenon results in a moderate, but still significant discrepancy between the number of vernacular names and that of botanical groups: of the 467 Old English complex botanical terms, 438 terms[34] cover 332 botanical taxa (see *Index of botanical names*). There are, on average, 1.3 Old English plant names to one botanical taxon.[35]

The semasiological analysis of the Old English botanical lexicon as a lexical field[36] frequently unveils cases of polysemy, but it also abounds with synonyms. *Synonymy* results on the one hand from variation within the diasystem of Old English, on the other hand from socio-historical and cultural realities. These have played a role in the creation of as many as fifteen terms for Alliums (in alphabetical order and including simplexes): *brādelēac* 'ramsons, etc.', *cīpe* 'onion', *cīpelēac* 'onion, etc.', *crāwanlēac* (*crāw(e)lēac*) 'crow garlic', *croplēac* 'common leek', *gārlēac* 'garlic', *hēafdeht por(r)* 'common leek', *hollēac* 'shallot', *hwītlēac* 'leek', *lēac* 'leek, plant, garden-herb', *por(r)lēac* 'common leek', *risclēac* 'rush garlic', *secglēac* 'rush garlic', *wudulēac* 'ramsons', *ynnelēac* 'onion'.The preponderance of Alliums in the Anglo-Saxon diet (Lee 2011: 151) and the high number of terms for Alliums can be explained by the cultural and nutritional importance of this plant, not only for the Anglo-Saxons, but for all Germanic people. Leeks and onions were used as ingredients for gruel, they provided flavour, had a high vitamin C content and could be kept for a longer period (Lee 2011: 151). Runic inscriptions on bracteates read the name of the leek and, in Old Norse, the royal sceptre is termed *ítrlaukr* lit. 'excellent leek'. (Birkhan 2012: 216). Shallots (Allium ascalonium L.) were very popular as a condiment in the later Middle Ages (Birkhan 2012: 159) and in Old English, the denotational range of *lēac* was extended to a more generic use, as the plant name denoted also a 'garden-herb, plant', i.e., was a life form term and appeared as such in many compounds (e.g. *lēactūn* 'garden', or *ynnelēac* 'onion'; see chapter 6.6. and Catalogue, s.v. *lēac*).

'Technical language' (diaphasic variation, i.e. variation according to the communicative setting, where different styles or registers are used)[37] is also the source of a number of synonyms. As the medico-botanical texts are the products of the intellectual elite of Anglo-Saxon England, the assumption that a number of Latin loan words or loan formations on Latin terms can be imputed to being part of the technical

---

34 The denotata of 14 complex plant names have not been identified yet and a further 15 of the complex botanical terms do not denote generic or specific taxa, but life forms such as *feldswamm* 'mushroom' lit. 'field-mushroom', *wæterbōg* 'sucker' lit. 'water-bough'.

35 Due to a number of uncertainties regarding the 'correct' identifications of some of the plants, the numbers will can vary when ambiguities in identification are eliminated. Still, they offer sufficient data to at least approximate the 'correct' ratio of complex vernacular plant names and botanical taxa. The ratio in Middle English (including the simplexes) is one to three: Hunt (1989: xi, xlviii ff.) counted about 1800 vernacular plant names, which cover about 600 species.

36 Unlike word fields, which are limited to morphologically simple lexical items, lexical fields can contain simple and complex terms (Lipka 2002: 167-8): "One of the reasons for doing this is the concept of 'lexical gap', which can only be proved to exist in the structure of simple words. If complex lexemes are admitted in such a field, the productive word-formation processes may close such a gap at any time" (Lipka 2002: 168).

37 The terms *diatopic variation* (Gk *τόπος* 'place') for regional variation, *diastratic variation* (Gk *στρατός* 'stratum') for variation according to social group, *diaphasic variation* (Gk *φήμη* 'talk') for variation according to communicative setting, *diamesic variation* (Gk *μέση* 'middle') for variation according to the medium of communication and *diachronic variation* (Gk *χρονος* 'time') relating to variation through time were introduced by Flydal (1951) and Coseriu (1981).

language of Anglo-Saxon medicine and botany suggests itself.[38] This was probably the case with the following pair of synonyms, *agrimonia*, a Latin loan (*DOE*, s.v. *agrimonia*), and *gārclīfe*, a native formation, which both denote the 'agrimony'. Where exactly the boundary between the basic botanical terminology and technical language can be drawn is, however, a matter of dispute. It must have been as fuzzy in Old English as it is today, with the areas of technical terminology and common vocabulary merging and interpenetrating. With regard to Modern German, English and French botanical jargons, Sauerhoff (2001: 31) not only assumes many plurivalent cases but also a diachronic development, an assumption that may well be transferred to the situation in Anglo-Saxon England:

> [Zum Gemeinwortschatz] gehört nicht nur der aktive Wortschatz des Durchschnittsprechers, der verhältnismäßig begrenzt ist, sondern auch sein passives Wissen von Lexikonwörtern, die jederzeit richtig rezipiert und in bestimmten Situationen durchaus auch richtig angewendet werden können. Der Gemeinwortschatz enthält aber stets auch einen beträchtlichen Anteil an Wörtern, die einer Fachsprache entstammen, schließlich allgemein bekannt und dann auch gemeinsprachlich verwendet werden.
>
> [It is not only the active vocabulary, relatively limited in terms of number, which belongs to the basic lexicon of the everyday speaker, but also his or her passive knowledge of learned terms, which can be received and employed correctly at any time. However, the basic vocabulary also contains a remarkable number of words that were, initially, part of a technical language, and then became widely known and used as parts of the basic vocabulary.]

Whether or not loans from Latin and loan formations of Latin lemmas were in actual use can hardly be assessed nowadays. Yet, it can be assumed that some of the plant names were restricted to the vocabulary of the trained elite and were never institutionalised, let alone lexicalised within the system of Old English. Given the fact that many loan formations occur as glosses, it may also be possible that some of them remained throughout the Anglo-Saxon period what they probably were intended to be initially: didactic interpretational bridges between Latin and Old English "[...] with the sole function of explaining Latin terms, respectively of making visible the compositional structure of the words they render" (Johnston 2011: 72).

38 However, Latin loan influence does not automatically entail synonymy. In most cases, foreign terms are introduced because of lexical gaps in the receiving language (Durkin 2009: 142).

2.2. SOCIOLINGUISTIC AND TEXTUAL SETTINGS

It is at this point that some observations on the limits of the textual corpus and, subsequently, on the limits within the corpus of Old English complex plant names, have to be made. Namely, the sociolinguistic settings of Anglo-Saxon text production and the specific nature of the medico-botanical texts account for certain restrictions in the botanical lexicon. These aspects will be addressed in the following.

As stated above, the intellectual elite of Anglo-Saxon England was among the first in Europe to ensure that a substantial number of texts dealing with theological, cultural or natural scientific issues, or engaged in literary pursuits, were available in the vernacular. To value a vernacular language as an adequate means of communicating complex scientific, literary, or religious ideas in writing is quite a modern attitude, and many European languages did not obtain full acceptance until the Renaissance period. However, English already began in the 7th century to spread into domains governed by communicative distance[39], and this process was accompanied by linguistic elaboration and refinement, by codification, and to a certain degree also by processes of standardisation of the language.[40] *Extensive elaboration*, i.e. the process of languages being codified graphically and expending into various spoken or written discourse types more or less as a 'natural' consequence causes languages to develop and be refined in line with the exigencies presented by the text and discourse types; this development, dubbed *intensive elaboration*, manifests itself in distinct lexical and grammatical usages (cf. Koch and Oesterreicher 2007: 364).

---

39 In a monograph and a number of essays, Koch and Oesterreicher (1985: 17-24; 1990: 8-12; 2001: 586; 2007: 351) differentiated between language of communicative proximity (G *Nähesprache*) and language of communicative distance (G *Distanzsprache*), arguing that, as the distinction between 'orality' and 'literacy' does not lead to satisfying explanations when it comes to explaining certain phenomena on the level of the *parole*, discourse analysis has to differentiate between a 'phonic' and a 'graphic' code and between conceptually 'spoken' and conceptually 'written' discourse types. Thus, a private chat with a friend, for instance (a prototypical example for communicative proximity), features a medially phonic communication and a conceptually spoken discourse. Communicative proximity is characterised by a high degree of privacy and intimacy between the communication partners, a face-to-face communication situation, cooperation and spontaneity. Linguistically, communicative proximity mirrors frequent use of contracted forms (e.g. *I've, I'm*), deictic expressions and paratactic syntax, to name but a few characteristics. By contrast, the parameters of communicative distance (e.g. the writing of a scientific treatise) include a fixed topic, hardly any communicative cooperation, itself the result of emotional, temporal and spatial distance between author and reader; then, a certain degree of publicity and the tendency to be used in the graphic code. Accordingly, communicative distance is characterised by a high degree of hypotaxis, frequently discernable features of cohesion and coherence, a lack of hesitation phenomena or repetition and a large and diversified lexicon.

40 Most of the extant Anglo-Saxon manuscripts "record the regulated, supra-regional variety commonly called 'late West Saxon'. This variety has often been dubbed 'Standard Old English' as it appears in manuscripts copied outside Wessex [...]. This does not mean, however, that it is legitimate to consider this Old English 'standard' identical to its modern written equivalents. We do not have enough data to prove its full supra-regional acceptance and it is furthermore evident that 'Standard Old English' allowed a good deal more variation than our modern fixed standard languages" (Lenker 2000: 227).

The botanical vocabulary of the Anglo-Saxons being preserved mainly in medico-botanical texts and in glossaries, respectively as glosses to Latin texts makes them part of 'scientific writing', i.e. a discourse type of communicative distance.[41]

2.2.1. THE TEXTS

With regard to their structure, texts of communicative distance betray a high degree of planning and reflectedness through their typically consistent and linear structuring of information (Koch and Oesterreicher 2007: 351). This is in contrast to communicative proximity, where information exchange does not require any pre-planning, where the communication partners change topics, digress and, at times, also interrupt each other. In relation to my subject of investigation, this linearity becomes obvious, for instance, in the ordering of entries in the medico-botanical texts.[42] Bald's *Leechbook* and *Leechbook 3*, for example, present the diseases and ailments in an *a capite ad calcem order*, i.e. beginning with the diseases of the head and proceeding downwards via the diseases of the body to the inflictions affecting the feet.

Another feature of reflectedness is the inclusion of a table of contents and the observance of the order of the table of contents in the main body of text. Of the six texts, three, namely, Bald's *Leechbook*, *Leechbook 3* and the Old English *Herbarium*, possess a table of contents. This not only renders the structure of the main body of text visible at once, but also offers readers the possibility of selecting the parts of the text they want to read, at the same time skipping other parts of the text without losing any vital information. The 'chapter' titles appearing in the table of contents are repeated in the main body of text where they precede the respective entries. In the Old English *Herbarium*, we can even find sub-chapters. Entry number 115, for instance, gives remedies based on the plant OE *hwerhwette* 'cucumber'[43] and reads as follows in the table of contents:

> CXV. Herba cucumeris siluatica þæt is hwerhwette
> 1. Wiþ þæra sina sare ⁊ fotadle.
> 2. Gyf cild misboren sy.
>
> (*Herbarium*, no. 115)

---

41 Texts of communicative distance are usually preserved in the graphic code – an obvious fact applying, as a matter of fact, to all texts before the invention of voice recorders in 1857. Phonautograms from 1857 by Édouard-Léon Scott de Martinville are the earliest known modern sound recordings (http://en.wikipedia.org/wiki/Sound_recording_and_reproduction>; accessed 29. January 2011).

42 For a brief introduction into the Anglo-Saxon medico-botanical texts, see chapters 4.3.-4.5.

43 "According to D'A[ronco], this cucumber is a wild species that grows near the sea in warm places" (van Arsdall 2002, 199, fn. 224); the scientific name is, then, Ecballium elaterium Rich., the Modern English name is 'squirting cucumber'.

[115. The plant *cucumeris siluatica* that is cucumber
1. For sore joints and for gout.
2. If a child is born prematurely or misshapen.]

In the main body of the text, the title is "CXV. Hwerhwette (c)ucumeris siluatica" and the sub-entries are marked as such, though the titles are not repeated:

> Ðeos wyrt þe man cucumeran siluaticum ⁊ oþrum naman hwerhwette nemneþ byþ cenned neah sæ ⁊ on hatum stowum.
> 1. Wið þæra sina sare ⁊ wið fotadle genim wyrtwalan þysse wyrte þe we cucumeran silfaticum nemdun, seoð on ele to þriddan dæle, [...].
> 2. Gif cild misboren sy genim ðysse ylcan wyrte wyrttruman to þriddan dæle gesodenne, [...].
>
> [This plant, called *cucumeris siluatica* and by another name cucumber, grows near the sea and in hot places.
> 1. For sore joints and for gout, take the root of this plant that we call *cucumeris siluatica*, simmer it down in oil by two thirds, (...).
> 2. If a child is born prematurely or misshapen, take the root of this same plant, simmered down by two thirds, (...).]

The Old English *Peri Didaxeon*, though lacking a table of contents, betrays a logical structure, too. The entries are not only arranged by diseases, but are headed by the diseases the ailment is thought to alleviate or cure. These 'chapter' titles are, at times, in Old English, in Latin or in a mix of the two. The titles of entries 7, 9 and 11, for instance, read as follows:

> 7. It*em.* Wid tobrocene heafod (*Peri Didaxeon*, 7).
>
> [Again. With a broken head.]
>
> 9. De cefalaponia (*Peri Didaxeon*, 7).
>
> [On headache.]
>
> 11. Ad jdem. Wid þæt ylca (*Peri Didaxeon*, 9).
>
> [On the same. On the same.]

As is typical for communicative distance, just one topic is addressed: the Anglo-Saxon medico-botanical texts are mainly concerned with the description of ailments and diseases in conjunction with remedies thought to alleviate them. The Old English

Herbarium provides, in addition, information such as on the procurement of the ingredients, on places whereof to find them and on methods for gathering and preparing them adequately.

The spatio-temporal distance between author/scribe and reader results from the fact that scientific writing is, in fact, an asynchronous communication act which does not necessarily require any personal contact, let alone the sharing of emotional experiences. On the contrary: since the medico-botanical texts do not address individuals, but the literate public in general, any elements of temporal, spatial or personal deixis are lacking. As Gärtner (2011: 113) reasons in her analysis of different scribal practices in the Middle Ages,

> [t]he dreaded and appraised effect of writing is of course that it dislocates the utterance from its speaker and from its direct, synchronic communication act so that it can prevail over its ephemerality and travel and stray to new readers and contexts [...].

The graphic codification, the elaborated structure, the formal style, the specific, limited number of topics and the asynchronous communication situation as characteristic parameters of communicative distance, are reflected in the language these texts evidence. Besides features such as frequent use of hypotaxis or the avoidance of anacolutha on the syntactic level and the absence of discourse markers on the pragmatic level (Koch and Oesterreicher 1985: 27), communicative distance manifests itself also in the vocabulary through avoidance of *passe-partout* words and expressive language on the one hand, or, in other words, through the use of precise terminology, a formal style and no, or hardly any, diatopic marking – exactly those features that are inherent to the Anglo-Saxon medico-botanical texts.

2.2.2. AUTHORS, COMPILERS & SCRIBES

With respect to the botanical lexicon, a high ratio of Latin or Latinate terms as well as a high ratio of non-common, technical botanical terminology are to be expected, as the texts were not only produced as texts of communicative distance, but also in very specific socio-linguistic settings, and by a very particular group of people. According to Jolly (1996: 104),

> [i]n a sense all medical texts were products of an elite because the reading and writing of texts was limited to that select, largely clerical, minority located in major centers (monastery, minster, or cathedral) with a scriptorium or library.

This literate elite, trained in copying and translating as well as producing Latin texts were certainly bound to rely on Latin, thereby taking over loan words, creating loan formations of various kinds and producing instances of loan syntax even when they wrote in the vernacular Old English.

As a matter of fact, its impact can be felt clearly in the medico-botanical vocabulary through the number of loan words and the number of loan formations recorded. The difficulty in interpretation is, however, that we do not know whether these loans and loan formations were used by a larger part of the Anglo-Saxon linguistic community, i.e. whether they were institutionalised, or whether their existence was restricted to the learned, intellectual monastic environment. Our insight into the extent of the Latin influence is narrowed by further non-linguistic events. According to Lenker (2000: 226),

> [t]he Anglo-Saxon manuscripts (including fragments, glossaries, continuous glosses etc.) which contain a substantial amount of Old English number fewer than 200, among them many translations whose language is often highly dependent on their Latin exemplar. The picture is even more distorted because we simply do not know what percentage survived the Viking raids, the Reformation and the ravages of time. In general, only the most valuable manuscripts were kept, mostly sumptuous codices recording a very formal style.

The details of the nature and the extent of the Latin influence on the botanical vocabulary of Old English are discussed in chapter 7.5. Suffice it to say for the moment that it is considerable with 36 percent of Latin elements, though it is hard to estimate the exact ratio of words with Latin origins in actual use, especially among the country people.

On the other hand, it seems too narrow an interpretation to confine these texts to a Latinate intellectual environment. For some decades, debates on Anglo-Saxon medico-botanical texts involved mainly the nature and the origins of the remedies appearing in the texts and the dichotomy 'learned and/or Mediterranean (including Greek)' vs. 'popular and/or Northern/Germanic medicine' (cf. chapter 3). As a side effect, this debate led to evaluatory statements on the value of the plant names recorded (cf. chapter 3.), certainly fuelled by the fact that large parts of the medico-botanical texts are based on Latin originals or go back ultimately to Latin and Greek texts. For a time, scholarship was convinced that neither the copying nor the compiling Anglo-Saxon clerk would have been capable of understanding the Latin and Greek originals, reasoning that, consequently, the Old English plant names were of no value either.[44] The debate has now come to a halt. While the influence of Latin texts on the Anglo-Saxon corpus remains undoubted, it is now a common consensus that the texts originated in learned contexts, but were also influenced by the people and the peoples involved in the production of these texts, i.e. by Anglo-Saxons and their culture. As will be shown in chapter 4.1., plants and the plant world were, indeed, integral parts of the everyday experiences of the populace – besides medicine, people experienced nature in craft and in alimentation as well as in housebuilding.

44 Prominent advocates of this idea were, for instance, Grattan and Singer (1952) and Bonser (1963); see chapter 3.

As parts of the Anglo-Saxon culture and society, those belonging to the intellectual elite of a country, however,

> [...] were not isolated within their libraries and scriptoria; rather, they interacted with their own culture, reflecting its needs and ideas in what they chose to read, copy, or write down, thus influencing popular practices by transmitting literate ideas (Jolly 1996: 104-5).

Accordingly, what have been preserved as the 'Anglo-Saxon medico-botanical texts' offer valuable insights into the language and thought of the Anglo-Saxon intellectual clerical elite, trained in copying and translating Latin texts, but also in compiling and interpreting common knowledge, intermixing popular and scientific wisdom and thus influencing both 'literate ideas' and 'popular practices' at the same time. Hence, the number of Old English complex plant names exhibiting Latin loan influence attests, indeed, to the significant impact a *lingua franca* like Medieval Latin had on the vernacular, but the ratio of more than half of the complex plant names of completely native origin serves to prove the intellectual independence of the Anglo-Saxons from Mediterranean learning.

2.3. SUMMARY

The chapter has presented the analysis of the Old English plant names as situated on the interface between different disciplines: linguistics, philosophy and botany. As 'descriptions in disguise', the Old English plant names encode botanical 'facts' such as the colour of the flowers or the efficacy of the plant against thunder and lightening, as well as real distinctions in nature, such as taxonomic super-ordination or sub-ordination. In their capacity as natural kind terms, however, they are not comparable to proper names, as their point of reference is not unique.

As concerns the socio-linguistic setting, the medico-botanical texts of the Anglo-Saxons were presented as being of a very specific text-type: Produced by the Anglo-Saxon intellectual elite, they are prone to contain many Latin and Latinate terms as well as many instances of loan influence in the form of loan translations, loan renditions and loan creations. On the other hand, our insights are always restricted by the limited number of texts and the very specific nature of the texts that survived from the earliest period in the history of the English language.

3. SURVEY OF PREVIOUS SCHOLARSHIP

One cannot research the botanical lexicon of the Anglo-Saxons without considering the historical, cultural and textual context in which the Old English plant names were transmitted. The history of the botanical lexicon in early medieval England is closely associated with the history of medicine and with the textual dissemination of medical knowledge from antiquity through the Middle Ages. As a result, the evaluation of the Old English botanical lexicon depends to a great extent on the value accorded to the Anglo-Saxon medico-botanical texts. Whereas the scholars who regard the texts as a means of transmitting real and practicable medical knowledge certainly believe in the benefit of plant identifications, for those who consider them unskilled translations of classical models mingled with a 'heathen gibberish' produced by uninformed translators, analysing the Old English plant names and identifying the plants must obviously seem a fruitless task. It is therefore most important to strip the texts of any kind of judgment, whether exaggeratedly positive or negative, in order to arrive at a correct and unbiased interpretation, taking at the same time into consideration the geographical, climatic, historical and cultural facts of Anglo-Saxon England. The following chapter provides a brief account of previous – biased as well as unbiased – scholarship on the Old English plant names. First, studies devoted to the Anglo-Saxon medical texts, to medical practices and to the botanical lexicon are presented. This is followed by an introduction into the information provided by the edition of the medico-botanical texts.

The negative view of modern scholarship towards Anglo-Saxon medical knowledge and practice originated in the prefaced edition and translation of the Old English botanical texts by Cockayne in the 1860s (*Leechdoms, Wortcunning and Starcraft of Early England*). It prevailed by and large until the late 1970s when Bierbaumer (1975-1979) published his investigation of the botanical vocabulary of a larger number of Anglo-Saxon medico-botanical texts, and Voigts (1979) decidedly challenged the idea of Anglo-Saxon medical texts as being uncritical and useless translations of classical models.

Cockayne's (1864-1866) translations of the Anglo-Saxon *materia medica*, executed in an archaising and quaint style, not only transformed "ancient texts on healing into literary oddities" (van Arsdall 2002: 36), his commentaries on the medical *status quo* of early medieval England also contributed decisively to the misconceptions that "medieval medicine was preposterous, that intelligent people could not have taken it seriously, and that it could not have worked much, if at all" (van Arsdall 2002: 35). Cockayne further promoted the view that the Anglo-Saxon doctors were "vainly seeking to comprehend classical remedies that were beyond their intellectual reach while chanting gibberish and saying nonsensical words reflecting native magic" (van Arsdall 2002: 35).

Cockayne's prejudiced attitude is explained by van Arsdall (2002) as being the result of his personal enthusiasm for the classics, which made medieval knowledge appear in a very unfavourable light (van Arsdall 2002: 12). This negative view was further supported by the state of medical knowledge and practices in Cockayne's time (van Arsdall 2002: 36ff.). Although Cockayne's professional failure and tragic suicide (van Arsdall 2002: 1-35) contributed, among other circumstances, to the non-acceptance of his works, and although he was not the only one to assess early medieval medicine negatively, he certainly laid the cornerstone for the disdain accorded early medieval medicine in England.[45]

Since Cockayne's aim was to edit and translate the Anglo-Saxon medico-botanical texts and not to provide a linguistic investigation of plant names (or of other parts of the vocabulary contained in the texts), we find only scant information on their etymology, their morphology or their semantics. Each of his three volumes is, however, equipped with a glossary listing all plant names; in addition, he gives in vol. 3 a separate list of "Saxon names of worts and trees from various sources, with some vegetable products" (*Leechdoms*, 3: 310-50). Both the glossaries and the list of plant names in the third volume provide primarily information on the identification of the plants. Since, however, attempts to identify them requires information on the etymology of the words, on the corresponding Latin and Greek plant names, and on the semantics (especially on the motivation) of the plants to be identified, Cockayne includes information of this kind where he feels the need to support his line of argumentation. For *bēowyrt*, for example, the reader is given the Latin and Greek equivalents as well as its motivation: this plant obviously was called *bēowyrt* because its pleasant smell attracts bees (*Leechdoms*, 2: 371-2).

Information on the word-formation of the plant names is given, if at all, only casually. Cockayne's use of the hyphen and blank spaces between words can be interpreted as a statement on the morphological status of some of the complex names, namely, whether he saw them as compounds or as syntactic groups. Since *bēowyrt* is written as one word, it was probably regarded as a compound (*Leechdoms*, 2: 371), while *wudu lectric* was regarded as a syntactic group because it is written with a space between the two elements (*Leechdoms*, 2: 411). In addition, Cockayne splits the plant names into their parts when he requires the individual parts as arguments in favour of the one or the other identification (cf., e.g., *clufþung* in *Leechdoms*, 2: 377, where Cockayne expounds the meanings of *cluf-* and *þung*). As was usual in his time, he indicates the genitive endings of the plant names in order to indicate the declension type of the lexeme.

Cockayne provides linguistic information casually, and he is neither consistent nor systematic. *Minte* for example is nowhere marked as a Latin loan word, while *cunelle* is: "a Latin word, cunila" (*Leechdoms*, 2: 377), and *bēowyrt* is not characterised explicitly as a compound, while *clufþung* is written as one word (*Leechdoms*, 2: 377). In the light of more recent research, and bearing in mind that Cockayne's estimation

45 See van Arsdall (2002: 41ff.). Van Arsdall (2002) describes very vividly not only the personal background of Cockayne, his failure to obtain the acceptance of his contemporaries as an Anglo-Saxon scholar and philologist, and his tragic suicide, but also his admiration for the classical languages, which he taught all his life in King's College School in London.

and interpretation of the Anglo-Saxon texts and plant names is biased by his addiction to the classics, it is important to approach his findings with caution, without, however, forgetting to acknowledge his achievements in the field of Anglo-Saxon medicine as the first to edit and translate the major Anglo-Saxon medico-botanical texts and glossaries.

In 1889, Hoops published his Ph.D. dissertation *Über die Altenglischen Pflanzennamen*, in which he was in search of the "Naturgefühl" ['feeling for nature'] of the Anglo-Saxons. Hoops draws a clear distinction between classical, Germanic and Christian elements in Anglo-Saxon medicine, characterising the remedies as a "durchweg buntes Gemisch von germanischen und antiken Elementen [...], äußerlich zusammengehalten durch das Band des christlichen Ritus und Glaubens" (Hoops 1889: 67) ['a colourful mixture of Germanic and antique elements throughout, kept together by Christian rites and Christian faith'], but this does not mislead him to assume a negative evaluation of anything which is not of classical origin (in contrast to, e.g., Cockayne 1864-1866, Grattan and Singer 1952, Bonser 1963). Hoops distinguishes three kinds of sources which provide information about the Anglo-Saxons' *Naturgefühl*:[46] the treatment of plants in poetry, the treatment of plants in texts containing superstition (here Hoops clearly refers to all Anglo-Saxon medical texts) and the plant names themselves, i.e. their semantics:

> Überall dagegen, wo die Pflanze oder ein Teil derselben mit irgend einem anderen ähnlich gestalteten Gegenstande verglichen wird, erkennen wir die Thätigkeit einer schöpferischen Phantasie und die ersten, wenn auch noch unentfalteten Keime einer ästhetischen Auffassung der Pflanzen (Hoops 1889: 17).
>
> [Yet, where a plant or a part of it is compared to some other object with a similar shape we recognise the work of a creative person and the first tiny buds of the idea that plants can have an esthetic value.]

The etymology and semantics of the plant names are important for Hoops, but only in order to position the plants within the cultural context. Since his aim is to describe the *Naturgefühl* of the Anglo-Saxons – whom he perceives as a purely Germanic people with a purely Germanic body of thought – he excludes from his corpus all translations of antique texts, as, in his view, they do not transport the "echt[en] altenglischen Nationalgeist[...]" (Hoops 1889: 14) ['real English national spirit'] and exclude all plant names which denote plants not native to Anglo-Saxon England (Hoops 1889: 14, 16-7). For quite a number of plant names isolated in the manner described above, Hoops provides throughout the whole work commentaries on their identification, on their etymology and on their motivation (cf. for example the list on pp. 17-9 and the discussion of *dweorgedwostle*, 1889: 49).

46 Johannes Hoops' ideas are strongly influenced by Romanticism, a movement that validated genuine ideas and creativity, and emphasised the importance of feeling and imagination.

Hoops announced the publication of a dictionary of Old English plant names (Hoops 1889: 5), which, however, never appeared. In his dissertation, he indicates the methods he planned to employ in the future dictionary. First, he lists all relevant plant names (more than 500; 1889: 68ff.) and then explains the procedure of identification. This was to be based on the Latin and Greek plant names glossed, on other Old English plant names, on cognates in other Germanic languages and on the semantics of the plant name. Then, he describes the etymologies of some of the plant names (Hoops 1889: 74f.), which he traces back to different stages: (a) to Indo-European (e.g., *æsc* 'ash'), (b) to Germanic (e.g., *apuldor* 'apple-tree'), (c) to West Germanic (e.g., *berwinde* '? lesser bindweed'), (d) to Anglo-Frisian (e.g., *secg* 'sedge') and, finally, (e) to Old English (e.g., *æscþrote* '? vervain, etc.'). Loan words and loan formations are treated separately: (a) before 400 AD, e.g., *cīpe* (< L *cēpa*) 'onion'; (b) between 400-450 AD, e.g., *bēte* (< L *bēta*) 'beet'; (c) between 450 and 600 AD, e.g., *box* (< L *buxus*, VL *boxus*) 'box-tree', (d) before 600 AD, e.g., *cymen* (L *cumīnum*) 'cumin' (Hoops 1889, 75-6). For some plant names, he assumes Celtic origins, such as, for example, for *cassuc* '? hassock, etc.'. He concludes his thesis with three specimen entries.[47]

Hoops was the first scholar to consider the etymology and semantics of Old English plant names in an unbiased and respectful way, but the information he gives is limited since his corpus is too restricted and he does not treat the plant names systematically. Unfortunately, the material he collected for the dictionary seems to have been lost (Bierbaumer 1, v).

Skeat's article (1896) is basically a list of circa 50 "Anglo-Saxon or Early English" (1896: 164) plant names formed on animal names, which he collected from the glossary of Cockayne's third volume (1866) and where he included, for example, *briddes nest* 'a Cretan plant', *kattes minte* 'catnip', *colts foot* 'colt's foot' and *cūslyppe* 'cowslip'. Since a large number of these plant names are Middle English (e.g., *briddes nest*, *kattes minte* and *coltesfōt*; see *MED*, s.v. *brid*, 5; s.v. *cat*, 3; s.v. *coltesfōt*), this list is of limited use for the present survey.

Payne's (1904) lectures on *English Medicine in the Anglo-Saxon Times* were intended to increase the interest of scholarship in the medicine in early medieval Britain. He saw European medicine in the period from the ninth to the eleventh century as being at its lowest level (1904: 57), but emphasised the fact that what seem to be superstitious elements could very well be remnants of "old-fashioned regular medicine which ha[ve] sunk down to the level of the unlearned, and ha[ve] sometimes put on a rustic dress" (Payne 1904: 108), since many of the remedies can be traced back to some classical source. On the other hand, Payne was also convinced that, with regard to herbal medicine and English plant names,

> [...] the Anglo-Saxons, before they translated Apuleius, had recognized and given English names to a large number of native plants. Similar evidence is supplied by the numerous English names occurring in the

47 Further articles by Hoops deal with individual plant names, e.g., Hoops (1898) and Hoops (1903).

> other Leech Books, and also in the 'glossaries' or lists giving English equivalents from Latin and Greek names of plants [...] (Payne 1904: 80).

In his opinion, a large part of the herbal medicine and knowledge of the plants was an original achievement of the Anglo-Saxon doctors, founded on empirical knowledge (Payne 1904: 61), and the remarkable number of 500 Anglo-Saxon plant names attested to the extensive knowledge of medicinal plants of the Anglo-Saxons (Payne 1904: 81).

In *Anglo-Saxon Magic*, Storms (1948) discussed magic under different perspectives, echoing, however, Payne's idea that pagan magic elements could be easily identified in early medieval texts and that it is even more difficult to separate pagan Germanic elements from pagan classical elements. Storms favoured the idea that the magic elements were part of a common pool of Indo-European beliefs and not characteristically Anglo-Saxon and that Anglo-Saxon and Greek or Roman superstition, therefore, need not differ from each other. He concludes his work with an edition and discussion of 86 charms. Storms is not concerned with plant names in particular, but in one longer passage he challenges Hoops' opinion that with the plants' names the superstition attached to them was imported, too. In the course of his argumentation, he explains that the Anglo-Saxons probably knew more plants than the number of native plant names suggests, a fact which is indicated by the co-existence of vernacular and loan plant names for the same plant (e.g., *agrimonie* and *gārclīfe* both denoting 'agrimony' and glossing L *agrimonia*). He further states:

> I cannot believe that a plant like mint growing throughout Western and Northern Europe was only imported into England after the establishment of the rule of Rome, nor that every medicinal and superstitious use of it was taken over from the Romans, by them from the Greeks, merey because mint is the only name by which it is known, and this name happens to be derived from Lat. *menta*, in its turn derived from Gr. μίνθη (Storms 1948: 124).

In other words, Storms' idea is that herbal knowledge and the medical tradition and practice of Anglo-Saxon England did not start from scratch with the first translations from Latin and Greek, as some of the subsequent scholars seem to believe, but were already a living knowledge and practice at the time the Anglo-Saxons became acquainted with the classical medical authorities. At the end of his work, Storms appends a glossary of plant names, "intended as a preliminary list of Anglo-Saxon plant names" (1948: 319). Whether he intended to enlarge this list himself at a later date is not mentioned.

Singer wrote a considerable body of literature on Anglo-Saxon medicine (e.g. Singer 1927, Singer 1928). In all of his papers, Singer expressed the conviction that the state of medical knowledge in early medieval England was poor and could not be taken seriously. This attitude clearly reflects Cockayne's (1864-1866) view and was further elaborated on in the joint publication with Grattan, *Anglo-Saxon Magic and Medicine* (1952). Grattan and Singer (1952) try to separate three strata of influence:

Greek or Byzantine medicine, Celtic and Teutonic insertions, and Christian piety. As more recent research has shown, this attempt is not always helpful for our understanding of the texts since the isolation of single strands is often not possible and very often blurs the view of the whole (van Arsdall 2002: 50). In addition, it misleads them into criticising all Anglo-Saxon efforts to adapt the ancient texts to their needs as

> [...] constant copying and re-copying, for over a thousand years, with enormous labour and technical skill, of a futile work with its unrecognizable figures and its incomprehensible vocabulary (Singer 1927: 104f.).

The fact that the Anglo-Saxon medical texts were useful tools in the hands of trained doctors and were indeed in use, as the enlargements and changes in the texts and the marginal commentaries prove,[48] unfortunately escaped Singer's attention.

As Singer informs us in the preface to *Anglo-Saxon Magic and Medicine*, Grattan had intended to add to the book "a special glossary on the forms of the plant-names based partly on the work of his pupil, Miss Fazackerley [which, unfortunately], was made impossible by printing and other difficulties" (Grattan and Singer 1952: vi). The treatment of Old English plant names in particular is, therefore, restricted to a chapter on the "Semantics of Anglo-Saxon Plant Names" (Grattan and Singer 1952: 80-91) (treating, in fact, not semantic aspects of the plant names but attempting an identification of the plants) and to the inclusion of the plant names in the general index at the end of the book.

Singer's contempt for the medical and botanical knowledge of the Anglo-Saxons pervades almost every line of his accounts, and the reader often wonders why Singer is concerned with the semantics of plant names at all: not only is the major part of plant names discredited as being mere translations or transliterations of Latin plant names, but the "resulting product in an A.S. manuscript can seldom represent more than an empty sound", the Latin plant names themselves being "mostly interpretations – or misinterpretations – of Greek names" (Grattan and Singer 1952: 81). Thus, only a small number of Anglo-Saxon plant names are considered as being worthy of any serious semantic analysis. And even for these, Singer finds only disdainful words: the men who wrote the Anglo-Saxon texts "were almost incapable of making exact statements, [they] had little conception of exactness by measure and weight, [their] colour discrimination was poor and their vocabulary for colours both meagre and vague" (Grattan and Singer 1952: 82). Accordingly, attempting to identify the plants "is a fruitless task" (Grattan and Singer 1952: 85), since all vernacular taxonomies concerned (Greek, Latin and Anglo-Saxon) are unstable and "early herbals were largely mere literary exercises" (Grattan and Singer 1952: 86). In the end, the reader is left with no additional knowledge at all – save the feeling that any attempt at finding something useful in Anglo-Saxon medical texts and their botanical vocabulary is of no value whatsoever.

48 For this aspect see, e.g. Siraisi (1990: 10) and d'Aronco (2007: 49ff.).

Grattan and Singer's (1952) ideas about the poor state of Anglo-Saxon medicine were repeated in 1963, when Bonser, a pupil of Singer, published *The Medical Background of Anglo-Saxon England.* Bonser not only further promoted the neat division of medical knowledge into its classical sources, the native or pagan elements, and Celtic traditions, but also echoed his predecessors' contempt for Anglo-Saxon herbal medicine, warning against "the waste of effort in attempting to identify the plants named in Anglo-Saxon texts" (Bonser 1963: 309).

Talbot's (1967) *Medicine in Medieval England* devotes his first chapter to Anglo-Saxon medicine, but it is not concerned with plant names at all. The other chapters deal with Arab and Salernitan medicine, with medical education at the universities, with hygiene, etc. As one of the first, Talbot took the opposite viewpoint of Singer and his followers, insisting upon the fact that "the Leech-Book is not barbarous, full of folk-lore, charms, incantations and the like [...]. In short, far from the Leech-Book being a tissue of folk remedies and irrational ideas, it embodies some of the best medical literature available to the West at that time" (Talbot 1967: 18f.). He is also the first to recognise Löweneck's false attribution of the *Peri Didaxeon* to the medical school of Salerno (Talbot 1967: 18; see also chapter 4.3.2.).

Grendon (1974) does not treat Old English plant names in particular in *Anglo-Saxon Charms.* He concentrates on the edition of one hundred and forty-six charms taken from various manuscripts and texts (Grendon 1974: 110ff.), clearly following in the footsteps of Cockayne (1864-1866) and Grattan and Singer (1952) in distinguishing between heathen and vernacular elements, Christian liturgical elements and gibberish (1974: 24).[49]

In 1975, Sandmann published his Ph.D. thesis on Old English charms. His intention was a careful analysis of fifteen charms with regard to the religious elements, the versification and the medical and botanical aspects, factored out by Storms (1948), Grendon (1974) and Grattan and Singer (1952).[50] His findings should be treated with caution: the first volume of Bierbaumer's work (1975) was published the same year, so that, unfortunately, Sandmann did not have access to this very important information. This is a problem, as he attempts to identify the plants, their healing properties and the diseases discussed mainly on the basis of Cockayne's and Storms' findings (Sandmann 1975: 8) and is, in addition, often misled by his eagerness to filter out pagan elements (e.g., Sandmann 1975: 3f.).

Although one must admit that the analysis of charms is very often speculative and that some of their secrets most probably will never be solved,[51] Sandmann sometimes

49 It was not until the 80s and 90s that the 'veil of magic' was striped off from Anglo-Saxon remedies: M.L. Cameron's (1983b) research was devoted to the text sources of Bald's *Leechbook* and in 1988, M.L. Cameron studied the Anglo-Saxon texts as 'medical works', i.e. focusing on the usefulness of their recipes as medical treatments as he believed that behind every seemingly magical prescription there could be a rational reason. He published a thorough and comprehensive study of the topic in 1993 (see also van Arsdall 2002: 74ff.).

50 He obviously did not know that Grendon had published the *Anglo-Saxon Charms* a year before.

51 With regard to the so-called *Nine Herbs Charm*, Pettit (2001, 2: 99f.) notes: "The history of scholarship on this charm is not marked by major advances in understanding. [...] This is unsurprising in view of the obscurity of almost all of the charm's allusions, the lack of any known written sources, the uniqueness and difficulty of much of its vocabulary [...]." Two studies of

overshoots the mark. Two plant names, *mucgwyrt* 'wormwood' and *wegbrāde* 'greater plantain, waybread' may illustrate his approach to botanical terms. Sandmann believes that OE *mucgwyrt* lit. 'midge-wort' belongs etymologically to G *mugeln* 'zeugen, to conceive' (Sandmann 1975: 204ff.) and proposes it as evidence for the "Zeugungspotenz *par excellence*" ['power of procreation *par excellence*'] (Sandmann 1975: 205) of the Germanic primordial being *Urwesengott*. He believes he has detected this *Urwesengott* encoded in the element *ūna*, which relates to *mucgwyrt* in the Old English text ("Una þu hattest, [...]" ['You are called una']; *Lācnunga*, 535) and which, according to his theory, should be emended to "**ūre* < preGmc **ūrōn* 'die zum Urwesengott Hinzugehörige' (Pflanze)" ['the plant belonging to the primordial being']. Thus, for Sandmann (1975: 204), **ūre* is a kind of *kenning* for *mucgwyrt*.[52] This rather unlikely emendation is, however, not necessary: According to Pettit, L *ūna* 'the only one', accompanying OE *mucgwyrt*, merely "reflects mugwort's antiquity and importance [...] and perhaps its position as the first plant in the incantation" (2001, 2: 124); Bierbaumer (1, s.v. *mucgwyrt*) and Marzell (s.v. *Artemisia vulgaris* L.) identify *mucg-* as a variant of *mycg* 'midge', G *Mücke* < Gmc **mukjōn*.

Sandmann believes that his etymology of *mucgwyrt* is further supported by other plant names, which are presumably derived from names of gods, too; e.g., OE *fornētes folm* lit. 'Forneots palm', and ON *baldrs brâ* lit. 'Baldur's brow' (Sandmann 1975: 204), Sw *Baldersbrâ*.[53] That these are formed using names of gods, is correct, but Sandmann neglects the structural and morphological differences between *fornētes folm* and the reconstructed form **ūre*: while *fornētes folm* and *baldrs brâ* are combinations of N_{gen}+N, **ūre* is a derivative, i.e. even if it existed, it would not fit at all into the above-mentioned group.

Sandmann's etymology of OE *wegbrāde* is hazardous, too: starting from the Germanic conception of the broadness of (mother) earth and from the literal meaning of OE *wegbrāde* "die Weg-Breite" ['way-broad'] (Sandmann 1975: 208), Sandmann speculates that OE *wegbrāde* refers not only to the plant Plantago major L., but also in a mythological way to 'mother earth', i.e. the plant names seem to reflect a kind of mythological knowledge about the earth:

> Beide Kompositionsbestandteile von *wegbrāde* haben sich als Bezeichnung für die Erde unter dem Aspekt der Ausgedehntheit erwiesen. Die gleiche Bedeutung dürfte daher auch für *wegbrāde* selbst anzusetzen sein (Sandmann 1975: 208).
>
> [Both parts of the compound *wegbrāde* have proven to denote the earth, alluding to its broadness. Hence, the same meaning can be assumed for *wegbrāde* itself.]

---

magic, with a special focus on medieval German charms were published by Monika Schulz (2000 and 2003).

52 Compare also Pettit (2001, 2: 124).

53 For a short account of *baldr* see Grimm (1878, 3: 78ff.).

This speculation is quite unnecessary and rather far-fetched. Plantago major L., denoted by OE *wegbrāde* is a common weed with broad leaves which, in general, is associated with its co-taxonym Plantago lanceolata L., in German *Spitzwegerich* lit. 'pointed *Wegerich*'. Both the German and Old English names of the Plantago-species indicate that these plants belong to the same group, one of them with broad, the other one with narrow leaves. Sandman could have noticed this parallel, since he gives the German names *Breitwegerich* and *Spitzwegerich* for the two species of Plantago L. The other pertinent feature of this plant (in addition to its broad leaves), namely, its durability on being constantly trampled down at the roadside (see also Catalogue, s.v. *wegbrǣde*), is neglected and, in an adventurous emendation of the Old English text, Sandmann instead attempts to prove that it is the plant's healing virtues which are described here (Sandmann 1975: 210). The nomenclatural similarity of G *Breitwegerich* and the OE *wegbrāde* seems to have escaped his attention.

It is obvious that, eager to extract the mythological component in charms isolated from their textual and intellectual context, Sandmann lost sight of the medical tradition as a whole and over-interpreted in a number of cases. According to van Arsdall, "by the early medieval period, practical medicine had fused classical and "barbaric" elements (which included Roman superstition) [...] [so that] the tradition must be seen as a whole to be understood, not broken into hypothetical parts that are allegedly Teutonic, Celtic, classical and the like" (2002: 50), and "the isolated pieces become fully comprehensible only when considered part of a larger medical continuum – essential pieces in a large and complex puzzle" (van Arsdall 2002: 75).

No investigation of Old English plant names can do without Bierbaumer's *Der Botanische Wortschatz des Altenglischen* (1975-1979).[54] For this three-volume work, the author systematically perused all major texts, glosses and glossaries containing botanical vocabulary, collecting and listing all plant names and related botanical terms in alphabetical order (in total c. 900 lemmas). Each plant name is supplied with information on its occurrences, with all certain, probable or doubtful identifications, and now and then Bierbaumer provides commentaries on the etymological origin, the word-formation and the semantics of the Old English plant name. In his second volume, Bierbaumer (2: iv) announced a systematic analysis of the etymology and the semantics of the plant names, which, however, was never published.

Voigts seminal article "Anglo-Saxon Plant Remedies and the Anglo-Saxons" (1979) initiated a reversal in the evaluation of Anglo-Saxon medical texts and knowledge. Voigts not only demonstrated a longstanding tradition of vernacular medicine in Anglo-Saxon England, but also a direct and good knowledge of most of the plants mentioned in the texts. The former is evidenced by the English origin of at least one part of the medico-botanical material and by an evident need for translations of Latin texts and plant names into Old English – obviously there were Anglo-Saxons who needed medical texts and knew the plants only by their vernacular names. The latter is suggested, among other reasons, because a large number of the Mediterranean and Eastern plants could be had in Europe by way of trade, and some of them could

---

54 The three volumes were digitised and are now (in part) available online (cf. *DOEPN*, at <http://oldenglish-plantnames.uni-graz.at>); see below.

most probably even be grown in England. Meteorological investigations showed that the average temperature around the year 1000 was circa 10 C° (50 F) higher than today, as the Anglo-Saxon period lay "partially in an era termed the 'Medieval Climate Optimum', a warm period that lasted from around AD 800 to AD 1300" (Lee 2011: 304). Voigts concludes that "[...] Anglo-Saxons valued healing plants, that they valued books about healing plants, and that they dealt with both intelligently" (1979: 266).

D'Aronco was the first scholar to deal systematically with linguistic aspects of the Old English plant names. In her article "The Botanical Lexicon of the Old English *Herbarium*" (1988), d'Aronco investigated the etymology of the vernacular plant names in the Old English *Herbarium*, being concerned in the main with the various influences of Latin. D'Aronco found that a large amount of the Old English plant names can be ascribed to Latin influence.

The first one to focus on different linguistic aspects of the Old English plant names in a more systematic way was Sauer. In a series of articles (1992b, 1995, 1996, 2000, 2003) which were based on different Old English medico-botanical texts, Sauer analysed the morphology (especially the word-formation), the semantics and the etymology of the plant names attested. Sauer (1992b) is concerned with the botanical lexicon of the Old English *Herbarium* and of the *Antwerp-London Glossaries* (excluding names for trees); (1995) looks more generally at the Old and Middle English names of flowers and herbs; (1996) examines the plant names in the thirteenth century *Trilingual Harley Vocabulary*; (2000) investigates the linguistic aspects of the Old English plant names in the *Épinal-Erfurt Glossary*, and (2003) presents an overview of morphological patterns of the Old English plant names.

Sauer demonstrated that different etymological strata can be distinguished in the plant name lexicon, the oldest plant names going back to Germanic and West Germanic origins (e.g. *āc* 'oak' < Gmc **aik-*), but also confirmed the considerable influence of Latin. Latin influence can be seen not only in loan words (e.g. *minte* 'mint' < L *ment(h)a*), but also in loan formations (i.e. loan translations, loan renditions and loan creations, e.g. *fīflēafe* 'fiveleaf, cinquefoil', formed on L *quīnquefolium*).

Morphologically, N+N compounds represent by far the largest group of plant names in his corpora; they are only outnumbered by the unanalysable simplexes. Other word-formation patterns include combinations of Adj+N and N_{gen}+N combinations (e.g. *dæges ēage* 'daisy', *foxes glōfa* 'foxglove, etc.'). A smaller number of complex formations are derivatives (e.g. *mistel* 'mistletoe' < {*mist*}+{*el*}). With regard to semantics, according to Sauer, the two largest groups of composite plant names in his corpora refer to the place of growth or express a relation of comparison or resemblance (Sauer 1992b: 395ff.; 1996: 148ff.).

Sauer also demonstrated some problems in the analysis. *Wermōd* 'wormwood', *colloncroh* 'white waterlily' and *þūþistel* 'sow-thistle', for instance, are cases in point. At first glance, *wermōd* seems to be a N+N compound (< *wer* 'man' + *mōd* 'mind'), but the semantic analysis casts doubts on the morphological analysis and leaves the interpretation of the etymology of this plant name open to speculation (for more details, see Catalogue, s.v. *wermōd*). Interestingly, the subsequent development of the plant's name into ModE *wormwood* by way of folk etymological processes indicates its complexity and the general desire to motivate the plant name. The elements of *þū-*

þistel and *colloncroh*, on the other hand, are neither morphologically nor semantically transparent; an interpretation of these plant names has to remain rather speculative (see Catalogue, s.vv. *colloncroh*, *þūþistel*).

Hankins' dissertation "Bald's 'Leechbook' Reconsidered" (1992) is a translation of the first part of Bald's *Leechbook* with an "extensive discussion of the primarily herbal remedies prescribed by the tenth-century physician", in which she concludes "that many of the prescriptions were efficacious" (Hankins 1992: iii). Hankins feels the need to retranslate the text and to discuss the medical material because, on the one hand, she wants to realign the negative picture of Anglo-Saxon medicine drawn by medical historians and on the other hand, because she regards Cockayne's translation as improper: "[...] because of his insistence on maintaining the original syntax, Cockayne adopted a pseudo-Germanic style which often seems archaic and clumsy" (Hankins 1992: 7). Hankins offers a translation of "the first volume into modern, readable prose" (Hankins 1992: 3). She includes the plant names occurring in this part of the *Leechbooks* in the appendix at the end of her work, giving their botanical names and their therapeutic values, but she is not concerned with their linguistic aspects. Unfortunately, her work on the medical efficaciousness of the herbs mentioned in Bald's *Leechbook* was not published and remained almost unnoticed.[55]

Another cornerstone in scholarship on the Anglo-Saxon *materia medica* was undoubtedly M.L. Cameron's *Anglo-Saxon Medicine* (1993).[56] Cameron's approach to the Anglo-Saxon *materia medica* (especially, chapters 11 and 12) is that of a natural scientist, meaning that instead of separating 'magical' from 'rational' elements he asks if the elements mentioned in the recipes – even if they seem to be magical at first sight – do actually have a rational, scientific basis, and often finds that they do (see, for example, the description of a remedy for sore eyes, M.L. Cameron 1993: 121). In discussing the recipes from a scientific point of view, he sometimes includes explanations on the semantics of the plant names and mentions taxonomic aspects of vernacular plant name nomenclatures (e.g. with regard to *net(e)l(e)* 'nettle' and *rēade net(e)l(e)* 'red dead-nettle'; M.L. Cameron 1993: 108f.). Since his treatment of plant names is, however, focused on the therapeutic value of the plants, his discussions on the semantics of the plant names are neither comprehensive nor systematic.

Pollington's (2000) approach to the Anglo-Saxon medical texts and their contents is an encyclopaedic one:

> This book is about a good many things: herbs and plants used in healing and cooking; Old English charms and runes; the increasing power of literacy and the English Church; omens, amulets, and dreams; gods, elves and dwarfs; the production and survival of Anglo-Saxon

55 To my knowledge, van Arsdall (2002: 46) is the only one of the more recent publications to mention Hankins' work.

56 Cf. also M.L. Cameron (1983b) on the compilation of Bald's *Leechbook* and M.L. Cameron (1988) on the therapeutical values of the Anglo-Saxon *materia medica*.

> manuscripts; the names the early English gave their plants and trees, and how we can track them down today; shamans and witches and cunning women; theories of magic and the contents of graves (Pollington 2000: 15).

Since Pollington tries to cover every possible subject connected to Anglo-Saxon healing, and includes three translations of Anglo-Saxon medical texts into modern English (the Old English *Herbarium*, the *Lācnunga* and *Lǣcebōc 3*), he cannot go into much detail. And with regard to the linguistic aspects of the plant names, he hardly provides the reader with any new information or insight into the subject.

Two chapters and one appendix are entirely devoted to herbs and trees, to their names and to related aspects: chapters 4 and 5 and appendix 6. Chapter 4, called "The cultivation of herbs in Anglo-Saxon England" deals mainly with the formation of the Old English plant names. Since Pollington draws his information primarily from Sauer (1992b) and d'Aronco (1988) and adheres in large parts to the information structure in Sauer (1992b), this chapter, unfortunately, not only fails to provide significant new information, but also the corpus analysed is the same as in Sauer (1992b) and d'Aronco (1988). In the second part of the chapter, Pollington presents a summary of Biggam's (1994) article on OE *hǣwenh(n)ydele* '? scurvy-grass'.

Chapter 5 is called "Old English Medical Materials" and is subdivided into the sections "The Plants Named in the *Lacnunga*, *Herbarium* and *Leechbook*" and "Other materials used in the Leechdoms". The aim of the first section is to provide information on the origin, the number of occurrences and the identification of circa 480 plant names, and to give some "notes on the names and other matters" (Pollington 2000: 95). Surprisingly, the headwords in this section are not the Old English plant names, but Modern English names, which, sometimes, are only Modern English renderings of the Old English names. Thus, we can find existing vernacular plant names such as *ash*, but also fictitious Modern English plant names like, for example, *athelfarthingwort* or *atterlothe* (Pollington 2000: 98), which are, clearly, 'modernisations' of the Old English plant names, but they do not exist in Modern English. The information he provides on these Modern English lexemes ranges from simply giving the Old English name (as is the case, e.g., for *wheat flour*, Pollington 2000: 165) to elaborate explanations on the use of the plant, on the semantics (e.g., for *waybread*, Pollington 2000: 164) or on the etymologies of the plant names (e.g., *pennyroyal*, Pollington 2000: 148).

In appendix 6, called "Tree Lore", fifteen selected tree names such as *alder*, *apple*, *birch*, *hazel* and *fir*, are discussed with regard to their etymology, taking into account mythological aspects which led to the formation of the Old English names, their motivation and examples of place names formed on the basis of these tree names (e.g., *apple*, Pollington 2000: 499f.). The principles on which these fifteen tree names were selected are, however, not explained.

Van Arsdall's *Medieval Herbal Remedies: The* Old English Herbarium *and Anglo-Saxon Medicine* (2002) consists largely of two parts. In the first part of the book, van Arsdall traces the life experiences and the tragic suicide of Oswald Cockayne and explains how the translation of texts in general, and of the Old English medical texts in particular, are always influenced by the personal and professional background of the translator(s). Subsequently, she demonstrates how Cockayne's method of translating and interpreting set the course for later scholars to denigrate Anglo-Saxon medical knowledge and the Old English texts on medical and botanical issues. Van Arsdall maintains the view that Anglo-Saxon monasteries and monks were part of a European-wide system of healing, on a par with Continental medical training and knowledge, and in their work markedly similar to today's *curandera* medical practice.[57]

Van Arsdall devotes the second part of her book to a new translation of the Old English *Herbarium*, which is

> [...] not a literal translation in the sense that Cockayne's is. [...] The most literal translation of a term or a phrase was used as far as possible. But where that literal meaning would confuse a modern reader, the closest approximation to the Old English was chosen when it had a clear meaning in modern English (van Arsdall 2002: 113).

Linguistic aspects of plant names are of no particular interest for van Arsdall. Her identifications of the plants rely on d'Aronco and Cameron (1998) (van Arsdall 2002: 116).

Recent papers related to linguistic aspects of the Old English plant names are Sauer and Krischke (2004) and Bierbaumer, Sauer, Klug and Krischke (2008) and Sauer and Krischke (2009) who present etymological, morphological and semantic patterns as introductory information to their work on the *Dictionary of Old English Plant Names* (*DOEPN*). The *DOEPN* is being prepared under the direction of Peter Bierbaumer (University of Graz) and Hans Sauer (University of Munich). So far, the three volumes of Bierbaumer's *Der botanische Wortschatz des Altenglischen* have been digitised and (partially) revised. Linguistic information on the plant names is not yet available.

Rusche (2008) is a thorough investigation into the sources for plant names in several Anglo-Saxon glossaries, with a focus on the *Laud Herbal Glossary*. By analysing meticulously parallel entries in the various glossaries, Rusche shows not only that the Anglo-Saxon compilors had access to a number Anglo-Saxon and continental medical sources, but also demonstrates how frequent compiling and re-compiling of different sources can blur the relation between Latin (and, originally, Greek) lemma and the Old English plant name.

Krischke (2009) is an analysis of the compound plant names in the sections *nomina herbarum* and *nomina arborum* in abbot Ælfric's *Glossary*. The focus is on the semantic aspects of these plant names. It is argued that complementing pre-structuralist and structuralist semantics by using notions and concepts from cognitive semantics and from *gestalt* psychology helps broaden our explicative possibilities, especially when it comes to describing metaphorical and metonymic compounds and syntactic groups. The distinction between 'motivation' and 'associative relation' is established and then applied to the corpus of 40 compound plant names.

57 *Curanderismo* is a Mexican healing tradition based mainly on plants and herbs (van Arsdall 2002: xv, 80-6).

Apart from more general works on Anglo-Saxon medicine, the various editions of the Anglo-Saxon medico-botanical texts provide linguistic information on the Old English plant names. In most cases, however, this information is scattered in the commentaries on the identification of the plants or in the appendixes and glossaries, and it is more occasional information than a systematic treatment of linguistic issues.

Three editions give some information on linguistic aspects of the plant names, in most cases trying to reconstruct their etymological history: Lindheim's edition of the *Durham Plant Name Glossary* (1941), Pheifer's edition of the *Épinal-Erfurt Glossary* (1974) and de Vriend's edition of the Old English *Herbarium* and the *Medicina de Quadrupedibus* (1984). Both Lindheim and Pheifer comment on the Latin lemmas and on the Old English plant names listed in the glossaries.[58] De Vriend's explanatory notes on the plant names start from the Linnéan nomenclature, giving, then, the vernacular Latin (and/or Greek) plant names translated, the Old English renderings and their Modern English equivalents. This is followed by the basic assumptions of his identifications, including, if it seemed necessary to him, comments on the motivation of the plant names and on their etymology.[59]

Olds edited and translated *Leechbook* 3 (1984). Obviously, she did not know Bierbaumer's *Botanischer Wortschatz*, because her "Glossary of plant names", where she informs us about the identification of the modern English plant names, about the scientific botanical name and about the medicinal properties of the plant names, relies largely on identifications given by Cockayne (1864-1866), Storms (1948), and Grattan and Singer (1952). Linguistic aspects play no role in her discussion of the Old English plant names.

D'Aronco and Cameron's facsimile edition of the Old English *Herbarium* (1998) contains a chapter on the identification of the plants and one on their medicinal efficacy, but no information on the etymology or related issues is included.

In 2001, Pettit published an edition and analysis of the *Lācnunga*, one of the four major Anglo-Saxon medical texts. The second volume, concerned with comments on the text and providing a bibliography, treats various plant names with regard to aspects of their linguistic structure, but always in view of the identification of the plants. There is, for example, a discussion on the morphological status of *brādbisceopwyrt*, which Pettit assumes to be a compound (and not a syntactic group), based on the fact that the first element is not inflected in the accusative, the case demanded by the syntactic structure the plant name is embedded in (Pettit 2001, 2: 12). The edition of *Lācnunga* also contains in an appendix (no. 6) a list of Old English plant names including plant products, where plant names not found in Hoops (1889) are flagged.

---

58 Cf., for example, the discussion of the etymology of OE *colloncroh* (*Durham Plant Name Glossary*, no. 11), or the notes on *popeg* and *gundaesuelgiae* in Pheifer (1974: 76, no. 253 and 125, no. 976). Since the *Épinal-Erfurt Glossary* is not confined to plant names, the notes do not deal exclusively with plant names.

59 See, for example, the notes to OE *feltwyrt* (de Vriend 1984: 305, no. 73), or the notes to OE *dweorgedwosle* (de Vriend 1984: 310, no. 94).

The brief summary of previous approaches to the Old English plant names demonstrated that scholarship findings must not be regarded in isolation: As often postulated in the philosophy of science, observations of any kind are theory-driven as well as theory-loaded (Schurz 2011: 16, 50). Scientific objectivity is an ideal that can never be achieved – it can only be approached by way of scientific intersubjectivity, achieved when a number of competent people (i.e. researchers) agree upon the findings of other researchers on the basis of the data presented (Schurz 2011: 27). This, in turn, implies that, as each researcher is necessarily biased, being a subject with his own theoretical predisposition and his own moral values, all kinds of research results have to be regarded against the theoretical background of the researcher, the values of his time as well as his personal moral values. On the basis of these premises, it was shown that much of the early research on the Old English plant names was biased, and produced, to say the least, problematic research results in consequence of the conviction that the Old English medico-botanical texts were merely dumb translations of the Latin originals (e.g. Grattan and Singer 1952; Bonser 1963). In contrast, more recent research, eventually, overcame the biased interpretation, often approaching the plant names interdisciplinarily, e.g. by proposing rational, scientific solutions for the question of 'magical' practices in preparing drugs and potions (e.g. M.L. Cameron 1993), or by analysing the plant names in relation to botanical taxa (Bierbaumer 1975-1979).

4. PLANT WORLD, PLANT NAMES & THE ANGLO-SAXON MEDICO-BOTANICAL TEXTS

> The texts are not intended to be instructional, but are like cookbooks for experienced cooks.
>
> (van Arsdall 2002: xiii)

It is not possible to describe the semantics, the morphology and the etymology of a part of the lexicon without looking at the conceptual context implied: When we accept that words are "meaningful units" (Marchand 1969: 1), we must ask ourselves, to whom and why they are meaningful. Any lexical neologism is created with a certain purpose. Concepts with no importance for a speech community never ever get a name.

It is, on the other hand, a well-known fact that the Anglo-Saxon medico-botanical texts derive from Latin and Greek source texts (see chapter 4.2.-4.5.) and that a number of plant names are loan formations on Latin models or are even loan words from Latin. Some of the plants the Anglo-Saxons had names for were not native to Anglo-Saxon England and when they were grown, they survived only in sheltered places, in some secluded herb gardens of monasteries. Therefore, a linguistic analysis of the Old English plant names, especially with regard to semantic and cognitive aspects, is in danger of being perceived as a rather fruitless task, since not Anglo-Saxon, but Latin (and ultimately Greek concepts) could be concealed behind apparently Old English plant names. I would like to address this problem briefly by drawing attention to the impact of the plant world on the life of the Anglo-Saxons (chapter 4.1.). Chapters 4.2.-4.5. look at the textual tradition of medico-botanical issues in the British Isles and show that very early, even earlier than the first surviving documents are dated, the Anglo-Saxons were familiar with a large body of medical texts and, probably, had several copies of them at hand.

4.1. THE PLANT WORLD AND ANGLO-SAXON ENGLAND

The Anglo-Saxon people lived in a world where plants and plant derivatives pervaded practically every area of life. Archaeo-botanical evidence shows that plants played a major role in most areas of human activity: food and drink, medicine, fuel, litter and thatch, building, textiles and rituals (Hall 2003: 108).[60]

60 For recent studies into the "material culture of daily living" in Anglo-Saxon England, see Clegg Hyer and Owen-Crocker, eds. (2011).

As in all agricultural societies, the Anglo-Saxon farmers and their families were the backbone of the community. In fact, most Anglo-Saxons worked in agriculture or in cattle farming. The metrical calendar of Tiberius (which survives in three complete copies from the 11th c.; namely, in manuscript London, British Library, Tiberius B. v; Gneuss *Handlist*, no. 373; in manuscript London, British Library, Cotton Galba A. xviii; Gneuss *Handlist*, no. 334; in manuscript London, British Library, Cotton Julius A. vi; Gneuss *Handlist*, no. 337 and in a calendar with selected verses, in manuscript Oxford, Bodleian Library, Junius 27 (5139); Gneuss *Handlist*, no. 641; cf. McGurk 1983: 44) shows people working the soil, ploughing, sowing, and harvesting. Others are depicted felling trees or caring for the cattle, or, for instance, tending sheep. Apart from agriculture, the Anglo-Saxons had a number of occupations in which they depended on and worked with the goods nature provided. Bakers used all kinds of cereals; carpenters used wood; merchants dealt with all kinds of natural and processed goods. In short, plants and their products were the working material for almost all professions.

Plants were ubiquitous in Anglo-Saxon private life, too. According to Banham (2003: 119f.), plant foods provided the bulk of the diet, even for the aristocracy. Cereals were certainly the most important plants, but fruits like strawberries, apples, pears and plums, and vegetables like beans, peas, carrots and parsnips, beets and leeks were important eatables. Wine, made of grapes, and ale and beer, made of hops, were every-day beverages.

Architectural evidence indicates that Anglo-Saxon dwellings were wattle and daub or timbered huts (for the common people) with the halls for the more wealthy (the royal family and the aristocracy) made of woods and covered with straw or reed-thatched roofs (Fernie 1983: chapter 1). Stone as a building material was reserved for churches, but should stone not be readily available, wood would be used for churches, as well. Thus, a fair amount of wood was needed, for building purposes, but also for preparing meals or for heating during colder times of the year. In order to provide an ever-renewing resource, woodlands were carefully managed (Pollington 2000: 37).

In addition, prominent trees and sometimes major shrubs served as boundary markers, as is evidenced by the Anglo-Saxon charters (Hooke 2003: 17-39). The entry 542 in the *Charters* ((Birch), 2: 162), "tó æsc ápaldre leage norðeweardne" ['to the northern meadow of service-trees'], has already been mentioned as attesting not only to the borders of the real estate granted to the monks of Worcester by Ceolwulf, king of Mercia, but also to a plant name not yet dealt with in the existing dictionaries, namely, OE *æscapuldre* lit. 'ash apple tree', the service tree.

Trees also often inspired the names of newly founded towns and villages. Compare for example the place names ModE *Appleford* < OE *Æppelford* based on OE *æppel* 'apple', ModE *Mapleton* < OE *Mapleton(e)* based on OE *mapul-* 'maple', or ModE *Boasley* < OE *borslea* based on the botanical term *bors* 'spiky or bristly plant'.[61]

61 The latter attests to a plant name not recorded elsewhere in Old English texts: OE *bors* (Hough 2003: 65ff.; not an entry in the *DOE*).

Plants also played a major role in curing people. The remedies and potions the Anglo-Saxons took for ailments were – as is indicated by the medico-botanical literature – based mainly on herbs and to a much smaller part on minerals or products of animals such as blood or faeces:

> Although plants accounted for only 70-80 per cent of drugs, they made up well over 90 per cent of the ingredients in the recipes, so that for most purposes Anglo-Saxon medicine was a herbal medicine (M.L. Cameron 1993: 101).

To come back to the issue raised at the beginning of this chapter, namely, the question of whether the Anglo-Saxons were in command of the botanical lexicon preserved in the Anglo-Saxon texts, it is obvious that a society depending so much on nature and natural goods must have had a good knowledge of the form, the habitats and different growth stages of the vegetables, grains, fruits and other plants. A dish can only be prepared if one knows which parts of which plants can be used and which are unpalatable. Some trees qualify for heating while others suit building purposes better. The parts of the lexicon the people needed daily we can definitely assume were at their command.

With regard to medicine and the medico-botanical texts, the picture is somewhat different. Two strands of knowledge can be assumed to have existed: first, the learned, monastic knowledge based on texts that were partly derived from classical sources. They are indications of the high level of education in early English monasteries, and, indeed, place the Anglo-Saxons on a par with other European medical schools of Early Medieval times, all of them joining to produce a "pan-European, early medieval medical tradition" (van Arsdall 2002: 35) with a common corpus of texts, remedies, and practices.

The second strand of medical knowledge is independent of schooling and textual tradition: a good portion of knowledge is accumulated by personal experience and passed on orally from generation to generation. Even today, with medicine and healing at a very high level, people have a general knowledge of medical and botanical issues. Many adults today know, for instance, about the therapeutical values of the essential oils emanating from hot onions for earache, or of the antiseptic agency of camomile tea and treat their children accordingly. In times when most people were without immediate access to physicians, as in the Anglo-Saxon period, those with injuries or illnesses were forced to cure themselves. The Roman historian Tacitus, for example, described how Germanic warriors, who were wounded during battle, went to their mothers and wives to be attended: "[...] ad matres, ad coniuges vulnera ferunt; nec illae numerare et exigere plagas pavent, [...]" ['to the mothers, to the wives they bring their wounds, and these are not afraid to count and to examine the bruises'] (*Tacitus*, 7, 4 ; cf. also Payne 1904: 11).

Finally, (mass) literacy is a rather new phenomenon. For centuries, medical and botanical knowledge was passed on orally, from generation to generation, from midwife to midwife and from doctor to doctor:

> When one particular part of an herb, say a root, was found as being effective for some specific action, this information was orally transmitted whenever and wherever men communicated and one generation taught the other. This process takes place independently of literary transmission (Riddle 1974: 165).

This view is also supported by van Arsdall (2002), who argues that what is missing from the Old English medical texts is the practical wisdom, which, however, must have been transmitted in a kind of apprenticeship:

> It appears that lay and monastic practitioners were part of the early medical tradition, with some of the information transmitted in texts, some orally. [...] It is also the way a good deal of practical wisdom is today transmitted, together with textbook information, by modern herbologists (van Arsdall 2002: 95).

The German monk Walahfrid Strabo (*c. 808-†849) explains in the *Liber de cultura hortorum* (or *Hortulus*) where his knowledge about plants and their use comes from. One portion of knowledge he acquired by learning from books; the other by talking to people, and the third by personal experience, hard labour and the power of observation:

> Haec non sola mihi patefecit opinio famae
> vulgaris, quaesita libris nec lectio priscis;
> Sed labor et studium, quibus otia longa dierum
> Postposui, expertum rebus docuere probatis.
> (*Hortulus*)
>
> [This I have learnt not only from common opinion
> And searching about in old books, but from experience –
> Experience of hard work and sacrifice of many days
> Whence I might have rested, but chose instead to labor.][62]

As pointed out above, medicine is only one aspect of Anglo-Saxon life in which botanical knowledge and terminology was needed. Cooking, heating, building, planting and harvesting all required botanical knowledge and terminology. As all this knowledge was necessarily shared with the other members of the community, relatives and children, we can certainly assume that the botanical lexicon they were in good command of comprised at least the plants, herbs and bushes they had to live in, live on, and live with every day.

62 Text and translation are taken from Voigts (1979: 268).

4.2. ANGLO-SAXON TEXTS CONTAINING PLANT NAMES

Old English plant names are recorded in many kinds of texts and contexts: they form large sections in the Old English glossaries and can be found as glosses to Latin texts. They survive in charms and in scattered prescriptions written on fly-leaves[63] as well as in remedies noted down in the margins of otherwise unrelated texts.[64] Old English plant names are constitutive elements of Old English place names and are frequently listed in the Old English charters, where they indicate estate and other boundaries (cf. Hooke 2003), and in three of the manuscripts containing plant names, illustrations of the plants are provided, as well (manuscripts London, British Library, Cotton Vitellius C. iii, fols. 11-85; Ker *Catalogue*, no. 219; Gneuss *Handlist*, no. 402; manuscript Oxford, Bodleian Library, Ashmole 1431 (S.C. 7523), fols. 3-43; Ker *Catalogue*, no. 289; Gneuss *Handlist*, no. 527; and manuscript Oxford, Bodleian Library, Bodley 130; Ker *Catalogue*, no. 302; Gneuss *Handlist*, no. 549; see also Ohlgren 1986). The two major sources yielding Old English plant names, however, are the medico-botanical texts and the Anglo-Saxon glosses and glossaries.[65]

The following section is intended to show that the extant medico-botanical texts, glosses and glossaries, impressive as they are in being among the first in a European vernacular other than Latin and Greek, and recording the large amount of Old English plant names, are based on an even larger tradition of source texts that were copied, revised and recompiled in Anglo-Saxon England and that gave rise, eventually, to the major texts we can refer to nowadays when talking about medicine and botany in early medieval England.

4.3. MEDICO-BOTANICAL TEXTS

The corpus of Anglo-Saxon medico-botanical texts compiled in or translated into Old English consists of six texts: three medical compilations and one fragment in the

63 Fly-leaves are leaves at the beginning or the end of gatherings which remained blank until someone found them suitable for introducing short texts as, for example, medical recipes. So-called fly-leaf leechdoms can be found, e.g., on the manuscripts Lanhydrock, Bodmin, Collection of Lord Clifden, B. 12. 16, fol. 144 (Ker *Catalogue*, no. 98), and British Library, Cotton Domitian I, fols. 2-55 (Ker *Catalogue*, no. 146; Gneuss *Handlist*, no. 326). The last leaf of manuscript London, British Library, Royal 12 D. xvii (the manuscript of the *Leechbooks*), for example, is a blank fly-leaf.

64 For example, on manuscript British Library, Cotton Titus D. xxvi, fol. 16v and 17r (Ker *Catalogue*, no. 202; Gneuss *Handlist*, no. 380).

65 For a complete list of Old English texts, see A. Cameron (1973). For a slightly dated, but still the most comprehensive list of texts containing plant names up to date (and including glosses, glossaries, fly-leaves, etc.), see Bierbaumer (1: v-xiv, 2: iv-xiv, 3: vii-xxxiv). For plant names in English place names, see Smith (1956) and *The Vocabulary of English Place-Names*; for plant names recorded in the Old English charters, cf., e.g., Hooke (2003). Archaeo-botanical evidence from Anglo-Saxon England is discussed, e.g., in Hall (2003). No cookery books survive from Anglo-Saxon England. The earliest English cookery books date from the fourteenth century, i.e. are in Middle English (Banham 2003: 120). For the natural world as described in Old English poetry, see, e.g., Neville (2003, with further references).

vernacular Old English – Bald's *Leechbook*, *Leechbook 3*, *Lācnunga* and the *Omont Fragment*[66] –, and two translations of Latin texts – the Old English *Herbarium* and the Old English *Peri Didaxeon*. All these texts are of major importance not only for the history of English medicine and botany, but also for the history of European medicine as they represent the oldest medical writings in the vernacular to survive from Northern and Western Europe.

4.3.1. COMPILATIONS IN OLD ENGLISH

Bald's Leechbook *and* Leechbook 3

Manuscript London, British Library, Royal 12 D. xvii (Ker *Catalogue*, no. 264; Gneuss *Handlist*, no. 479) contains three books concerned with medical issues (hereafter referred to as *Leechbook* 1, 2 and 3).[67] *Leechbook* 1 has eighty-eight chapters and deals with external diseases; the main body of text is preceded by a table of contents. The arrangement of the chapters follows by and large the traditional order of ancient medical books, namely, the *a capite ad calcem* or 'from-head-to-foot' order. *Leechbook* 2, which is preceded by a table of contents, too, consists of sixty-seven chapters and deals mainly with internal diseases. A Latin colophon in verses at the end of *Leechbook* 2 discloses the presumed owner of the book, a certain 'Bald', and informs us that he ordered the book to be written (? or compiled) by a certain 'Cild' "Bald habet hunc librum Cild quem conscribere iussit" (*Lǣcebōc*, 91/1). Since the colophon, which is written in the same hand as the texts of *Leechbook* 1 and 2, is positioned at the top of one leaf (fol. 109r) and immediately precedes the table of contents of *Leechbook* 3, it is fairly safe to assume that

> [...] the scribe of Royal 12 D. xvii had before him two manuscripts, one containing Books I and II and concluding with the Latin verses and the other made up of what is now called Book III, and that having copied the first manuscript he went straight on with the second (Wright 1955: 14).

The organisation of contents supports this assumption: external and internal diseases are treated in *Leechbook* 1, respectively *Leechbook* 2, and a mixture of both in

66 The *Omont Fragment* was either a fly-leaf or, as was suggested by M.L. Cameron (1993: 31), the last leaf of a larger medical compendium.

67 Bald's *Leechbook* and *Leechbook 3* were edited by Oswald Cockayne (1864-1866) in *Leechdoms, Wortcunning and Starcraft of Early England*, vol. 2, and by Günter Leonhardi (1905) in *Kleinere angelsächsische Denkmäler* on pp. 1-120 (the edition used in the present work, quoted as *Lǣcebōc*); in addition, Bald's *Leechbook* was published in a facsimile edition by Cyril E. Wright (1955) as *Bald's Leechbook: British Museum, Royal Manuscript 12 D. xvii*; *Leechbook 3* was edited by Barbara Olds (1984) as "The Anglo-Saxon Leechbook III: A Critical Edition and Translation".

Leechbook 3.[68] The first and the second book are generally known as Bald's *Leechbook*, while the third book is termed *Leechbook 3*.

Like the first two parts, *Leechbook 3* has a table of contents for its seventy-six chapters, which outlines the content of the main body of text. The remedies are organised in the *a capite ad calcem* order, as they are in the first two books, but, in contrast to the first two parts, *Leechbook 3* is much simpler in structure and content, lacking almost completely discussions of symptoms or diagnoses; in addition, it contains a higher proportion of magical elements, charms and the like than Bald's *Leechbook* (M.L. Cameron 1993: 35-42). All three *Leechbooks* were intended as manuals of instruction. Most of the remedies are based on herbal material; the others rely on the use of animals, blood and faeces, as well as on rites, charms and the like.

There are indications in the *Leechbooks* informing that these texts were not alone in conveying medical knowledge in early medieval England. Two Anglo-Saxon practitioners are mentioned in *Leechbooks* 1 and 2, one called *Oxa* (*Lǣcebōc*, 37/4), the other *Dun* (*Lǣcebōc*, 88/30). On another occasion other medical compendia are mentioned: *Leechbook* 2 starts one of its section as follows: "Sume bēc lærað [...]" ['Some books teach us'] (*Lǣcebōc*, 85/29). The existence of medical books, which, unfortunately, did not survive to our day, and the approximate number of manuscripts to which the three *Leechbooks* are indebted can be further specified by an examination of their dating and origins.

Date and origin

The manuscript containing the three *Leechbooks*, manuscript London, British Library, Royal 12 D. xvii (henceforth referred to as *Royal*) is generally dated c. 950. It was probably written in the same scriptorium, presumably in Winchester, and certainly by the same scribe as the annals for the years 925 - 955 of the *Parker Chronicle*.[69]

M.L. Cameron (1993: 30, who follows Wright 1955: 16-7), suggests that *Royal* may be "a copy of a lost exemplar which may have been composed about fifty years earlier in the last years of Alfred the Great" since a contemporary of King Alfred, Elias,

68 According to Rubin (1974: 56), "the third leechbook is purely a collection of prescriptions, instructions and advice for a variety of diseases, and is sufficiently similar in content and style to the first two books as to give the appearance of a natural supplement to them."

69 Wright (1955: 22-3) questions the Winchester provenance of the manuscript: "My only remarks by way of conclusion would be to recall that the outstanding character of the handwriting of A 925-955 and of the Leechbook is just this quality of roundness in which we may perhaps see a reflection of a reaction natural to a revival of native tradition and national consciousness, both of which would be implicit in a movement inspired by Dunstan and Æþelwold and fostered by the royal patronage of Edgar and his Queen". What Wright suggests is that due to the troubling events caused by the Benedictine reform movement, there was a break in the scribal tradition. This is evidenced by the distinct hands for the entries 925-955 of the *Parker Chronicle* and the preceding ones. Consequently, even though the hands of the annals for 925-955 in the *Parker Chronicle*, of manuscript Royal 12 D. xvii (the *Leechbooks*) and of London, British Library, Cotton Otho B. xi (parts of which correspond to remedies in the *Leechbooks*, see Ker *Catalogue*, no. 180, art. 11), go back to one and the same scribe, and even if we know that the entries of the annals for 925-955 are of Winchester origin, written in the Old Minster, "it does not follow that the scribe had not been brought from another house and the Leechbook, if it is correctly identified as his work, may have been written therefore elsewhere" (Wright 1955: 23).

Patriarch of Jerusalem from about 879 to 907, ordered that some of the remedies be told to the king: "þis eal het þus secgean ælfrede cyninge domine helias patriarcha on gerusalem" ['all this commanded Helias, the patriarch of Jerusalem, to be said to King Alfred'] (*Lǣcebōc*, 88/8). Considering that king Alfred succeeded to the crown in 871 and that he died in 899, the compilation of Bald's *Leechbook* can possibly be dated to the period between 879 and 899.

However, the evidence of another manuscript in London, British Library, Cotton Otho B. xi (henceforth referred to as *Otho*; beginning of the 11th century, Winchester; Ker *Catalogue*, no. 180; Gneuss *Handlist*, no. 357), shows that the date of origin must be pushed even further back. In the attempt to trace the stemmatic relationship between *Royal* and *Otho*, Torkar (1976) dated the compilation of one part of Bald's *Leechbook* to a pre-Alfredian period. A small collection of remedies identical in large parts to remedies in Bald's *Leechbook*, originally written on the now destroyed last leaves of *Otho*, survived in a transcript carried out by Laurence Nowell in 1562.[70] Some of the recipes are unique, i.e. they do not occur elsewhere in the corpus of Old English texts, others correspond fairly clearly to remedies recorded in Bald's *Leechbook*. A direct comparison of the corresponding remedies in the two manuscripts showed that *Royal* could not have been the model for *Otho* or vice versa. *Otho* contains remedies not recorded in *Royal*, while archaic graphemes in these remedies in *Otho*, in contrast to more modern orthography in *Royal* (e.g. *eburf*earn in *Otho* versus *eoforfearn* in *Royal*), seem to point to the fact that *Otho* was compiled or copied earlier than *Royal* (Torkar 1976: 327).

Linguistically and paleographically, both manuscripts exhibit Anglian features masked by West-Saxon dialectal and paleographic peculiarities. In terms of ratio, Anglian features are prevalent in *Otho*, and this is striking, in particular with regard to the remedies identical in *Otho* and in *Leechbook* 1 of *Royal* (Torkar 1976: 323-5). In conclusion, *Otho* is either a copy of an already existing compilation of two texts of different ages (one older, with more archaic features and the other one younger, with a more modern orthography), or the scribe of *Otho* compiled the text using two distinct manuscripts, one older and one younger.

Specific for *Leechbook* 1 is, for example, the writing convention for [v] and [f]: in *Leechbook* 1 the digraph <uu> is used for the allophone [v]. Since this is an orthographic feature which can be found only in the early annals of the *Parker Chronicle* (465, 544, etc.) and which was already obsolete by the end of the 9th century (Torkar 1976: 326), it is highly probable that at least the original compilation of *Leechbook* 1 can be dated back to pre-Alfredian times, i.e. before c. 850 AD (Torkar 1976: 326). This dating is further supported by the use of <b> for [f], an orthographic convention which reflects a stage of Old English where Primitive Old English still preserved the distinction between the labial spirants *f* and *ƀ* (which fell together when *f* was voiced in intervocalic position and *ƀ* was unvoiced in word-final position). According to Blumbach (1974: 25), this is an orthographic convention which had its

70 For further details concerning *Otho* and Nowell's transcript (London, British Library, Add. 43703, fols. 261r-264v) see Torkar (1976), and (Ker *Catalogue*, no. 180). The contents of Bald's *Leechbook* are complemented by manuscript British Library, Harley 55, fols. 1-3, dated to the beginning of the 11th century (Ker *Catalogue*, no. 225, art. 1; Gneuss *Handlist*, no. 412).

heyday in the 8th and 9th centuries, Pheifer (1974: lxxx-lxxxi; lxxxix) dates it to the end of the 7th and the beginning of the 8th centuries with regard to the common ancestor of the *Épinal-Erfurt* glossaries, which both retain this distinction.

From this evidence, we may trace the following history of *Royal*: *Leechbook* 1, *Leechbook* 2, and *Leechbook* 3, originally, were three different and independent compilations. The original compilation of *Leechbook* 1 dates to pre-Alfredian times, i.e. at least before c. 850: diachronically, it shows archaic, and diatopically, Anglian features, which, however, were smoothed out in *Royal* at the latest. The compilation of *Leechbook* 2 must have been carried out during the reigning years of King Alfred, i.e. between c. 871 and 899. *Leechbook 3* was compiled c. 950 at the latest. Storms suggests that "there is sufficient internal evidence to put it down as a later compilation" than *Leechbook* 1 and 2 (Storms 1948: 16), without, however, revealing this internal evidence, but M.L. Cameron (1993: 35, 43) suggests a date of origin even earlier than the compilation of *Leechbook* 1 and 2, on the basis of the fact that *Leechbook 3* is hardly influenced by Mediterranean sources.

There are no indications of when exactly Bald ordered Cild to copy (or compile) Bald's *Leechbook*. This event must have taken place in any case between c. 900 (the compilation of *Leechbook* 2) and 950 (the dating of *Royal*). Bald's *Leechbook* and *Leechbook 3* were, eventually, combined to *Royal*, either by the scribe of *Royal* directly (Wright 1955: 14) or by a scribe who created an archetype, which was then copied by the scribe of *Royal*.

As the history of *Otho* shows, there is further evidence for the fact that another exemplar of Bald's *Leechbook* must have existed, different from the one leading to *Royal*. According to Torkar (1976: 327-8), *Otho* was copied from the archetype of Bald's *Leechbook*. The differences in the ratios of archaic and dialectal features in *Otho* and *Royal* are explained as being based on the use of different sources:

> Das für Ot [Otho B. xi] benutzte, auf x zurückführende Exemplar bot – im Vergleich zu Ot – angl. Dialektalismen sowie Archaismen in ebenso hohem oder in noch stärkerem Maße. In dem zu R führenden Redaktionsstrang wurden solche Formen abgebaut; in R selbst treten sie nur noch relikthaft auf (Torkar 1976: 327).

Consequently, it must be supposed that two different sources were available: one that led to *Otho*, the other one that led to *Royal*.[71] All in all, it is particularly noteworthy that what survives as a solitary text in a single manuscript, namely, as *Royal*, and fragmentarily in the single leaf of *Otho* must have existed in at least five copies in Anglo-Saxon England, and that one of these copies can possibly be dated earlier than c. 850.[72]

---

71 This view is supported by Wright (1955: 24), who suggests the existence of another version of Bald's *Leechbook* at the beginning of the 11th century, based on evidence of a piece of marginal writing on Royal 12 D. xvii.

72 See also M.L. Cameron (1993: 91), who insists on the existence of different, now lost sources for the Anglo-Saxon medical texts.

The Lācnunga

The *Lācnunga*, which survives only in manuscript London, British Library, Harley 585, ff. 130-193 (Ker *Catalogue*, no. 231, art. 2; Gneuss *Handlist*, 421), is a collection of herb recipes and charms based on various classical Greco-Roman and northern sources (Pettit 2001, 1: xxvii-xxxii).[73] The manuscript was dated to the end of the 10th, beginning of the 11th century, i.e. around the year 1000 in England, but no details about the origin and/or the provenance are known (Gneuss *Handlist*, no. 421).[74] M.L. Cameron is inclined to believe that "there is indeed no good reason to suppose that it was a planned work, but that it was always a commonplace book in which things were entered higgledy-piggledy" (1993: 46).

Date and origin

Basing their argument on the nature of the remedies, Grattan and Singer (1952: 20-2) believe the text to have been the work of several scribes:

> An Anglo-Saxon leech, working in the ninth century, put together a collection of recipes. He first adopted the ancient and recognized plan [...]. Then, *Lacnunga* **a** came into the hands of a second leech who was only very superficially Christianized [...].
>
> The combined collection of *Lacnunga* **a** and *Lacnunga* **b** came, in due course, into the hands of a third compiler. [...] Finally, this rebound copy was copied shortly after the year 1000 by the scribe who produced the existing manuscript and who added some passages at the end which might be treated as a fourth stratum, or as a fifth, if the entries which we designate as **x** be given a stratal significance (Grattan and Singer 1952: 20-1).

According to Ker (*Catalogue*, no. 231) and M.L. Cameron (1993: 46), only two scribes or compilers were at work, one, who copied the first two-thirds of the text and a second one who took over in the middle of fol. 179r, l.11.[75]

Since the text displays Anglian, Northumbrian, Kentish, and late West Saxon traits, Grattan and Singer (1952: 212) and Leonhardi (1905: 158) suggest that it is a West Saxon translation of a Northern original. The most recent investigation of *Lācnunga* by Pettit (2001), challenges this interpretation, arguing that both the specifically early

73 The *Lācnunga* was edited by Oswald Cockayne (1864-1866) in *Leechdoms, Wortcunning and Starcraft of Early England*, vol. 3, by Günter Leonhardi (1905) in *Kleinere angelsächsische Denkmäler* on pp. 121-59; by J.H.G. Grattan and Charles Singer (1952) in *Anglo-Saxon Magic and Medicine: Illustrated Specially from the Semi-Pagan Text* 'Lacnunga' on pp. 96-205, and by Edward Pettit (2001) as *Anglo-Saxon Remedies, Charms, and Prayers from British Library MS Harley 585: The* Lacnunga, vol. 1 (the edition used in the present work and quoted hereafter as *Lācnunga*). A facsimile edition of the entire manuscript was published in the *Anglo-Saxon Manuscripts in Microfiche Facsimile* by A.N. Doane and Phillip Pulsiano (1994-), vol. 1.

74 Fols. 179/10-fol 193 are dated to the beginning of the 11th century (Ker *Catalogue*, no. 231, art. 2).

75 For a detailed account and discussion of the scribes and the script of Harley 585, see Pettit (2001, 1: 134-5, 138-40).

(archaic) and the specifically Anglian and Kentish features have been thought to be much more numerous in the manuscript than they actually are (Pettit 2001, 1: 201-2).

In contrast to the general body of text of the *Lācnunga*, the Old English gloss to the *Lorica of Laidcenn*, known formerly as the *Lorica of Gildas* (no. 65 in *Lācnunga*) exhibits specifically Anglian and a few Mercian forms, in addition to some archaic features such as the use of <b> for [f] for example in *deoblu*, *neabulan* or *li[b]re*, which points to the fact that this gloss probably existed as early as 8th c. AD as an Anglian (Mercian) text (Pettit 2001, 1: 201-2).[76]

The Omont Fragment

In 1977, a single leaf, containing on the *recto* side a number of remedies in Old English, was edited for the first time by Schauman and A. Cameron (1977). This leaf, manuscript Louvain, Bibliothèque Centrale de l'Université, Section des manuscrits, Fragmenta H. Omont no. 3 (Gneuss *Handlist*, no. 848), called the *Omont Fragment* "might be dated c. 850-900 and placed in a centre under Mercian influence, with a slight preference for a date nearer to 850 [...]" (Schauman and A. Cameron 1977: 301; see also Ker *Catalogue*, no 417).

Since the *Omont Fragment* may have been the last leaf of a much larger medical compendium, as has been suggested by Schauman and A. Cameron (1977: 295) and M.L. Cameron (1993: 31), and since the origin of the manuscript has been dated earlier than that of the other Anglo-Saxon medico-botanical manuscripts,[77] we can assume that "there were other earlier collections similar to the surviving *Leechbooks*, of which this leaf is a lone surviving witness" (M.L. Cameron 1993: 31). Parallels between recipes in the *Omont Fragment* and in *Leechbook* 2 (no. 64 and 66; cf. *Lǣcebōc*, 87-90) and between the *Omont Fragment* and *Lācnunga* (no. 49) further support this assumption, especially as a direct copying from either one or vice versa is excluded. To conclude, the original text from which the *Omont Fragment* was copied must be dated earlier than c. 850.

76 "Such clusters of linguistic features untypical for LWS – combined with the haphazard ordering and presentation of the text in the MS – lend support to the common opinion that the compiler(s)/scribe(s) of *Lacn.* were working from more than one (and probably many) exemplars, which may well have had distinguishing linguistic features. Numerous linguistic features characteristic of LWS and late OE in *Lacn.* which are in accord with the MS date are to be set against the very few possibly early features. [...] The dialectal origins of *Lacn.* (excluding for the moment *LL.*) are for the most part unclear [...]. However, decisively Anglian (as opposed to possibly Kentish, EWS, or late OE) forms are not numerous – and appear even less so if we exclude instances in the verse [...] and recognize the presence of occasional specifically Kentish forms. The disproportionate number of non-WS forms in *LL.* may suggest that this situation does not result from WS regularisation by the latest *Lacn.* scribe(s) at least, but that the bulk of *Lacn.* had prior (not necessarily primary) existence in the WS dialect, though distinctively EWS forms are few" (Pettit 2001, 1: 201-2).

77 "From the discussion of the verso it is clear that the earliest writing on the fragment is the unplaced Continental semi-uncial of the 7th or 8th century" (Schauman and A. Cameron 1977: 299).

4.3.2. TRANSLATIONS FROM LATIN

The two medico-botanical texts translated from Latin into Old English are the Old English *Herbarium* and the *Peri Didaxeon*. In contrast to the compilations in Old English and to the *Peri Didaxeon*, the *Herbarium* is arranged according to plants; it discusses their properties, their habits of growing and the diseases they were assumed to cure. The Old English *Herbarium* is outstanding as it is the first translation of a medico-botanical text into a European vernacular (d'Aronco 2008: 113).

***The Old English* Herbarium**

The Old English *Herbarium* survives in three manuscripts from the Old English period and in one from the Early Middle English period: (1) in manuscript London, British Library Harley 585, fols. 1r-129v, which also contains the *Lācnunga* (c. 1000, Ker *Catalogue*, no. 231; Gneuss *Handlist*, no. 421); (2) in manuscript London, British Library, Cotton Vitellius C. iii, fols. 11-85 (c. 1050, Ker *Catalogue*, no. 219; Gneuss *Handlist*, no. 402); (3) in manuscript Oxford, Bodleian Library, Hatton 76 (c. 1050, Ker *Catalogue*, no. 328b, art. 1); (4) and in manuscript London, British Library, Harley 6258B, fols. 1r-51r (c. 1200, Ker *Catalogue*: xix).[78]

Despite its title – *Herbarium* on fol. 19v of Cotton Vitellius C. iii ("[H]ERBARIU[M] / APVL[EI] [P]LAT[ONICI] / QUOD AC[CE]PIT AB E/SCOLAPIO ET [A]LCH[I]RONE / CENTAVRO MAGISTRO / ACHILLIS"; d'Aronco and Cameron 1998: 23, plate III) – it is not a work on botanical issues, but a medical text containing remedies based mainly on herbal ingredients. The Old English version of the enlarged *Herbarium* consists of basically three texts (*De herba vettonica liber*, *Herbarium* and *Liber medicinae ex herbis femininis*), which contain a collection of remedies based on herbs, and is followed by the Old English version of the *Medicina de quadrupedibus*, a collection of remedies based on animal products (de Vriend 1984: xiv).[79] Both are made up of different source texts. According to d'Aronco, the remarkable work of fusing six different source texts into one was carried out in 10th century Anglo-Saxon England:[80]

> [...] in an Anglo-Saxon centre, some scholars had the idea of putting together six Latin treatises, transmitted separately in their respective

---

78 The Old English *Herbarium* was first edited by Oswald Cockayne (1864-1866) in *Leechdoms, Wortcunning and Starcraft of Early England*, vol. 1; then, by H.J. de Vriend (1984) as *The Old English Herbarium and Medicina de Quadrupedibus* (this is the edition used in the present work and quoted hereafter as *Herbarium*); further, it was published as a facsimile edition by Maria Amalia d'Aronco, and Malcolm L. Cameron (1998), *The Old English Illustrated Pharmacopoeia: British Library Cotton Vitellius C III*, and, recently, it was translated by Anne van Arsdall (2002), *Medieval Herbal Remedies. The* Old English Herbarium and Anglo-Saxon *Medicine.*

79 For the complete inventory of the codex, see de Vriend (1984: xiv-xv), Gneuss *Handlist* (no. 402).

80 For the source texts, see d'Aronco (2007: 38-48), who disagrees with M.L. Cameron (1993: 68), who believes that the Old English manuscripts translate a fifth-century Latin compilation.

> textual traditions. They selected the texts carefully, translated them faithfully and even conserved the illustrations found in the respective Latin sources (d'Aronco 2007: 56).

All four extant manuscripts reproduce the same Old English original, which may have been produced in one of the Benedictine monasteries in Winchester (d'Aronco 2007: 46). Thus, the production of the *Herbarium* is, just as that of the *Leechbooks*, closely related to the Benedictine reform movement. The Old English *Herbarium* provides information on the medicinal plants, the ailments these plants cure and directions concerning their gathering and their rendering into remedies. In contrast to the three pre-conquest manuscripts, the herbs in the 12th century manuscript are arranged in the alphabetical order of the Latin plant names.

The Old English Peri Didaxeon

The Old English *Peri Didaxeon* survives in a single manuscript from the second half of the 12th century, manuscript London, British Library, Harley 6258b, fols. 83v-98v (cf. Ker *Catalogue*: xix), where it is preceded by the late version of the Old English *Herbarium*. Because of the transitional character of its language, which exhibits characteristics of Old English as well as of Middle English, the manuscript was included in Cockayne's *Leechdoms* (vol. 3) and, ever since, has been regarded as a Late Old English text. It is a collection of medical recipes translated from a Latin text entitled *Practica Petrocelli*, which was erroneously ascribed to an unidentified, eleventh-century Salernitan teacher Petrocellus by the first editor of the Latin text, Salvatore de Renzi (1856). The first editor of the Old English *Peri Didaxeon*, Löweneck (1896), published his edition after Cockayne's.[81] He did not question de Renzi's dating of the text and was probably tempted to remedy the negative assessment of Anglo-Saxon medicine by connecting the Old English text to the more respected medicine of 12th century Salerno, as Glaze points out in her account of text production and the medical training in early medieval England:

> If, he [Löweneck] reasoned, Anglo-Saxon medical knowledge was as rational and respectable as that of Salerno, the center where medical education of a professional variety found its start in the 12th century, then some of Cockayne's detrimental impressions might be undone (Glaze 2007: 473).

Unfortunately, by linking the *Peri Didaxeon* to the school of Salerno, de Renzi and, later on, Löweneck 'deprived' the Anglo-Saxons of a text which shows that, at least one century before the medical school of Salerno was founded, the Anglo-Saxons were already at the height of medieval medical knowledge. According to Maion (2007: 502), "the Latin exemplar of the *Tereoperica* used by the Old English translator

81 *Peri Didaxeon*, ed. by Max Löweneck (1896) (henceforth referred to as *Peri Didaxeon*); Sanborn's unpublished edition "Peri Didaxeon" (1983) is not available; neither is the most recent edition by Danielle Maion (1999), "Eduzione, Traduzione e Commento del Peri Didaxeon", part of her unpublished dissertation "Cultura e traditioni letterarie del mondo germanico antico e medievale".

belonged to the textual tradition of the oldest witnesses of this Latin work".[82] The oldest extant copy of the text is manuscript Paris, Bibliothèque Nationale de France, lat. 11219 (Maion 2007: 498), dated to the ninth century; the Old English *Peri Didaxeon* was probably translated between the end of the 11th and the beginning of the 12th century (Maion 2007: 506).[83]

The *Peri Didaxeon* consists of sixty-six chapters of prescriptions and recipes ordered according to the diseases they are believed to cure. Occasionally, Greek words appear: e.g., *glaucomata* in the entry "[*þæt* grecas nemniađ] glaucomata" (*Peri Didaxeon*, 15).

4.4. GLOSSES AND GLOSSARIES

Another important source for Old English plant names are glosses and glossaries. Some of the glossaries contain only a handful of botanical terms, others are replete with botanical vocabulary (e.g. the *Épinal-Erfurt Glossary*), or even dedicated entirely to plants and plant names (e.g. the *Durham Plant Name Glossary* or the *Laud Herbal Glossary*). Similar to the Anglo-Saxon medico-botanical texts, the Anglo-Saxon glossaries are of different ages and were conceived to meet different purposes. While many of the Latin lemmas were more or less easily identified, at times, the compilers or translators had to 'invent' Old English names for plants they did not know or for plants they knew, but had no name for. This, certainly, triggered the creation of new names as well as the borrowing and subsequent 'Anglicising' of Latin plant names; there are also instances where the glossator preferred to leave a gap in the manuscript. Other glossaries, most notably the *Antwerp-London Glossaries* and Ælfric's *Glossary*, were conceived as school-books, as manuals of instruction for young people, and are therefore less likely to contain unusual, arcane or learned plant names, let alone gaps. In the following, I would like to highlight briefly some important aspects.

The Épinal-Erfurt Glossary *and the* Leiden Glossary

The *Épinal-Erfurt Glossary* is one of the earliest witnesses to the Old English language (but see 'The *Antwerp-London Glossaries*' below) It is a glossary in *a*-order, arranged according to the Latin lemmas. It contains about 1100 Old English glosses to the more than double as many Latin lemmas and survives in two manuscripts which are largely identical (some leaves are missing in *Épinal*), but which are not copies from one another. The *Épinal* manuscript (Épinal, Bibliothèque Municipale 72(2), fols. 94-107; Ker *Catalogue*, no. 114; Gneuss *Handlist*, no. 824) is dated to the end of the 7th and the beginning of the 8th centuries,[84] and was, apparently, written by an Anglo-Saxon scribe in England. Manuscript Erfurt, Codex Amplonianus f. 42, fols. 1-14v

82 The compilation known as the *Peri Didaxeon* is the translation of the part of the *Practica Petrocelli Salernitani*, which is known as the *Tereoperica* (Maion 2007: 500).

83 M.L. Cameron (1993: 71) suggested that the *Petrocellus* be the work of an English compiler, since there are a number of similar arrangements in the *Petrocellus* and in Bald's *Leechbook*.

84 See also A. Cameron (1973: 248, no. D 7) and Gneuss (1981, no. 824). Pheifer dates the mansucript c. 725 (1974: lxxxix).

(Ker *Catalogue*, App. no. 10)[85] is a slightly shorter copy of a common ancestor of *Épinal* and *Erfurt*, executed by a single scribe in the minuscule used at the Cathedral School in Cologne in the late eight and early ninth centuries (Pheifer 1974: xxvi). Since the scribe often makes mistakes, misinterpreting, for example, <a> for <cc> or <f> for <p> (Pheifer 1974: xxvi), it may be assumed that the scribe did not know Old English and was copying *literatim*.[86]

According to Rusche (2008: 130-1), the plant names in the *Épinal-Erfurt Glossary* can be traced to the so-called *Hermeneumata Pseudo-Dositheana*,[87] a Greek-Latin schoolbook used in language teaching, and made up of an alphabetical glossary, a subject glossary and some colloquies (Rusche 2008: 130-1). As he further argues, at least one copy of the subject glossary must have reached England by the end of the seventh century, as large parts of the plant name section of the *Hermeneumata* are found, equipped with Old English glosses, in the *Épinal-Erfurt Glossary*, and a smaller amount of plant names in the *Leiden Glossary* (Rusche 2008: 131).[88] This assumption is contested by Porter (2011b), who argues instead that the *Leiden*, the *Épinal-Erfurt*, the *Corpus*, the *Cleopatra*, the *Brussels* and the *Antwerp-London* glossaries derive from a now lost, 7th century bilingual Latin-English glossary, which was compiled on the basis of Isidore's *Etymologiae* (see 'The *Antwerp-London Glossaries*' below).

The Antwerp-London Glossaries

The *Antwerp-London Glossaries* in manuscript Antwerp, Plantin-Moretus Museum, 47 and manuscript London, British Library, manuscript Add. 32246 (Ker *Catalogue*, no. 2; Gneuss *Handlist*, no. 775 and *Handlist*, second add., no. 775), are made up five components: Latin-Latin and Latin-Old English glossaries, written in the margins and at the end of the Latin grammar *Excerptiones de Prisciano* and containing 4,225 entries (Porter 2011b: 155).[89] While the manuscript is from the 11th century, the order of the entries and frequent parallels to the *Épinal-Erfurt Glossary*, the *Leiden Glossary* and to Isidore's *Etymologiae* point to the fact that the *Antwerp-London Glossaries*

---

85 See also A. Cameron (1973: 251, no. D 36).

86 The two glossaries were published by Wright and Wülcker (1884) in *Anglo-Saxon and Old English Vocabularies*; by Wallace M. Lindsay (1921) in *The Corpus, Épinal, Erfurt and Leyden Glossaries*; by Sweet and Hoad (1978) in *A Second Anglo-Saxon Reader* and by Jason D. Pheifer (1974) as *Old English Glosses in the Épinal-Erfurt Glossary* (the edition used and quoted as *Épinal-Erfurt Glossary*); a facsimile edition was published by Bernhard Bischoff et al. (1988) in *The Epinal, Erfurt, Werden, and Corpus Glossaries*.

87 The extant copies of the *Hermeneumata* were edited in vol. 3 in Georg Goetz, ed. (1888-1923), *Corpus Glossariorum Latinorum a Gustavo Loewe inchoatum*.

88 Manuscript Leiden, Rijksuniversiteit, Vossianus Lat. Q. 69, fol. 24 was edited by John H. Hessels (1906) as *A Late Eighth Century Latin-Anglo-Saxon Glossary preserved in the Library of the Leiden University (MS Voss. Qo. Lat. no. 69)*.

89 The first editions of the glosses in the Antwerp and London manuscripts do not cover all glosses. A part of the glosses are printed by Meritt (1945), no. 22; another by Förster (1917); then by Wright and Wülcker (1884), nos. 4 and 5. The three longest lists of glosses are edited by Lowell Kindschi (1955) "The Latin-Old English Glossaries in Plantin-Moretus MS 32 and British Museum MS Additional 32, 246" (henceforth used, and quoted as *Antwerp-London Glossaries*). For the most recent edition, containing all glosses and glossaries, see David W. Porter (2011a), *The Antwerp-London Glossaries*. Unfortunately, it was published after the submission of my thesis.

were, in fact, much older than *Épinal-Erfurt* and *Leiden*, and even contributed to them (Porter 2011b: 155). Porter's (2011b) analysis of the textual origins of the *Antwerp-London Glossaries* suggests that the *Antwerp-London* class list and its close sibling, Ælfric's *Glossary* (see below), are based on a lost seventh-century glossary based on Isidore's *Etymologiae* and produced at Canterbury, probably under the direction of Theodore and Hadrian. This Latin-Old English class glossary was, according to Porter (2011b), the basis from which the *Leiden Glossary*, the *Épinal-Erfurt Glossary*, the *Corpus Glossary*, the *Cleopatra Glossaries*, the *Brussels Glossary* and Ælfric's *Glossary* descended, and whose influence can be felt in Aldhelm's *De virginitate* and *Enigmata*, as well.[90]

The Corpus Glossary

Manuscript Cambridge, Corpus Christi College 144 (Ker *Catalogue*, no. 36; Gneuss *Handlist*, no. 45), dated to the late eighth or early ninth century, consists of two glossaries originating in South-West England, one shorter (fols. 1-3v), the other one longer (fols. 4-64 v), known as the *Corpus Glossary*.[91] According to Pheifer (1974: xxix), it seems as if it derives from the same *Hermeneumata* as *Épinal-Erfurt*, being an important witness to their common ancestor, as it may be regarded as "virtually a third text of the Épinal-Erfurt Glossary, and one which derives from the archetype independently of the common exemplar of Épinal and Erfurt". Rusche (2008: 132), in contrast, argues that the *Corpus Glossary* is basically a compilation of the *Épinal-Erfurt* glossary with additional material, while Porter (2001b: 168-70) classifies the *Corpus Glossary* as a compilation from many sources, ultimately descending from the now lost Canterbury ancestral glossary (see 'The *Antwerp-London Glossaries*' above).

The Cleopatra *glossaries and the* Brussels Glossary

In the 10th century, the *Corpus Glossary* was combined, possibly, with the *Hermeneumata* (Rusche 2008: 132) to form two further derivatives known as the *Cleopatra Glossaries* (manuscript London, British Library, Cotton Cleopatra A. iii; Ker *Catalogue*, no. 143; Gneuss *Handlist*, no. 320). The volume is made up, in fact, of three glossaries (an alphabetical glossary on fols. 5-75v; a subject glossary on fols. 76-91v and a glossary with interlinear glosses on fols. 92-117v), of which only part of the entries are entirely devoted to plants and trees, as they are part of the subject glossary.[92]

---

90 In addition, Porter (2011b: 159) suggests that the Old High German glossaries (ed. Steinmeyer and Sievers 1879-1922) were derived from the same source.

91 The two glossaries were printed by J.H. Hessels (1890) as *An Eight-Century Latin-Anglo-Saxon Glossary*; by Wright and Wülcker (1884) in *Anglo-Saxon and Old English Vocabularies* as number 1; by Wallace M. Lindsay (1921) in *The Corpus, Épinal, Erfurt and Leyden Glossaries*; by Sweet and Hoad (1978) in *A Second Anglo-Saxon Reader*, pp. 1-101 (the edition used and quoted in the present work as *Corpus Glossary*); a facsimile edition was published by Bernhard Bischoff et al. (1988) in *The Epinal, Erfurt, Werden, and Corpus Glossaries*.

92 The *Cleopatra* glossaries were first edited by by Wright and Wülcker (1884) in *Anglo-Saxon and Old English Vocabularies* as numbers 8, 11 and 12. William Stryker (1952), "The Latin-Old English Glossary in Ms Cleopatra A iii" edited the first of the glossaries (fols 5-75v); his edition is used when the first of the *Cleopatra* glossaries shall be referred to, henceforth as *Cleopatra*

Closely related to the first of the *Cleopatra Glossaries*, and possibly copied from it, is a glossary from the 10th/11th c. in manuscript London, British Museum, Cotton Otho E. i (Ker *Catalogue*, no. 184), which was edited by Voss (1996).[93]

Manuscript Brussels, Bibliothèque Royale, 1828-30 (185) (Ker *Catalogue*, no. 9; Gneuss *Handlist*, no. 807), dated to the beginning of the 11th century, contains five Latin-Old English class lists, again probably based ultimately on the Greek-Latin *Hermeneumata* (cf. Rusche 2008: 132; Ker *Catalogue*, no. 9) and closely related to *Antwerp-London* (Porter 2011b: 173-4), known as the *Brussels* glossaries. Fols. 94v-95v of the manuscript hold the plant name glosses, which form the fifth of the glossaries, beginning "Nomina herbarum grece et latine" ['The Greek and Latin names of the plants'].[94]

The **Durham Plant Name Glossary** *and the* **Laud Herbal Glossary**

Two glossaries are entirely devoted to plant names: the *Durham Plant Name Glossary* and the *Laud Herbal Glossary*. Fols. 82-84v of Durham, Cathedral Library, manuscript Hunter 100, contain a list of plant names in Latin glossed by Old English lexemes (plus one plant name on fol. 106v, see Ker *Catalogue*, no. 110).[95] According to Lindheim (1941: 1-6), the manuscript is a 12th century copy made by a Norman scribe of an Anglo-Saxon exemplar. Since some of the Latin and Greek lemmas are not accompanied by Old English glosses, Lindheim (1941: 5-6) suggested that the basis for this glossary was a trilingual Latin-Greek-Old English glossary, connected to a Latin-Old English glossary of the 7th century based on Greek-Latin hermeneumata – the archetype for the *Corpus Glossary* and the *Épinal-Erfurt Glossary* – which, however, have not survived. According to Rusche (2008: 139-41), however, the *Durham Plant Name Glossary* is rather a derivative of a glossary made from the Old English *Herbarium*, which has not survived either, but which, very probably, served as a model for a part of the *Laud Herbal Glossary*, too, as a number of similarities and corresponding entries seem to suggest. In addition, the *Durham* glossary made heavy use of glosses related Dioscoride's *De materia medica*, which was probably brought to Canterbury in the 7th century (Rusche 2008: 140).

Glossary (Stryker); see also Voss (1988) for addenda and corrigenda. John J. Quinn (1956) edited the second and the third of the glossaries; his edition is referred to when the second or the third of the glossaries are quoted as *Cleopatra Glossary* (Quinn); see also Voss (1989). All three glossaries are edited by Rusche (1996) "The Cleopatra Glossaries" (unpublished dissertation). A brief overview is provided by Lendinara (1999: 22-7); Kittlick (1998) provides analyses of the phonology, the inflectional morphology and dialectal features of the glosses.

93 The entries in Otho not found in the first *Cleopatra* glossary were published by Meritt (1961). For addenda and corrigenda to Meritt (1961), see Voss (1996).

94 The *Brussels* glossaries were published by Wright and Wülcker (1884) in *Anglo-Saxon and Old English Vocabularies* as number 9 (henceforth quoted as *Brussels Glossary*); a facsimile edition and a description was published by Rolf H. Bremmer and Kees Dekker (2006) in *Manuscripts in the Low Countries*.

95 The glossary was edited by Oswald Cockayne (1864-1866) in *Leechdoms, Wortcunning and Starcraft of Early England*, vol. 3, and by Bogislav von Lindheim (1941) as *Das Durhamer Pflanzenglossar: Lateinisch und Altenglisch* (the edition used and quoted in the present work as *Durham Plant Name Glossary*).

The *Laud Herbal Glossary* (manuscript Oxford, Bodleian Library, Laud Misc. 567; Ker *Catalogue*, no. 345) was written in the 12th century in England; apart from the glossary, the manuscript contains several medical texts.[96] According to Rusche (2008: 135-42), it shows many parallels with other Old English glossaries and derives from different sources. The origins of the entries to letter A, for instance, can be traced to a Greek-Latin plant name glossary that has not survived, to the *Hermeneumata*, to glosses related to the Old English *Herbarium* (and another Greek-Latin glossary that has not survived), to the same glossary that survives in the *Durham Plant Name Glossary* and to two further, unattested, but independent sources.

The* Harley Glossary*, and other related glossaries

A number of glosses and glossaries stand in relationship to the ones discussed above. Important for the present study are the *Harley Glossary* of the late 10th or early 11th century (London, British Library, manuscript Harley 3376; Ker *Catalogue*, no. 240; Gneuss *Handlist* 436)[97]; further, the glossaries in manuscript Oxford, Bodleian Library, Bodley 730 (Ker *Catalogue*, no. 317; 12th/13th c.)[98] and the glossary in Bodley 163 (manuscript Oxford, Bodleian Library, Bodley 163 (2016) on fols. 250-251; Ker *Catalogue*, no. 304, art b; Gneuss *Handlist* 555; middle of the 11th century), whose origins can be clearly traced back to the *Leiden Glossary* (Lendinara 1999: 330).[99]

***A topical glossary: Ælfric's* Glossary**

Ælfric's *Glossary*,[100] is a manual intended to be used in schools in the teaching of Latin. It is somewhat exceptional, because it probably was the first class glossary of English terms according to Hüllen (1999: 65), while Porter (2011b) assumes close correspondence with the *Antwerp-London Glossaries* arguing for the existence of a now lost, 7th century Canterbury glossary that served as the model for a number of glosses and glossaries (see 'The *Antwerp-London Glossaries*' above). According to Hüllen's classification (1999: 62ff.), it is a universal topical glossary, a precursor of the modern thesauri.[101] The eight subject fields covered by Ælfric's *Glossary* are titled *nomina* (general terms relating to the world, e.g., "*deus omnipotens*, þæt is god ælmihtig" 'God', "*caelum* heofen" 'heaven' and "*stella* steorra" 'star'), *nomina membrorum* (names of the parts of the body, including other terms, such as those describing kinship relationships, for which no title is provided; cf. Zupitza 1880: 299-306), *nomina avium* (the names of birds), *nomina piscium* (the names of fish), *nomina ferarum* (the names of animals), *nomina herbarum* (the names of plants), *nomina arborum* (the names of trees), and *nomina domorum* (the names of buildings). All

96 The *Laud Herbal Glossary* was edited by Richard Stracke (1974).

97 The glossary was edited by Robert Oliphant (1966), *The Harley Latin-Old English Glossary*.

98 Edition by Waltraut Ziegler (1981), referred henceforth as *Bodley 730*.

99 The glossary was edited by Lendinara (1988-1989) and reprinted in Lendinara (1999), on pp. 347-55.

100 The glossary was edited by Zupitza (1880).

101 Hüllen's (1999) study describes the development of English dictionaries from 800-1700. For descriptions and classifications of glosses and glossaries, see further, Lendinara (1999), and the overview in Sauer (2009).

these sections contain a number of more or less unrelated terms, too. The two sections of special interest in the present study are the *nomina herbarum* and the *nomina arborum*, as these provide us with the names for plants and trees.[102]

Ælfric composed the *Glossary* to complement his *Grammar*, to which it is attached as many as seven times. *Grammar* and *Glossary* provide the young not only with a comprehensive and contrastive grammar of Latin, but also with the vocabulary for what Ælfric seems to have considered to be the most important concepts of the society, the religion, the flora and fauna, the geographical realities, etc. Both *Grammar* and *Glossary* enjoyed a great success, even far beyond the Anglo-Saxon period, as is attested by the remarkable number of manuscripts in which they are preserved: the *Grammar* survives in fifteen manuscripts, the *Glossary* in seven of them.[103]

Despite the fact that Ælfric was one of the most learned men of his times and that we would perhaps expect a larger number of learned Latin or Latinate terms in his glossary, Ælfric's *Glossary* "[does] not contain any rare or obscure 'glossary' words [...]. It is a subject glossary, well-suited for use in constructing a 'subject colloquy' of the type which pupils in Anglo-Saxon schools used to practise their skills" (Lendinara 1999: 226) and the plant names it records probably belong to the core lexicon of the Anglo-Saxons. The plant name entries in the *Glossary* are topic-related and the items arranged largely in taxonomic order, starting with the higher taxa and descending to the lower (Krischke 2011: 95-7) while related concepts are grouped together.

The present overview demonstrated that the glosses and glossaries not belong to the earliest writings to survive from the Old English period, but also testify to even older compilations, which, unfortunately, have not survived to our days. As regards the plant name sections in the glosses and glossaries, their origins can be traced to a number of sources; among them the Greek-Latin *Hermeneumata*, an Isidore-based 'Canterbury glossary', and the Old English *Herbarium*. In 2003, Rusche (2003: 191) even argued that they may descend from a common archetype based on Dioscorides' *De materia medica*.

> At some point in time, by the mid- to late-seventh century, whether in England or elsewhere, plant-names were collected out of the Greek text of Dioscorides' *De materia medica*; at first, these were written down in textual order and supplied with Latin glosses [...]. This glossary was then, at least partially, alphabetized. By this time, the glossary must have come to England, if it was not there from the beginning, for a number of Old English glosses were added to it. [...] At some point between these early glossaries and the later surviving witnesses, the copies of the Dioscorides glossary diverged, some being heavily alphabetized, some having many more Old English glosses added (Rusche 2003: 191).

102 A large number of glosses in Antwerp, Plantin Moretus Museum, manuscript 47 and London, British Library, Add. 32246 overlap with Ælfric's *Glossary* (Porter 2011b: 155-8; see also Lendinara 1999: 224-6).

103 For a list of manuscripts containing Ælfric's *Grammar* and *Glossary* see Gneuss (2001a: iv ff.) and Gneuss (*Handlist*, p. 150).

Further research into the stemmatic filiation of the Anglo-Saxon glossaries, their Old High German counterparts and the prose texts may even elucidate, in the words of Porter (2011b: 177), "the birth of written English at the Canterbury school" of Theodore and Hadrian in the 7th century.

4.5. SOURCE TEXTS AND ANGLO-SAXON COMPILATIONS IN LATIN

The sources of the Anglo-Saxon compilors and translators were the medical texts of the Mediterranean cultures, which had been transmitted constantly in the form of summaries and re-workings, and which incorporated knowledge and practices from the Roman, the Greek, North African and Byzantine, but also from the Germanic and Celtic cultures. Both the translations into Old English and the Anglo-Saxon compilations in Old English, whether fragmentary or complete, depended to varying degrees on medical and botanical compendia in Latin (see below). There is no indication that the Anglo-Saxons could handle Greek well enough to understand Greek medical texts (Feulner 2000: 4-5), but, as by the end of the 6th and 7th centuries, most of the Greek texts had been translated into Latin or had been integrated into compilations in Latin in the form of quotations, of summaries or of re-workings (M.L. Cameron 1993: 65), this was not a major problem.

The glossaries (and the glosses) of the Anglo-Saxons derived from a plethora of sources, as well. In view of the compilation of the plant names in the *Laud Herbal Glossary*, Rusche (2008: 142) states that the heterogeneous character of the glossary

> suggest that whatever center compiled the Laud Glossary had access to a large variety of earlier Old English materials, both lexical and medical, and to numerous Greek-Latin medical/herbal glossaries that had been circulating on the continent since at least the ninth century.

The Anglo-Saxons produced a number of medico-botanical texts in Latin, as well. Two large compilations in Latin contain medical texts: one is known as the *Canterbury Classbook* (manuscript Cambridge, University Library, Gg. 5. 53; middle of the 11th c.; origin and provenance Canterbury; cf. Gneuss *Handlist*, no. 12);[104] here we find a number of verses on medical terminology and excerpts mainly from Pseudo-Soranus and Petrocellus. The other is a manuscript dated to the beginning of the 12th century and known as the *Ramsey Scientific Compendium* (manuscript Oxford, St John's College, 17 and British Library, Cotton Nero C. vii, fols. 80-84; cf. Ker *Catalogue*, no. 360; Morgan 2010). On fols. 1v-2v and 175r-177v are medical texts dealing with the qualities of the four humours, with bloodletting and with proper dieting in order to keep the humours in balance.

Besides them, a large a number of medical texts in Latin must have circulated. Manuscripts London, British Library, Sloane 475, fols. 125-231, and Sloane 2839

104 Rigg and Wieland (1975) give a description of the compilation as a whole and of its contents in detail. The songs on fols. 432–41 were edited by J.M. Zielkowski (1994) as *The Cambridge Songs*.

(both 11th/12th c.; Gneuss *Handlist*, no. 498.1 and 498.9), and Cambridge, Peterhouse 251, fols 106-91 (11th/12th c.; Gneuss *Handlist*, no. 145), for instance, testify to the fact that texts such Galen's *Epistola de febribus* and *Ad Glauconem de medendi methodo*, the *Liber Esculapii de chronicis passionibus*, Isidore's *Etymologiae*, and the 'Petrocellus' were well-known texts in Anglo-Saxon England and circulated widely.[105]

It is not known whether the medical text of the classical and post-classical antiquity first came to England in the 7th century with Theodore of Tharsus, reputed of having a good medical knowledge (Rusche 2008: 133), but the Anglo-Saxon compilations in Old English and Latin and the glossaries suggest the availability of numerous classical and post-classical sources. As the varied and complex stemmatic filiations of the Anglo-Saxon medico-botanical texts (including the glosses and the glossaries) and their sources have been analysed in detail in the respective editions and the secondary sources, I will only give a brief overview, intended to demonstrate their variety.[106]

Among the classical and post-classical authors to whom the Anglo-Saxon texts paid direct or indirect lip service are Pliny's *Naturalis historia*, the prognostics of Pseudo-Hippocrates (cf. Ker *Catalogue*, no. 102) and Galen's texts (see above), Dioscorides' *De materia medica*, Pseudo-Pliny's *De re medica* and Oribasius' *Synopsis*. Circulation can also be attested for texts such as Odo of Meung's *De naturis herbarum* and for Isidore of Seville's *Etymologiae sive origines*, which draws on Cassius Felix of Numidia's *De medicina ex Graecis Logicae Sectae auctoribus liber translatus* and was the source for a number of Anglo-Saxon glosses. Further, Bede quotes from Vindicianus' *Epitome altera* and from the *Epistula Vindiciani ad Pentadium nepotem*, which had a wide circulation in the Middle Ages (M.L. Cameron 1993: 67), as did the work of the Roman Theodorus Priscianus. Marcellus of Bordeaux' *De medicamentis*, was probably one of the sources of *Leechbook 3*. In the 6th century, Alexander of Tralles authored, among others, a medical work in the *a capite ad calcem* order, which survived – altered and supplemented – as the well-known *Practica Alexandri*. The *Practica Petrocelli* and the *Passionarius Galeni*, in turn, yielded the sources for the Old English *Peri Didaxeon*.

4.6. SUMMARY

From all these texts, medical compendiums and Latin-Old English glossaries, we learn that beginning with the earliest Anglo-Saxon period and up to the Norman Conquest in the 11th century and even after, some of the Anglo-Saxons were not only familiar with these texts but were also interested in them and used them extensively. The extant manuscripts provide clear evidence of a considerable number of original texts in Old

105 For a list of manuscripts containing medical texts, see Gneuss (*Handlist*, p. 173).

106 An overview (which is also the basis of my brief account) is provided by M.L. Cameron (1993, chapter 9). For the editions of the texts, see chapters 4.3.1, 4.3.2, 4.4. and 4.5.. In addition, the following secondary sources may be consulted: for Bald's *Leechbook* and *Leechbook 3*: M.L. Cameron (1983a, 1983b); for the Old English *Herbarium*: d'Aronco (2003, 2007, 2008, 2009), Hofstetter (1983); for the glossary in Bodley 163 Lendinara (1999); for the *Peri Didaxeon*: Glaze (2007) and Maion (2007); for the *Laud Herbal Glossary* Rusche (2008).

English and of Latin-Old English glossaries and glosses, which must have been in circulation, and which were constantly copied, recopied, revised and recompiled. In these texts, the remarkable number of c. 900 plant names were collected – plant names that yield a vivid picture of the way the Anglo-Saxons viewed their natural world, experienced it and coined it into words. Despite the differences in the origins and nature of the texts, it is important to keep in mind that they share the feature of being outstanding in medieval Western Europe: The Old English *Herbarium* is the first translation of a medical text into a European vernacular and Bald's *Leechbook* is the first compilation of a medical text in a European vernacular.

The number and the variety of source texts in Latin that circulated in Anglo-Saxon England demonstrate the great interest in medicine and the vigorous activity in the field of medicine, reflected by the intense text production, translation, copying and glossing related to this field of learning. In addition, by their mere existence, the Anglo-Saxon medico-botanical texts in Old English prove that in early medieval England certain groups of the population were sufficiently literate in Old English to read and understand the often detailed and sophisticated instructions, even though they may not have been so in Latin.

5. MORPHOLOGY AND WORD-FORMATION OF THE OLD ENGLISH COMPLEX PLANT NAMES

> sume naman synd PRIMITIVA, þæt synd frumcennede oððe fyrmyste, swâ-swâ ys *schola* on englisc scôl, *mons* dûn, *ciuitas* ceaster. sume synd DIRIVATIVA, þæt synd, ðâ ðe cumað of ôðrum namum: *scholasticus* sê ðe on scôle ys, [...].
>
> (Ælfric's *Grammar*, ed. Zupitza 1880: 11)

5.1. PRELIMINARIES

The first step towards a linguistic analysis of the Old English plant names is a formal one: The present chapter looks at the morphological make-up of the complex plant names. The morphological analysis of the plant names provides us with the basis for the ensuing semantic analysis, as it uncovers the elements that carry semantic information. In the absence of a comprehensive description of Old English word-formation,[107] the morphological analysis of the Old English complex plant names in the present survey is based on Bauer (1983) and on Marchand (1969). As will be seen in the following, the two approaches to word-formation, on the one hand the traditional strand of analysis, represented by Bauer (1983), and, on the other, the structuralist one, represented by Marchand (1969), complement each other in various respects, but are different in the evaluation of others. In these cases, the differences are pointed out and their explicative value is considered with regard to the corpus of plant names. This is supplemented by Kastovsky (1992), who follows Marchand's (1969) structuralist methodology in his approach to word-formation in Old English.

Applying modern theories of word-formation to a language no longer spoken raises a number of problems, and some are addressed in the following as they are of a more general nature.[108] According to Marchand (1969), word-formation is the "branch of science of language which studies the patterns on which a language forms new lexical units, i.e. words. Word-formation can only be concerned with composites which are analysable both formally and semantically" (Marchand 1969: 2).[109] This definition immediately raises several fundamental questions.

107 As pointed out earlier, a comprehensive description of the word-formation of Old English is, still, a desideratum; this was noted, for instance, by Kastovsky (1992: 408), Gneuss (1993: 139) and, recently, by Johnston (2011: 67).

108 Compare Kastovsky (1992: 357-61). More specific problems are pointed out in the analytical sections (especially in chapters 5.4. and 5.5.).

109 Simplexes, e.g., *āc* 'oak' or *trēow* 'tree', by definition *monemes* (or *monomorphemes*), are on principle excluded from the morphological analysis. Loan words, e.g., *finul* 'fennel', which are not morphologically or semantically analysable within the system of the receiving language are

To begin with, it is the question of productivity, which has to be considered. It is not possible to test the productivity of a word formation pattern reliably without (living, human) informants. We can, of course, come to conclusions about the productivity of a word-formation pattern on the basis of the ratio of extant lexemes formed according to one pattern in contrast to the rest, or on the basis of the continued productivity of that pattern in subsequent periods (Kastovsky 1992: 357), but as the textual evidence of Old English is, to say the least, problematic (cf. chapters 2.2. and 5.4.), statements with regard to productivity always have to be made cautiously. We can state on the basis of the extant corpus of Old English complex plant names, for instance, that the majority of the Old English complex plant names are N+N compounds (5.3.3.1.), but this does not preclude that other word-formation patterns were, perhaps, even more productive, because the limited number and specific nature of the texts surviving from Anglo-Saxon times may have failed to record these plant names.

Second, as languages are in a state of constant flux and development,[110] diachronic processes may affect the morphological and semantic transparency[111] of the lexemes, thus distorting the possibility of clear-cut formal and semantic analyses. This can affect several aspects of the analysis. One of them is the problem of delimiting word-formation and inflectional morphology. *Morphology* "deals with the internal structure of word-forms" (Bauer 1983: 33) and is divided into *inflectional morphology* and *word-formation*. This distinction implies that there is a discrete boundary between inflectional morphology and word-formation: while inflectional morphology deals with the combination of bound grammatical morphemes with lexical morphemes and produces word-forms (e.g. {*trēow*} 'tree' + {Pl.} > *trēowu*, pl. of 'tree'), word-formation treats the combination of free or bound lexical morphemes with other free or bound lexical morphemes and produces new lexemes (e.g. {*pīn*} 'pine' + {*trēow*} 'tree' > *pīntrēow* 'pine-tree'). This assumed boundary is, however, often transgressed because, for instance, the morphological status of affixes changes through time. The second element of OE *fīflēafe* 'fiveleaf', for instance, is a case in point: the simplex *lēaf* 'leaf' is inflected according to the strong declension when used alone but according to the weak declension (*n*-declension) when used as a second element in the compound *fīflēafe*. The question of whether to classify the ending *-e* as a word-formation element – as a bound lexical morpheme – or as an inflectional element – a bound grammatical morpheme – can be answered in different ways: Some linguists classify the *-e* at the end of *fīflēafe* as a derivational suffix, while others(e.g., Kastovsky 1992: 383; 2006: 153) regard it as an inflectional element (cf. chapter 5.3.3.4.). This may seem a minor point, at first, but it touches upon an important question of word-formation pattern. A compound is defined as the combination of two

classified as polysyllabic monemes (cf. Marchand 1969: 3, 5-8). Note that Marchand's notion of *lexical unit* is not the same as Cruse's (1986: 76-80); see, further, fn. 160 in the present chapter.

110 "The fact of language change is a given; it is too obvious to be recorded or even listed among the assumption of our research" (Labov 1994: 9).

111 "A lexeme is said to be transparent if it is clearly analysable into its constituent morphs and a knowledge of the morphs involved is sufficient to allow the speaker-listener to interpret the lexeme when it is encountered in a context" (Bauer 1983: 19).

stems "that has not subsequently been subjected to a derivational process" (Bauer 1983: 29) – his counter-example is the denominal adjective *school-master/ish*.[112] The question of whether OE *fīflēafe* is an exocentric, metonymic compound with the form {*fīf*}{*lēafe*}, that is, with the *-e* classified as an inflectional element, or a suffixation with a complex modifier and a suffix as head ({*fīf*}{*lēaf*}/{*e*}) is, then, difficult to decide, and both analyses are possible.

Occasionally, plant names are extant in more than one morphological shape (cf. chapter 5.5.). OE *crāwlēac* 'crow garlic' lit. 'crow leek', is a N+N compound made up of {*crāw*}{*lēac*}, while OE *crāwanlēac* 'crow garlic', lit. 'crow's leek' is a N_{gen}+N formation: {*crāw*}{*an*}{*lēac*}: As the first element of the latter formation is in the genitive while the other is not, the two plant names instantiate different word-formation patterns. Yet, the difference between them is one of structure and not of denotation, as the denotata are the same:[113] They are morphological variants of one and the same plant name. Similarly, *blindnetele* (with an uninflected first element; cf. Zupitza 1880: 311/4) and *blinde netele* (with an inflected first element, cf. *Lǣcebōc*, 21/1) are morphological variants of the same plant name denoting the dead nettle. Between the first pair of plant names and the latter yet another difference pertains: while *crāwlēac* and *crāwanlēac* are both compounds[114], *blindnetele* and *blinde netele* are not: *blindnetele* is a compound, while *blinde netele* is an idiomatised syntactic group (for details and criteria of classification, see chapter 5.3.1.).

Third, the morphological and semantic analysability of linguistic signs does not automatically entail their transparency. On the one hand, lexicalisation processes (chapter 5.3.1.) can impair the transparency of lexemes, on the other, reverse processes of folk etymology (chapter 7.5.) – though aiming at (re-)establishing a transparent relationship between the constituents of the lexeme and other words of the lexicon – do not necessarily constitute a meaningful, secondary motivation of the lexemes in question.

When a lexeme is synchronically simple this does not mean that it was not complex at its origins and vice versa. Indeed, simple Old English plant names may have been complex at their origins: For example, OE *īfig* 'ivy' was probably a compound in Germanic, that is to say, **īf-heȝ* (*Kluge*, s.v. *Efeu*; *OED*, s.v. *ivy*). In contrast, synchronically complex Old English plant names may have been simple at their origins: OE *lufestice* 'lovage', for example, was originally, as a loan from Latin, a moneme within the system of Old English. The Old English form *lufestice* is, however, a compound made up of {*lufu*} 'love' + {*stice*} 'stitch' – the result of the folk etymological change of MedL *levisticum* in analogy to OE *lufu* 'love' and OE *stice* 'stitch' (see Catalogue, s.v. *lufestice*). The fact of diachronic processes shining through in the morphology of the complex plant names is allowed for by a diachronic-synchronic approach touching, for instance, upon questions such as lexicalisation (see chapter 5.3.1.) and the reversal, folk etymology (chapter 7.5.), as well as on questions

112 *Schoolmasterish* is, according to Bauer (1983: 29), a derivative (suffixation) from the complex (compound) noun *schoolmaster*.

113 Cf. also Gardner (1968: 15) with regard to the difference between OE *folccwen* and *folces cwen*, both 'queen of the people'.

114 For the status of genitive combinations, see 5.3.3.2.

such as stem variability (cf. chapter 5.5.). For all these plant names, whether they are clear and unequivocal in their morphological make-up, or whether some of their aspects require a specific discussion and analysis, an airtight classification is not what the present study aims at. Instead, in the following, it is attempted to unveil the general and the more specific morphological patterns inwrought in the complex Old English plant names and the results are intended to contribute to the more general aim of this survey, namely, to an understanding of the plant-naming processes as a whole.

5.2. THE CORPUS' FUZZY BOUNDARIES

Traditionally, complex lexemes are divided into two major groups: compounds and derivatives.[115] Compound lexemes are made up of (at least) two free lexical morphemes, respectively two full linguistic signs, while derivatives are made up of a free lexical and a bound lexical morpheme (Bauer 1983: 28-30).[116] As mentioned in the introduction (1.1., 1.3.), from the entire botanical lexicon of Old English the compound plant names, those made up of two lexemes were chosen as the corpus to be investigated in the present survey: their complexity in morphological and semantic terms promises to produce interesting insights into the early English ways of coining new terms for plants.

Despite the theoretical differentiation between *inflection* and *word-formation* on the one hand (5.1.), and between *compounds* and *derivatives* on the other, quite a number of Old English plant names transgress the boundaries between the categories and have to be analysed individually: here, the categories' fuzzy boundaries become visible. Some plant names combine composition with inflection. These plant names are N_{gen}+N group formations such as *hundeshēafod* 'small snapdragon' ({*hund*} 'dog' + {*es*} + {*hēafod*} 'head'), or Adj+N formations, such as *blind(e)net(e)l(e)* 'dead nettle' ({*blind*} 'blind' + {*e*} + {*net(e)l(e)*} 'nettle'), and their analysis raises the important issue of delimiting compounds from syntactic groups, as the first elements of compounds – in contrast to those of syntactic groups – are usually not inflected. Inflection of the first element can, thus, be an indicator that the complex expression is a syntactic group (phrase) and that, in turn, suggests that the complex expression cannot be part of a morphological analysis with regard to word-formation, but is a 'mere' syntactic construction with a modifier/head relationship.

---

115 Apart from composition and derivation, which act on the morphemic level, clipping and back-formation (which operate beyond morpheme boundaries) were also productive in the corpus of Old English plant names; they can be traced in some of the loan words (cf. chapter 7.4.2.1.).

116 Marchand's classification (1969) differs from Bauer's in that he distinguishes in the category *reduced grammatical syntagmas* (i.e. combinations of full linguistic signs as opposed to non-grammatical syntagmas, e.g. blendings, acronyms, etc.) between *expansions* (compounds and prefixations) on the one hand and *derivations* (suffixations) on the other. Expansions are combinations of two morphemes (A and B); with the second morpheme (B) being free and lexical, modified by the first (A, which can be free or bound; Marchand 1969: 11). In derivations, B is not a free lexical morpheme, but a bound lexical morpheme that transposes element A (a free lexical morpheme) from one word class into another, or from one lexical class into another (e.g. the suffix *-ly* transposes the adjective *clear* into an adverb *clearly*; cf. Marchand 1969: 12-3).

Because of this impact, dictionaries sometimes treat plant names with an inflected first element differently from plant names uninflected first element. In the *DOE*, for example, compounds are entries in their own right, while syntactic groups on principle do not appear as separate entries, but are subsumed to the modifier or the head of the plant name, which is the headword in the dictionary. Thus, idiomatised[117] plant names, such as the N_{gen}+N formation *foxes fōt* 'unbranched bur-reed, etc.' lit. 'fox's foot' or *blind(e)net(e)l(e)* 'dead nettle' lit. 'blind nettle', 'disappear' in the entries to *fox* 'fox', respectively, to *blind* 'blind' (*DOE*, s.v. *blind*, s.v. *fox*), while the lexical items *cistbēam*, *cistelbēam* and *cistenbēam*, all denoting the 'chestnut-tree' (all three lit. 'chestnut tree'), are given as separate entries although they can be classified as 'mere' morphological variants of one plant name (cf. chapters 5.5., 7.4.2.1.). Thus, one can get the impression that the chestnut was an important tree in medieval Britain, but *cistbēam* and *cistelbēam* are attested only once; *cistenbēam* is attested more often: 5 occ. according to the *DOE* (s.v. *cisten-bēam*). And here, again, it is only with caution that we can draw quantitative conclusions from the vocabulary recorded in the extant Old English texts, because, on the one hand, the texts are representative of the language use of the educated elite but not necessarily of the Anglo-Saxons in general, and on the other hand, because the corpus of extant Old English texts is certainly not the complete corpus of texts produced in Anglo-Saxon England (see chapter 2.2.).

Fuzzy boundaries also exist between different morphological types within composition. The aforementioned 'crow garlic', for instance, appears as the N_{gen}+N compound *crāwanlēac* lit. 'crow's leek' and as the N+N compound *crāwlēac* lit. 'crow-leek', and synthetic compounds such as OE *grundeswelige* 'groundsel' can be analysed as compounds and derivatives at the same time (see chapter 5.3.3.6.). Here, decisions can be taken as to whether the two morphological variants are regarded as reflecting diachronic processes, such as the coalescence of syntactic groups into compounds, e.g. from *crāwanlēac* lit. 'crow's leek' to *crāwlēac* lit. 'crow leek', with continuous morphological univerbation (cf. Durkin 2009: 39), or whether the variants are regarded as independent, semantically parallel but morphologically differing formations. A detailed analysis of texts from different periods that could give us insightful results is not possible since we do not have many early witnesses of early Old English. Fuzzy boundaries can also be claimed for the plant names in which morphological, phonological and spelling variants are not easily distinguished (see chapter 5.5.).

5.3. COMPOSITION

In accordance with the definition given above, compounds are complex lexemes made up of two (or more) full linguistic signs, i.e. combinations of at least two free lexical morphemes. According to the word-classes of the lexemes resulting from composition, compounds can be sub-divided into compound nouns, compound adjectives,

117 For discussion of lexicalisation, idiomatisation and institutionalisation, see chapter 5.3.1.

compound verbs, etc. Since plant names are by definition nouns, I will be concerned with the classification and the further analysis of compound nouns only.

5.3.1. COMPOUND OR SYNTACTIC GROUP? LEXICALISATION, IDIOMATISATION & INSTITUTIONALISATION

As mentioned above, one of the major problems one faces when dealing with a corpus of complex lexemes is the delimitation of compounds from syntactic groups (or phrases; cf. also Durkin 2009: 37-8). A large number of Old English plant names can be assigned unequivocally to one of these groups, such as the compound *eoforfearn* 'polypody', {*eofor*} 'boar' + {*fearn*} 'fern'; or *brēr, þe hiopan on weaxaþ* 'dogrose', which is a noun phrase with a post-modifying relative clause, lit. 'thorn-bush on which hips grow'. Yet, an equally large number of complex expressions do not fit neatly into one of these two categories. They raise questions such as: Is this composite a compound or is it a (semantically lexicalised) phrase? Where do syntactic processes stop and where does the lexical process of compounding begin (and the other way round)? Is it necessary to look for botanical identifications, that is, do these expressions denote species of plants, or are they merely descriptions of plants made up of, for instance, a noun with a pre-modifying adjective? How do we treat compounds with different morphological shapes and/or structures, which denote, however, the same plants? Which criteria shall we apply?

These questions, all relating to the distinction between compounds as the results of lexical processes and syntactic groups as the results of syntactic processes, have been the object of much controversy and many different criteria have been proposed[118]. It is, however, still one of the major problems one faces when dealing with complex expressions in general and with the Old English plant names in particular. None of the classification principles proposed so far has been accepted universally. In addition, testing different criteria against the same composites often leads to different and sometimes even contradictory results.

And yet it is important to discuss the differences between compounds and syntactic groups, as this distinction is often a criterion for lemmatising lexemes in the dictionaries (see also chapter 5.2.). The highly idiomatised plant name *foxes fōt* 'unbranched burr-reed, etc.' lit. 'fox's foot' mentioned already above, for instance, is regarded as an idiom with a specific denotation, though, morphologically, as a syntactic group by the *DOE* and is, hence, not an entry on its own: *foxes fōt* 'unbranched burr-reed, etc.' is subsumed to the lemma *fox* 'fox'. Similarly, the semantically rather transparent plant name *brādþistel* 'milk thistle, etc.', is not a lemma on its own in the *DOE*, although a comment to the adjective *brād* 'broad' (s.v. *brād*, 1.a.i.), states: "brad þistel 'broad thistle', a kind of thistle (the form may alternatively be read as a compound)", pointing to the fact that, as a compound, the plant name would have been lemmatised by the editors of the *DOE*. And, indeed, a

118 For different criteria of demarcation see, for example, Marchand (1969: 20-9), Koziol (1972: 48f.), Faiss (1978: 19-28), Kastovsky (1982: 176-9), Kastovsky (1992: 369f.), Sauer (1992a: 57-124), Plag (2003: 159-60), Booij (2005: 81-3), and Bauer (2006: 483-94).

large number of the Old English complex plant names are not morphologically isolated; still, they denote specific plants and exhibit specific morphological and semantic features. They are semantically lexicalised (or 'idiomatised'), and, since their semantic heads lie outside the constructions as with exocentric compounds (cf. chapter 5.3.2.2.2.), it seems justified to include them in the morphological analysis.

It is in many cases extremely difficult to decide whether a plant name should be regarded as a compound or as a syntactic group. As will be shown in the following, morphological and semantic isolation are not absolute values: for many plant names a definite position in favor of one or the other group cannot be taken and alternative concepts must be employed. It seems appropriate to employ a concept that covers all morphologically different expressions: simplexes, composites, derivatives and syntactic groups (all with various degrees of idiomaticity). Wunderli's (1989) concept of the *lexie* can serve as a super-ordinate term for all the complex plant names here, since it is not defined morphologically, but, largely, semantically: *lexies* are defined as auto-semantic units[119] that can be morphologically simple, morphologically complex, or even syntactic groups. The indispensable pre-condition for a word or a combination of words to be called a *lexie* is that it is equipped with a specific meaning, i.e. that it is *idiomatised*, and that it is an integral and universally accepted part of the lexicon of a speech community, i.e. that it is *institutionalised* (Wunderli 1989: 15).

The terms *institutionalisation* and *idiomatisation* as well as *lexicalisation* have been defined differently by different scholars[120]. Following Lipka (2002: 113), *lexicalisation* is "the phenomenon that complex lexical items, through frequent usage, may lose their syntagmatic nature and tend to become formal units with specific content", or to put it in the words of Durkin (2009: 49): "Lexicalised words are opaque – in meaning, or composition, or both." Lexicalisation and idiomatisation overlap as – since *idiomatisation* is the process whereby formally complex linguistic expressions develop semantically in a way that "their meaning is not derivable from that of their constituents" (Lipka 2002: 112) – both lexicalisation and idiomatisation result in 'specific meanings' of the complex expressions.[121] Thus, noun phrases with postmodifying relative clause, e.g. *bānwyrt sēo þe hæbbe croppan* lit. 'bonewort which has flowers' denoting the 'common centaury' (*Lǣcebōc*, 80/3-4) or *ompre sīo þe swimme* lit. 'dock that is swimming' denoting the 'yellow water-lily' (*Lǣcebōc*, 98/32), are not lexies, despite denoting species of plants: they are not idiomatised nor lexicalised; they are descriptive expressions. In contrast, the simplexes *āc* 'oak', *lēac*

119 Semantically speaking, there are three kinds of lexical items: syn-semantic lexical items whose function is to link lexical elements to larger syntactic units (i.e. grammatical, free morphemes, e.g. *in*, *with*, *to*, *while*, etc.), deictics, i.e. lexical items which can establish anaphoric and cataphoric relations between different elements of speech (e.g. *tomorrow*, *yesterday*, *here*, etc.) and auto-semantic lexical items, i.e. free, lexical morphemes (e.g. *dog*, *to run*, *to be frightened*, etc.) (Blank 2001: 5).

120 Compare, for instance, Lipka (2002: 110ff.) with an overview of different approaches; see, further, Herbst, Stoll and Westermayr (1991: 508), Sauer (2000b: 127ff.), and Durkin (2009: 49-56).

121 For Bauer (1983: 50-61), idiomatisation is sub-ordinated to lexicalisation as a semantic form of lexicalisation and is, thus, on a level with morphological lexicalisation, phonological lexicalisation and syntactic lexicalisation.

'leek, plant', *wyrt* 'plant, wort', the complex formations *pīnbēam* 'pine tree', *bisceopwyrt* 'marshmallow' and the phrases *basu popig* 'common red poppy' or *blæc pipor* 'black pepper' are considered lexies.[122]

Lexical items can be lexicalised in a number of different ways: morphologically, syntactically, phonologically and semantically (Bauer 1983: 50-61). One criterion which is often used to distinguish lexicalised from not lexicalised items in Modern English is spelling (Marchand 1969: 21). Compounds are generally written as one word or with a hyphen, while phrases are written as two (or more) distinct words. Spelling, however useful it may be in some cases, may be misleading in others. The Modern English spelling system often lacks uniformity in its treatment of compounds and syntactic groups: The reader of Modern English may, for instance, encounter three different graphemic options for *flower pot*: *flower pot*, *flower-pot* and *flowerpot* – all of them correct (Quirk et al. 1985: 1569).[123] The orthographical situation in Old English is by no means clearer: besides the fact that early medieval orthographical practice may vary diachronically, regionally, and idiosyncratically (i.e., from scribe to scribe), syllabification and word division were neither standardised nor did they depend on words or syllables as they do today, but on accent groups.[124] An example from the corpus of medico-botanical texts may illustrate this: in the clause "Ðeos wyrt þeman [...] feld wyrt nemneþ […]" ['This plant, which is called *feld wyrt*'] (Old English *Herbarium*, facsimile edition of British Library, Cotton Vitellius C. iii, fol. 27r, by d'Aronco and Cameron 1998) the distance between *feld* and *wyrt* is big enough to be interpreted as a gap left on purpose to indicate the distinction of two different words, while the writing of the relative particle *þe* and the indefinite pronoun *man* without a gap as if they formed one word, suggests a compound to modern readers.[125] Hence, spelling is not a trustworthy indicator. As Sauer (1992a: 62ff.) informs us, hyphens were known in Early Medieval England, but were not used very often. Morphological and semantic criteria are more reliable and shall be discussed in the following.

Morphological criteria

The most important condition which has to be fulfilled in order for a combination to be counted as compound is, according to Marchand (1969: 21), the isolation of the compound from the parallel syntactic group on morphological grounds, i.e. the *morphological isolation*. Different paths can lead to morphological isolation:

122 *Institutionalisation* is "the integration of a lexical item, with a particular form and meaning into the existing stock of words as a generally acceptable and current lexeme" (Lipka 2002: 112).

123 Not at all helpful is Jones' statement (1956: 101, fn. 6): "By a compound word is meant here a word made up of two words written in conventional spelling as one, with or without a hyphen".

124 An overview of writing, spelling conventions and word division in Old English is given by Campbell (1959: §§ 23-9) and study on the orgins of spaces and dots between words provides Saenger (1997); word accent and sentence accent in Old English are discussed in Campbell (1959: §§ 71-98).

125 Cf. also Pheifer (1974: xxiii) for similar examples from the *Épinal-Erfurt Glossary*.

(a) Prosodic pattern:[126] While compounds are marked, in general, by morphophonemic fore stress, " ´ / ` " (e.g. ModE *bláckbìrd*), phrases have level stress " ´ / ´ " (e.g. ModE *bláck bírd*). This condition seems to be met by many Modern English combinations.[127] In the analysis of the OE plant names, this criterion is, on the one hand, easy to handle because, owing to Germanic stress patterns, there are good reasons to postulate fore stress for all compounds before Early Modern English.[128] On the other hand, stress as a criterion is more problematic for Old English than for Modern English because in prose texts stress can be hardly ascertained and in verse texts word stress often was overlapped by verse stress patterns (Campbell 1959: §§ 71-98). Thus, in many cases, we simply cannot reconstruct the Old English stress patterns. With plant names it is even more difficult because a large number of the plant names are recorded in glosses and glossaries and there are hardly any plant names in verse texts (cf. Neville 2003).

(b) Morphological shape and structure: Combinations with determinants which could not act as modifiers in phrases are morphologically isolated (Marchand 1969: 22; Kastovsky 1982: 79; Sauer 1992a: 58f.). In view of the Old English plant names this criterion applies to N+N combinations, such as *brōþorwyrt* 'pennyroyal', {*brōþor*} 'brother' + {*wyrt*} 'plant' as long as the first element is not inflected (compare, in contrast, the N_{gen}+N formation *hundeshēafod* 'small snapdragon', {*hund*} 'dog' + {*es*} + {*hēafod*} 'head'),[129] then to the synthetic Pron+N compounds (e.g. *selfhǣle* 'wood march' {*self*} 'self' + {*hǣl-*} '(to) heal' + {*e*}) and to V_{stem}+N compounds (e.g. *springwyrt* 'caper-spurge' {*spring-*} '(to) jump' + {*wyrt*} 'wort, plant').[130]

(c) All synthetic compounds and all *bahuvrihi* compounds are morphologically isolated: synthetic compounds because their morphological structure cannot be ascertained (see chapter 5.3.3.6.) and *bahuvrihi* compounds because their morphological shape does not correspond to the expected morphological structure (cf. chapter 5.3.2.2.2., Bauer 1983: 57): as the semantic heads of *bahuvrihi* compounds lie outside the construction, the denotata of *bahuvrihi* compounds and of parallel syntactic groups are not the same; compare, for

126 For Bauer (1983: 50-1), prosodic and segmental features are separate, phonological criteria.

127 However, there are cases where stress cannot be taken as an indicator for the morphological status of a combination; for examples and their discussion see e.g. Koziol (1972: 49), Faiss (1978: 21f.) and Sauer (1992a: 68f.).

128 See Sauer (1992a: 69f.) with further references.

129 According to Sauer (1992a: 148), this is not always clear: "Der Typ Sb/Sb ist im Ae. und Me. morphologisch nicht isoliert, weil manche Verbindungen aus Sb+Sb auch Genitivfügungen darstellen können, vor allem wenn das Erstelement auf *-e* endet" ['The type Sb/Sb is not isolated in Old and Middle English because some of the Sb+Sb formations can, theoretically, be genitival formations, especially when the first element ends in *-e*'].

130 This criterion is problematic for the modern stages of English, because, due to the loss of inflectional markers and markers of word class, adjectives and nouns very often cannot be distinguished formally, e.g. (the) *brave*, noun and *brave*, adjective. Cf. also Koziol (1972: 48) and Sauer (1992a: 98-101).

instance, the denotatum of the plant name *fīflēafe* (Num+N) 'cinquefoil' with that of the noun phrase *fīf lēaf* 'five leaves' (n., a-decl.).

(d) Morphological isolation can be ascertained for combinations where the first element of the compound is not inflected, but should be in the corresponding syntactic group in order to ensure agreement (Sauer 1992a: 75-6), as for example in N+N compounds (e.g. *crāwlēac* 'crow garlic', {*crāw*} 'crow' + {*lēac*} 'leek') or in Adj+N compounds (e.g. *blæcber(i)(g)e* 'whortle-berry, etc.' {*blæc*} 'black' + {*ber(i)(g)e*} 'berry'). Accordingly, N_{gen}+N combinations are morphologically not isolated: with the first element in the genitive, the compound does not differ formally from the corresponding phrase. Compare for example the lexicalised Old English compound *hræfnes fōt* 'crowfoot, buttercups' lit. 'raven's foot', and the noun phrase *hræfnes fōt* 'raven's foot, foot of the raven'.

(e) Combinations where one of the elements is not used independently are said to be morphologically isolated, too (Sauer 1992a: 84-5). Compounds that meet this condition are either synthetic (chapter 5.3.3.6.), compounds the elements of which are affected by processes of attrition (e.g. in *leloþre* '? meadow-burr') or compounds with presumably *blocked* elements (cf. chapter 5.4.). The problem with this criterion is that we cannot claim the uniqueness of the morph in question as our knowledge of Old English is restricted to the texts that have survived to the present, and these are limited in number and scope, too, as was already stressed. Thus, no matter how clearly the blocked status of elements can be ascertained for some of the modern English combinations (cf., e.g., Schmid 2005: 31), it is unclear for earlier stages of the languages. For instance, despite the fact that *-hige*, the second element of the plant name OE *haranhige* 'hare's foot clover', is not attested elsewhere in the extant Old English texts, we cannot claim that it was not in use in an independent form. It may simply have happened that the text documenting the use of *-hige* got lost in the course of the centuries from the Anglo-Saxon period to our times (see Catalogue, s.v. *haranhige*).

(f) Morphological isolation can also be assumed when the compounds take forms which the simplexes do not have (e.g. *lēaf* 'leaf' with a strong declension as a simplex, but *-lēafe* with a weak declension in compounds, e.g. in *fīflēafe* 'cinquefoil' lit. 'fiveleaf').[131]

Other criteria for delimiting syntactic groups from compounds are of a more syntactic nature and can hardly be verified for Old English as we lack native speakers to inform us about whether certain constructions are perceived as idiomatic or not. To these belong, for instance, (a) introduction of a modifier without change of sense, which would imply that compounds cannot be split while syntactic groups can (e.g. ModE *new black market* vs. **a black new market*; cf. Sauer 1992a: 80), then, (b) modification by an adverb (e.g. ModE **the very darkroom* (compound) vs. *the very dark room* (noun phrase)), and (c) comparison (ModE *a hothouse*, but **a hotterhouse*). There are, however, exceptions to this rule: adjectives that are part of lexicalised phrases (idioms)

131 V_{ing}+N combinations are, further, problematic since their morphologic status has changed throughout the history of the English language (cf. Sauer 1992a: 73ff.). However, this morphologic pattern does not occur in my plant name corpus.

cannot be modified by an adverb and cannot be compared either **the very black market*, **the blacker market* or **the very Old English*, **Older English*. Other criteria are, e.g. reversed word-order (e.g. *flý-físhing* 'fishing with flies', Kastovsky 1982: 178) and the question of whether articles, pronouns and adjectives placed in front of compounds refer to the combination as a whole or not (**a brown black bird* is nonsensical, while *a brown black-bird* does make sense: female blackbirds have brownish feathers, while male blackbirds have black feathers).

Semantic criteria

Where morphological isolation 'fails', semantic isolation can be helpful. *Semantic isolation* (or *idiomatisation*) refers to the phenomenon that the meaning of complex expressions cannot or can only partly be derived from the meaning of the two (or more) constituents. This is an inherent feature of compounds (cf. Schmid 2005: 121f.), but also of idiomatised syntactic phrases. In most cases, unidiomatised expressions are marked by additional semantic elements, e.g. [habitually], or [professional], etc., so that they often have a more specific meaning than the not idiomatised syntactic group.

As idiomatisation is not limited to compounds, a large number of complex expressions are highly lexicalised semantically, sometimes even up to the point of not being transparent any more (of being *opaque*; Bauer 1983: 19-20). The syntactic group OE *hēah bēam*, for instance, literally denotes a 'high tree', but its lexicalised meaning is 'gallows'. In a similar way, a number of plant names are semantically lexicalised to various degrees: Who would recognise without previous knowledge that *openærs* lit. 'open arse', *eorþgealla* lit. 'earth gall' and *lēonfōt* lit. 'lion's foot' are the Old English plant names denoting the medlar, the knapweed and the lady's mantle?

Semantic isolation can lead to the analysis of the complex expressions as compounds although there may be no morphological isolation. Thus, the expressions *lēonfōt* 'lady's mantle' or *dæges ēage* 'daisy' are classified as compounds despite their inflected first element. Once a complex formation is lexicalised, it can be the base for new complex expressions. OE *wīngeard* 'vineyard', for instance, adopted a second sense – 'vine, grapes' – as a semantic loan from polysemous L *vīnea* 'vine, vineyard' (see Catalogue, s.v. *wīngeard*); this second meaning is the one active in the plant names formed on *wīngeard* 'vine': *blæc wīngeard* 'black bryony', *hwīt wīngeard* 'vine, etc.' and *wilde wīngeard* 'black bryony'. In the same way, the metaphorical element in the compound *wulfescamb* 'wild teasel' lit. 'wolf's comb' and the metonymic element in *wegbrǣde* 'waybread' lit. 'way-broad (plant)', for instance, are neutralised in the plant names that derived from these lexicalised items: *brāde wulfescamb* 'wild teasel' lit. 'broad wild teasel', *rūwe wegbrǣde* 'hoary plantain' lit. 'hairy waybread' and *smēþe wegbrǣde* 'greater plantain' lit. 'smooth waybread'. Semantic isolation can be established on the basis of:

(a) Semantic specification: Semantic specification is extant in formations based on the metaphorical or the metonymic use of language (Jackendorff 1975: 657).[132]

132 This is rejected by Bauer (1983: 57-8) on the grounds that the semantic heads of exocentric espressions are not necessarily lexicalised: exocentric formations can be used figuratively, e.g., when ModE *redskin* 'American indian' (*OED*, s.v. *redskin*) is used to advertise a specific species of apples.

Since the semantic head is not expressed on the surface of these combinations, they differ from the corresponding syntactic groups: in metaphorical and metonymic formations, the compounds are not semantic subgroups of the second elements, while in syntactic groups they are (see also chapters 5.3.2.2.1., 5.3.2.2.2.).

(b) Compounds with morphemes of unclear origins and whose meanings cannot be identified (cf. chapter 5.3.3.9.), such as *hǣwenh(n)ydele* 'scurvy-grass', where the denotatum of *-h(n)ydele* is unknown (see Catalogue, s.v. *hǣwenh(n)ydele*), or compounds where one (or both elements) are affected by morphological and/or phonological attrition such as the element *le-* in OE *leloþre* 'dock' (see Catalogue, s.v. *leloþre*) are semantically isolated: the semantic content of these elements cannot be derived unambiguously from the form. This does not mean that the denotation of the whole compound is a puzzle. We know, for example, that *hǣwenh(n)ydele* denotes the 'scurvy-grass', but what exactly the second element *-h(n)ydele* denotes and why the combination of the two elements is used for the scurvy-grass has not been resolved, yet.

Morphological and semantic isolation: a continuum

Drawing a neat demarcation line between compounds and syntactic groups is not an easy task: in many cases more than one criterion applies, and the consideration of different aspects can easily lead to contradictory results. Morphologically isolated compounds, for example, may not be isolated at all semantically (cf., e.g. OE *āctrēow* 'oak(-tree)' lit. 'oak tree') and the other way round: semantically highly idiomatised expressions may not be isolated morphologically; cf., e.g., *openærs* 'medlar' lit. 'open arse'.[133] Following the deliberations made above, composites where the following criteria apply can be safely regarded as compound plant names:

(a) Combinations with determinants that could not act as modifiers in syntactic phrases, such as, for example, N+N combinations, e.g. *crāw(e)lēac* 'crow garlic' lit. 'crow leek' or V_{stem}+N combinations, e.g. *springwyrt* 'caper-spurge' lit. 'jump-wort'.
(b) Synthetic compound, e.g. *g(r)undeswelige* 'groundsel' lit. 'ground-devouring (plant)'.
(c) Metaphorical and metonymic compounds, e.g. *dæges ēage* 'daisy' lit. 'day's eye' or *ælfþone* 'dogwood' lit. 'elf tendril'.
(d) Semantically and/or morphologically opaque compounds, e.g. *wermōd* 'wormwood' lit. 'man-mood' or *leloþre* 'dock' lit. 'meadow-burr', and combinations where the complex lexeme may assume a form which the simplex does not have, e.g. *fīflēafe* 'cinquefoil' lit. 'five-leaf' (cf. chapter 5.3.3.4.).

For the plant names where these four criteria do not apply, individual decisions must be made. OE *englisc more* 'wild carrot' lit. 'English carrot', and *fūla bēam* 'black alder tree' lit. 'foul tree', for instance, are semantically but not morphologically isolated: Their denotation is more than the sum of the meanings of its parts, i.e. these are cases of semantic lexicalisation (or idiomatisation; cf. 5.3.1. and also Bauer 1983:

133 For a discussion of this problem see also Kastovsky (1982: 178) and Sauer (1992a: 117-23).

58; Schmid 2005: 79). Criteria such as absence or presence of inflection, stress or spelling must also be taken into account as far as possible. Thus, *englisc more* and *fūla bēam* are considered to be semantically lexicalised syntactic groups: the editions render *englisc more* as "englisce moran" (*Lǣcebōc*, 12/38; Cockayne, *Leechdoms*, 2: 38, no. 2), suggesting that the plant name was regarded as consisting of two separate intonation units; in addition, the adjective is inflected. *Fūla bēam* is edited as "fulanbeames" (*Lǣcebōc* 24/29; Cockayne, *Leechdoms*, 2: 78, no. 32), i.e., as one intonation unit, while the adjective is inflected.

For other plant names, especially for those attested with different morphological shapes that betray diachronic developments, the answer can be only tentative. As mentioned already, the plant name *blind(e)net(e)le*, for instance, is attested as a compound with an uninflected first element (Zupitza 1880: 311/4) and as a syntactic group (*Lǣcebōc*, 21/1). Since the plant name appears as a compound in Middle English (cf. *MED*, s.v. *blind*, 5 (a)), we can assume an intermediate stage in the development of the plant name between a low and high morphological isolation for the Old English period – a development whose intermediate stages are fully visible in the extant Old English texts – as it were, a diachronic development *in actu*.

However, morphological and semantic isolation are not absolute values: isolation is a matter of degree and depends on the number and reliability of the criteria that can be applied. Some criteria are more stringent and lead to a high degree of isolation, such as for example the figurative use of language (*dæges ēage* 'daisy', for example, cannot be mistaken for a transparent syntactic group 'day's eye') while other criteria are less stringent and lead to a lower degree of isolation, such as for example the absence of inflection in the compound e.g. *hwītlēac* 'leek' lit. 'white leek'. Yet, morphological isolation does not automatically imply semantic isolation or vice versa and the two can even be contradictory.

The correlation of the grade of morphological isolation with the grade of semantic isolation may indicate whether a complex lexeme is a morphologically and/or semantically isolated compound, a semantically isolated syntactic group or a transparent syntactic group. Fig. 2 shows exemplarily the correlation of morphological and semantic criteria for some of the Old English complex plant names. The two extreme ends are *transparent* and *opaque* (for semantic isolation) and *compounds* and (syntactic) *phrases* (for morphological isolation).

The degree of morphological isolation for *pīnbēam* 'pine-tree' lit. 'pine tree', for example, is high, since it is a combination of a noun premodified by an uninflected noun, while its degree of semantic isolation is low: *pīnbēam* is semantically transparent. *Hræfnes fōt* 'buttercups' lit. 'raven's foot' and *dæges ēage* 'daisy' lit. 'day's eye', on the other hand, are semantically highly isolated since the meaning of the composites may not be inferred from the meaning of the parts. Their degree of morphological isolation, in contrast, is rather low, as the use of nouns in the genitive, premodifying other nouns, is frequent in noun phrases.[134]

134 According to Booij (2005: 83), N_{gen}+N combinations "function in the same way as compounds [since] these expressions instantiate a lexicalised syntactic pattern that functions to create new labels." Booj argues that this kind of combination should be regarded as an idiomatic pattern between compounds and syntactic groups (form: *N's N*), serving to create new lexical expressions

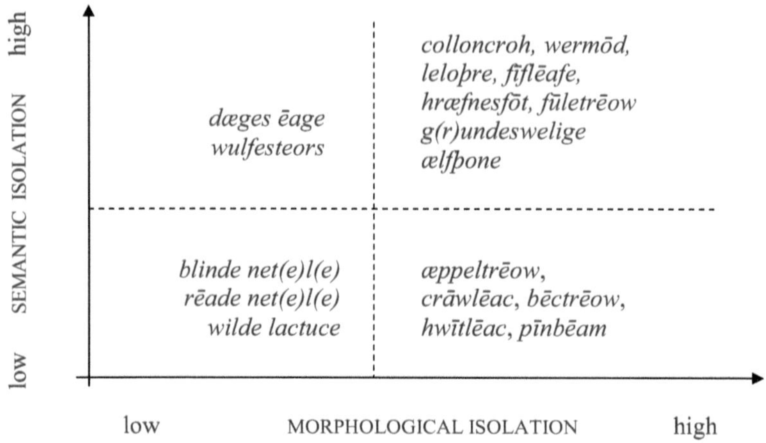

Fig. 2: Morphological and semantic isolation: cross-classification, with examples

Fīflēafe 'cinquefoil' lit. 'five-leaf' is semantically isolated because it is a metonymic formation (through figurative use of *lēaf(e)*). However, the second element *-lēafe* belongs to the word field of plants and of parts of plants, so that the semantic transfer from the simplex to the complex word is not from one domain of knowledge to a totally different one (as is the cases with metaphorical compounds), but takes place within one domain. Hence, its degree of isolation is lower than that of metaphorical compounds such as *dæges ēage*. Its morphological isolation rests on the change of declension paradigm of *lēaf* as simplex vs. *-lēafe* as a second element of complex words. As a second element it belongs to the weak declension paradigm (neuter, n-decl.), as a simplex, *lēaf* belongs to the strong declension (neuter, a-decl.). *Hwītlēac* 'leek' lit. 'white leek' is not particularly isolated as the meanings of noun phrases made up of pre-modifying adjective and a noun are transparent; so is the meaning of this Adj+N compound.

For some of the plant names attested in the nominative only, as often happens with lemmas in glossaries, the status cannot be identified since the plant names are not isolated morphologically or semantically, and as they are not attested in a syntactic position that would require an inflected form, their status cannot be ascertained syntactically, either. In the *Lǣcebōc* (16/39-17/1), for instance, the syntactic position of *ofersǣwisc rind* 'cinnamon' as part of the subject requires the use of the nominative: "Wyrc þus toþsealfe, ofersæwisc rind & hunig & pipor meng tosomne, lege on" ['Make a salve against toothache as follows: Mix *ofersǣwisc rind* and honey and pepper, apply.']. As *ofersǣwisc rind* is a hapax legomenon, the question of whether this plant name is a compound or a syntactic group cannot be answered. It

in English; he calls them *constructional idioms*: "A constructional idiom is a fixed syntactic pattern in which some positions may be filled by all kinds of words of the right category, whereas other positions are filled by specific morphemes or words" (Booij 2005: 83).

may seem, therefore, inappropriate to analyse syntactic groups by giving them labels such as 'endocentric, copulative: attributive'. Yet, we would miss much of the structural and semantic information about the Anglo-Saxon way of expressing ideas on plants if we excluded the syntactic groups from the morphological analysis.

5.3.2. APPROACHES TO THE MORPHOLOGICAL ANALYSIS

According to Marchand (1969: 53ff.), a comprehensive description of compounds includes the discussion of

(1) the morphological shape,
(2) the morphological structure, which indicates the determinant/determinatum relationship between the elements of the compound,
(3) the grammatical deep structure (syntactic-semantic structure), preceded by the formulation of a paraphrase, and
(4) the type of semantic content (reference-type), based on the grammatical deep structure.

Apart from the morphological shape, which yields purely formal information, all levels of description provide semantic information on the compounds in question.[135]

5.3.2.1. MORPHOLOGICAL SHAPE & MORPHOLOGICAL STRUCTURE

The morphological shape indicates the morphological status of the constituents of the lexies, i.e. whether they are free or bound, or grammatical or lexical morphemes, and describes their word-class affiliation (cf. Marchand 1969: 54). Thus, the analysis of the morphological shape produces in equal measure all inflectional and derivational morphemes of the complex linguistic sign. To give an example: the plant name *āctrēow* 'oak' lit. 'oak tree' is, morphologically, made up of two free, lexical morphemes, namely, of {*āc*} and {*trēow*}. The complex plant name *dæges ēage* 'daisy' lit. 'day's eye', in contrast, is made up of three morphemes: {*dæg*}{*es*}{*ēage*}: that is, of the two free, lexical morphemes {*dæg*} 'day' and {*ēage*} 'eye', and of the bound, grammatical morpheme {*es*}, which is the genitive ending of the first morpheme.

The analysis of the morphological structure, in contrast, indicates the modifier/head relationship (Bauer 1983: 30), resp. the determinant/determinatum relationship (Marchand 1969: 54-5) of complex linguistic signs.[136] Whereas the analysis of the

---

135 Since the analysis of semantic aspects of the complex Old English plant names in Ælfric's *Glossary* (Krischke 2009) required some preliminary thoughts about morphologic aspects, parts of the line of argument presented here were also discussed.

136 The modifier/head principle describes the principle that grammatical entities such as, e.g., noun phrases, consist of a dependent, modifying element and an autonomous head. The terms determinant/determinatum relate basically to the same principle (with slight modifications; cf. fn. 140 and 142), but, as 'modifier' and 'head' are more established than 'determinant' and 'determinatum', I prefer their use.

morphological shape produces a description of all morphemic parts of the linguistic sign investigated, the analysis of the morphological structure is analytical insofar as it produces the functional elements of the linguistic sign – the two (potential) stems. While in terms of the morphological shape more than two elements can be identified, in terms of structure, there is always a bipartite result: irrespective of the number of morphemes involved, complex linguistic signs can be reduced to one (simple or complex) element modifying a (simple or complex) head. To take the examples used above: in the compound *āctrēow* the element *āc* is the modifier, while *trēow* is the modified element (*āc/trēow*); similarly, the analysis of *dæges ēage* produces the structure *dæges/ēage* (with a morphologically complex modifier {*dæg*} {*es*}).[137]

The complex plant names are, structurally, made up of two stems but in terms of morphological shape, many of them are made up of more than two morphemes. In these cases one or both stems – either the modifier or the head or both – are complex:[138] e.g., *sūrmilscapuldor* 'apple-tree, etc.' lit. 'sour-sweet apple-tree', which can be split into {*sūr*} {*milsc*} {*apul*} {*dor*}; *ofersǣwisc rind* 'cinnamon' lit. 'overseas bark', consisting of {*ofer*} {*sǣ*} {*(w)isc*} {*rind*}; *sūþerne bisceopwyrt* and *sūþerne wermōd* with the complex elements *bisceopwyrt* 'marsh mallow' and *wermōd* 'wormwood' denoting the *genus proximum*, and the first, equally complex elements ({*sūþ*} {*ern*} {*e*}) the *differentia specifica*: *sūþerne bisceopwyrt*, for instance, is seen as a 'southern' kind of *bisceopwyrt*, i.e. as taxonomically sub-ordinated to *bisceopwyrt*, while *sūþerne wermōd* is taxonomically sub-ordinated to *wermōd*, and denotes a 'southern' kind of *wermōd* (see Catalogue, s.vv. *sūþerne bisceopwyrt*, *sūþerne wermōd*).

5.3.2.2. MORPHOLOGICAL STRUCTURE

The Indian grammarian Pañini (4th c. BC) identified three main types of compounds in Sanskrit (Gardner 1968: 17): (i) *dvandva* 'pair', (ii) *tatpurusa* 'of whose servant' (with the subgroups *karmadhāraya* 'action-bearer' and *dvigu* 'a pair of oxen') and (iii) *bahuvrihi* 'having much rice'. *Dvandvas* are additive copulative compounds (e.g., ModE *bitter-sweet*, *actor-director*), *tatpurusa* are compounds where the main semantic element is the head of the formation (e.g., ModE *police-dog*) and *bahuvrihi* are defined as adjectival compounds where the semantic head is outside of the composition.[139] This classification system, which combines grammatical with semantic information,

137 The analysis of *dæges ēage* as a metaphorical, endocentric compound, which follows Marchand's approach (1969) at this point of the survey, is questioned in chapter 5.3.2.2.2. It is argued that metaphorical formations such as *dæges ēage* are exocentric, too, and should be grouped together with metonymic formations such as *fīflēafe* (according to Marchand 1969: 13-5, pseudo-compounds with a complex modifier and zero head, i.e. *fīflēafe/Ø*), as the semantic head of both lies outside the formation.

138 For details on stem variants and linking vowels, see chapter 5.5.

139 Cf. Carr (1939: xxv); Gardner (1968: 17). Originally confined to attributive formations such as G *barfuss* 'having bare feet', the category of *bahuvrihis* was later extended to also include imperative formations, e.g., ModE *pickpocket*; cf. Carr (1939: xxvii).

has survived – slightly modified – into the present and is used to ascertain the morphological structure of compounds. Accordingly, compounds are grouped into *dvandvas* (additive compounds, e.g., *Alsace-Lorraine*), appositional compounds (e.g., ModE *maid-servant*), into endocentric compounds (determinative compounds) and into exocentric compounds (compounds with the semantic head outside the combination, including *bahuvrihis*, e.g., ModE *paleface, redskin* and imperative formations, e.g., ModE *pickpocket*; Bauer 1983: 30-1).[140]

The description of the morphological structure reflects the binary modifier/head relationship of complex words (cf. chapter 5.3.2.1.), and, since it depends on whether the semantic head of the combination is encoded in the compound, it is not only a morphological description, but also provides information on the semantics of the complex lexeme: "Compound nouns can be further subdivided into four groups according to semantic criteria" (Bauer 1983: 30). In *endocentric compounds*, the head indicates the grammatical and the semantic class of the combination and can be substituted for it; the head is a super-ordinate – a hyperonym – of the composite (Bauer 1983: 30; e.g., *dog* in ModE *police-dog*, or *bēam* 'tree' in OE *āctrēow* 'oak' lit. 'oak tree'). The modifier is semantically more dominant than the head since it provides the newly-given information, specifying the head and, usually, preceding it in the Germanic languages.

Dvandvas and *exocentric compounds*, on the other hand, do not have the modifier/head relationship endocentric (determinative) compounds have. In *dvandvas*, both elements are semantically and morphologically on equal terms (e.g., *actor-director, bitter-sweet*). In *exocentric compounds*, e.g., *paleface* (*bahuvrihi*) and *pickpocket* (imperative compound; Marchand 1969: 380)[141], the semantic head lies outside the composite: a *paleface* does not relate to a certain kind of face, but to a person having a pale face. This means that exocentric compounds are not sub-classes of the items denoted by one of the elements.[142]

There are no *dvandva* compounds in the corpus of Old English plant names. This is not surprising, because this type of compound was not productive in Old English (cf. Kastovsky 1992: 365). Consequently, two major types of compounds remain to be discussed: the endocentric and the exocentric compounds. Both can be further sub-classified.

140 Marchand (1969: 11-3) differs in that he distinguishes in the first step between expansion (composition and prefixation) and derivation (suffixation, zero-derivation); cf. also fn. 116 in this chapter. In his categorisation, exocentric formations are zero-derived with a complex modifier (hence termed pseudo-compounds) and not compounds. Sauer (1992a: 389ff.) subsumes the *dvandvas* to the copulative compounds; for Bauer (1983: 31), '*dvandva*' and 'copulative compounds' denote the same morphological pattern.

141 Imperative compounds do not appear in Old English (Marchand 1969: 381).

142 The morphologic head of exocentric compounds, however, still passes on morphosyntactic features such as word-class affiliation, gender or number; cf. e.g., Bauer (2003: 42f.). Marchand (1969: 13) classifies exocentric compounds as derivations with a zero-morpheme (of the structure AB/Ø), i.e. for him, they are pseudo-compounds.

5.3.2.2.1. ENDOCENTRIC COMPOUNDS

The structure of endocentric plant name compounds can be reduced to the formula AB = B ('AB is (a kind) of B') or, more precisely, AB ⊂ B (AB is a subset of, is included in B),[143] since the second elements of these complex lexemes are attested as plant names themselves. In this formula, A refers to the first element of the compound, B to the second element. Examples of endocentric plant names are *blind(e)net(e)l(e)* 'dead nettle, etc.' lit. 'blind nettle', *bōctrēow* 'beech' lit. 'beech tree', *cederbēam* 'cedar of Lebanon' lit. 'cedar of Lebanon tree', *clifwyrt* 'great burr' lit. 'cleav(ing) plant', *crāwanlēac* 'crow garlic' lit. 'crow's leek', *dweorged(w)os(t)(l)e* 'pennyroyal' lit. 'dwarf's *dost*', *elebēam* 'olive-tree' lit. 'oil-/olive-tree', etc. Syntactic groups such as *(sēo) læsse bisceopwyrt* 'betony' lit. 'lesser, smaller betony', are by nature endocentric.[144]

Endocentric compounds can be further sub-classified. The decisive criterion is the nature of the verb used in the syntactic paraphrase of the compound (Marchand 1969: 40). If the verb used is the copula verb *(to) be*, the compounds are called *copulative compounds* (for examples, see chapter 5.3.2.2.1.1.); if a verb other than *(to) be* is inserted, the compounds are called *rectional compounds* (for examples, see chapter 5.3.2.2.1.2.). This classification is, again, not only a morphological one: it includes additional semantic information, since the classification of all rectional compounds depends heavily on the semantic and morpho-syntactic features of the verb in the paraphrase. The main features of copulative and of rectional compounds will be discussed in the following.

5.3.2.2.1.1. COPULATIVE COMPOUNDS

Copulative (or copula) compounds can be paraphrased by constructions using the copula verb *(to) be*, e.g. OE *āctrēow* 'tree which is an oak'. Three subgroups are distinguished: attributive compounds, subsumptive compounds and compounds based on comparisons (cf. Kastovsky 1992: 365ff.; Sauer 1992a: 389ff.).

(a) In *attributive copulative compounds*, the first element denotes a specific property of the second element. To this category belong for example *blind(e)net(e)l(e)* 'dead nettle, etc.' lit. 'blind nettle' and *hārhūne* 'hoarhound' lit. 'hoary hoarhound'. Attributive copula compounds are, often, made up of an adjective and a noun.

(b) The main feature of *subsumptive compounds* is that the modifier or determinant denotes a subclass of the head (or determinatum): the head denotes the *genus proximum*, the modifier the *differentia specifica* (cf. Sauer 1992a: 396f.) The conditions AB ⊂ B, A ⊂ B and AB = A are met for instance by *bōctrēow* 'beech' lit. 'beech tree', *cederbēam* 'cedar of Lebanon' lit. 'cedar of Lebanon tree', *fīctrēow* 'fig-tree' lit. 'fig tree', *lāwerbēam* 'laurel' lit. 'laurel tree', *mōrbēam*

143 The notions AB = B, etc., is adapted from Marchand (1969).

144 The status of complex formations where the first element is inflected in the genitive, however, is disputed, and especially for those that are used in a figurative way (see chapter 5.3.1.).

'mulberry tree' lit. 'mulberry (tree) tree' and *pīntrēow* 'pine (tree)' lit. 'pine tree'. As the first constituents of these compounds mark the *differentia specifica* and denote the same significates as the compounds attested independently, we can assume that the information provided by the second element is and was perceived, in fact, as redundant. These types of compounds are therefore called *pleonastic* (cf. Kastovsky 1992: 366f.). Most of the pleonastic compounds in the Old English plant name corpus, strikingly, are made up of a Latin loan word and a native second element.[145] Sauer (1992a: 396f.) therefore suggests that the second elements served as an explanation for the first elements which – being loans – were probably not understood by everybody and that the second elements were therefore by no means redundant.

A small number of the copulative subsumptive compounds seem not to be pleonastic but tautologic. The difference between pleonastic and *tautologic* compounds is that while pleonastic compounds are characterised by the fact that one of the elements is superfluous or redundant, e.g. the notion that a *pīnbēam* 'pine(tree)' is a *bēam* 'tree' – one of the necessary semantic features of *pīn* is its being a tree (seme [+ tree]) – tautologic compounds repeat the same notion in different words. Thus, the plant names *por(r)lēac* 'common leek', *cīpelēac* 'onion' and *ynnelēac* 'onion' could be called tautologic, as *por(r)* and *lēac*, as well as *cīpe* and *lēac* and *ynne-* and *lēac* are synonyms: all these terms can denote Alliums, i.e. leeks, onions. In fact, though, *por(r)lēac*, *cīpelēac* and *ynnelēac* are rather pleonastic than tautologic, which is clear when the meaning of the lexeme *lēac* is examined. As the leek was an important nutritious plant in the early Middle Ages (Biggam 2011: 32; Birkhan 2012: 72; see also chapter 6.6. and Catalogue, s.v. *lēac*), its meaning was extended to the very general meaning of 'plant'; this can be seen in compounds such as ON *laukagarþr* '(herb-) garden', OE *lēactūn* '(herb-)garden' and OE *lēacweard* 'gardener'. In combination with OE *por(r)*, *cīpe*, and *ynne-*, which are all three loans from Latin (L *porrum*, L *cēpa*, respectively, L *uniō*; see Catalogue, s.vv. *por(r)*, *cīpe*, *ynne-*), the element *lēac* serves to explain the first element. In these compounds, therefore, *lēac* is not a tautologic repetition of the (identical) notion, *por(r)*, *cīpe*, and *ynne-*, but explains these as being kinds of *lēac* 'garden-herbs, plants' (see also chapter 6.4. (1)).

(c) Compounds containing a *comparison* are sometimes subgrouped to the attributive compounds, sometimes to the subsumptive compounds (e.g. Kastovsky 1992: 366). Sauer's proposal (1992a: 401) to group them separately is most convincing to me, since these compounds are quite numerous and differ significantly from the two other types of copula compounds. In compounds containing a comparison, the modifier (determinant) denotes a concept to which the plant (or a part of it) is compared (AB $\subset$ B). In the present survey, they are called *partial metaphors* since the base of the comparison is not the whole plant, but one aspect of the plant only, such as the shape of the stem (e.g. in *gārclīfe*

145 For (some of) the pleonastic compounds seen from an etymological perspective, i.e. as hybrid formations, see chapter 7.4.2.1. For the pleonastic compounds analysed in terms of their semantics, see, e.g., chapter 6.3. (1), (5) and 6.6.

'agrimony' lit. 'spear-cleavers'), or its texture (e.g. *feltwyrt* 'great mullein' lit. 'felt wort').[146]

Strictly speaking, compounds containing a comparison are not copulative. A putative sentence paraphrasing *gārclīfe*, for example, should not amount to "The *clīfe* is (a) *gār*", but rather to "The *clīfe* is like/resembles a *gār*", which qualifies the compound basically as rectional. To grasp the crucial point, it is necessary to distinguish between the motivating (extra-linguistic) concept and the way to verbalise this concept. In *candelwyrt*, for instance, it is the shape of the plant that provokes the association of a candle and leads to the plant name *candelwyrt*, lit. 'candle-wort'. *Candelwyrt* and similar compounds are called rectional, partial metaphors.[147]

From the point of view of motivation, attributive copulative compounds such as *smælþistel* 'prickly cnicus' lit. 'small thistle' and partial metaphors such as *dweorged(w)os(t)(l)e* 'pennyroyal' lit. 'dwarf's *dost*' may be similar,[148] but the way they verbalise the motivating concept can still differ: the first elements of both *smælþistel* and *dweorged(w)os(t)(l)e* ('small', respectively, 'dwarf') refer to the fact that the plants are rather small. In *smælþistel*, this attribute is verbalised literally via the adjective *smæl* 'small', while in *dweorged(w)os(t)(l)e* it is verbalised metaphorically, via a comparison of the plant with a dwarf (see also Catalogue, s.v. *dweorged(w)os(t)(l)e*).

5.3.2.2.1.2. RECTIONAL COMPOUNDS

All compounds that cannot be paraphrased by a copulative sentence but that fulfil the condition for endocentric compounds (formula AB $\subset$ B), can be classified as rectional (Marchand 1969: 40). Subgroups to rectional compounds are, according to Marchand (1969), on the one hand, compounds containing a verbal element, that is, compounds "representing the entire verbal nexus in an overt form, including the predicate as the essential part of the sentence" (Marchand 1969: 18) – these are called *verbal nexus compounds*. Examples of verbal nexus compounds from the plant name corpus are *grundeswelige* 'groundsel' lit. 'ground-devouring (plant)' (synthetic compound) and *snīþstrēo* 'carline thistle' lit. 'cut-straw' (V_{stem}+N).

The other group of rectional compounds are the *nominal* ones, that is, "compounds, that do contain the syntactic relations, but only implicitly, while the semantic element

146 See also Sauer and Krischke (2009: 153).

147 Partial metaphors must be seen in contrast to metaphorical compounds, where the metaphorical relation holds between the concept denoted by the second element and the concept denoted by the compound as a whole. Compare, e.g., *candelwyrt* lit. 'candle-plant', where the first element is used metaphorically and the second element is used literally (a *candelwyrt* is a kind of *wyrt* 'plant') with *dæges ēage* lit. 'day's eye', where the metaphor extends over the whole compound, including the second element: a *dæges ēage* 'daisy' is not a kind of *ēage* 'eye'.

148 For this reason, compounds containing a comparison are sometimes subgrouped to the attributive compounds (see (a) above).

of the verb is not expressed" (Marchand 1969: 18). These are called *non-verbal nexus compounds*. Examples are *āctrēow* 'oak' lit. 'oak tree' (N+N), *halswyrt* 'throatwort' lit. 'throat-wort' (N+N) or *blindnet(e)l(e)* 'dead nettle' lit. 'blind nettle' (Adj+N). Compounds where the second elements are deverbal but attested independently as nouns, such as *gārclīfe* 'agrimony' lit. 'spear-cleavers', are counted as non-verbal nexus compounds; the element *clīfe* 'cleavers' in *gārclīfe* is a deverbal noun, deriving from the verb *clīfan* 'to cleave'.[149]

Although the syntactic analysis of rectional compounds, i.e. their paraphrasing, is in most cases fairly simple and straightforward, a considerable number of compounds – namely, the nominal compounds – are problematic: they often cannot be turned into fitting and clear-cut syntactic phrases. The problem with all non-verbal nexus compounds is that the linguist has to insert a verb into the sentence postulated as being at the base of the compound, because the verbal element is not expressed overtly (Marchand 1969: 18). Clear and easy as this seems to be at first sight, it is a problem, since it is often possible to find different sentences all matching one and the same compound.[150] Compare, for example, Bauer's (1983: 160) Modern English example *police-dog*. This compound can be paraphrased as follows:

(1)	*The dog*	*serves*	*the police.*	'B serves A'
	S	V	O_d	
(2)	*The police*	*use*	*the dog.*	'A uses B'
	S	V	O_d	
(3)	*The dog*	*works*	*for the police.*	'B works for A'
	S	V	O_{ind}	
(4)	*The police*	*work*	*with the dog.*	'A works with B'
	S	V	O_{ind}	

Not only can we not say which of these underlying sentences is valid and should be used for further analysis as all of them seem plausible, but also the semantic content, expressed in an abstract way by the type of reference – ascertained on the basis of the syntactic function of the determinatum (head) in the corresponding syntactic structure – is not clear, as the different verbs chosen are of different syntactic valencies and entail, therefore, different *semantic types* or *types of semantic content*. If we use paraphrase number (1), we would classify the compound *police-dog* as an S-type (Subject-type) since the head of the compound is the subject in the paraphrase; if we use number (2), we would classify the compound as O_d-type; if we use number (3), we would classify the compound again as S-type; and if we use number (4), we would classify *police-dog* as O_{ind}-type. On the same grounds, we can postulate (at least) four different semantic relations between the two elements *police* and *dog*: 'B serves A', 'A uses B', 'B works for A' and 'A works with B'.

149 For a discussion of the difficulties of distinguishing between N+N and $N+N_{deverbal}$ compounds in Old and Middle English see Sauer (1992a: 132-44).

150 Sauer (1992b: 397), therefore, regards compounds as, in principle, polysemous, and having more than one possible semantic relation between the constituents.

The possibility of multiple interpretations is, of course, not at all useful for the analysis of the nominal compounds: "The interpretation of the compound is determined, to some extent, by the grammatical pattern available for the verb" Bauer (2006: 494) notes, and therefore is to some extent arbitrary, it depends on the linguistic skills of the researcher.[151] In addition, this kind of syntactic analysis is not applicable at all to metaphorical compounds: ModE *monkshood* denoting the plant Aconitum napellus L., for example, cannot be paraphrased sensibly as *The monk wears a hood* just as OE *dæges ēage* 'daisy' lit. 'day's eye' cannot be paraphrased sensibly as *The eye opens at the beginning of the day*. Marchand himself indicates, however unconsciously, the weak point of this classification system by providing lists of model examples of syntactic paraphrases for different compound types, e.g. for the type *whetstone* (V_{stem}+N; cf. Marchand 1969: 72f.), but all of them refer to verbal nexus compounds. Nominal compounds are not equipped systematically with lists of syntactic paraphrases.

Formulating underlying sentences for Old English compounds – as we would be forced to do for the Old English plant names – confronts us with still another problem: As there are no native speaker informants, we are forced to use Modern English source sentences (as is often done for older stages of English), but this is an anachronistic approach. When we try to formulate Old English paraphrases, we can never be sure about the validity and idiomaticy of our Old English sentences.[152]

The value of syntactic phrases for describing the morpho-semantics of compounds was already questioned in the late 1930s. According to Carr (1939: 319), syntactic phrases are not able to describe the meaning of compounds comprehensively:

> Although the logical relationship between the parts of the compound may be defined and stated by a syntactic phrase, such a definition will not always give the meaning of the compound as a whole. It would, for instance, be impossible to define the meaning of the compound *Johanniswürmchen* by stating the logical relation of *Johannis* to *Würmchen*, and even if the meaning of a primary compound can be defined by a syntactic phrase, the compound cannot be identified with it and has certainly not arisen from it (Carr 1939: 319).[153]

Carr (1939: 319) further stressed that the impreciseness of nominal compounds, which results both from the fact that the relation between the parts is not expressed

151 Jespersen (1942: 143) remarks that "the number of logical relations between the two elements is endless", while Faiss (1978: 30) criticises the arbitrariness involved: "Eine gewisse Willkürlichkeit scheint der Konzeption des US immanent zu sein, wenn es unerheblich ist, ob für oak tree von "the oak is a tree" oder von "the tree is an oak" als US ausgegangen wird" ['A certain arbitrariness seems to be inherent to the concept of ultimate constituents when it is not important whether the ultimate constituents of oak tree are 'the oak is a tree' or 'the tree is an oak''] (Faiss 1978: 30).

152 Gardner (1968: 31-3), for instance, criticises Reibel (1963) for using Modern English source sentences to paraphrase Old English compounds.

153 By 'primary compounds' Carr (1939) understandes 'nominal compounds'; for Carr's (1939) distinction between primary and secondary compounds, see fn. 171 in this chapter.

morphologically and from the additional information attached to them as a result of lexicalisation processes, is not a flaw, but rather an advantage, since nominal compounds thus gain in flexibility and in suggestiveness.

In Marchand's (1969) classification system the formulation of syntactic paraphrases yields the semantic types the compounds belong to and, these, in turn, indicate the grammatical deep structure of the compounds. Levi (1978) and Bauer (1983; 2006) bypass establishing the grammatical deep structure and concentrate on the semantics of nominal compounds directly. In search of the semantics of nominal compounds, Bauer (2006: 494ff.) even rejects the formulation of source sentences and discusses Levi's (1978) proposal of capturing the relationship between the elements of the endocentric nominal compounds in a set of nine general predicates, three of which can be either active or passive: "cause" (e.g., *drug death*, *tear gas*), "have" (e.g., *picture book*, *feminine intuition*), "make", "use", "be", "in", "for", "from", "about".[154] This classification system seems to cover the majority of English compounds. A closer examination reveals, however, that even these rather abstract relationships, which have the advantage of leaving room for a wider interpretation, are not able to cover all kinds of compounds: for instance, compounds containing metaphorical elements such as *wisdom tooth* or *iron horse* resist Levi's classification principle.[155]

Bauer's solution is what he calls the *mnemonic device*. By the mnemonic device, Bauer (2001: 495) understands a very general formula: "A type of element-2 efficiently brought to mind by mention of element-1". The relationship between element 1 and element 2 is meant to be non-modal, positive and inherent or permanent. On the one hand, it seems as if the mnemonic device has solved all problems since it is applicable to all endocentric nominal compounds. In fact, however, the mnemonic device can just as easily be dismissed since it adds no new information to what we already know: namely, that there is, or seems to be, a kind of semantic relationship between the elements of nominal compounds. Schmid (2005: 100) concludes his section on different attempts at capturing the nature of the semantic relationship encoded in compounds as follows:[156]

> Es muß also etwas desillusioniert festgehalten werden, daß eine semantische Analyse von komplexen Lexemen, die über die morphologischen und syntaktischen Aspekte hinausgeht, weitgehend auf der Intuition der einzelnen Linguisten beruht.
>
> [It must, consequently, be asserted, with a certain disappointment, that the semantic analysis of complex lexemes, which exceeds the morphological and syntactic analysis, is based, largely, on the intuition of individual linguists.]

---

154 For a discussion of Levi's (1978) "recoverably deletable predicates" (abbreviated as RDP), see also Kastovsky (1982: 207ff.).

155 See also, e.g., ten Hacken (2000: 352).

156 See also Sauer (1992a: 36ff.).

This is certainly not a good starting point for a semantic (and morpho-semantic) analysis of the Old English compound plant names. All the same, coining neologisms always means encoding meaningful information in language. It is therefore worthwhile to look for alternatives to the formulation of underlying sentences; all the more, since structural morphology and semantics still face the challenge of analysing and classifying metaphorical and metonymic compounds.

5.3.2.2.2. EXOCENTRIC COMPOUNDS

As mentioned above (chapter 5.3.2.2.), exocentric formations are formations in which the composite is not a hyponym of its second element (AB $\not\subset$ B): in exocentric compounds, the semantic head is outside the composite. The so-called *bahuvrihi* compounds are typical exocentric formations. To them I would like to add the compounds based on a metaphorical relationship between the concept denoted by the second element and the plant denoted by the whole compound, which are counted among the endocentric compounds by Bauer (1983) and Marchand (1969). This is because I believe them to fit much better into the group of exocentric compounds, as I will explain in the following.[157]

The Modern English plant names *bluebell* and *monkshood* are both plant names where the composite is not a hyponym of its head (AB $\not\subset$ B): the *monkshood* is not a kind of *hood*, nor is a *bluebell* a kind of *bell*. Marchand (1969: 68) classifies *monkshood* as an endocentric but metaphorical compound:

> [...] the combination [is], however, not understood literally but figuratively. They [i.e. *monkshood* and other similar compounds] are metaphorical names of a plant, an animal, a concrete or abstract thing [...]. [...] a *monkshood* is a 'monk's hood' in appearance and form, though not in reality: a *monkshood* is a plant.

In his argumentation for the classification of the compound *redskin* as an exocentric compound (in his terms a "pseudo-compound", 1969: 13), Marchand states:

> This is the difficulty Lees has not been able to solve for the type *redskin*: *redskin* cannot be embedded in the matrix sentence *this is a skin* as a

157 To give some examples from the corpus of Old English plant names: *bahuvrihi* and metaphorical compounds are, for instance (in alphabetical order), *dæges ēage* 'daisy' lit. 'day's eye', *eorþgealla* 'knapweed, etc.' lit. 'earth-gall', *hennebelle* 'henbane' lit. 'hen's bell', *hræfnes fōt* 'buttercups, etc.' lit. 'raven's foot', *cicena mete* 'chicken-weed' lit. 'chicken's food', *fīflēafe* 'cinquefoil' lit. 'five-leaf', *hoclēaf* 'common mallow' lit. 'mallow-leaf', *sundcorn* 'saxifrage' lit. 'sea-corn', *strēawber(i)(g)e* 'strawberry' lit. 'straw-berry', *sūðerne wudu* 'southernwood' lit. 'southern wood', *wegbrǣde* 'waybread' lit. 'way-broad (plant)', etc. Metaphorical and metonymic compounds can be identified on the basis of the relation between the concepts denoted by the second elements and the whole compound. When the metaphorical relationship is between the first element and the concepts it denotes (as in, e.g., *candelwyrt* lit. 'candle-wort', *goldbloma* lit. 'gold-blossom'), we speak of partial metaphors (cf. chapter 5.3.2.2.1.1., (c)).

> substitute for *skin*. By a tour de force Lees embeds it in the matrix sentence *this is a man* for *man*, which is not in accordance with his own principles. If *redskin* is a compound, it must be a nominalization of sentence *the skin is red*, and by Lees' explanation it should be embedded for *skin*, but not for a substantive of an entirely different semantic class, such as *man* 'living human being' (Marchand 1969: 14).

It is clear that, following Marchand's argument, as *redskin* cannot be substituted for the semantic class *skin*, it is an exocentric compound (or, in Marchand's (1969: 13-5) terms, a pseudo-compound).

Let us look again at the example *monkshood*. Can *monkshood* be substituted for *hood* in the sentence *This is a hood*? It is obvious that this is not possible – just as it was not possible for *redskin*: The Modern English compound *monkshood* must be embedded for a substantive of the semantic class 'plant' and not for 'hood', since a *monkshood* is, in reality, a plant, as Marchand himself states (see quote above); the formula that applies is $AB \not\subset B$; $AB = X$ [plant]. Nevertheless, and not in accordance with his own argumentation, Marchand groups the metaphorical *monkshood* together with the non-figurative N_{gen}+N combinations, i.e. classifies it as endocentric compound.

Metaphorical compounds are not included in Bauer's systematics either. The exocentric compounds he mentions, such as for instance *redskin* and *highbrow* (1983: 31), or *egghead* and *highbrow* (2003: 43f.), are *bahuvrihis* (based on a relation of metonymy). Metaphors are, according to Bauer, meaning extensions of already existing lexemes with non-figurative meanings: for instance, *headhunter* 'one who recruits executives for a large corporation' is explained as the metaphorical extension of the literal meaning of the compound, which is "person who hunts heads, i.e. people" (1983: 63), and not as a semantic type on its own.[158] Schmid (2005: 124) concludes in line with Marchand (1969) and Bauer (1983):

> Beispiele wie *backbone*, *nutshell* und *ponytail* geben Anlaß, hervorzuheben, daß Determinativkomposita metaphorische Komponenten enthalten oder in einer wörtlichen und einer metaphorischen Bedeutung lexikalisiert sein können. In diesen Fällen weist natürlich nur die wörtliche, nicht aber die übertragene Bedeutung die Hyponymiebeziehung zum *head* auf.
>
> [Examples like *backbone*, *nutshell* and *ponytail* prompt us to stress that determinative compounds can contain metaphorical components or can be lexicalised with a literal and a metaphorical meaning. In these cases, it is only the literal, not the metaphorical meaning, which has a relationship of hyponymy with the head.]

---

158 *OED* (s.v. *head-hunter*): "(a) one who practises head-hunting; (b) orig. U.S., an employment agent or agency specialising in the recruitment of managers and other skilled personnel by identifying and approaching preferred candidates; one who recruits using these methods."

In other words, a lexeme with two meanings, one literal and one metaphorical, can be explained – at least with regard to the literal meaning – as an endocentric (determinative) compound. In its metaphorical meaning, the composite does not exhibit any relation of hyponymy to the head.

It is certainly possible to accept that it is sufficient to categorise compounds like e.g. *ponytail* by taking into account their literal meaning only. However, consulting the *OED* with respect to the meanings of, for example, ModE *ponytail*, *bluebell* and *monkshood*, we discover that these lexemes are not polysemous: the only meaning recorded is their figurative one (*OED*, s.vv. *ponytail*, *blue bell/blue-bell*, *monkshood*). The question is, then, how to deal with their classification when a literal meaning does not exist?

This is the crucial point: when compound lexemes with a metaphorical meaning are not hyponyms of their second elements, they cannot be classified as endocentric. But when these lexemes are not exocentric either because they do not express a HAVE-relationship as required of exocentric – *bahuvrihi* – compounds, these (metaphorical) expressions are, from the point of view of classification, in no man's land.[159]

The solution to this paradox seems to be to allow for the parallels in the structure of metaphorical and metonymic compounds by treating them as equals: both types of compounds are exocentric, as the semantic heads of both are not expressed morphologically on the surface of the construction. Therefore, metaphorical compounds should not be treated as endocentric (rectional) compounds but should be included in the group of exocentric compounds. The difference between metaphorical and metonymic compounds, then, is not a structural one, but one of associative relation, i.e. a cognitive, and not a formal criterion. Compare, for instance, the metonymic compound *whitethorn* and the metaphorical compound *bluebell*: both are Adj+N compounds (same morphological shape) with the semantic head lying outside the composite (same morphological structure). The difference between the two is that *whitethorn* is based on a relation of metonymy (metonymic compound) while *bluebell* on a metaphorical relation (metaphorical compound). In other words, metaphorical and metonymic compounds display the same morphological structure, namely, AB $\not\subset$ B; AB = X [plant], but they are based on different associative relations: on relations of (metaphorical) similarity, respectively, on relations of contiguity.[160]

As I have attempted to show, the morphological structure provides the first kind of information on the compound type. Yet, the question of how to describe both endocentric and exocentric compounds systematically with regard to the relationship

159 Sauer (1992a: 243) already recognised this paradox, but preferred to stick to the traditional classification. See also some pages earlier (1992a: 147f.), where he rightly asked whether the metonymic or metaphoric use of words must be treated within word-formation or should not rather be regarded as shifts of meaning.

160 To capture the morphologic and semantic ambiguities in polysemous compounds, one can use Cruse's (1986: 76-7) notion of *lexical unit*: for the compound ModE *headhunter* for example, one can choose between the lexical unit *headhunter*$_1$ 'person who hunts heads' (non-figurative meaning), which is an endocentric compound, and the lexical unit *headhunter*$_2$ 'one who recruits executives for a large corporation' (figurative meaning), which is an exocentric (metaphorical) compound.

of their elements is still unanswered. We have seen that a great part of the endocentric compounds, namely, the rectional nominal compounds, and all metaphorical and metonymic compounds still fall through the grids structural morphology and semantics have been able to provide: neither the indication of the morphological structure nor Levi's predicates or the formulation of an underlying sentence are able to encompass all compounds and certainly cannot pin down the semantic content of the compounds. Lipka's (2002: 138) proposal to go beyond the level of language-immanent analysis for the semantics of metaphorical and metonymic compounds, is, therefore, the only possible solution:

> For both [i.e. for metaphor and metonymy], the boundaries of a language-immanent structuralist approach must be transcended and a semiotic model must include referents and the extra-linguistic world. It is the contiguity of the referents or denotata of linguistic signs in reality, or rather our cognitive models of it, as well as similarity in our view of the world, which are involved here.

It is, of course, the extra-linguistic, encyclopaedic knowledge that helps us identify a *whitethorn* or a *bluebell* as plants, and not as thorns or as bells.[161]

The morphological analysis of the compound plant names is therefore based in the first place on formal aspects, namely, on the morphological shape of the compounds. The ensuing analysis of the morphological structure makes use of the traditional classification into endocentric and exocentric compounds, but it takes into account the semantics of the plant names as something that cannot be described morpho-syntactically, but has to be regarded on the semantic-cognitive level, using the notion of associative relations holding between the motivating concept and the concepts encoded in the plant names. A detailed analysis of the semantic aspects of the compound plant names, providing also a semiotic model and a discussion of the associative relations, is presented in chapter 6.

5.3.3. ANALYSIS OF THE PLANT NAMES ACCORDING TO THEIR MORPHOLOGICAL TYPES

In the following, the Old English complex plant names are discussed according to their morphological types. The *morphological types* are established on the basis of the morphological structure of the complex formations according to the word-class of the modifier and the head. That is, when for instance *blind(e)net(e)l(e)* is classified as an Adj+N formation, the type corresponds to the formal analysis of the plant name as *blind* (adj.) 'blind' + *net(e)l(e)* (noun) 'nettle', not taking into consideration the fact

161 This is why Gardner (1968: 15) distinguishes between two levels of meaning: the relationship between the content of the compositional elements (which I have called the *signification*; see chapter 6.4., fn. 216) and "the relationship between the compound and its referent, described in terms of how the referent is approached (literally, metaphorically, etc.)."

that, in terms of morphological shape, the composite is even more complex, as the first element is inflected ({*blind*}{*e*}) and the second element derived (suffix *–el* ({*net*}{*el*}).

In the corpus of Old English complex plant names, the following morphological types are present: first, N+N compounds (e.g. *āctrēow* 'oak tree'); then, combinations of N_{gen}+N (e.g. *hundestunge* 'dog's tongue' lit. 'dog's tongue'); combinations of Adj+N (e.g. *blind(e)net(e)l(e)* 'dead nettle' lit. 'blind nettle'), Num+N compounds (e.g. *fīflēafe* 'cinquefoil' lit. 'five-leaf'), V_{stem}+N compounds (e.g. *springwyrt* 'caper-spurge' lit. 'spring-wort'), synthetic compounds (e.g. *grundeswelige* 'groundsel' lit. 'ground-devourer', *sunfolgend* 'common marygold' lit. 'sun-following (plant)'), combinations of N+N_{deadj} (e.g. *īsenhearde* 'knapweed' lit. 'iron-hard (plant)'), $PropN_{es/ing}$+N compounds (e.g. *simǣringwyrt* 'mallow' lit. '*Simǣring's* wort'), followed by cases, which seem to be complex, but where an exact analysis is not possible as one or both elements of the plant names are not transparent (e.g. *gasdie* 'upright pellitory' lit. 'glass-?' with unclear second element).

The analysis concludes with an account of stem variancy (chapter 5.5.) and a brief discussion of the explicative value of the notions 'hapax legomena' and 'blocked morphemes' (chapter 5.4.) with regard to the Old English plant names.

5.3.3.1. NOUN + NOUN (e.g. *æppeltrēow*)

Noun + noun compounds (henceforth N+N) are made up of two lexemes; compare, e.g., *ācmistel*, which consists of the element *āc* 'oak' and *mistel* 'mistle'. N+N compounds are verbal nexus compounds (e.g., *gārclīfe* with the element *clīfe* 'cleavers' derived from the verb *clīfan* '(to) cleave') or non-verbal nexus (or nominal) compounds (e.g., *feltwyrt* made up of *felt* 'felt' and *wyrt* 'plant, wort'). 251 of the Old English complex plant names in my corpus are made up of N+N. Examples are *fīc/trēow* 'fig-tree' lit. 'fig-tree', *hals/wyrt* 'throatwort, etc.' lit. 'throat-wort', *dweorge/dwos(t)(l)e* 'pennyroyal' lit. 'dwarf's *dost*', *plūm/trēow* 'plum-tree' lit. 'plum-tree', and *wealh/more* 'common parsnip' lit. 'foreigner carrot'.

For a number of compounds included in the group of N+N formations the analysis does not yield clear-cut results. This is sometimes due to the textual situation, which does not allow conclusive statements about the status of (in-)dependence of some of the morphemes (cf. chapter 5.4.), or sometimes to morphological and/or phonological processes, which changed the morphological make-up of the elements (see chapters 5.5. and 7.3.1.2.). Sometimes, the elements of N+N formations are not attested outside the plant names though there are indicators suggesting their classification as independent elements. Thus, compounds where one of the elements is not attested independently in Old English (e.g. *-dwos(t)(l)e* in *dweorge/dwos(t)(l)e*) are included when the existence of the element under investigation is attested in cognate languages: the occurrence of G *Dost* permits to regard *-dwos(t)(l)e* as a noun and to group OE *dweorged(w)os(t)(l)e* with the N+N compounds (see Catalogue, s.v. *dweorge-*

d(w)os(t)(l)e).[162] In a similar way, the first elements of *cist/bēam* lit. 'chestnut tree' and *ynne/lēac* 'onion' lit. 'onion-plant' are still recognisable as going back to Latin, albeit in a modified form (see Catalogue, s.vv. *cist-*, *ynne-*); nevertheless, *cist-* and *ynne-* are not attested outside these combinations.

Verbal nexus compounds are included in the category N+N when the lexeme derived from a verbal element is attested as a lexicalised noun in Old English, such as *clīfe* 'cleavers' in *gār/clīfe* 'agrimony', lit. 'spear-cleavers'. If it is not, such as *-sūge* (from *sūgan* '(to) suck') in *hunig/sūge* 'honeysuckle', the compound is listed and analysed as a synthetic, $N+N_{deverbal(-e)}$ compound (see chapter 5.3.3.6.). It is, of course, not easy to decide about the status of the deverbal element as this question touches, again, upon the problem of the 'incomplete' and not necessarily representative corpus of Old English texts (cf. chapter 2.2.). This problem shows up also with plant names consisting of a primary noun and a deadjectival noun, e.g., *wegbrǣde* 'waybread' lit. 'waybroad'. When the deadjectival noun is attested independently, the compound is counted as N+N compound; e.g. *āt(t)or/lāþe* 'betony' lit. 'poison-harm' with *lāþe* 'pain, harm'. In contrast, as deadjectival *-brǣde* is not attested independently, *wegbrǣde* is classified as a $N+N_{deadj}$ compound (see also, for instance, Catalogue, s.v. *wudu(h)rōfe* and chapter 5.3.3.7.).[163]

Plant names in which the morphological status of one (or both) of the elements cannot be ascertained on the basis of the extant texts or through comparison with other languages, as for example, the element *-fuge* and *þelne-* in *fēfer/fuge* 'feverfew' lit. 'fever-[drive-away]' and *þelneþung* 'celery-leaved crowfoot' lit. '?-poisonous plant', are listed and discussed separately (chapter 5.3.3.9.). On the other hand, lexemes like *wermōd* 'wormwood' lit. 'man-mood' and *lufestice* 'lovage' lit. 'love-stitch', where the nouns *wer* 'man' and *mōd* 'mood', respectively *lufe* 'love' and *stice* 'stitch', can be recognised clearly, however unclear their semantic motivation may be, are included in the group of N+N compounds. These formations are of a special nature as their semantic motivation is secondary (cf. chapter 6.2.) because of folk etymological processes: the morphological and semantic make-up of these plant names was established on the basis of largely formal, phonological criteria, in the attempt of the speakers to establish some kind of formal relationship between these originally and presumably unmotivated plant names and other established terms in their lexicon (see chapter 7.5.).

As mentioned before, some plant names are attested with different morphological shapes. While the forms of the plant names mentioned in chapters 5.1. and 5.2. transgressed the boundaries of morphological types, in the following examples, the morphological shapes of the stems do not always change the morphological type of the compound. For the chestnut-tree, for example, not only the simplex *cisten* 'chestnut tree' (loan word) is attested, but also the complex N+N formations *cist/bēam* (with clipped first element), *cistel/bēam* (with suffix substitution in *cistel*) and *cisten/bēam* (cf. also chapter 7.4.2.1., and 7.4.2.1.2.). Other examples are *bisceop/wyrt* 'marsh

162 Old English *-dwostle* is attested in many variant forms (see Bierbaumer 1, 2, 3, s.v. *dweorge-dwostle*) and is a combination of *-d(w)ost* + suffix {*el*}; For the suffix family *-el*, *-il*, *-ol*, *-ul*, see Sauer (2001).

163 For a postulation of a zero-suffix compare the explanation to *fīflēafe* (cf. chapter 5.3.3.4.).

mallow' and *bisceop/wyrtel* or *bisceopwyrt/el* 'marsh mallow' – once without, once with suffix {*el*}. As *bisceop/wyrtel*, the plant name is a compound with a derived second element; as *bisceopwyrt/el*, it is a derivative with a complex modifier. Two morphologically different heads are parts of the compounds *clufþung* and *clufþunge*, where the former plant name is in the masculine (*a*-declension), and the latter in the feminine (ending in *-e*, weak *n*-declension). The plant names *æscþrote* and *æscþrotu* 'goutweed, etc.' lit. 'ash-throat' ending differently, are both feminine.

Endocentric and exocentric N+N compounds are in a ratio of one to three: 198 formations face c. 46 exocentric formations, including metonymic (e.g., *clātecrop* 'great burr' lit. 'burr-umbel'; or *furhwudu* 'pine' lit. 'furrow-wood'; see Catalogue, s.vv. *clātecrop, furhwudu*) and metaphorical compounds (e.g., *coltetræppe* 'caltrop' lit. 'colt's trap' and *cūslyppe* 'cowslip' lit. 'cow-dung, cow-slime'; see Catalogue, s.vv. *coltetræppe, cūslyppe*), including partial metaphors such as, e.g., *æscþrote* 'goutweed' lit. 'ash-throat', *furhwudu* 'pine' lit. 'furrow-wood', *goldblōma* 'marigold' lit. 'gold-blossom', where the metaphorical relationship affects the first element of the compound (cf. chapter 5.3.2.2.1.1., (c), and see Catalogue, s.vv. *æscþrote, furhwudu, gold/blōma*).

The majority of the endocentric compounds are rectional such as, e.g., *ācmistel* 'oak mistletoe' lit. 'oak mistletoe', *clufþung* 'celery-leaved crowfoot' lit. 'clove(d) poisonous plant' and *mōderwyrt* 'water-mint, etc.' lit. 'motherwort', some of which are partial metaphors such as, e.g., *candelwyrt* 'great mullein' lit. 'candle-wort' and *cnēowholen* 'knee-holly' lit. 'knee-holly'.

For a number of exocentric plant names the morphological status cannot be ascertained, as their motivation is opaque. The plant name *gōtwōþe*, for instance, is a case in point, as it denotes an unidentified plant, and a literal translation cannot be given since the constituting elements are not transparent (see Catalogue, s.v. *gōtwōþe*). The morphological status of *lufestice* as a N+N compound, on the other hand, is problematic in that it is valid from the point of view of folk etymological processes changing the intransparent Latin loan *ligusticum* on the basis of phonological similarity to the Old English nouns *lufu* 'love' and *stice* 'stitch', but semantically, there is no transparent relationship between the denotatum of the compound and the denotata of the elements of the compound (see Catalogue, s.v. *lufestice*). On the same grounds, the morphological status of the abovementioned *wermōd* cannot be established: it is very probably a compound that acquired its shape by folk etymological processes, but no transparent relationship between the denotatum of the compound and the denotata of its constituents can be ascertained (cf. chapter 7.5.; Catalogue, s.v. *wermōd*).

A noteworthy subgroup of the copulative compounds are the pleonastic ones as exemplified by, e.g., *cedertrēow* 'cedar of Lebanon' lit. 'cedar of Lebanon-tree', *fīcbēam* 'fig-tree' lit. 'fig-tree', *cīpelēac* 'onion' lit. 'onion-plant' and *gærswyrt* 'grass, hay' lit. 'grass-plant' (cf. chapters 5.3.2.2.1.1. (b)). As most of them are made up of a non-native first element and a native second element, they demonstrate that even loan words more or less fully integrated in the language system such as *cīpe* 'onion', at least at one point in their history, benefited from a semantically supportive element that positioned them clearly into the domain of plants.

5.3.3.2. NOUN$_{GEN}$ + NOUN (e.g. *hundestunge*)

Noun$_{gen}$+Noun compounds (henceforth N$_{gen}$+N) are made up of (at least) two full linguistic signs. The first element is inflected in the genitive, e.g. *cyningeswyrt* 'marjoram' lit. 'king's wort': {*cyning*}{*es*}{*wyrt*} 'plant, wort'. Since the first element is inflected, it has often been suggested that genitive combinations have developed from syntactic groups by way of *univerbation* (Durkin 2009: 39; G *Zusammenrückung*; Koziol 1972: 60f.). As such, they are non-verbal nexus compounds. But for a considerable number of compounds it seems far-fetched to assume the existence of a syntactic phrase as precursor. Old English *dæges ēage* 'daisy' lit. 'day's eye', for example, is a case in point: since there is a metaphorical relationship between the source concepts EYE and DAY and the target concept DAISY, the compound is semantically isolated from the syntactic group. It is therefore reasonable to assume that compounds such as OE *dæges ēage* or OE *foxes glōfa* 'foxglove' lit. 'fox's glove' may have been formed directly, and have not developed from syntactic phrases (see also Sauer 1992a: 152).

The status of combinations with the first element inflected in the genitive is disputed. Some classify them *a priori* as syntactic groups, as, for instance, the *DOE* (see below).[164] Others, such as, for example, Kastovsky (1992), posit a classification as compounds:

> For Noun (genitive) + Noun combinations it cannot always be decided with absolute certainty whether a given combination should be treated as a syntactic group or a genitive compound ('secondary compound' in Carr 1939: 309ff.), but it would seem unjustified to deny the existence of genitive compounds (see Nickel *et al.*, 1976: II, 20) in view of the behaviour of words such as *domesdæg, cristesboc* (Kastovsky 1992: 369).

The *DOE* as a rule treats combinations with the first element in the genitive as syntactic groups; they are sub-ordinated to one of the elements (Healey 2009: 102). This principle is undermined, however, by cases such as *dæges ēage* 'daisy', a N$_{gen}$+N combination which appears as a headword on its own in the *DOE* (s.v. *dæges ēage*), and this seems inconsistent in view of examples such as the N+N compound *crāwlēac*, which bears the same idiomatised meaning as the N$_{gen}$+N combination *crāwanlēac* but is – in contrast to the latter – a headword on its own in the *DOE*. *Crāwanlēac*, on the other hand, appears as a sub-entry to the headword *crāwe*, "as an element in plant names" (*DOE*, s.v. *crāwe*).

Drawing a demarcation line between N$_{gen}$+N compounds and corresponding syntactic groups on the basis of the arguments presented earlier (5.3.1.) is not an easy task, as some of the otherwise 'safe' criteria, such as transposition of the first noun, or

---

164 For a discussion of the problems in distinguishing compounds from syntactic groups, especially with regard to Old and (Early) Middle English genitive combinations, see Kastovksy (1992: 369-70) and Sauer (1992a: 57ff.).

non-inflection of the first element are not applicable.[165] What is more, since some genitive combinations may have had their origins in morphologically identical syntactic groups, they do not differ from the latter in shape or in structure (an exception are the metaphorical and the metonymic compounds). As concerns the status of *-(e)s-* between two constituents of a complex form, it must be kept in mind that, while an *-s-* in Modern English is generally regarded as a linking element (which, originally, may have been a genitive marker) so that the combination is classified as a compound (or at least as an idiomatised syntactic group) – the *officer's mess* is not the same as the **mess of the officer* (Marchand 1969: 27; Kastovsky 1982: 177f.) –, in Old English, the situation is different. Here, intermediate *-(e)s* is in most cases a genitive marker (gen. sg., *a*-declension) and not a general linking element. Besides other variants of the genitive morpheme (see below), there are also several linking elements such as intermediate *-e-* (e.g. *hild* 'war, combat' > *hilderinc* 'warrior') and the absence of nominative singular endings (e.g. *cyrice* 'church' > *cyricbōc* 'churchbook') marking composition (cf. Sauer 1992a: 81-4). In some complex formations, intermediate *-s-* is inherent to the stem of the word and not a genitive a marker nor a linking vowel, e.g. *ælmesmon* (see below; cf. Sauer 1992a: 156). The most decisive and often the only criterion applicable to delimit compounds from syntactic groups with N_{gen}+N composites is semantic isolation, i.e. when the meaning of the whole combination cannot be derived from the meaning of its constituents. Thus, the metaphorical and metonymic genitive formations *dæges ēage* 'daisy', *foxes fōt* 'unbranched burr-reed', *mūsepise* 'tufted vetch' and *cicena mete* 'chicken-weed', all of them semantically isolated, can be considered compounds. Isolated are, further, compounds which owe the form and meaning of their elements to secondary motivation (e.g. *-teors* 'penis' in *wulfesteors* 'wolf's teasel', lit. 'wolf's penis'; see chapter 7.5., and Catalogue, s.v. *wulfesteors*).

With regard to their morphological structure, 19 of the 43 combinations[166] of N_{gen}+N are endocentric, with the structure AB $\subset$ B: e.g. *crāwanbēam* 'goat's beard' lit. 'crow's tree', *foxesclāte* 'greater burdock' lit. 'fox's cleavers', *gēaces sūre* 'wood sorrel' lit. 'cuckoo's sorrel', *hræfnes lēac* 'wild orchid' lit. 'raven's leek', *hundescwelcan* (pl.) 'guelder rose' lit. 'dog's ?', *toscanlēac* 'horsetail' lit. 'toad's plant', and *witmǣres wyrt* 'unidentified plant' lit. '*Witmǣr's* wort'. The other half (24 plant names) are exocentric: AB $\not\subset$ B, AB = X (plant). The majority of exocentric formations are metaphorical compounds (17 plant names): e.g. *dæges ēage* 'daisy' lit. 'day's eye', *dracanblōd* 'dragon's blood tree' lit. 'dragon's blood', *fornētes folm* 'cuckoo flower' lit. '*Forneot*'s hand', *oxanslyppe* 'oxlip' lit. 'ox-pat', *oxantunge* 'alkanet' lit. 'ox-tongue' or *wulfescamb* 'wild teasel' lit. 'wolf's comb'; the rest (7 plant names) is metonymic: e.g. *cicena mete* 'chicken-weed' lit. 'chicken's food', *mūsepise* 'tufted vetch' lit. 'mouse-pea', or *sunnancorn* 'common gromwell' lit. 'sun's corn'.

---

165 Cf. also Marchand (1969: 20-9), Koziol (1972: 48f.), Quirk et al. (1972: 1019f.), Faiss (1978: 19-28), Kastovsky (1982: 176-9), Bauer (1983: 42-61), and Sauer (1992a: 57-124, 152-5).

166 This number includes *witmǣres wyrt*, which is counted twice: as combination of a proper name and a noun and as combination with inflected first element.

Forms of the Old English genitive and some problems

The genitive endings in Old English amount to a handful of forms: they can be *-a, -an, -e, -es, ∅* or *umlaut* of the root vowel in the singular, and *-a, -ena, -ra* in the plural, depending on the declension paradigm the noun belongs to. Not all of these forms occur in the Old English plant names.[167] The most often occurring ending is *-es* (gen. sg.), as, for example, in *cyningeswyrt, foxes fōt, hundestunge*, and *wulfescamb*. The other groups are not so numerously represented: the second biggest group is the one ending in *-an* (gen. sg.), e.g. *crāwanlēac, oxanslyppe* and *sunnancorn*, followed by those ending in *-e* (gen. sg.), e.g. *gātetrēow, mūsepise* and *hennebelle*, and those ending in *-ena* (gen. pl.): *cicena mete* 'chicken-weed' and *oxnalybb* (from *oxena* with syncope of the *-e*; Campbell 1959: 249).

A number of uncertainties are connected with the identification of genitive inflectional endings. One of them concerns the polysemous (inflectional) suffix *-e*: "The intermediate vowel could be interpreted either as a genitive marker, a 'linking vowel' or *Fugenelement* (empty morph), or as the stem formative of the determinant" (Kastovsky 1992: 369f.).[168] To this can be added the *-e* representing a *Schwa* as a result of weakening and reduction of ending (due, probably, to Germanic fore stress, which caused the last syllables in polysyllabic words to weaken and to be reduced; Campbell 1959: §§ 368-93). Eventually, the *Schwa* was lost at the end of the Old English period, respectively during the Middle English period. This reduction affected all genitive markers with the exception of the *-(e)s* of the masculine nouns of the *a*-declension,[169] and this can be observed in the different forms of the Old English compound *crāwanlēac* 'crow garlic' lit. 'crow's garlic' (N_{gen}+N), which was obviously reduced via *crāwelēac* to *crāwlēac* lit. 'crow garlic' (N+N) – at least, if we believe that the different forms depend on each other and are not due to polygenesis (cf. chapter 5.5.; Catalogue, s.vv. *crāwanlēac, crāw(e)lēac*).[170]

The reverse occurred with N+N compounds developing linking vowels in the joint between the stems of the compounds, especially to support the pronunciation of heavy consonant clusters. Even though the origins of these compounds were nominal,

167 For further details on genitive forms and functions, see Brunner (1960-1962, II: 2-10, 27-40); Campbell (1959: 222-60) and Hogg (1992c: 124-38).

168 For further details, see also Sauer (1992a: 155f.).

169 Sauer (1992a: 127, 155) excludes all compounds where the genitive is not explicitly marked by an *-(e)s* from his group of genitive compounds in Early Middle English, explaining: "[...] weil diese Formen des Gen. im Me. aber gegenüber dem s-Genitiv auf dem Rückzug sind, und auslautendes *-e* im Dt zudem oft verschieden interpretiert werden kann [...], wird hier ebenfalls nur der formal eindeutige Typ Sb+s/Sb eigens aufgeführt. Die übrigen Verbindungen werden unter Sb/Sb eingeordnet." ['because these forms of the genitive are retreating in comparison to the *s*-genitive and because final *-e* in the determinant often can be interpreted in different ways [...] here, it is also only the type N+*s*+N which is listed separately. All other formations are classified as N/N formations'].

170 Interestingly, the Modern English cognate of OE *crāw(an)lēac* is a genitive combination again, but now – due to processes of analogy – with the etymologically 'incorrect' *-(e)s* genitive: *crow's leek*. Non-botanical examples are, e.g., the compounds *lādman, winterdæg*, which turned into ModE *loadsman* and *winter's day*.

synchronically, they are classified as secondary compounds, i.e. compounds with the first element inflected in the genitive (see Carr 1939: 310).[171]

Scribal errors can be sources for faulty interpretations, too. It is, for example, not clear, whether *fleaþorwyrt* (*Durham Plant Name Glossary*, no. 252) is really a scribal error for *flēaþan wyrt* lit. 'water-lily' (with a genitival *-an*; see Catalogue, s.v. *flēaþe wyrt*), as was suggested by Lindheim (*Durham Plant Name Glossary*, 62), or whether it is a variant of an Adj+N compound *flēaþe wyrt* lit. 'floating plant' as was suggested by Bierbaumer (1, s.v. *flēaþe wyrt*). Therefore, it must always be kept in mind that, however clear the morphological shape may seem, some of the plant names classified as N+N compounds may have been genitive combinations at their origins, and that some of the genitive combinations, originally, may have been primary N+N compounds. Seventeen N_{gen}+N compounds are attested only once and a number of them only two or three times. Among them are, e.g. *crāwanbēam* 'goat's beard, etc.' lit. 'crow's tree', *foxesclāte* 'greater burdock' lit. 'fox's clivers' and *gātetrēow* 'common dogwood', lit. 'goat's tree' etc.

5.3.3.3. ADJECTIVE + NOUN (e.g. *blind(e)net(e)l(e)*)

128 plant names are made up of an adjective and a noun (henceforth Adj+N); 67 of the combinations are hapax legomena. Many of the Adj+N compounds were, originally, syntactic groups that coalesced into compounds (*univerbation*); this explains why some of the plant names still display the remains of inflection as, for example, *hāra(n)sp(r)ecel* 'cudweed' lit. 'hoary wort' and *hāranwyrt* 'cudweed' lit. 'hoary wort', where some of the forms attested have inflectional *-n-* (cf. Catalogue, s.v. *hāra(n)sp(r)ecel*). As for the plant names recorded only in glossaries, it is extremely difficult to decide whether the combination is morphologically lexicalised or not (cf. chapter 5.3.1.), since the lemmas are seldom inflected in glossaries.

Endocentric Adj+N formations are not morphologically isolated and, in some cases, the plant names appear both as a syntactic group and as a compound: e.g., *þā brādan bisceopwyrt* (in *Lācnunga*, 697-9) and *brādbisceopwyrt* 'unidentified plant' lit. 'broad marsh mallow' (in *Lācnunga*, 41).[172] Morphological isolation can be assumed when

---

171 Carr (1939: 309ff.) distinguishes between 'primary compounds', i.e. compounds whose first members appeared originally in the stem form without an ending, and 'secondary compounds', i.e. compounds whose first member are case-forms of nouns. The latter is a younger type of compound than the former, but developing independently in the languages belonging to the Indo-European family. The development of secondary compounds is the result of the univerbation of a syntactic group noun + attributive genitive (or inflected adjective; with both elements bearing a main stress) to a combination of these two elements under one main stress which then was on the attributive genitive. The isolation of meaning of these combinations distinguishes them clearly from the corresponding syntactic groups.

172 In their subsequent history, Adj+N combinations sometimes changed their 'status': the OE compound *hwīteclǣfre* 'white clover', for example, was decomposed into the Modern English idiom *white clover*, and, on the other hand, the phrase *sūþerne wudu* 'southernwood' turned into the Modern English compound *southernwood*. The Old English phrase *rēad clǣfre* 'red clover', in contrast, is still an idiomatic phrase today, namely ModE *red clover*.

the semantic head is not expressed on the surface of the combination, i.e. in metaphorical compounds as, e.g., in *openærs* 'medlar' lit. 'open ass', or *hāra(n)sp(r)ecel* 'viper's grass' lit. 'hoary/whitish spot', and in metonymic compounds as, e.g., in *wīelisc* [*wudu*] 'southernwood' lit. 'foreign [wood]'. Compounds that owe their forms to folk etymological processes such as, e.g., *gōdæppel* 'quince' lit. 'good apple' (< OE *coddæppel* 'quince' lit. 'husk-apple'; cf. chapter 7.5. and Catalogue, s.v. *gōdæppel*), are morphologically isolated as well.

Whereas most of the adjectives used in Adj+N compounds are in the positive, some deviate from the 'norm'. In four of the Adj+N combinations, the adjective is in the comparative. These are the plant names using *læsse* and *māra* (or *māre*): The comparative of *lȳtel* 'small, little', namely, *læsse*, appears in *læsse bisceopwyrt* 'betony' lit. 'smaller/lesser marsh mallow', in *hierdewyrt sēo læsse* 'common centaury' lit. 'lesser centaury' and in *curmealle sēo læsse* 'centaury' lit. 'lesser centaury'. One plant name uses *māra* the comparative of *micel* 'much': *curmealle sēo māre* 'knapweed' lit. 'bigger centaury'. In addition, the positions of the adjectives in *curmealle sēo læsse, hierdewyrt sēo læsse, curmealle sēo māre* and *wermōd se hāra* 'sea wormwood' lit. 'hoary wormwood', are inverted. As Mitchell (1985, 1: 78) points out, "in OE, as in ModE, the order adjective + noun is the norm" for attributive adjectives, but postposition of adjectives in attributive position is common as well (Mitchell 1985, 1: 61-2).[173] While attributive, predicative and appositive use are often not easy to distinguish (Mitchell 1985, 1: 49), noun phrases like *Sidroc eorl se gioncga* lit. 'earl Sidroc the young' (*Chronicle* (Bately), 48, [871]) suggest that our four plant names exhibit appositive use.[174]

Appositional groups in Old English usually consist of one (or more) nouns, adjectives, demonstratives, possessives, numerals or personal pronouns (Mitchell 1985: 1, 599), and the structure noun + demonstrative + adjective occurs, albeit a relatively young structure in Old English (Mitchell 1985, 1: 611, § 1463), quite often. Of the five semantic types of the elements in apposition (not including sentence apposition), namely, 'augmenting proper names', 'clarifying', 'enumerating', 'repetitive' and 'partitive' (Mitchell 1985, 1: 607-10), it is probably the second and the last one which can be invoked for the Old English plant names in question, though neither is exactly fitting. While the group of clarifying elements is characterised as "explaining the first [...] by translating it [...] or by making it more specific" (Mitchell 1985, 1: 608) with examples such as *upp on ane dune* 'up on a hill' for the specifying function, partitive apposition "occurs when one element (usually the second) limits the other by specifying a particular part of it, e.g. [...] *Hamtunscir se dæl se hierre behinon sæ was* ['the part of Hamptonshire that was close to the sea'] (Mitchell 1985, 1: 609). The elements *sēo læsse, sēo māre* and *se hāra* have, indeed, a limiting function as they indicate that it is only a part of the group of *curmealle, hierdewyrt,* and *wermōd*, respectively, which is denoted by the plant name, and they are clarifying,

173 For the relative position of qualifying elements in Old English noun phrases, see Mitchell (1985, 1: 68ff.) and Brunner (1960-1962, II: 52-73).

174 On the appositive use of adjectives, see Mitchell (1985, 1: 63); on apposition in general, see Mitchell (1985, 1: 597-612).

which part of the kind is meant, but they are more than limiting or clarifying in the descriptive way as the examples in Mitchell (1985) imply.

Beyond the fact that the semantic implication is not entirely graspable, as these plant names are lexicalised after having started out as syntactic groups, they are, further, characterised by the fact that the adjectives are not, as usual with most of the Adj+N plant names, in the positive but in the comparative. While, according to Mitchell (1985, 1: 81), adjectives in the comparative are used either attributively or predicatively and thus, apparently, not in an appositional function, Brunner (1960-1962, II: 71-2), at least does not exclude appositional use of adjectives in the comparative, when he implies that structures as present in the modern English noun phrase *John the Elder* go back to Old English. As it is not possible to distinguish clearly whether these four plant names are lexicalised nouns + appositional demonstrative and adjectives, or lexicalised nouns that are postmodified by attributive demonstrative and adjective, they will be labelled 'postmodified noun + demonstrative + adjective' and it is clear that the use of the postmodified elements is restrictive rather than non-restrictive: the elements following the noun are specifying a taxon subordinated to that denoted by the first element when used alone.

The group of nouns postmodified by demonstrative + adjective is conspicuously small in comparison to the rest of plant names made up of a noun and an adjective, and *curmealle sēo māre*, which glosses *centauria maior* lit. 'the bigger centaury' and *curmealle sēo læsse*, which glosses *centauria minor* lit. 'the smaller centaury', as well as *hierdewyrt sēo læsse*, which denotes the same plant as *curmealle sēo læsse*, are loan translations or assumed loan translations of the Latin terms: they are clearly inspired by the Latin order of the elements, as is also, probably, the unidentified hapax legomenon *wermōd se hāra* (see 7.4.3, and Catalogue, s.vv. *hierdewyrt sēo læsse*, *curmealle sēo læsse*, *curmealle sēo māre* and *wermōd se hāra*).

It is sometimes difficult to decide whether the first element of the combination is an adjective or a noun. It is, in general, quite easy to identify Old English adjectives – either due to their specific morphology (stem + adjectival suffix) which is often accompanied by *i*-mutation (e.g. *wealh* 'foreigner' – *wīelisc*, *wȳlisc* 'foreign', *gold* 'gold' – *gylden* 'golden'), or because of case, number and gender agreement with the accompanying article, the noun and verb of the sentence. The presence or absence of inflection, therefore, is a criterion to distinguish Adj+N compounds from parallel syntactic groups, and presence, respectively, absence of inflection of the adjective is the element indicating that *brādbisceopwyrt* (*Lācnunga* 41) is a compound, while (*þā*) *brādan bisceopwyrt* (in the accusative, sg., fem., *Lācnunga* 697-8) is not. Sometimes, however, the three criteria – word-formation, *i*-mutation and inflection – are neutralised. Especially with lexemes denoting materials it is difficult to take a decision for one word-class or the other since here, the noun and the adjective are sometimes homonyms and one could argue that the adjective was derived from the noun by zero-derivation. OE *ellen*, for example, can be used both as a noun ('elder') and as an adjective ('of elder') and forms compounds such as *ellentrēow* 'elder' or *ellenwyrt* 'dwarf-elder'. Since I had to take a decision as to which group they should be included in, I have classified them as N+N compounds, i.e. regarded the first elements as nouns in analogy to similar plant names where the status of the first element is not ambiguous, like, for example, in *bēctrēow* 'beech' (in comparison to the syntactic group *bēcenan trēow*, attested in the noun phrase "on becenan treowes wyrttruman" ['on the roots of a beechen tree'] in *Herbarium*, no. 78) or *pīnbēam* 'pine tree', with *pīn* 'pine (tree)' attested as the noun denoting the pine (see Catalogue, s.v. *pīn*).

The translations of *wealhhnutu* as 'foreign nut, walnut' by *BT* (s.v. *wealhhnutu*) and of *wealhmore* as 'foreign root, carrot, parsnip' (*BT*, s.v. *wealhmore*) characterise the combinations as Adj+N compounds (see also Sauer 1992b: 398 and Bierbaumer 1, s.v. *more*, who both translate the plant name as '*welsche Möhre*' i.e. lit. 'foreign carrot'). There is, however, an alternative interpretation that is based on morphonological criteria: as the Old English adjective *wīelisc* 'Welsh, foreign', derived from *wealh* by suffixation with the adjectival suffix *-isc* (twice in the botanical lexicon: in *wīelisc* [*wudu*] 'southernwood' and in *wīelisc moru* 'wild parsnip'), clearly displays phonologically conditioned allomorphy, the element *wealh* in the composite *wealhhnutu* is most probably the noun and not the adjective, so that the compounds *wealhhnutu* 'walnut' and *wealhmore* 'common parsnip, wild parsnip', are N+N and not Adj+N compounds.[175]

Indications of directions such as *ēast* 'east' *norþ* 'north', *sūþ* 'south' and *west* 'west' are either adverbs or adjectives when used independently. As adjectives they are often in the comparative or in the superlative (Sauer 1992a: 106-7). Far more often than that, however, they were used in a derived form such as, e.g., *ēastern*, *sūþern*, etc. The suffix *-ern* (< Gmc **-rônjo-*) is an adjective termination form denoting 'a place, towards (a place), coming from (a place)' (Koziol 1972: § 543) – and this is the form we find in the complex plant names *ēasterne hnutbēam* 'almond' lit. 'eastern nut tree', *sūþerne cymen* 'cumin' lit. 'southern cumin' or *sūþerne wudu* 'southernwood' lit. 'southernwood'.

The large majority of the Adj+N plant names are endocentric (119 plant names); only 5 plant names are exocentric and a further 4 plant names are postmodified noun phrases of the type noun + article + adjective, which probably translates a Latin model (see above). With the exception of *sculdþȳfel* 'reed' lit. 'shallow bush', which is a rectional formation, all Adj+N plant names are copulative (attributive), but, as pointed out above (see chapter 5.3.2.2.1.2.), some of the analyses can be questioned, as formulating paraphrases for non-verbal nexus compounds is problematic: lacking a verb, we can create quite a large number of plausible and fitting paraphrases. This problem is obvious in many of the Adj+N combinations: *affricanisc æppel* 'pomegranate', for instance, can be circumscribed as 'The apple is African' or 'The apple comes from Africa'. Following the first paraphrase, *affricanisc æppel* would be a copulative, attributive formation, following the second, a rectional formation. The same difficulty can be encountered in *brādbisceopwyrt*: the paraphrase 'The bishopswort is broad' turns the plant name into a copulative, attributive compound; the more complex paraphrase 'The bishopswort has leaves which are broad' into a rectional compound. This demonstrates all the more that language structures and language use are not clear-cut and easily classifiable, but that, often, the boundaries

175 According to *ClH* (s.v. *wealh*), *wealh* is a noun, and denotes 'foreigner, stranger, slave, etc.'.

between the categories are transgressed, and, in these cases, more than one interpretation is valid.

Of the exocentric compounds, two are metaphorical (*openærs* 'medlar' lit. 'open arse' and *hāra(n)sp(r)ecel* 'viper's grass' lit. 'hoary speck') and three are metonymic (*līþelēaf* unidentified lit. 'mild leaf', *sūþerne wudu* 'southernwood' lit. 'southernwood' and *wīelisc* [*wudu*] 'southernwood' lit. 'foreign wood').

5.3.3.4. NUMERAL + NOUN (e.g. *fīflēafe*)

Numeral + Noun formations (henceforth Num+N) are made up of two lexemes; they are non-verbal nexus compounds.[176] There is no general consensus as to the status of numerals: According to Greenbaum and Quirk (1990: 16, note [a]) and Ungerer et al. (1984: §§ 354-60), numerals form a word class of their own. At at a later point, however, Greenbaum and Quirk (1990: 127-8), list both cardinal and ordinal numerals as "pronouns without a person contrast". ModE *half* is listed as one of the numerals by Greenbaum and Quirk (1990: 127); OE *healf* is labelled 'adjective' by *ClH* (s.v. *healf*) and as 'prefix' by Koziol (1972: 98f.).[177] For Sauer (1995: 312) *healfwudu* is a Num+N compound.[178]

Some classify Num+N compounds as representing an independent word-formation pattern (e.g., Sauer 1992a: 171-4), while others subgroup them to the Adj+N compounds. For Jespersen (1942: 158), they belong to the Pron+N compounds, while for Marchand (1969: 386-9), formations such as ModE *five-finger*, and OE *fīflēafe* are not compounds at all, but zero-derivatives (or pseudo-compounds, also called *bahuvrihi*-substantives).

Six of the plant names in my corpus qualify as Num+N compounds; these are *fīffingre* 'fiveleaf' lit. 'five-finger', *fīflēafe* 'fiveleaf' lit. 'five-leaf', *healfwudu* 'calamint' lit. 'half-wood', *seofonlēafe* 'common tormentil' lit. 'seven-leaf', *twīlēafa* 'large butcher's broom' lit. 'two-leaf' and *þrilēfe* 'wood sorrel' lit. 'three-leaf'. With the exception of *healfwudu* lit. 'half-wood', they form a semantically and formally homogeneous group, as all are paraphrasable as 'plant which has two (three, five, seven) leaves'. *Fīffingre* is slightly different from the others because there is a metaphorical relationship between the concept denoted by the second element and the plant; all the other Num+N sequences, including *healfwudu*, form metonymic compounds.

Fīflēafe, seofonlēafe, and *þrilēfe* are loan translations of Latin plant names; OE *twīlēafa* is based on Gk *δίγλωσσος* lit. 'two-tongued', which means that it is not a loan translation but a loan rendition (see Catalogue, s.v. *twīlēafa*), since it renders *-γλωσσος* 'tongue' by OE *-lēafe* 'leaf'. However, it has probably been formed in analogy to *fīflēafe*, etc., so that it may be assumed that not only the models for single plant names

176 For an overview of the history of this word-formation type and for further references, see Sauer (1992a: 171-4).

177 G *halb* is labelled as an adjective in the *Kluge* (s.v. *halb*).

178 For a recent investigation into the nature of Old English cardinal numbers, see Mengden (2010).

have been taken over from Latin, but that the pattern produced further plant names without Latin models.[179] *Fīffingre*, *healfwudu* and *þrilēfe* are hapax legomena.

The Num+N compound plant names are morphologically and semantically isolated from the corresponding syntactic groups since their head is not expressed on the surface of the combination as all of them are exocentric formations. Morphological isolation is, further, apparent in compounds that assume a form which the simplex does not have. The example of *lēaf* (strong declension) and *(fīf-)lēafe* (weak declension) has already been mentioned: The final *-e* in *fīflēafe* is a weak noun extension and not a derivational suffix. According to Kastovsky (1968, 2005a, 2005b and 2006), the Germanic stem formative (which was attached to the root and preceded the inflectional ending) could act either as a derivational element (producing deverbal nouns or denominal verbs), or as an inflectional marker (indicating the declension paradigm, cf. also Kastovsky 2005a: 289). Due to the re-structuring of the morphological system from Germanic to Pre-Old English (and again from Pre-Old English to Old and Middle English), the stem formative "had become an inherent feature of the stem" (Kastovsky 2006: 165), so that the "endings [e.g. in *cum-a* n.< *cum-an* vb., *hunt-a* n. < *hunt-an* vb.] are part of the inflectional system, and thus do not have any derivational function" (Kastovsky 2006: 153). Therefore, formations of this kind (in simplexes and in compound words) must be regarded, synchronically, as zero-derived (resp., as conversed).[180]

All six Num+N plant names are exocentric formations: the compounds are not hyponyms of their second elements (AB $\not\subset$ B, AB = X [plant]). OE *þrilēfe* lit. 'three-leaf' for example, is not a kind of 'leaf', but a plant. With the exception of *fīffingre*, which is a metaphorical compound, all are formations based on a relation of contiguity between the concept of the second element and the plant; *fīflēafe*, *seofonlēafe*, *twīlēafa* and *þrilēfe* are *pars pro toto* compounds, i.e. connected by a relation of meronymy; see chapter 6.3. (7)).

5.3.3.5. VERB$_{STEM}$ + NOUN (e.g. *springwyrt*)

Verb$_{stem}$+Noun (henceforth V$_{stem}$+N) compounds are made up of a verbal stem and a noun; examples are *clifwyrt* 'great burr' lit. 'cleav(ing) plant', *pricþorn* 'holly' lit. 'prick-thorn', *springwyrt* 'caper-spurge' lit. 'jump-wort' and *sticwyrt* 'greater stichwort' lit. 'stitchwort'. According to Carr (1939: 175-93), this compound type was not very common in Old English and did not exist at all in Germanic or West Germanic; it may have developed from complex formations where the deverbal noun was, by chance, identical with the verb stem (Kastovsky 1992: 370-1). Seven plant names from the corpus are V$_{stem}$+N compounds, and two of them, namely, *brȳsewyrt* 'comfrey, etc.' lit. 'bruise-wort' and *pricþorn* 'holly' lit. 'pricking thorn', are hapax legomena.

179 See, for example, Sauer (1992b: 391); Sauer and Krischke (2004: 186).
180 Cf. also Carr (1939: 164, 237).

V_{stem}+N compounds are morphologically isolated since verbal stems cannot act as modifiers of nouns. The verbs *clīfan* '(to) cleave', *(æt)clīðan* '(to) adhere', *prician* '(to) pierce', *springan* '(to) spring', and *stician* '(to) stitch', which provide the first elements in the plant names *clifwyrt* 'great burr', *cliþwyrt* 'great burr', *pricþorn* 'holly', *springwyrt* 'caper-spurge' and *sticwyrt* 'greater stitchwort', also form independent nouns, namely, *cliðe* 'great burr', *clife* 'great burr', *stice* 'stitch' and *prica* masc., *price* fem. 'prick' – good grounds on which the analysis of these formations as V_{stem}+N could be questioned, and an alternative analysis as N+N compounds could be proposed.[181] Yet, since it seems unusual to eliminate the final *-e* or *-a* of nouns in compounds when what is left is a consonant cluster which is not as easily pronounced as if the final *-e* or *-a* had been retained, the compounds are rather grouped with V_{stem}+N compounds.

With the exception of *snīþstrēo* 'carline thistle' lit. 'cut-straw' which is an exocentric, metaphorical compound, all V_{stem}+N compounds are endocentric and rectional; compare, for instance, *brȳsewyrt* 'comfrey' lit. 'bruise-wort', *springwyrt* 'caper-spurge' lit. 'jump(ing) plant', and *clifwyrt* 'great burr' lit. 'cleav(ing) plant' (see Catalogue, s.v. *brȳsewyrt, springwyrt, clifwyrt*).

5.3.3.6. SYNTHETIC COMPOUNDS (e.g. *grundeswelige*)

A number of complex plant names consist of a primary noun or pronoun and a deverbal noun; in terms of morphological shape they combine two free lexical morphemes with a bound grammatical morpheme: e.g. *grundeswelige* 'groundsel': {*grunde*} 'ground' + {*swelig(ean)*} '(to) swallow' + {*e*}; *sunfolgend* 'marigold': {*sun*} 'sun' + {*folg(ian)*} '(to) follow' + {*end*}.[182]

As the deverbal nouns forming the second elements are not attested independently in Old English, the analyses of the shapes of the plant names on the one hand, and of the structures, on the other, produce contradictory results:[183] On the one hand, it is not possible to classify these composites as compounds because the major condition for a lexeme to be called a compound – namely, that the composite is a combination of two independent lexemes – is not fulfilled, since the second elements, for example *-swelige* or *-sēce* in *grundeswelige* 'groundsel' and *sōlsēce* 'common marigold' are not attested as independent items. On the other hand, it is not possible to classify these kinds of composites as derivatives because the complex verbs one is forced to form – e.g. **grundeswelig(e)an* or **sōlsēc(e)an* – are not attested verbs of Old English (the so-

181 According to Carr (1939: 103) for instance, "the first part of [*springwyrt*] is not a verbal stem: see OHG. *springa*, wild caper."

182 With regard to the status of the final *-e*, I follow Kastovsky (e.g. 2006: 165), who argues that since the stem formative had become an integral part of the stem in Pre-Old English, endings, e.g. final *-e* or *-a* in deverbal nouns, e.g. *hunt-a* 'hunter' < *hunt-an* 'to hunt', were inflectional and had no derivative force (see also 5.3.3.4. above).

183 When the nominalisations of the verbal constituents are attested independently, such as *clīfe* 'cleavers' in *gārclīfe* 'agrimony', or *-stice* 'stitch' in *lufestice* 'lovage', the compounds are classified as N+N compounds (5.3.3.1.).

called *bracketing paradox*; cf. Schmid 2005: 136). Compounds of this kind are called *synthetic*.[184]

The Old English synthetic plant names are without exception agent nouns, ending in *-e* or in *-end* (a highly productive suffix forming deverbal agent nouns; Kastovsky 1992: 385). It is obvious that, since plants do not usually perform actions, the number of synthetic compounds is rather small and the actions verbalised in the plant names refer chiefly to the creeping and winding features of these plants (see further below and chapter 6.4., (8)). In total, 16 plant names are synthetic, of which seven are hapax legomena, namely, *berwinde* 'lesser bindweed', *menlufigend* 'cleavers', *nǣdrewinde* 'bistort', *sunfolgend* 'common marigold' and *selfhǣle* 'wood march', *weoþobend* 'woodbine' and *wudubindel*. Synthetic compounds are morphologically isolated because the second elements are not attested as independent lexical items.

The second element of some synthetic compounds is *-winde* (< *windan* 'to wind, curl, twist', *ClH*, s.v. *windan*), an element which indicates that these plants are winders, i.e. climbing plants: *berwinde* 'lesser bindweed' lit. 'barley-winder', *nǣdrewinde* 'bistort' lit. 'adder-winder', *wiþowinde* 'small bindweed' lit. 'rope-winder'. As *-winde* is not attested independently in Old English, and since it still has a strong verbal force, compounds using this element are included in the group of synthetic compounds. Alternatively, one could argue that, in analogy to *-d(w)os(t)(l)e* (in *dweorged(w)os(t)(l)e* 'pennyroyal'), which is not attested in Old English, but in German, namely as *Dost*, formations with *-winde* can be classified as N+N compounds, as the German plant names *Ackerwinde* and *Winde* 'winder, climbing plant' suggest that the second element may have existed in Old English as well and the fact that it is not attested in Old English may well be due to an incidental lack of recording. However, while *-d(w)os(t)(l)e* is purely nominal, *-winde* is clearly deverbal through its being derived from the verb *windan* '(to) wind' (see Catalogue, s.v. *dweorged(w)os(t)(l)e*).

The compounds *weoþobend* (*wiþobend*) 'woodbine', *wiþowinde* 'small bindweed, etc.', *wudubinde* (*wudubend*) 'woodbine', and *wudubindel* 'woodbine' combine three Germanic elements which were apparently easily confused: (i) Gmc **widu-*, *wudu-* 'tree, wood', cognate to ON *við* 'wreath, rope' and also recorded in OHG *witewinda*; (ii) Gmc **windō-* 'winder, creeping plant', also recorded in, e.g., MLG *wedewinde*; and (iii) Gmc **bindō-* 'winder, binding plant', also recorded in, e.g., Dan *vedbende*. According to Carr, it is not clear whether, for example, OE *wiðewinde*

> [...] represents a modification of the original compound with OE. *wudu* and *bindan*, or whether it is an independent formation [...]. Certain considerable changes have taken place, for whilst the ON. compound meant 'woodbine', the Mod. Icelandic *viðvindill* denotes 'ivy' (Carr 1939: 53).

184 For Marchand (1969: 15-7), all combinations with deverbal second elements whose verbal bases form a direct nexus with the first elements must be called synthetic, independently of whether the deverbal noun is attested or not. For criteria of classification in Early Middle English and for a discussion of unclear and problematic cases, see Sauer (1992a: 132-41). See further Kastovsky (1992: 367-8) and Schmid (2005: 135-7).

Apparently, the phonological similarity of the elements **widu-*, *wudu-*, **windō-* and **bindō-* led to frequent confusion and to the assimilation of these forms which results in the coalescence of forms and denotations of these elements.

Describing the morphological structure of synthetic compounds is problematic, as already discussed. Neither the analysis as a compound nor the analysis as a derivative is entirely satisfactory. In the present survey, synthetic compounds are regarded primarily as compounds and the second element is considered as a potential lexeme with an independent meaning, even if it is not attested independently. For speakers of German, for instance, this is quite easy to relate to some of the plant names, such as for those ending in *-winde* (< *windan* '(to) wind, curl, twist') and *-bind(e)* (< *bindan* '(to) bind') as these plant names exist in German, too: e.g., in G *Ackerwinde* 'bearbind'. It is, however, more problematic to relate the Old English plant names ending in *-hwette* (< *hwettan* '(to) whet, sharpen') or *-sēce* (< *sēcan* '(to) seek'), since these endings do not immediately evoke the images of plants and are intuitively rejected as potential independent plant names – at least to the speaker of Modern English, who is, however, not necessarily a good informant on Old English idiomaticity.

All second elements of synthetic compounds are potential agent nouns, i.e. nominalisations of action verbs such as for example *swelig(e)an* '(to) swallow', *hwettan* '(to) whet', *sūgan* '(to) suck' or *folgian* '(to) follow'. The first elements in these putative derivatives denote concepts which are conceptually contiguous to the plants. For example *grundeswelige* 'groundsel', a creeping plant, seems to swallow the ground it grows on, hence the second element *-swelige*; and *sōlsēce* 'common marigold, etc.' turns its head following the movements of the sun, i.e. seems to seek the sun, hence the second element *-sēce*. The potential plant names *-sēce*, *-winde* or *-bind(e)* are, thus, taxonomically super-ordinated (cf. chapter 6.3. (1)) to *hwerhwette* 'cucumber', *sōlsēce* 'common marigold, etc.', *berewinde* 'lesser bindweed' and *wudubinde* 'woodbine' but at the heart of their formation lies a conceptual contiguity (cf. chapter 6.3. (8)): a *-winde*, for instance, is a winding plant; a *-sēce* is a (sun-) seeking plant (see Catalogue, s.vv. *grundeswelige*, *sōlsēce*, *berewinde*, *wudubinde*).

Subgroups of synthetic formations can be established on the basis of the morphological shape of the formations. OE *grundeswelige* 'groundsel' and *sōlsēce* 'common marigold', for instance, are $N{+}N_{\text{deverbal}(-e)}$ compounds. The plant name *wudubindel* is a morphological variant of *wudubinde* as the second element is derived with the suffix *-el* instead of *-e* (see Catalogue, s.vv. *grundeswelige*, *sōlsēce*, *wudubinde*; cf. also s.vv. *bisceopwyrt* and *bisceopwyrtel*).[185]

Two plant names are $N{+}N_{\text{deverbal}(\textit{-end})}$ formations: *sunfolgend* 'common marigold' and *menlufigend* 'cleavers'. The *-end* is polysemous, i.e. it can have two functions: it can form the present participle of verbs (e.g. in the sentence OE *he wæs huntiende* 'he was hunting') or deverbal agent nouns (e.g. OE *biddend* 'petitioner' < *biddan* '(to) beg, ask, pray' or *lǣrend* 'teacher' < *lǣran* '(to) teach'). As a suffix deriving deverbal

185 For a recent publication on Old English plant names with suffixes, with a special focus on the suffix *-el* and its various functions (e.g. as diminutive, as suffix forming agent nouns and instrument nouns, see Sauer (2012). The suffix (and its variants *-il, -ol, -ul*), apparently, was not very productive in the Old English plant name corpus with 27 simplexes (which occur as part of compounds, as well; see Sauer 2012: 217).

agent nouns, *-end* was with *-ere* (e.g. *leornere* 'disciple', *scipere* 'sailor') one of the most productive elements in Old English (Kastovsky 1992: 385-6).

Two synthetic plant names have a different shape, as they are Pron+$N_{deverbal(-e)}$ formations: *selfǣte* 'common groundsel' lit. 'self-eat(ing) plant' and *selfhǣle* 'wood march' lit. 'self-heal(ing) plant'.

5.3.3.7. NOUN + NOUN$_{DEADJECTIVAL}$ (e.g. *īsenhearde*)

Noun + $Noun_{deadjectival}$ (henceforth N+N_{deadj}) compounds are made up of two nouns of which the second is deadjectival. N+N_{deadj} compounds are non-verbal nexus compounds. N+N_{deadj} compounds do not represent a word-formation type on their own in accounts of Old English and Middle English word-formation (e.g. Kastovsky 1992, Sauer 1992a), nor do they in surveys on Modern English word-formation (e.g. Marchand 1969, Bauer 1983). This may be due to the fact that many of the deadjectival nouns occurred independently: for instance *brǣd* 'breadth' and *heard* 'hard object' are independent lexemes of Old English (*ClH*, s.vv. *brǣd*, *heard*), and the compounds they are part of can be analysed as N+N compounds. Since some of the deadjectival elements in the complex plant name corpus do not occur independently (*-rīfe*, *-rife* and *-(h)rōfe*),[186] the N+N_{deadj} formations are listed separately in the present survey.

Five plant names qualify as N+N_{deadj} compounds: *giþrīfe* 'corn-cockle', *hegerīfe* 'cleavers', *īsenhearde* 'knapweed', *wegbrǣde* 'greater plantain' and *wudu(h)rōfe* 'woodruff-aperule'. All of them are exocentric formations as the second elements denote contiguous qualities of the plants. *Wegbrǣde* 'waybread' lit. 'weg-broad', for instance, is a plant with broad leaves (see Catalogue, s.v. *wegbrǣde*) and *īsenhearde* 'knapweed' lit. 'iron-hard (plant)' has stems so hard that they are compared to iron (see Catalogue, s.v. *īsenhearde*).

5.3.3.8. PROPER NOUN*ES/ING* + NOUN (e.g. *simǣringwyrt*)

Proper $Noun_{es/ing}$ + Noun (henceforth PropN+N, $PropN_{es}$+N, or $PropN_{ing}$+N) compounds are made up of two lexemes, e.g. *simǣringwyrt* 'mallow' lit. '*Simǣring*'s wort'.[187] As the first element is a proper noun, the medial *-ing* is seignoral, i.e. it indicates descent from the person denoted by the proper name; the intermediate *-es* in *witmǣreswyrt* is genitival.

Three Old English plant names are $PropN_{(ing)}$+N formations, namely, *æþelferþingwyrt* 'greater stitchwort' lit. '*Æþelferþ's* wort', *simǣringwyrt* 'mallow' lit. '*Simǣring's* wort' with the form $PropN_{ing}$+N and *tunsingwyrt* 'white hellebore' lit.

---

186 As to the problems arising from the textual situation, see 2.2.

187 Plant names containing proper nouns can be of two kinds: either $PropN_{ing}$+N compounds or N_{es}+N compounds. The semantic difference between these two kinds is based on the perception that medial *-ing* is seignoral while the genitive ending indicates possession (cf. Bandle, Insley and Andersson 2005: 350).

'*Tunsing's* wort', and all three are endocentric, rectional compounds. In contrast, *witmǣreswyrt* has the proper name inflected in the genitive (see 5.3.3.2.); structurally, it is an endocentric, rectional compound, as well.

5.3.3.9. UNANALYSABLE FORMATIONS (e.g. *gasdie*)

For a small number of Old English complex plant names, the synchronic and the diachronic analysis do not yield clear results: one element, at least, in what appears to be a compound is not analysable, and, in rare instances, both. It cannot be connected to any known element of the Old English lexicon, nor to those of cognate languages (a counter-example, where this can be done is *dweorged(w)os(t)(l)e*; see Catalogue, s.v. *dweorged(w)os(t)(l)e* and chapter 5.3.3.1.). The first element of *fēfer/fuge* 'feverfew', for instance – an early Latin loan word (< L *febrifugia*) – was changed in analogy to OE *fēfer* 'fever' due to the phonological similarity of OE *fēfer* and L *febri-* (from L *febris* 'fever'), supported by the fact that the plant helps with fever. In this process of folk etymology, the second element, *-fuge*, was not accomodated and remains semantically and morphologically opaque. It was possibly understood as 'drive-away', so that *fēferfuge* was understood as a 'plant that drives away the fever' (Sauer 1996: 146) by the educated Anglo-Saxons who were used to reading Latin and probably knew the plant name or at least the Latin verb *fugāre* '(to) drive away, (to) put to flight', but whether the non-educated Anglo-Saxons actually knew the meaning of the Latin lexeme is doubtful. The fact that *-fuge* was opaque, i.e. in no transparent relationship to other Old English lexemes, hampers the morphological and also the semantic analysis; in the further history of the word, this fact probably led to subsequent changes on its way to Modern English, as it developed into ME *fewerfue* and ModE *feverfew* by processes of folk etymology, thus changing the Old English element *-fuge* in analogy to OE *fēawe* > ME *fewe* > ModE *few*. With regard to the attempt of a morphological analysis of this plant name, the second element of *fēfer/fuge* cannot be assigned to any of the categories presented above (i.e. it is not a noun, a deverbal noun or a deadjectival noun) – this element is not analysable (see Catalogue, s.v. *fēferfuge*).

Elements of complex plant names that are unanalysable are so for different reasons. Borrowing, as in the case of *fēferfuge*, is one reason. The other major reason is morphological and phonological attrition, that is, changes in the word form, typically triggered by the operation of sound changes. As the early stages of languages are poorly documented or even not documented at all in writing, however, the original forms of the plant names and their constituing elements remain obscure. For example, both elements of *leloþre* 'dock' and the first element of *þūþistel* 'sow-thistle' are morphologically and semantically opaque: there is no transparent relationship between the elements of the plant names and the plant names themselves. According to Pheifer (1974: 99, no. 606), *le-* can be connected to OE *lēah* 'meadow', and *þū-* to *þūf* 'tuft(ed)' (1974: 98, no. 601), but *loþre* is completely opaque (see Catalogue, s.v. *leloþre*). Though OE *þūþistel* can be connected to OHG *dudistel* (see Catalogue, s.v. *þūþistel*), this information does not help very much as the origin of *du-* in OHG *dudistel* is unclear, as well (*Holthausen*, s.v. *ðuðistel*; Björkman 1901: 226; 1902: 291). In modern English morphology, these elements are called *cranberry*, *blocked* or *unique* morphemes,[188] but it is doubtful whether this is a category that is applicable to early stages of languages with few written and no spoken documents (see chapter 5.4.).

Other formations with unclear or problematic elements are: *gasdie* '? upright pellitory' (both elements), *geormanlēaf* 'common mallow' (the element *georman-*), *hǣwenh(n)ydele* 'scurvy-grass' (the element *-h(n)ydele*), *þēfanþorn* 'common buckthorn' (the element *þēfan-*) and *þelneþung* 'celery-leaved crowfoot' (the element *þelne-*). In total, 8 formations belong to this group. *Gasdie* and *þelneþung* are hapax legomena.

5.4. HAPAX LEGOMENA AND BLOCKED MORPHEMES

Hapax legomena of Old English are words that occur only once in extant Old English texts (Waldorf 1953: 2). A number of complex Old English plant names (c. 245) qualify for this category as they are attested only once. In a number of cases the decision as to whether a lexeme should be included or excluded from the category is not easy to make. When a word occurs in two manuscripts which contain basically the same text, such as for instance compilations of glossaries based on the same sources, these instances are regarded as as hapaxes (Waldorf 1953: 2) although the fact that, frequently, the copying of texts was not carried out *literatim*[189] and that compilations often changed the original texts to a considerable degree led to a sometimes pointed difference between the end products and between them and the original texts. When a plant name occurs in two more or less different contexts in the same text, the two instances are counted as two occurrences of the same word in the present survey. Thus, when it is recorded twice in the same context, or, even more, in the same sentence, as *grēne helde* 'cinquefoil' in "Nim grene helda [...], and gif nabbe grene helda [...]" ['Take cinquefoil, and if you do not have cinquefoil'] (*Peri Didaxeon*, 39/1, 39/7) – it is not counted as a hapax legomenon.

Hapax legomena have to be distinguished from spelling variants of the same plant name. The plant names *ananbēam* and *wananbēam*, for instance, each attested only once, are two different entries in *ClH* and thus qualify as two individual hapax legomena; in the present survey, *ananbēam* is, however, considered a spelling variant of *wananbēam* 'prickwood' lit. '? flexible tree' (following Bierbaumer 1, s.v. *ananbēam*). Likewise, *mircapuldor* is a variant of *milscapuldor* 'apple-tree' lit. 'sweet apple-tree', and *hārewinta* a variant of *hāreminte* 'wild mint' lit. 'hoary mint', and do not represent two different plant names.

Morphological variants of the same plant name, on the other hand, must be recorded separately despite their identical denotata. Compare, for instance, the aforementioned variants of the Old English name for the chestnut tree *cistbēam*, *cistelbēam* and

188 For a brief introduction into the concept of *cranberry* morphemes (or morphs), see Durkin (2009: 56-8).

189 More insights into different scribal practices may be gained from, e.g. McIntosh (1973), Laing (2004), and Gärtner (2011).

cistenbēam (chapter 5.2.), or the plant name *brādbisceopwyrt* lit. 'broad marshmallow', which is attested once as a compound and once with an inflected adjective, i.e. as a syntactic group *þā brādan bisceopwyrt* (see chapter 5.3.3.3.). A few plant names show attestations with different genders, as for instance, *clufþung* and *clufþunge*, the 'celery-leaved crowfoot' and for some of the plant names, a clear-cut analysis is not possible at all. The word forms *curmille*, *culmille*, *curmealle*, *curmelle*, for instance, are regarded as spelling variants of the plant name *curmealle* 'centaury' following Bierbaumer's suggestion (1, 2 and 3, s.v. *curmealle*; cf. *DOEPN*, s.v. *curmealle*), but the attempt to connect the first elements of the plant names *eolhsecg* and *ilugsecg* 'sedge, etc.' to *eolete* 'water' demonstrates the limits of analysis: even if we assume a connection between the three items *eolh* 'elk', *ilug-* '?' and *eolete* 'water', the nature of the relationship between these items remains largely obscure. The *DOE* (s.v. *eolh-secg*), in contrast, connects the plant name to L *helix*, *helice* (see Catalogue, s.v. *eolhsecg*).

The question of hapax legomena does not only affect complex formations as a whole; the elements of the plant names themselves, i.e. modifier and head, can also be attested only once, namely, in the complex formation itself. Surveys of Modern English morphology and word-formation refer to this phenomenon in Modern English as to the phenomenon of *blocked* (or *unique*, or *cranberry*) morphemes: blocked morphemes are elements of complex formations that do not occur independently but, as they appear in conjunction with free lexical morphemes, they can also be attributed the status of lexical morphemes.[190] These elements are not productive (Marchand 1969: 2; Sauer 1992a: 340) and, in many cases, their semantic contribution to the complex word is unclear, too (Kortmann 2005: 88-9). Their origins are varied. Some of them were originally native words which died out in independent use, others go back to loan influence (e.g., *Satur-* in ModE *Saturday*, from L *Sāturnus*, the name of the planet Saturn; *OED*, s.v. *Saturday*) or to loan influence followed by folk etymological processes[191] (e.g., *cray-* in ModE *crayfish* < OF *crevisse, crevice*; *OED*, s.v. *crayfish*) and some may have developed due to processes of phonological and/or morphological attrition (e.g., *Fri-* in ModE *Friday*, from *Frija*, the genitive form of the Germanic goddess *Frij* (ModE *Friday* < OE *Frīgedæg*); *OED*, s.v. *Friday*). In the corpus of the present study a number of plant names do feature indeed elements that are not attested independently: *-hige* in *haranhige* 'hare's foot clover' lit. 'hare's foot', for instance, is an element in question; so is *-d(w)os(t)(l)e* 'dost' in *dweorged(w)os(t)(l)e* 'pennyroyal' lit. 'dwarf's dost' and many more examples (see Catalogue).

The crucial question with *hapax legomena* and with *blocked morphemes* is the question of whether we can speak of hapax legomena and blocked morphemes in Old English at all. It is closely connected with the question of the productivity of word-formation patterns and with the question of the nature of the lexical corpus of a language under investigation. Thus, the high number of hapaxes and blocked morphemes in the Old English plant name corpus is striking on the one hand and not

190 See, e.g., Marchand (1969: 2), Kastovsky (1982: 71-2, who subsumes them to the formatives), Lipka (2002: 87), Bauer (2003: 48, 50), and Schmid (2005: 31).

191 Folk etymological processes are discussed in chapter 7.5.

easily explained on the other. Namely, one of the problems is that although the number of words that have survived until today is impressive – between roughly 24,000 (Scheler 1977: 14) and 35,000 (Healey 1997: 47, fn. 4) – we must also assume the existence of words which are lost today. Some may never have been written down: the Anglo-Saxon period stretches over roughly 600 years, but the period of written texts only over some 250 years. Furthermore, not all texts written in Anglo-Saxon England have survived up to our day: as parchment leaves were expensive, texts were often obliterated and the parchments re-written. Other texts were destroyed unintentionally – due to unfavourable storage conditions – or deliberately, for example during Viking raids in the Anglo-Saxon period or during the dissolution of the monasteries in the 16th century. Thus, if a word is recorded only once in extant Old English texts, it does not automatically follow that it was rarely in actual use. It may have been used much more often without, however, leaving any traces in written documents.

On the other hand, we must also not forget that some of the hapax legomena were coined in order to render Latin words. This means that they were probably not used in day-to-day communication but were learned book-words or even "kind[s] of contrastive linguistic tool[s] that make visible the specific characteristics" of two languages (Johnston 2011: 72) – an explanation, perhaps, of why some of the terms, such as the hapaxes *coltnægl* 'coltsfoot' lit. 'colt's nail' ← L *ungula caballi* and *candelwyrt* 'great mullein' lit. 'candle-wort' ← L *herba lucernaria*, were written down only once.[192]

The case of OE *hwītlēac* indicates how meagre the textual basis for research sometimes is: Because this composite occurs only four times, where it glosses three times *alba cepa* lit. 'white leek' and once *poleus*, Bierbaumer (3, s.v. *hwītlēac*; 2003: 157) considered it to be an erroneous gloss. In his interpretation, *poleus* had to be regarded as a variant of L *propoleos* 'wax', which would have turned the other two lemmas into *alba cera* lit. 'white wax'. Following this interpretation, the Old English glossator translated L *alba cēpa* as 'white leek' into Old English *hwītlēac*, although, following Bierbaumer, this plant name did not exist in Old English. However, as recent evidence from Old West Norse studies revealed, *hwītlēac* existed, indeed, as a plant name – not only in Old English, but also in Old Norse as *hvítr laukr* (see Catalogue, s.v. *hwītlēac*).

As the elements in the now following examples are in a transparent relationship to existing Old English morphemes, they are not exactly *blocked*, though they are not attested independently: the elements *-hige*, *-winde* and *-hǣle* in *haranhige* 'hare's foot clover', *berewinde* 'lesser bindweed' and *selfhǣle* 'wood march' can be related to the verbs *higian* '(to) hurry', *windan* '(to) wind' and *hǣlan* '(to) heal', and, thus, they can be taken as independent, self-contained potential stems, differing from non-transparent elements such as *-loþre* in *leloþre* 'dock' and *-die* in *gasdie* 'upright pellitory' (see Catalogue, s.vv. *haranhige*, *berewinde*, *selfhǣle*). There is, on the other hand, no reason to assume an independent meaning and standing for *beg-* or *cist-*, which are clipped from OE *ber(i)(g)e*, respectively of OE *cisten* (< L *castanea*), and thus

192 For the loan translations, loan renditions or loan creations in the Old English plant name corpus, see 7.4.3.

allomorphic variants of *ber(i)(g)e* and *cisten* in *begbēam* 'mulberry, etc.' and *cistbēam* 'chestnut tree', respectively (see below, and Catalogue, s.vv. *begbēam*, *cistbēam*).

5.5. STEM VARIANTS

Some elements of the complex plant names occur only once but can be related to other, independent morphemes by way of their spelling, their morphology or their phonology. Variants that go back to spelling differences are, for instance, *ferdwyrt* and *feldwyrt* 'felwort' (see Catalogue, s.v. *feldwyrt*), or *ananbēam* and *wananbēam* 'prickwood' (see Catalogue, s.v. *wananbēam*). As the Old English spelling system was, though quite regular, not standardised, different spellings of the same words were not unusual.

Variants resulting from morphological processes are, for instance, *beg-* in *begbēam* 'mulberry, etc.' which is clipped from *beger*, itself a variant of *ber(i)(g)e* 'berry' and *mapul-* in *mapultrēow* 'maple tree', which is probably a back-formation of *mapuldor* 'maple tree' (see Catalogue, s.v. *mapul-*). *Cist-* and *cistel-* and *cisten* in *cistbēam*, *cistelbēam* and *cistenbēam* are all variants of a loan from Latin: OE *cisten* goes back to L *castanea*; *cist-* is clipped from *cisten* and in *cistel*, the putative suffix *–en* was replaced by the Old English suffix *–el* (Campbell 1959: § 518; see also chapter 7.3.1.2.). Morphological variancy is also expressed where the elements belong to different word classes, e.g. the heads in the compounds *clufþung* masc. and *clufþunge* fem. (cf. chapter 5.3.3.1.)

The crow garlic with its three different morphological shapes *crāwanlēac* (N_{gen}+N), *crāwelēac* and *crāwlēac* (N+N) (Catalogue, s.v. *crāwanlēac*, *crāw(e)lēac*) can be interpreted in two ways. We can either assume (i) that these plant names reflect a transitory language situation, where the inflection of the first element was on the point of being eliminated, and where both the (semantically isolated) syntactic phrase and the compound were still in use, or (ii) that these are incidents where two (or three) independent word-formation processes happen to have the same denotatum as a point of reference (i.e. a case of polygenesis). Variants whose shapes are the result of borrowing at different points of time are, for instance, *leahtric*, *-lectere* and *lactuce* from L *lactūca*, or *cāwel* and *caul* from L *caulis* (see Catalogue, s.vv. *leahtric*, *cāwel*; chapter 7.4.2.1.2.).

In some cases it is difficult to decide whether two (or more) elements should be seen as variants of another and in many cases this question cannot be answered conclusively. In some cases it even leads to different morphological and semantic interpretations. To resume the example introduced in 5.5.: Depending on whether the attested terms *fleaþan wyrt* and *fleaþor wyrt* are interpreted as variants of one and the same plant name (*BT* and *ClH*, s.v. *flēaþe wyrt*) or as two plant names (*DOE*, s.vv. *flēaþe wyrt*, *flēaþor-wyrt*), depending on whether the first element is analysed as a noun (as by *BT*, *ClH*, *DOE*, all s.v. *flēaþe wyrt*), or as an adjective (Bierbaumer 1, s.v. *flēaþe wyrt*), three possible analyses come up: *flēaþe wyrt* as N+N formation (with *fleaþor wyrt* as scribal error for *fleaþan wyrt*), *flēaþe wyrt* as Adj+N formation (with *fleaþor wyrt* as scribal error for *fleaþan wyrt*) or *flēaþe wyrt* as N+N formation and *flēaþorwyrt* as an independent N+N compound (hapax legomenon) (see Catalogue, s.v. *flēaþe wyrt*).

Linking Vowels

Throughout the corpus of Old English complex plant names we find examples where the first and the second elements are combined via a so-called *linking vowel* (also *bridge vowel* or *glide vowel*; G *Fugenelement*): for example, the intermediate *-e-* in *coltetræppe* 'caltrop', in *dweorged(w)os(t)(l)e* 'pennyroyal', in *grundeswelige* 'groundsel', and in *brādelēac* 'ramsons'. According to Carr (1939: 268), in Early Germanic, the first elements of compounds were in the pure stem form without any case endings, but very early exceptions to this principle appeared and vowels were introduced for various reasons. In Old English, a vowel between two elements of a compound could be the result of (sporadic) introductions of vowels into the bridges of compounds (composition vowels); they could also be inflectional endings, especially with feminine nouns or in genitive combinations and, in many cases, the linking vowels were simply reduced forms of stem vowels.[193]

The vowels in *dweorged(w)os(t)(l)e* and *grundeswelige*, for instance, are relics of the old stem vowels (Carr 1939: 282), whereas the *-e-* in *coltetræppe* is probably a glide (Carr 1939: 283). *Brādelēac*, on the other hand, probably has retained the *-e-* as an adjectival declension relic (Carr 1939: 283). *Sticwyrt*, which is classified as V_{stem}+N compound in the present paper, is seen as a N+N compound by Carr (1939: 285), who argues that the stem syllable *-e-* was simply dropped. In compounds with *u*-stems as first elements, such as *wudu* or *hagu*, the *-u* was either retained (e.g. *haguþorn* 'common hawthorn'), reduced to *-e* (e.g., in *wudemerce* 'celery, etc.', which existed alongside *wudumerce*; see Catalogue, s.v. *wudumerce*), or fell out completely (e.g., *hægþorn* 'common hawthorn' which existed alongside *haguþorn*, see Catalogue, s.v. *haguþorn*).[194]

5.6. SUMMARY

This morphological analysis of the Old English complex plant names was undertaken on the basis of two approaches to word-formation developed for Modern English: Marchand (1969) and Bauer (1983), with the former being 'a synchronic-diachronic approach', as it traces the word-formation patterns back to their origins (cf. Marchand 1969: 8). The questions related to the difficulties that an analysis of a language no longer spoken or written can bring about were, then, discussed and it became clear that

193 For a discussion of different explanations with examples see Carr (1939: 281ff.) and Sauer (1992a: 81-4). In *CONE*, non-inflectional vowels that appear between the first and second elements and that can be of different origins (e.g., stem-final vowel of the first element, linking vowel, otiose element or the result of epenthesis) are labelled 'intermorphemic e'.

194 For the classification of consonants and vowels in the bridge of compounds with the first elements inflected in the genitive see, further, chapter 5.3.3.2. Sauer (1992a: 81-4) provides a detailed discussion of composition vowels with many Early Middle English examples.

little can be said about the productivity of word-formation patterns or about the validity and idiomaticity of paraphrases of complex expressions when no native speakers of the language in question are at hand. The questions of whether elements not attested independently in the extant manuscripts can be called *blocked* and of whether we can assume that the composites attested only once were not used very often, can be answered only tentatively, as well. Here, the special textual situation, arising as a result of the limited number of the Old English texts and their very specific nature as the products of a minority elite have to be respected.

As regards the morphological shape and the morphological structures of the complex plant names, it was observed that the boundaries between the categories are often transgressed by the Old English plant names. Some names are attested as compounds, others as idiomatised syntactic groups, and a number of plant names are attested with different shapes and have to be classified as different morphological types, such as, for instance, *crāwlēac* (N+N), *crāwelēac* (N+N, with intermediate *-e-*) and *crāwanlēac* (N_{gen}+N), all 'crow garlic'; or *cistbēam*, *cistelbēam* and *cistenbēam* (all N+N: the first, with clipped first element, the second, with suffix substitution in the first element, and the third, with the first element developed 'regularly' from L *castanea*).

As to the distinction between endocentric and exocentric compounds, it was proposed to consider metonymic and metaphorical compounds as structurally similar, since, in both cases, the semantic heads of the compounds are not encoded on the surface of the expression, but have to be established on the basis of encyclopaedic knowledge. Owing to this and to problems in the paraphrasing of non-verbal nexus compounds, the analysis of the semantic content of the composites was transferred to the area of semantics, which is to be treated in the next chapter (chapter 6).

As far as the morphological shapes and the morphological structures are concerned, the analysis of the corpus of Old English complex plant names produced the following morphological types: The majority of the plant names are made up of two nouns (N+N; 251 plant names); e.g. *bōctrēow* 'beech' lit. 'beech-tree', *cederbēam* 'cedar of Lebanon' lit. 'cedar of Lebanon tree', *feltwyrt* 'great mullein' lit. 'felt-wort'. 43 plant names are made up of a noun in the genitive + noun (N_{gen}+N), e.g. *foxes glōfa* 'foxglove, etc.' lit. 'fox's glove', *wulfestǣsel* 'wolf's teasel' lit. 'wolf's teasel' or *lambescærse* 'touch-me-not, etc.' lit. 'lamb's cress'. 128 plant names are composed of an adjective and a noun (Adj+N), as, for example, *blind(e)net(e)l(e)* 'dead nettle' lit. 'blind nettle' and *hārhūne* 'hoarhound' lit. 'hoary horehound'. Six plant names contain a numeral and a noun (Num+N), e.g. *fīflēafe* 'cinquefoil' lit. 'five-leaf', and seven formations are V_{stem}+N compounds, such as for instance *clifwyrt* 'great burr' lit. 'cleav(ing) plant'. 16 plant names are synthetic, e.g. *grundeswelige* 'groundsel' lit. 'ground-devouring (plant)', *selfǣte* 'common groundsel' lit. 'self-eating (plant)' or *sunfolgend* 'marigold' lit. 'sun-following (plant)'. Five plant names are made up of a noun and a deadjectival noun ($N+N_{deadj}$), e.g. *īsenhearde* 'knapweed' lit. 'iron-hard (plant)'. Three plant names are $PropN_{(es/ing)}$+N compounds, e.g. *simǣringwyrt* 'mallow' lit. '*Simǣring's* wort' and for eight plant names a morphological shape cannot be ascertained, e.g. *gasdie* 'upright pellitory' lit. 'glass-?'; they are called unanalysable.

6. THE SEMANTICS OF THE OLD ENGLISH COMPLEX PLANT NAMES

> sume synd GENERALIA, þæt synd gemǣnlîce: [...] *arbor* ælces cynnes trêow. sume syndon SPECIALIA, þæt synd synderlîce, þâ ðe bêoð tôdǣlede fram þâm gemǣnelîcum. [...] gemǣnelîce *arbor* trêow; synderlîce *uitis* wîntrêow, *laurus* lawerbêam, *corilus* hǣsel, *abies* æps, *quercus* âc, *malus* apuldre.
>
> (Ælfric's *Grammar*, Zupitza 1880: 14).

According to Lipka (2002: xiii), "combinations of linguistic signs, as the result of extremely productive processes in the lexicon, may be motivated or transparent to a greater or lesser degree, and are traditionally the concern of the discipline of *word-formation*". One of the main conclusions in the previous chapter has, however, been that a systematic description of the semantic relationship of the elements of both endocentric and exocentric compounds cannot be carried out comprehensively within the framework of word-formation: A large part of the endocentric compounds – the non-verbal nexus, rectional compounds and all metaphorical and metonymic compounds – fall through the grids of the syntactic-semantic analysis. The indication of the morphological structure of compounds, the formulation of syntactic paraphrases or the application of Levi's (1978) Recoverably Deletable Predicates did not prove comprehensive enough. This called for a different approach to the semantics of Old English complex plant names.[195]

Within the area of lexical semantics, two paths can be followed: a structural one and one which approaches semantics from a cognitive point of view. Each is extremely valuable in yielding particular results and insights, but neither is entirely exhaustive. In fact, as is argued, the meanings of words can be described best when words are approached through structural and through cognitive semantics in equal measure. Before focusing on the semantic structures in the corpus of the Old English complex plant names proper, a short overview of structuralist and cognitive approaches will, therefore, be given and the most important terms and concepts introduced. This will be followed by a new approach to the description of meaning. This approach, which was developed by Blank (1997) as a framework for the analysis of semantic change, presents a model of the linguistic sign that integrates structuralist and cognitive views on the meanings of simple and complex lexemes, and which will serve as the theoretical background for the description of the semantics of the Old English complex plant names in the present chapter.

195 See also Krischke (2009).

Structural and cognitive semantics: A brief introduction

Trier's (1931) research on specific German word-fields heralded the era of *structural semantics*, with the identification of the micro and macro-structural organisation of the lexicon as its focus. Based on the Saussurean (1916) concept of the linguistic sign, structural semantics centres on the identification of semantic features (semes) that form the sememes (bundle of necessary, distinctive features) of words. The sememe represents the internal semantic structure of the lexeme in question and helps to position the lexeme within the micro and macro-structures of the entire lexicon of a language (cf., e.g., Kastovsky 1982: chapter 4). Lyons' (1977) development of compositional analysis – a method for describing the meanings of words on the basis of the paradigmatic and syntagmatic relations they enter into with other lexemes – complements structural semantic analysis.

Its many values and useful insights notwithstanding, one of the shortcomings of the structuralist approach, it has been argued, is its failure to pinpoint the actual meanings of words. Just as the meanings of complex lexemes are not comprehensively described by the addition of the meanings of their elements, the meanings of simple and complex lexical items are not comprehensively described by the addition to sememes of their distinctive, semantic features. Hence the warning by Seuren that

> [...] the total meaning of a lexical item may be more than the value resulting from the functional calculus of its parts (or, as it is often put in vulgarizing terms, more than the sum of its parts) (Seuren 1988: 174).[196]

As a result, *cognitive semantics* developed *inter alia* in consequence of the attempt to overcome the limitations of structural semantics, and focused on the rapport between extra-linguistic reality, human conceptualisation of extra-linguistic reality and its verbalisation – the lexicon and its semantic structure.[197] One of the basic assumptions of cognitive semantics is that the way the lexicon is organised semantically reflects the human way of experiencing and categorising reality, but does not reproduce reality as such: it can indeed differ from the impartial, scientific categorisation and organisation of the world (Blank 2001a: 36).[198] Single lexical items are considered to be linguistic representations of cognitive categories: the English lexeme *bird*, for example, represents the cognitive category BIRD. Cognitive categories are the product of the mental classification of extra-linguistic reality (Ungerer and Schmid 2006: 8) and bear several features: They are arranged prototypically, i.e. they are made up of a prototypical core and more or less marginal members, and have fuzzy boundaries. The meanings of lexemes, which are equated with the cognitive category the lexeme represents (although not in a one-to-one relationship as most lexemes are polysemous;

196 This is the so-called "Mehrwert der Komposition" ['additional value of composition'] (Blank 1998: 22); see also Kastovsky (1982: 164ff.) and Schmid (2005: 100).

197 Recently, the *Handbook of Cognitive Linguistics*, eds. Geeraerts and Cuyckens (2007).

198 For example, G *Walfisch* is perceived as a kind of fish in German folk taxonomy but, in fact, it is part of the group of mammals and not of the group of fish. Likewise, the *jellyfish* belongs to the Cnidaria, a group of animals which is part of the corals and sea anemones, and is by no means a kind of fish as its name implies.

cf. Ungerer and Schmid 2006: 23), are made up of a set of theoretically indefinite salient features called *attributes*, which form an open inventory and include linguistic, idiosyncratic and culture-specific information. This flexibility of the cognitive categories contrasts significantly with the rather rigid model of the sememe in structural semantics, which is limited to the necessary and distinctive features of the lexemes in question (Blank 2001a: 37).

One of the major achievements of cognitive semantics is to stress the importance of the *saliency* of specific attributes in the description of semantic meaning as opposed to their *consistency*, which is important in structural semantics (Blank 1997: 81). The highly salient attribute 'has green leaves' of the category PLANT, for instance, does not pertain to all members of the category in equal measure: conifers have needles instead of leaves,[199] and mushrooms do not have green leaves at all; still, most people would posit that one of the most prominent features of PLANTS is that they have 'green leaves'. The principle uniting mushrooms, conifers and plants with green leaves such as, e.g., leeks as members of the cognitive category PLANT, is labeled *family resemblance* in cognitive semantics. This term describes a network of overlapping similarities which are not all of them category-wide, but, as Rosch and Mervis (1975) point out, a network where "each item has at least one, and probably several, elements in common with one or more items, but no, or few, elements are common to all items" (quoted in Ungerer and Schmid 2006: 29) – just as not all plants have 'green leaves'.

Cognitive categories are organised in frames (or scripts)[200]. Frames are global contexts of knowledge based on everyday experience; they are described as a kind of stereotyped situations, that can be determined with the help of questions such as, for instance, 'Who belongs to the frame?', 'What is done?', 'What is affected?' (cf. Ungerer and Schmid 2006: 207ff.). Frames are based on relations of contiguity and have prototypical structures.

> Ein Frame ist, mit anderen Worten, ein globaler, gestalthafter oder kulturspezifischer Wissenskontext, eine "coherent schematization of experience", auf dessen Basis wir Alltagssituationen aller Art bewältigen. Da Frames das allgemein Rekurrente vielfältiger Einzelerlebnissen speichern, haben auch sie eine prototypische Struktur und lassen evtl. bestimmte randständige, nicht primär erwartbare, aber mögliche Variationen zu (Blank 2001a: 55).
>
> [A frame is, in other words, a global, culture-dependent context of knowledge with a specific *gestalt*, a "coherent schematization of experience" on the basis of which we cope with various situations of

---

199 Although needles, strictly speaking, are, indeed, leaves, but of a very special nature: With the exception of those of the larch, needles are perennial, i.e. they do not fall in autumn. This, again, shows that the difference between attributes in cognitive semantics – which can be culturally specific or even of an idiosyncratic nature – and scientifically proven features of plants, which are universally valid is not a problem: "To get through everyday life, laypersons do not need scientifically correct models, but functionally effective ones" (Ungerer and Schmid 2006: 57).

200 *Frame theory* was pioneered by Fillmore (1975), among others.

> everyday life. Since frames contain recurrent aspects of various individual experiences, they have a prototypical structure, too, and allow, at times, the inclusion of marginal, not originally anticipated, but possible variations.]

Not all elements of a frame are focused on evenly – some are highlighted against others, which remain in the background. The elements that are in focus are, generally, called *figure*, the elements against which the figures stand out are called *ground*.

Leaving ongoing debates on the value of structuralist and cognitive semantics aside, the following discussion will introduce an integrative approach, which combines both structural and prototype semantics. When it comes to describing Old English complex plant names, several layers of meaning emerge, and, as the next chapter will indicate, an analysis that is restricted to one of the two major approaches to semantics would lack the comprehensive scope the subject demands. Instead, it is argued, a view should be adopted in which both structuralist and cognitive methods are on a par – allowing for the semantic description of lexemes from different perspectives, while nevertheless acknowledging that they yield not contradictory, but complementary results.[201]

6.1. BLANK'S SEMIOTIC MODEL OF THE LINGUISTIC SIGN

When both language-immanent and extra-linguistic information are addressed in a semantic description of words, a semiotic model is needed which is able to represent this diverse information in balanced terms. According to Blank (1997: chapter 2; 2001a: 7ff.), both the structuralist dualistic model of de Saussure (1916: 67) and Ogden and Richards' (1956: 11) semiotic triangle are insufficient for this purpose. In the Saussurean model of the linguistic sign, human cognition and extra-linguistic knowledge are not represented at all since the *signifié* (the sememe) is established on the basis of relations to other lexemes (such as synonymy, hyponymy and antonymy), which means, on a level of knowledge that is language-dependent. While Ogden and Richard's model includes extra-linguistic reality in the form of the *thought* and the *reference*, it provides no place for intra-linguistic, word-specific knowledge such as information on the word class, on collocations or on paradigmatic relations within the lexicon. Cognitive models describing meaning (via cognitive categories), in his view, seem likewise inadequate, since, here, words are equated with extra-linguistic concepts, notwithstanding the fact that even though their extra-linguistic reality may be roughly the same, each language community can exhibit its own unique linguistic interpretation and organisation of reality (Blank 1997: 82-3).

In order to overcome these insufficiencies, Blank (1997: 96ff.) proposes a semiotic model that attempts to integrate all the different kinds of knowledge that could possibly be connected to and associated with lexemes, structural and cognitive

201 Cf. Lipka (2002: 74): "In my opinion, both the notion of 'prototype' and of 'semantic feature' [...] are necessary and useful. [...] The two approaches to lexical meaning are complementary and there is a division of labour".

included. Fig. 3 presents Blank's (1997: 102) model,[202] and will be briefly introduced in the following:

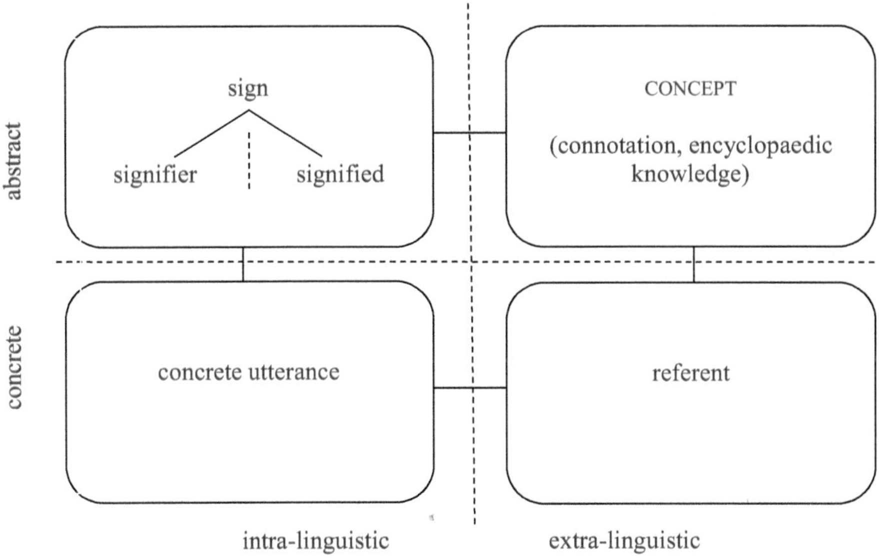

Fig. 3: Blank's semiotic model (cf. Blank 1997: 102).

Three kinds of knowledge are at the heart of Blank's semiotic model: (1) language-immanent, sememic knowledge about the lexeme under investigation, (2) language-immanent, lexical knowledge about the lexeme under investigation and (3) extra-linguistic, encyclopaedic knowledge about the concept(s) denoted by the lexeme under investigation.[203]

Language-immanent, *sememic* knowledge is encoded in the *signified* and includes all distinctive features (semes) that are needed to delimitate one word meaning from another. They can be deduced by componential analysis and help to establish relations of hyponymy, synonymy and antonymy in the lexical field. Language-immanent, *lexical* knowledge (encoded in the *sign* as a whole) includes information on the word class of the lexeme in question, on the syntagmatic relations it may enter as well as information on the diasystem, i.e. on the diatopic, diachronic, diamesic and diaphasic varieties to which the particular lexeme belongs. Extra-linguistic, *encyclopaedic* knowledge includes all the approaches to meaning developed in cognitive linguistics,

202 The model is insofar adapted as I changed Blank's (1997) labels *Designat* 'designate' to *concept* (which corresponds to his *Konzept* 'concept' – the label employed in Blank 2001: 9) and *konkrete Lautung* 'concrete pronunciation' to *concrete utterance* without, however, meaning to change his general outlines and the implications of the model.

203 For a more concise explanation of his semiotic model and the different levels of knowledge see Blank (1997: 96ff. and 2001a: 7-12).

with the notions of prototypes and frames being of specific importance in the present context. Encyclopaedic knowledge can reflect scientific findings, but it can also be in contrast to it (see fn. 198 and 199) and depends on individual experiences and/or on general, cultural knowledge.[204] While the language-immanent, abstract level can best be described in structuralist terms, the language-external, abstract level can best be described in terms of cognitive semantics. Within this semiotic model, intra- and extra-linguistic information is connected via different associative relations.

Before the associative relations are explained, however, I will attempt to locate in Blank's semiotic model exemplarily different kinds of knowledge that we have on one of the Old English plant names. A good example is OE *bānwyrt* 'daisy, common centaury, etc.', the plant name introduced in the introductory chapter (1.1.): On the intra-linguistic, abstract level of knowledge, there is the phonological knowledge about OE *bānwyrt* as having a specific texture, which can be transcribed broadly as [ba:nwyrt]. This information is part of the signifier. As a concrete utterance (on the intra-linguistic, concrete level of knowledge), the idiosyncratic pronunciations of the speakers of Old English (level of parole) were, probably, slightly different.

With regard to the signified, *bānwyrt* can be described as being polysemous: it can denote the daisy, the common centaury, the comfrey and the violet. As regards the lexical knowledge about *bānwyrt* (encoded in the sign), we know that it is a feminine noun belonging to the *i*-declension and that it is attested 23 times in Old English texts (*DOE*, s.v. *bān-wyrt*). We do not know anything, however, about its distribution in terms of diatopic, diastratic or diaphasic variety, and our knowledge of this aspect of the Old English language is, in general, very limited (Kastovsky 1992: 292-3), but we know that *bānwyrt* is mainly recorded in medical recipes and glossaries (*DOE*, s.v. *bān-wyrt*).

Taking the denotation 'comfrey' as a starting point for the componential analysis of the signified (still on the intra-linguistic, abstract level of knowledge), we can say that comfrey is 'a plant' (as opposed to trees), that it is a perennial (as opposed to plants which are not perennial), that it 'has a turnip-like root' (as opposed to plants with, e.g. tap roots), that it has violet flowers (as opposed to flowers of other colours), that its flowers are rather small and bell-shaped (as opposed to plants with large flowers of other shapes), etc. The sum of all these semes forms the sememe of *bānwyrt*, encoded in the signified.[205]

All the features mentioned are sememic, which means that they are necessary on the intra-linguistic level of knowledge to distinguish *bānwyrt* from other kinds of *wyrt* such as from, e.g., *brōþorwyrt* 'pennyroyal', *candelwyrt* 'great mullein', or *ceasterwyrt* 'Christmas rose'. On the other hand, they are anchored on the encyclopaedic, extra-linguistic level of knowledge, in the concept of what a *bānwyrt* is in extra-linguistic reality, and this can be described on the basis of cognitive category and frame theory. The difference from language-dependent semantic information is that encyclopaedic knowledge on whatever item one chooses to investigate can be vast

---

204 The association of crowds of doves populating St Mark's square with the concept of VENICE, for example, is generalised even for people who have never been in Venice.

205 This analysis is probably not entirely exhaustive but it serves its present purpose of illustrating the different kinds of semantic information encoded in the plant name.

and is not necessarily always relevant for the language system – it is not sememic. Thus, very idiosyncratic, personal feelings, for instance, do add to the 'meaning' of items or events (the so-called *experiential aspects* of meaning; Ungerer and Schmid 2006: 2), but they do not add to the 'meaning' of items within the system of language.

In the end, the concept COMFREY can be encountered in the form of a concrete referent (on the concrete, extra-linguistic level of knowledge), and this can deviate in its general form from the idealised concept: the leaves of the plant at the end of summer are, probably, withered and not green any more and, still, the plant is a *bānwyrt* with the same denotational meaning.

One of the most salient attributes, the one that was singled out by the language community to determine the plant's name, is its healing virtue. Since antiquity, comfrey has been said to heal bone fractures and other diseases of the skeleton and has been recommended accordingly. One of the texts advises, for instance:

> Gif scancan sare synd genim giþrifan & dolgrunan & hamorwyrt & betonican & banwyrt & linwyrt & wudumerce & eorðgeallan & brunwyrt, seoþ on buteran, smire mid.
>
> Gif scancan synd forode nim banwyrt, gecnuwa, geot æges þæt hwite, meng tosomne scancforedum men.
>
> (*Lǣcebōc*, 21/13-7)

> [When the shanks are sore, take corn-cockle, and pellitory of the wall and Christ-hellebore and betony and comfrey and line and celery and knapweed and water-betony, seethe in butter, smear therewith.
>
> When the shanks are worn out, take comfrey, pound it, take the white of egg, mix [them] together [and apply it] on the man with the worn-out shanks.]

This pain-relieving feature of the plant is situated on the conceptual, extra-linguistic level of knowledge and, while it may be a distinctive feature on the level of language structure, it does not have to be sememic.

What is more important than the question of whether the features of the plants encoded in the plant names are system-relevant (from a structuralist point of view), category-wide, or attributes of family resemblance (the latter two from a cognitive point of view), is the question of how these attributes are put into words and which processes were involved: The question of why it is exactly these elements that are constituents of the Old English complex plant names, and why they are put together in the way they are will be the focus of the semantic analysis.

As we have seen in chapters 5.3.2.2.1.2. and 5.3.2.2.2., paraphrases of the complex lexemes often fail to grasp the information central to the plant name (true especially for formations used metaphorically), and for a large part of the complex lexemes at least two different paraphrases can be formed, while a decision as to which one is the 'right' or 'better' one cannot be taken. Here, the idea of 'associative relations' holding between the different kinds of knowledge is of great help (cf. chapter 6.3.). Beforehand, however, it is important to introduce briefly the concept of *motives* and *motivation* to the analysis, since these two, as they are understood in the present survey, refer to the most salient attributes of the plants – the features that attracted so much attention or that were of some (and sometimes of utmost) importance in medical or nutritional terms that they triggered and determined the form and structure of the Old English complex plant names.

6.2. MOTIVES AND MOTIVATION

All new lexemes are codes for human thought, created out of the need and desire of people to impart what concerns them to other members of their community: their knowledge about the world, their ideas and their feelings. Thus, each instance of lexical innovation automatically seems to imply some kind of motivation: except for creations *ex nihilo* and morphologically or semantically intransparent words, all forms of lexical neologism can theoretically be traced back to some kind of motive (Boretzky 1987: 54). This, of course, holds for plant names, too. According to Biggam, plants were named by their

> notable visible characteristics or invisible properties or associations, and ascertaining these concepts often explains, for example, why a certain name refers to more than one plant. Such information may support a definition of an Old English plant-name or suggest an anomaly (Biggam 2003b: 206).

Motivation refers to the fact that lexemes are not arbitrary labels for things and concepts but that they are semantically and often also morphologically transparent expressions. Most obviously motivated are complex lexemes as their meanings can often be inferred from the meanings of their constituents (Kastovsky 1982: 19-20). According to Ullmann (1962: 81ff.), lexemes may be motivated phonetically (onomatopoeia), morphologically (word-formations) or semantically (metaphors and metonymies), but motivation can also be mixed. Phonetic motivation, for instance, is not only extant in onomatopoetic formations such as OE *æspe*,[206] to give an example from the plant name corpus, but also in complex formations that owe their shapes to processes of folk etymology triggered by the phonological similarity of the sound patterns of two otherwise unrelated lexemes (or parts of lexemes): e.g., *lufestice* 'lovage' lit. 'love-stitch', which was changed from CL *ligusticum* (see Catalogue, s.v. *lufestice*).

Morphological motivation refers to the notion that the meanings of complex formations can be deduced from the meanings of their elements and from the semantic relation between the elements, which is or can be expressed by a verb or a verbal construction on the basis of these elements (see chapters 5.3.2.2.1.2. and 5.3.2.2.2.). However, as was demonstrated, paraphrasing complex formations on the basis of their

---

206 According to Biggam (2003b: 204ff.), *æspe* was formed in allusion to the famous whispering of the poplars.

elements is often not sufficient, since additional semantic components cannot be pinned down by this method and since the verbal element defining the relation between the constituents depends heavily on the linguist's perception of reality and on his or her language skills. Especially with metaphorical and metonymic compounds, a morpho-syntactic analysis that remains on the (surface) level of language is problematic because it does not provide paraphrases apt to render the semantic implications of the compounds in question.

Where morphological and semantic motivation coincide, the concept of *mixed motivation* can be applied: "It must be pointed out [...] that very often metaphor, metonymy and word-formation are combined as in *bigmouth*, [...] *egghead*" (Lipka 2002: 138; cf. also Lipka 2002: 109-10).[207] Given the fact that morphological motivation has proven to be inadequate, mixed motivation must be advanced for all kinds of complex lexemes, except for those where 'phonetic texture', i.e. the sound pattern of a word – a purely formal feature – prompts secondary, phonological motivation, which can be accompanied by secondary, semantic motivation. This in turn means that motivation always involves at least one or more semantic components (with the exception of phonological motivation, as pointed out above), which I call the *motives*. Thus, with regard to Old English complex plant names, I understand by the term 'motivation' all the attributes of the plants that were singled out and verbalised in the plant names – be it the colour, the shape, the growing habits or healing virtues, etc. of the plant (called *motives*). As some of the plant names owe their morphological make-up to processes of folk etymology, which, in turn, are triggered by the formal, phonetic texture of otherwise intransparent plant names (or parts of it), 'sound pattern' can also be a motive singled out to determine the plant's name, albeit, of course, not a semantic one.[208]

Motivation can be *first* or *secondary*. While the motives that are grounded in extra-linguistic aspects of the plants can be summarised as the first motivation of the plant names, phonetic texture (sound pattern) triggering the change of intransparent plant names to transparent ones is always secondary. Secondary, phonological motivation can be supported by secondary, semantic motivation. The following example, describing the motives of *grundeswelige* and of *gundeswelige*, will present exemplarily both kinds of motivation. It has been chosen because, in contrast to many plant names changed by processes of folk etymology, the first motivation of these plant names is still traceable: The Old English plant name *grundeswelige* 'groundsel' lit. 'ground-devour(ing plant)' is a creeper. The plant name encodes the motive of 'creeping' by expressing that the plant 'swallows the ground'. The plant name is, however, recorded two times as *gundeswelige* lit. 'pus devour(ing plant)'. Sauer (2000: 30), therefore, assumed that, originally, the plant's name was 'pus-devourer' (*gunde* 'pus' + *sweligean* '(to) devour') – as groundsel is a medical plant used to remedy pus

207 For Saussure (1916: 67), linguistic signs are on principle arbitrary, i.e. non-motivated, with the exception complex lexemes, which are 'relatively' motivated (1916: 181). Bally (1944: 129) distinguishes between motivation by the *signifié* (e.g. *poirier* 'pear-tree'), motivation by the *signifiant* (e.g. *croquer* 'to crack) and motivation by both of them together (e.g. *croque-mort* 'grave-digger').

208 Processes of folk etymology are explored in chapter 7.5.

– and that the formal, phonological similarity between *gunde* 'pus' and *grund* 'ground' led to the secondary phonological motivation of the plant name, which was supported by the secondary, semantic motivation, namely, that the plant is a creeper. To sum up: the fact that the plant groundsel was used to cure pus is its first, semantic motivation, while the phonetic textures of *grund* and *gunde* embody the secondary, phonological motivation, which is supported by secondary, semantic motivation (see Catalogue, s.v. *grundeswelige*).

As Biggam (2003b: 206) observed, all plants or plant species have individual, eye-catching or culturally relevant features that are so salient that they trigger the formation of a name for the plant. In most of the complex plant names two motives can be isolated since each element can encode a feature of its own.[209] The plant named *gārclīfe* 'agrimony' lit. 'spear-cleavers', for instance, was, apparently, associated with a spear through the element *gār* 'spear', and with other kinds of *clīfe* 'cleavers' through its second element: The association of the first element is initiated by the plant's stems that look like (tiny) spears; the second element encodes the fact that agrimony was seen, at least in Anglo-Saxon folk taxonomy, as a sub-species of *clīfe* (see Catalogue, s.v. *gārclīfe*).

The motives triggering the formations of Old English plant names are varied: they are reaching from eye-catching colours (e.g. the first element in *basu popig* 'common red poppy' lit. 'purple poppy'; see Catalogue, s.v. *basu popig*), to particular shapes (e.g. the combination of the first and the second element in *openærs* 'medlar' lit. 'open arse'; see Catalogue, s.v. *openærs*), preferred habitats (e.g. the first element in *fenampre* 'water-dock' lit. 'water dock'; see Catalogue, s.v. *fenampre*), healing virtues (e.g. the first element in *lǣcewyrt* 'rib-wort, etc.' lit. 'doctor's wort'; see Catalogue, s.v. *lǣcewyrt*), origins (e.g. the first element in *affricanisc æppel* 'pomegranate' lit. 'African apple'; see Catalogue, s.v. *affricanisc æppel*) and time specifications (e.g. the first element in *dæges ēage* 'daisy' lit. 'day's eye'; see Catalogue, s.v. *dæges ēage*).

Single motives such as 'red', 'spear-shaped' or 'prefers moist ground' can be grouped into larger units, the *motivation categories*, as I have termed them. The category 'size', for example, includes motives such as the breadth of the leaves (e.g. *brādelēac* 'ramsons' lit. 'broad leek'; see Catalogue, s.v. *brādelēac*) or the height of the plant (e.g. *dweorged(w)os(t)(l)e* 'pennyroyal' lit. 'dwarf's *dost*'; see Catalogue, s.v. *dweorged(w)os(t)(l)e*); the category 'colour' references to all kinds of colours (e.g. 'red' in *rēad sealh* 'purple loosestrife' lit. 'red sallow'; see Catalogue, s.v. *rēad sealh*, or 'white' in *hwīt popig* 'opium poppy' lit. 'white poppy'; see Catalogue, s.v. *hwīt popig*). Often, motives relate to 'parts of plants' (e.g. *æppeltrēow* 'apple-tree' lit. 'apple-tree'; see Catalogue, s.v. *æppeltrēow*), to their nutritional value (e.g. *meteswamm* 'edible mushroom' lit. 'food-mushroom'; see Catalogue, s.v. *meteswamm*) or verbalise information on the plants' medical virtues (e.g. *bānwyrt* 'daisy, etc.' lit. 'bone-wort'; see Catalogue, s.v. *bānwyrt*). The motivation categories are presented in chapter 6.4.

Before the catalogue of motives is presented, however, the notion of *associative relation* must be discussed, as this is a concept crucial for the understanding of the

209 Integral metaphors and integral metonyms are different; see 6.3. below.

encoding of the motives into the forms of the plant names. It is claimed that cognitive, psychological processes, that are identical with those postulated to be at the heart of semantic change, can account for the way the motives 'enter' the plant names.

6.3. ASSOCIATIVE RELATIONS

Human categorisation of reality as described by *gestalt* psychology is based on three associative relations that, according to Aristotle, governed the process of remembering, namely, *similarity*, *contiguity* and *contrast*:

> So we track down the sequence [of our ideas] by starting from the present moment or from something else and from something similar or opposite or close [to it]. That is the way remembering comes about (Aristotle, *De memoria et reminiscentia*, 451b: 18-22, quoted in and translated by Koch 1999a: 141).

Similarity, contiguity and contrast combine to create certain universal principles or laws that are at the heart of human cognition and the perception of reality: the principle of holism (G *Übersummativität*), the principle of good *gestalt* or *prägnanz* (depending on proximity, similarity, closure and continuation, cf. Ungerer and Schmid 2006: 36) and the principle of figure and ground. Thus, different mental operations are used to categorise reality: similar and contrasting things are related to each other; bigger scenarios are fragmented into smaller entities; similar and contiguous concepts are combined into bigger entities and prominent features or concepts are profiled against less prominent backgrounds (cf. Blank 2001a: 38ff.).

The theoretical framework of *gestalt* psychology was recognised as being extremely important in linguistics and, eventually, became the background of cognitive semantic theory. Namely, the basic notions in cognitive linguistics – cognitive categories, prototypes and frames – are grounded in the three associative principles similarity, contiguity and contrast:

> Frame und Prototyp sind in dieser Sichtweise komplementäre, einander ergänzende Konzeptionen zur Erfassung der Welt: Die nach dem Similaritätsprinzip organisierten prototypischen Kategorien werden erst in Frames aktualisiert; sie sind ansonsten Abstraktionen, die auf Erfahrungen der Wirklichkeit beruhen, aber nicht in konkrete Wissenskontexte eingebaut sind. Die nach Kontiguitäten organisierten Frames auf der anderen Seite sind selbst prototypisch, insofern sie 'normale', typische Lebenssituationen erfassen und zentralere Komponenten oder Teilhandlungen enthalten sowie randständigere (Blank 2001a: 59).
>
> [Frame and prototype are, in this view, complementary concepts that supplement each other and help to grasp reality: The prototypical

categories, organised according to the principle of similarity, are realised primarily in the frames; they are otherwise abstractions based on experiences of reality, but which are, however, not included in concrete contexts of knowledge. Frames, on the other hand, organised by relations of contiguity, are themselves prototypical in that they capture 'normal', typical, life situations, and contain more central components or fractions of action as well as more marginal ones.]

To sum up, the internal structures of categories and frames, of concepts and linguistic signs are best described in terms of relations of similarity, of contiguity and of contrast.

The relations of similarity, contiguity and contrast can be located on all levels of knowledge in Blank's (1997, 2001a) semiotic model. The application of the three associative relations, similarity, contiguity and contrast to the various kinds of information stored on the different levels of knowledge as presented in the semiotic model produces a set of ten associative relations (Blank 1997: 137ff.; 2001a: 43ff.). These organise the mental lexicon, create polysemy, are responsible for semantic change and can be used both to describe the semantics of neologisms as well as that of syntactic structures. In the following, the ten associative relations shall be introduced briefly and shall be illustrated by examples taken, largely, from the Old English complex plant name corpus:

(1) metaphorical similarity

Metaphorical similarity (Blank 2001a: 43) is one of the major associative relations. It is based on a relation of similarity between two concepts that cannot be compared directly to one another as they belong to different frames of knowledge on the extra-linguistic, abstract level: two otherwise unrelated concepts – a source and a target concept – are seen as similar since they have one feature in common, which is the *tertium comparationis*. To give an example: the source concept HEAD 'upper, directive part of the body' is conceptually similar to the HEAD as 'chief, directive part of a company'; the basis for this similarity – the *tertium comparationis* – is the notion of 'upper, principle' (the same idea is present in G *Hauptmann* 'captain'). When metaphorical similarity affects simple items, which are, eventually, lexicalised with both meanings, it produces polysemous lexemes, such as in the example just mentioned, or, to name another classic example in many textbooks, *mouse* as an 'animal' and as a 'tool used with computers'.[210]

In compounds, metaphorical similarity can affect both the first and the second element; in the first case, it yields *partial metaphors* (chapter 5.3.2.2.1.1. (c)). Take, for example, the complex plant name *gārclīfe* mentioned above, an endocentric compound: the first element *gār* 'spear' refers to the shapes of the leaves of garlic, which, as the botanical morphology of the plant tells us, are pointed and similar to spears – a fact that is verbalised metaphorically in the first element of the plant name

210 Metaphorical extension of meaning is also used as one criterion to distinguish between polysemy and homonymy (Lipka 2002: 153-7).

(see Catalogue, s.v. *gārclīfe*). When the second elements of complex lexemes are used metaphorically, they produce *metaphorical compounds* or *total metaphors* (chapter 5.3.2.2.2.). To pick up the example *hræfnes fōt* 'crowfoot' lit. 'raven's foot': the similarity between the shape of the leaves of crowfoot and a raven's foot, for example, gave rise to the metaphorical Old English plant name *hræfnes fōt*, lit. 'raven's foot' (see Catalogue, s.v. *hræfnes fōt*).

(2) co-taxonomic similarity
Co-taxonomic similarity is a relation of similarity holding between two concepts that are related paradigmatically, i.e. between two taxa on the same hierarchic level (Blank 2001a: 43). The plants denoted by *elebēam* 'olive-tree' and *mōrbēam* 'mulberry tree', for instance, are sub-ordinated to *bēam* 'tree', but are, taxonomically, on the same hierarchic level. Co-taxonomic similarity can lead to semantic change, but this is restricted to cases in which people with a specific cultural background enter a new cultural background and, as it were, 'change the frame'. One example from Blank's corpus is the animal name F *caméléon* 'chameleon', which was changed to Creole *kamaléõ* 'lizard' on the island Réunion (Blank 2001a: 86), when the French colonisers were confronted with a fauna they did not have names for.

In the corpus of Old English plant names, co-taxonomic similarity triggered the polysemy of *more*: When the parsnip was introduced to the Anglo-Saxons, the Old English lexeme *more*, which originally referred to the carrot, became polysemous and denoted henceforth the parsnip, too, as, at that time, the parsnip and the carrot looked quite similar and were not easily distinguished: both were whitish and rather thin (see Catalogue, s.v. *more*). Co-taxonomic similarity is, however, not restricted to cases where people, quasi, 'change the frame'. Co-taxonomic similarity is a relation that is extant in many of the Old English polysemous plant names. The plant name *bēowyrt*, for instance, denotes both the common balm and the black horehound. The Old English plant name *bēowyrt* lit. 'bee-wort' was, initally, motivated by the most salient feature of common balm, namely, that it attracts bees. The denotation as black horehound, in turn, though, is based on the outward similarity of the two plants (common balm and black horehound), while the polysemy of the plant name results from this co-taxonomic similarity (see Catalogue, s.v. *bēowyrt*; for another example, see Catalogue, s.v. *blind(e)net(e)l(e)*).

As, however, the Anglo-Saxons invading England did not enter a territory climatically different, or with large differences in the flora and fauna from the one they had settled before, there are no cases of co-taxonomic similarity in the Old English complex plant names resulting from contact between the Celts and the Anglo-Saxons, nor from the contact between the Norse people and the Anglo-Saxons (cf. chapter 7.3.). The only climatic difference relevant to the corpus of Old English plant names is that between the Mediterranean and Anglo-Saxon England. Many plants were imported from the Mediterranean, and together with them also the plant names. Most of them were integrated into the Old English lexicon, and some plants are marked explicitely as foreign (e.g. *wealhhnutu* 'walnut' lit. 'foreigner-nut'; see chapter 6.4. (10)). Plant names originating in co-taxonomic similarity as a result of different climatic conditions, such as *more* 'carrot, parsnip' rare.

(3) taxonomic super-ordination

Taxonomic super-ordination is a relation of similarity between two concepts and can be described on the intra-linguistic, sememic level of knowledge (Blank 2001a: 43).[211] The sememes of the two words are identical but for one distinctive feature: one of the concepts is taxonomically super-ordinate to the other. OE *bēam* 'tree', for example, is super-ordinate to *mōrbēam* 'mulberry tree'; *lēac* 'leek, plant, herb' is super-ordinate to *crāwanlēac* lit. 'crow garlic, etc.'.[212]

Taxonomic super-ordination accounts for the polysemy of OE *lēac*: initially denoting the 'leek', a second meaning, the denotation 'plant, garden-herb' developed, because leeks were culturally and nutritionally highly important as well as taxonomically central and among the most important plants used by the Germanic peoples (see Catalogue, s.v. *lēac*). Then, it also accounts for the polysemy of *brēmel* and *brēr*, both denoting specifically 'bramble' and more generically 'thorn-bush' (see Catalogue, s.vv. *brēmel, brēr*), of *cærse* as 'cress' in *brōccærse* lit. 'brook-cress' and as 'plant' in *hole cærse* lit. 'hollow plant', and of *æppel* as 'apple' in *æppeltrēow* and as 'fruit' in *coddæppel* 'quince' and *eorþæppel* 'mandrake, etc.' (see Catalogue, s.vv. *brōccærse, hole cærse, æppel, æppeltrēow, coddæppel, eorþæppel*; see also chapter 6.6.).

(4) taxonomic sub-ordination

Taxonomic sub-ordination is the opposite relation of (3) above (Blank 2001a: 44). Taxonomic sub-ordination leads to semantic specialisation. OE *dēor* 'animal', for example, developed the very general meaning 'ruminant quadrupeds, with horns or antlers' (*OED*, s.v. *deer*, 2 (a)) in Modern English; *hund* 'dog' developed the meaning "dog kept or used for the chase, usually one hunting by scent" (*OED*, s.v. *hound*, 2 (a)). Taxonomic sub-ordination is the relation that holds in many Old English plant names between the concept denoted by the second element and the plant. Thus, *bānwyrt* 'daisy, comfrey' lit. 'bone-wort' is taxonomically sub-ordinated to *wyrt* – it is a 'kind of *wyrt*' (see Catalogue, s.v. *bānwyrt*).

(5) conceptual identity

Conceptual identity results from the association of the same, identical concept (Blank 2001a: 44). Compounds formed on the basis of conceptual identity are generally known as tautologies. Kastovsky (1992: 367) cites the examples *holtwudu* 'wood' and

211 Often, intra-linguistic taxonomy reflects extra-linguistic, scientific taxonomy as in the case of *elebēam* and *bēam*; in several cases, the two are contradictory: G *Walross* 'walrus', which is not denote a kind of *Ross* 'horse', but a marine mammal (see also fn. 198 and 199). Taxonomic super-ordination is a frequently employed mechanism leading to semantic change. For instance, OE *bridd* 'young bird' adopted the meaning of its super-ordinate OE *fugol* 'bird (young and old)' in the course of the development from Old English to Modern English: the denotation of ModE *bird* is not restricted to young birds anymore (*OED*, s.v. *bird*, 1 (a), 2).

212 Traditionally, this test frame is applied to establish relations of hyponymy within a word field. Cruse (1986: 163 and 1988: 82f.) argued in favour of a distinction between hyponymy (as a word-specific, intra-linguistic relationship) and taxonomy (a relation between concepts); relations of taxonomy belonging to our encyclopaedic knowledge of concepts, whereas hyponymy, on the other hand, describes relations between word-specific meanings.

wuduholt 'wood', both lit. 'wood-wood', *deaþcwealm* lit. 'death-death' and *feorhlīf* lit. 'life-life' as examples of tautologies, but it seems more likely that different meanings of the lexemes are activated: in *holtwudu*, for instance, the element *wudu* denotes the 'wood, forest', in *wuduholt* the trees that make up the wood 'forest'.

Conceptual identity is extant in many Old English complex tree names, especially, in hybrid pleonastic formations. The first element of *fīctrēow* 'fig-tree', for instance, a loan from Latin, is attested independently in Old English as *fīc* 'fig-tree', so that the relation between the synonyms *fīc* 'fig-tree' and *fīctrēow* 'fig-tree' is one of conceptual identity. The fact that this happens, as just stated, with many plant names where the first element is a loan from Latin, is very conspicuous and can probably be explained by the fact that, as originally Latin words, they were probably not understood by everyone (cf. chapter 5.3.2.2.1.1. (b)).

(6) formal, or phonological similarity

The *formal* or *phonological*[213] *similarity* (Blank 2001a: 44) of two *signifiants* can lead to folk etymological changes. L *hibiscum*, for example, was remotivated after OE *bisceop-* 'bishop' in *bisceopwyrt*, and remotivation is also responsible for the development of MedL *levisticum* into OE *lufestice* lit. 'love-stitch'. Remotivation on the basis of phonological similarity does not depend on any semantic relation between the source and the target concepts; it is a strictly language-internal process, though it can be supported by a semantic relation (cf. for example the relation between *gundeswelige* lit. 'pus-devour(ing plant)' and *grundeswelige* 'ground-devour(ing plant)'; see Catalogue, s.v. *grundeswelige*; cf., further, chapters 6.2. and 7.5.).

(7) conceptual contiguity

Conceptual contiguity describes a relation of contiguity between different concepts that belong to the same frame of knowledge (Blank 2001a: 44). Relations of contiguity can be, for example, of a temporal, spatial or causal nature and belong – together with metaphorical similarity – to the most important structuring principles of our conceptual and linguistic system. Conceptual contiguity is the principle governing the creation of many complex plant names. *Blind(e)net(e)l(e)*, for example, verbalises in *blinde* a contiguous feature of the plant, namely, the fact that it does not burn when touched (see Catalogue, s.v. *blind(e)n(e)tel(e)*). OE *hylwyrt* verbalises with *hyl* 'hill' the preferred habitat of the plant and thus a spatial relation (see Catalogue, s.v. *hylwyrt*).

Metonymic compounds are all based on relations of contiguity. The loan-translation *fīflēaf(e)* lit. 'five leaves' ← L *quinquefolium*, for example, relates to the fact that the plant Potentilla reptans has leaves that are divided into five parts. To the layperson it appears as though there are five leaves, and this characteristic is, apparently, so eye-catching that it stands for the whole plant. A special sub-type of metonymy, present in *fīflēaf(e)*, is meronymy. Meronymy refers to part-whole relationships (*pars pro toto*) that hold between words on different hierarchic levels (Kortmann 2005: 204). To take

213 I prefer to call Blank's 'formal similarity' the 'phonological similarity' in order to narrow it down to the motive 'sound texture', in contrast to the morphological make-up, which can also represent a 'formal' aspect of a lexeme.

the example *fīflēaf(e)*: *lēaf* 'leaf' is a meronym of *wyrt* 'plant' (which is the holonym to *lēaf*) and when *fīflēaf(e)* denotes a plant, the relation between the second element of the plant name and the plant fiveleaf is one of contiguity (sub-type 'meronymy'; cf. Catalogue, s.v. *fīflēaf(e)*). Another example relates to the first element of the complex plant name: the relationship of *æppel* 'apple' and the plant *æppeltrēow* 'apple-tree' is a meronymic relation, as apples are parts (products, the fruits) of apple-trees (see Catalogue, s.v. *æppeltrēow*).

Many elements of the complex plant names developed additional meanings on the basis of a relation of conceptual contiguity. Thus, OE *þorn* 'thorn' developed the meaning 'thorn-bush' based on the meronymic relation between thorns and the bushes they grow on (cf. Catalogue, s.v. *æppelþorn*); *wudu* 'wood' developed 'wooden', relating to the texture, in *wuduæppel* 'crab apple' and *wuducunille* 'wild thyme'.

(8) co-taxonomic contrast

Co-taxonomic contrast produces antonyms, i.e. co-hyponyms belonging to the same frame of knowledge (Blank 2001a: 44). Co-taxonomic contrast can be used to produce irony or can mark registers, such as for instance when ModE *bad* 'not good' is used with the meaning 'excellent' in slang varieties. This relationship is not present in the plant name corpus investigated.

(9) antiphrastic contrast

Antiphrastic contrast (Blank 2001a: 44) holds between concepts belonging to different frames of knowledge that are used in a kind of reverse metaphor: for example, F *pensionnaire* meaning 'boarder' takes over the meaning of 'prisoner, convict', i.e. an involuntary 'boarder' in French slang (cf. Blank 2001a: 44). This associative relation is not present in my plant name corpus.

(10) syntagmatic contiguity

Syntagmatic contiguity is a language-immanent relationship holding between linguistic signs on the concrete linguistic level of knowledge; these two (or more) lexemes form a syntagm (Blank 2001a: 44). It can lead to elliptical formations as in *daily* from *daily newspaper*, i.e. when the second element of the syntagm *daily newspaper* is dropped and the first element, *daily* 'every day, day by day', takes on the meaning of the original complex formation, becoming *daily* 'daily newspaper'. This relation is not extant in the Old English complex plant name corpus.

As pointed out above, human categorisation and the conceptualisation of reality depends to a large extent on mental operations, such as the creation of associative relations, or the classification of concepts into cognitive categories. It is, however, not only human thought, but also the encoding of thought into language that is decisively governed by associative relations. According to Blank (2001b), associative relations form the groundwork of any lexical innovation, be it a derivation or a compound, a new lexical unit resulting from semantic change or from conversion, an onomatopoeic formation, a loan word or a semantic loan.

> Damit ergibt sich für die Lexikologie die nicht ganz selbstverständliche Erkenntnis, daß alle lexikalischen Verfahren, mit denen Sachverhalte versprachlicht werden können, eine semantische Komponente besitzen. Diese semantische Komponente beschreibt die Relation zwischen dem zu versprachlichenden Konzept (dem Zielkonzept) und dem Konzept oder den Konzepten, die hinter den Wörtern und Affixen stehen, welche an der Bildung beteiligt sind (den Quellkonzepten) (Blank 2001a: 124).
>
> [Thus, the outcome is the insight that all lexical methods which can be used to verbalise facts have a semantic component; and this insight can by no means be taken for granted. This semantic component describes the relationship between the concept to be verbalised (the target concept) and the concept or the concepts behind the words and affixes that take part in the creation of the new lexeme (the source concepts).]

In other words, all kinds of lexical innovation, even if based on the most disparate verbalisation strategies, can be put down to a common denominator: the associative relations holding between source and target concepts. Composition, conversion (or zero-derivation) and semantic loan – distinct from the point of view of verbalisation strategy – can be analysed cognitive-semantically on the basis of the same basic psychological facts (Blank 2001b: 6). The relation of similarity, for example, produces new lexical units not only by semantic change (e.g. ModE *headhunter* 'someone who hunts heads' > 'someone who recruits people for a company'), but also new compounds when combined with word-formation processes (e.g. OE *hræfnes fōt* lit. 'raven's foot' > 'crowfoot (plant)'), and folk etymologies (based on the formal (phonological) similarity of the *signifiants* between, e.g., MedL *levisticum* and OE *lufestice* lit. 'love-stitch (plant)'). Using the notion of associative relations, therefore, seems the most obvious solution in the analysis of the relation between the denotata of the elements of the complex formations and the denotata of the complex formations as a whole. Not only can we avoid using syntactic paraphrases and the problems they raise, we can also leave the strictly language-immanent approach and describe all plant name compounds, including the metaphorical and metonymic ones, where they are rooted: on the interface between the extra-linguistic level of knowledge and the language-dependent level of knowledge.

It has already been stated that not all of the ten possible associative relations are present in the Old English plant name corpus: taxonomic super-ordination could be stated for some of the simplexes that are part of the complex plant names (e.g., *lēac*), but there is no specimen in the plant name corpus that displays this kind of associative relation; likewise, antiphrastic contrast, co-taxonomic contrast and syntagmatic contiguity are not present in the corpus investigated. When the motives encoded in the plant names cannot be ascertained, it is clear that the appropriate associative relations cannot be established either. This is the case, for instance, with *wermōd* 'wormwood' – a frequently recorded plant name whose semantic and formal origins are unknown (see Catalogue, s.v. *wermōd*).

It is important to stress that the associative relations do not hold between the two elements of the complex plant names, respectively, between the concepts they denote, but between the concepts denoted by the elements of the plant names and the concepts denoted by the plant names as a whole. To give an example: The endocentric plant name *gārclīfe* is constructed of the modifier *gār* 'spear' and the head *clīfe* 'cleavers. As a whole, *gārclīfe* denotes 'agrimony' and can literally be rendered as 'spear-cleavers'. The relation between GĀRCLĪFE as 'agrimony' and its first source concept, GĀR 'spear', is one of metaphorical similarity, as the shape of the stem (long and pointed) resembles the shape of a spear. The relation between the concept of GĀRCLĪFE 'agrimony' and its second source concept, CLĪFE 'burdock', is one of taxonomic subordination since, in Old English folk taxonomy, *gārclīfe* is seen as a kind of *clīfe* (as both plants cleave to the garments of passers-by; cf. also Krischke 2009: 251-3).

The relations in exocentric compounds are more intricate – a complexity which is mirrored in their morpho-semantic structure AB $\not\subset$ B. While there is a language-immanent relation of determination between A and B (B is the *grammatical* head of exocentric formations), the relation between the concept of the plant and the two elements of the plant name is a conceptual, extra-linguistic one. In *fīflēaf(e)* lit. 'fiveleaf', for example, the numeral delimits the second, nominal element: the plant is characterised by five, and not by six or seven, etc., leaves. In the metaphorical plant name *hræfnes fōt* 'crowfoot, etc.' lit. 'raven's foot', the choice of animal sets limits to comparison of the shape of the leaves of the plant with feet: the plant is not compared to just any foot, but to a raven's foot – and not to a dog's or a cat's paw. The relations that hold between the concepts of the individual elements and the plants as denoted by the complex lexemes are (i) metaphorical similarity for *hræfnes fōt*, and (ii) conceptual contiguity in the *pars pro toto* (or meronymic) relation between the plant *fīflēaf(e)* and the leaves of the plant.

Metaphorical similarity between the denotatum of the first element and the plant (or a part of it) results in *partial metaphors*, while the metaphorical similarity between the second element and the plant results in a *metaphorical compound* (cf. chapter 5.3.2.2.2.). Because of the interrelation of the denotations of the elements of metaphorical and metonymic compounds on the extra-linguistic level of knowledge, Blank (1998) called them *integral metaphors* and *integral metonyms*, and postulated that each is based on one concept only. He relativised this statement with regard to metonymies, when he stated that for them the postulation of two associative relations is possible, but not decisive (Blank 1998: 23), and, indeed, the analysis of the corpus of Old English plant names revealed that, while for many of the plant names two individual motives and, hence, two associative relations can be postulated – one for each element –, this is not pivotal: not for total metaphors, nor for metonymic formations.[214] Of the 90 exocentric plant names, 51 are metonymic, and 39 are metaphorical formations. Integral metaphors, respectively, metonyms, do exist, but in different proportions.

214 In Krischke (2009: 242, fn. 52) I was more strict about this point, when I stated that "In my opinion, however, two lexical morphemes encode always two concepts, even if these are interdependent".

A number of the integral metaphors conceptualise the similarity of the plant or of parts of it to animal body parts. As the example *hræfnes fōt* 'crowfoot, etc.' lit. 'raven's foot' analysed above shows, the two elements of this metaphorical plant name form, conceptually, one integral motive: a 'raven's foot' that the leaves of the plant are compared to. Further examples are *hundesheāfod* 'small snapdragon' lit. 'dog's head', *lēonfōt* 'lady's mantle' lit. 'lion's foot', and *wulfesteors* 'wolf's teasel' lit. 'wolf's penis'. In total, c. 21 metaphorical plant names can be reduced to one integral motive, i.e. about half of the metaphorical compounds. The other half, in contrast, conceptualises two different motives. The plant name *ēalifer* 'water-hemp, etc.' lit. 'water-liver', for instance, refers to the fact that waterhemp needs moist grounds in the first element, *ēa* 'river, water', while the shape of the plant's leaves, which are compared to a liver, is alluded to metaphorically in the second element of the plant name. Here, the first element *ēa* cannot be related directly to the second element, *lifer*, in the way it was possible to relate *hræfn* and *fōt* in the example mentioned above (see Catalogue, s.vv. *hræfnes fōt*, *hundesheāfod*, *lēonfōt*, *wulfesteors*, *ēalifer*).

Not all plant names can be reduced to clear-cut semantic 'results'. The plant name *eorþgealla* 'knapweed, etc.', for instance, encodes two motives: a conceptually contiguous feature of the plant, namely, that it grows in the earth, in the first element, and the bitterness of the plant, alluded to metaphorically in the second element. On the other hand, however, as *gealla* 'gall' is not a concept that belongs to the domain of plants, the first element is perhaps not directly linked to the plant itself, but is rather used to transfer the allusion to the body fluid gall from the domain of 'body' to the domain of plants (see Catalogue, s.v. *eorþgealla*).

Regarding the metonymic formations, the ratio is different. Of the 51 plant names, only seven can be called integral metonyms. Examples of integral metonyms are *fīflēafe* 'fiveleaf' lit. 'five leaf', *seofonlēafe* 'common tormentil' lit. 'seven-leaf' and *þrilēfe* 'wood sorrel' lit. 'three-leaf', where the first element determines the second and is not necessarily linked independently to the plant. In contrast, the two concepts denoted by the elements of the plant name *hæþber(i)(g)e* 'whortleberry' lit. 'heath-berry' can be linked independently to the concept of the plant: the first element refers to the habitat of the plant, the second element expresses a *pars pro toto* relationship between the plant and the concept denoted by the second element of the plant name (see Catalogue, s.vv. *fīflēaf(e)*, *seofonlēafe*, *þrilēfe*, *hæþber(i)(g)e*).

In this regard, we can revert to the morphological analysis and present a semantic explication for phenomena that had emerged in the formal analysis of the mophological structure of the complex plant names. Namely, the fact that there is a difference between exocentric (metaphorical and metonymic) compounds and endocentric compounds in morphological terms is mirrored in the associative relations that emerge in the semantic analysis of the plant names: The relationship between modifier and head of metaphorical and of metonymic compounds cannot be pinned down conclusively on the language-immanent level of knowledge, since there is no morpheme expressing the semantic head on the surface of the combinations ($AB \not\subset B$; AB is not a kind of B; while it can be for endocentric compounds; cf. $AB \subset B$; AB is a kind of B). Quite the contrary: The semantic relation between the denotata of the elements of complex formations and the denotata of the complex formations as a whole, are best explained on the conceptual, cognitive level, where they are based on the motivating concepts (the motives), which are 'translated' into the plant names with the help of the different associative relations. The associative relations and their functions in the naming of plants introduced, the motives encoded in the plant names, will be presented in the following section.

6.4. MOTIVATION CATEGORIES

Some of the motivation categories can be grouped into larger semantic units such as plant morphology (categories 'shape', 'part of plant', 'colour' and 'size'), or medical values (categories 'illness', healing virtue' and 'pathogen) but others, such as 'provenance' (e.g. in *sūþerne wudu* lit. 'southern wood') or 'person' (e.g. in *tunsingwyrt* lit. 'Tunsing's wort') are so varied that they can hardly be assigned to groups; and 'sound pattern', which refers to the phonetic texture, is not a semantic category at all. They indicate, however, the diversity of motives that triggered the creation of the Old English plant names.

When the denotata of the plant names are known and the plant names are transparent, the motives can be reconstructed with the help of botanical and encyclopaedic works. Botanical information helps when the plant names allude to, e.g. the morphology of the plants, or to their preferred habitats, while encyclopaedias and etymological dictionaries (e.g., the etymological dictionary of Latin plant names by Genaust, 2005) help with motives that relate to material culture (e.g. the products of plants), to (assumed or real) medical virtues or to religious or magical beliefs about plants.[215]

The motives encoded, for instance, in *æppeltrēow* 'apple-tree' – a transparent plant name made up of the elements *æppel* 'apple' and *trēow* 'tree' – can easily be reconstructed: *æppeltrēow* denotes a tree that bears apples. When the plant names are highly idiomatised or where phonological and/or morphological attrition led to loss of substance, and/or to (subsequent) folk etymological changes, the analysis of the motivation is difficult or even impossible. The plant name *hǣwenh(n)ydele* 'scurvy-grass, etc.' for instance, is transparent as far as the first element, the adjective *hǣwen* 'grey, green, blue', is concerned. The second element *-h(n)ydele* is not attested independently in Old English, which means that neither the morphology nor the motivation of the plant name can be ascertained (see Catalogue, s.v. *hǣwenh(n)ydele*). In contrast, the plant name *lufestice* 'lovage', for instance, is transparent, as the elements *lufu* 'love' and *stice* 'stitch' can be discerned clearly. Since the plant name is a folk etymological change of, originally, L *ligusticum*, however, there is no semantic

---

215 The identifications are based mainly on Bierbaumer (1, 2, 3), the *DOEPN* and the *DOE* (cf. also chapter 1.3.). As for literature of an encyclopaedic nature, I have recurred to Marzell's *Wörterbuch der deutschen Pflanzennamen* (1943-1979), the *Brockhaus* encyclopaedia, and to information provided by etymological dictionaries, such as *Holthausen*, *Genaust*, the *OED* and the *ODEE*, besides works dealing with specific aspects of the Old English language and Anglo-Saxon culture, such as Feulner (2000) on the Greek loan words or Clegg Hyer and Owen-Crocker (2011) on *Material Culture* in Anglo-Saxon England.

connection between the denotata of *lufu* and *stice* and the denotatum of the plant name *lufestice*, lovage; the only connection is made by the phonetic texture of the Latin plant name and the phonetic textures of the two Old English lexemes, *lufu* and *stice*, which led to the secondary, phonological motivation (see Catalogue, s.v. *lufestice*).

As the example of *lufestice* demonstrates, there is, sometimes, no obvious connection between the denotatum and the features encoded in the elements of the plant name. In the example of *hræfnes lēac*, in fact, the analysis of the motives points to the identification of a different plant than the one accepted until now. Old English *hræfnes lēac* 'wild orchid' lit. 'raven's leek' is glossed by *satyrion* in the Old English *Herbarium* (no. 16) and is equipped with a stylised picture of an orchis. This led Bierbaumer (2, s.v. *hræfnes lēac*), de Vriend (1984: 291, no. 16) and d'Aronco and Cameron (1998: 50) to identify *hræfnes lēac* as the wild orchid (Orchis L.). On the other hand, the wild orchid has no features which would suggest the association of a raven (Marzell, s.v. *Orchis* L.), but there is an Old English plant name that may help determine *hræfnes lēac*. The solution I propose is to regard *hræfnes lēac* as a synonym of *crāwan lēac* 'crowleek' lit. 'crow's leek', both in denotation and in signification.[216] This is because in Middle English, MedL *saturion* is glossed by ME *crowlek* (*MED*, s.v. *croue*). As MedL *saturion* often denotes the bluebell, which, in turn, strongly resembles the crowleek, it seems probable that OE *hræfnes lēac* was coined in analogy to OE *crāwanlēac*, with crows (OE *crāwe*) and ravens (OE *hræfn*) being birds that belong to the same species and that are quite similar. In this case, the motivation of the plant name *hræfnes lēac* would be, in analogy to *crāwanlēac*, the fact that the plant is a kind of leek, but in contrast to the leeks grown in gardens, this one is inferior. The inferiority is expressed metaphorically via the element *crāwan-* 'crow's', respectively, *hræfnes-* 'raven's'.

When the plant names refer to more than one species – a feature found in all folk taxonomies, and often a problem for the identification process (cf. Sauer 1992a: 383ff.) – the question is sometimes whether all of the proposed denotata can be used for the semantic analysis and if not, how the polysemy of the plant names can be explained. OE *bēowyrt*, for example, was identified by Bierbaumer (3, s.v. *bēowyrt*) as:

A: Melissa officinalis L. / balm (common ~, lemon~) / Zitronenmelisse
B: Ballota nigra L. / black horehound / Schwarzer Andorn
C: ? Acorus calamus L. / sweet flag / Kalmus
D: ?? Onopordon acanthium L. / Scottish thistle / Eselsdistel
D: ?? Acanthus mollis L. / bear's breech / Pracht-Akanthus or any other ?? thistle-species or thistle-like plant / Distelart oder eine distelähnliche Pflanze

Several scenarios are possible. One of them is, of course, that the plant name could denote all identifications proposed. The indication of different denotata does, however,

216 By *signification*, I understand with Lipka (2002: 57) "the content of the words or the relation between the contents of the words" in contrast to the denotation, which is the relation between the linguistic signs and their extra-linguistic denotata.

not automatically mean that the plant name is polysemous: it is, sometimes, not clear which of the alternative plant names are, exactly, denoted, as the texts, the contexts, or the Latin lemmas that are glossed are ambiguous. It is, for instance, quite obvious that, as a hapax legomenon, *ēowuhymele* lit. 'ewe-hop', cannot denote both hops and bryony at the same time and in the same text. Here, the analysis of the motivation (see Catalogue, s.v. *ēowuhymele*) can help, as it indicates that, probably, the plant name was coined originally to denote the hops, but that the outward similarity of the two plants contributed to the extension of meaning, as the name of hops was used to denote the bryony, too.

When polysemy is assumed, the question has to be answered of whether all identifications can be connected to the denotata of the elements of the plant name. In the case of *bēowyrt*, for instance, I concentrated on the botanical information of the plants most probably referred to by the Old English plant name. Since the question marks given by Bierbaumer (1, 2, 3) indicate degrees of probability, two question marks – as in our example given for meaning D – seemed to mark rather improbable identifications; I excluded, therefore, the identifications with two question marks. Then, since Bierbaumer's comment on identification C says that this interpretation is probably based on a misreading of the Latin text, C could also excluded. Thus, the plants identified in A and in B were left and could be considered in the semantic analysis. On the identifications A and B Bierbaumer notes:

> Marz. (3, 128) erwähnt, daß fast alle alten Namen von M. officinalis auch für andere ähnliche, von Bienen besuchte Lippenblütler gelten; z.B. für Lamium L., Taubnessel, die B. nigra (Marz. 1, 535) sehr ähnlich ist. (Bierbaumer 3, s.v. *bēowyrt*).
>
> [Marz[ell] (3, 128) notes that almost all old names of M. officinalis can be used for similar labiates, frequented by bees as well; for instance, for Lamium L., dead-nettle, which is very similar to B. nigra (Marz. 1, 535).]

We can, therefore, infer that OE *bēowyrt* originally denoted very probably Melissa officinalis L, the common balm, as bees (OE *bēo*) are strongly attracted by the scent of the common balm. The polysemy of *bēowyrt*, then, i.e. the fact that it could also denote the black horehound (Ballota nigra L.), resulted from the outward similarity of Melissa officinalis and Ballota nigra – a relation of co-taxonomic similarity (see Catalogue, s.v. *bēowyrt*).

Regarding the motivations of loan translations, loan creations or loan renditions (see chapter 7.4.3.), it is evident that it is not necessarily Anglo-Saxon views on the plants that we find encoded. When an Old English plant name is a loan translation of a Latin plant name, the original motivation goes back not to Anglo-Saxon, but to Roman origins and the people that created the original name were not the Anglo-Saxons but the Romans, or even other peoples, when the plant name is, in Latin, a loan word, too. Thus, the motive verbalised in *hundeshēafod* 'small snapdragon' lit. 'dog's head', a translation of L *canis caput* (see Catalogue, s.v. *hundeshēafod*), namely, the fact that the seed vessels look like skulls of humans or animals, pertains to the Latin name and was taken over into the Old English plant name as this is a translation of the Latin term. The notions 'translation', 'creation' or 'rendition' do not, however, automatically imply a dumb translation practice on the part of the glossators, with hardly any insight into the subject. Mapping foreign composites in one's own language requires the conscious recognition of the morphological and semantic make-up of the foreign models. Whether the Anglo-Saxon translators recognised the original motivation of the foreign (mostly Latin) plant names because they knew the concepts that lay behind them, or whether they translated "worde be worde",[217] i.e. mechanically, cannot, however, be definitively affirmed. After a thorough study of the Old English *Herbarium*, d'Aronco and Cameron are convinced that it

> is evident that the translator's choice of the name was not determined mechanically by the Latin model but depends directly on his experience and knowledge of plants and their classification. The fact that some names were left untranslated does not signify that the translator did not know what herb he was discussing but only that he did not know the corresponding name in his own language. His sense of professional duty and thorough, serious attitude prevented him from coining new names *ad hoc*, exposing users of the herbal to the potentially very dangerous consequences of mistaken identification (d'Aronco and Cameron 1998: 46).

In the following, the motivation categories are listed in descending order according to the numbers, are then explained, and, for each category, some examples are given (of which several are discussed in more detail). For further information, whether on the plants, the motives or the naming process, or on further reference work, the reader is advised to consult the Catalogue *sub voce* the plant name(s) in question.

(1) botanical information

Botanical information is by far the most often used motivation category in the corpus of Old English complex plant names. It is encoded in three quarters of all terms (c. 350 times) and includes (i) botanical information, where the element providing the information denotes a plant, as the second elements do in, e.g., *ācmistel* 'oak mistle-toe' lit. 'oak mistle-toe', *brōccærse* 'water-cress' lit. 'brook cress' and *fenampre* 'water-dock' lit. 'fen-dock'; (ii) botanical information where the element providing the information denotes a life form such as the second elements in, e.g., *æppelþorn* 'crab-tree' lit. 'apple-thornbush', and *bēctrēow* 'beech' lit. 'beech-tree'; (iii) botanical information, where the element providing this motive denotes a part of a plant, such as in *cornæppel* 'pomegranate' lit. 'corn-apple' or 'pip-apple', *coddæppel* 'quince' lit. 'husk-apple' or 'husk-fruit', *wæterbōg* 'sucker' lit. 'water-bough'. Taking botanical information as a motive means using the naming of a plant or a life form indicating taxonomic rank. In terms of associative relations, botanical information is, in most

217 The quotation is from King Alfred's letter prefixed to the *Cura Pastoralis* (Whitelock 1994: 5).

cases, connected to taxonomic sub-ordination and indicates where the plant is (probably) placed in Anglo-Saxon folk taxonomy.

Folk taxonomies are assumed to have five levels (Berlin et al. 1974: 25f.; Ungerer and Schmid 2006: 68): a level of (0) unique beginner, represented in Old English by *plante* and *wyrt* 'plant'; a level of (1) life form taxa, such as *þorn* 'thorn-bush' and *trēow* 'tree'; a level of (2) generic taxa, such as *lēac* 'leek'[218] or *bisceopwyrt* 'mallow'; a level of (3) specific taxa, e.g., *gārlēac* 'garlic', *por(r)lēac* 'leek' and *ynnelēac* 'onion' and a level of (4) varietal taxa, e.g. *micel worþignet(e)l(e)* 'great nettle'. Taxonomic ranking in folk taxonomies is not absolute and unambiguous (see chapter 2.1.), although it can be quite regular, and is identifiable on the basis of amounts of semantic and morphological information encoded in the plant names (cf. Krischke 2011). Thus, a large number of plant names can be identified as super- or sub-ordinate to other taxa on the basis of their morpho-semantic shape. For instance, plants denoted by endocentric plant names, where the second element denotes, in turn, a plant or a life form, are sub-ordinated to the denotatum of the second element; examples in question are *æppelþorn* 'pomegranate' lit. 'corn-apple' or 'pip-apple' and *bēctrēow* 'beech' lit. 'beech-tree', for instance, which are taxonomically sub-ordinated to the life forms *þorn* 'thorn-bush' and *trēow* 'tree'; *feldminte* 'horse-mint' lit. 'fen-mint' is sub-ordinated to *minte* 'mint', as is *wuducerfille* 'cow parsley' lit. 'wood-chervil' to *cerfille* 'garden chervil'.

As just mentioned, however, there are cases that are not entirely unambiguous. The compound *wealhmore* 'parsnip' lit. 'foreigner-carrot', for instance, is a co-taxonym of OE *more* 'carrot', while the morphological shape and the structure of the compound *wealhmore* (AB $\subset$ B) in relation to *more* suggest that one is the hyponym of the other, and that the compound reflects a relation of taxonomic sub-ordination, respectively super-ordination of the taxa in extra-linguistic world. This, however, is not the case: while *more* is simultaneously both a super-ordinate generic name and sub-ordinate specific plant name – i.e. vertically polysemous[219] –, taxonomically, *wealhmore* and *more* are on the same taxonomic level: they are co-taxonyms. Perhaps to circumvent this potential source of error, *more* is sometimes premodified by *englisc* (*englisc more* 'tansy' lit. 'English cost(mary)') so that the taxonomic weight of *more* in comparison to *englisc more* and *wealhmore* is more balanced.

Botanical information that relates to hierarchical levels in Anglo-Saxon folk taxonomy is encoded in the second elements of the complex plant names, as the examples mentioned so far illustrate. A number of plant names encode botanical information in the first element, as, for instance, *bēctrēow* 'beech' lit. 'beech-tree', *bōctrēow* 'beech' lit. 'beech-tree', *cederbēam* 'cedar of Lebanon' lit. 'cedar of Lebanon tree', *fīctrēow* 'fig-tree' lit. 'fig-tree', *hoclēaf* 'common mallow, etc.' lit. 'mallow leaf', *lāwerbēam* 'laurel' lit. 'laurel-tree', *mōrbēam* 'mulberry tree' lit. 'mulberry (tree) tree'. As these examples nicely illustrate, the concepts denoted by the first elements are not sub-ordinated taxonomically to the concepts denoted by the plant

218 However, *lēac* is polysemous; see chapter 6.6. and see Catalogue, s.v. *lēac*.

219 Blank (2001a: 19). The relationship of vertical polysemy and metonymy was investigated by Koskela (2005). This is comparable to the case of ModE *man*, which has two meanings: (1) 'human (male and female)' and (2) 'man (male)'.

name as a whole (though the concepts denoted by the second element are, with the exception of *hoclēaf*, super-ordinated to the concepts denoted by the complex formations) – they are identical: A *bōctrēow* 'beech' is a *bōc* 'beech', a *cederbēam* 'cedar of Lebanon' is a *ceder* 'cedar of Lebanon', and so forth. The associative relation holding between the botanical information encoded in the first elements and the denotatum of the complex plant names is one of conceptual identity. This is mirrored in the morphological make-up of these plant names, as most of them are endocentric and pleonastic (see chapter 5.3.2.2.1.1. (b)) – in contrast to the plant names where the botanical information is linked via taxonomic sub-ordination: these plant names display a wider range of morphological shapes.

(2) habitat

The plants' preferred habitats can be expressed directly by naming the geographical entities the plant lives near, i.e. a pond, the earth, a fence, water, etc., and indirectly. Direct reference to the preferred habitat can be found, for instance, in the element *ēa* 'river, water', which appears, for instance, in *ēacærse* 'water-cress', *ēadocce* 'water-dock' or *ēarisc* 'rush'. Indirect reference is expressed, e.g., when 'habitat' is indicated by the typical flora or fauna the plant lives in: *hindheolaþ* 'water-hemp' lit. 'hind-elecampane' for instance, grows preferably where deer typically live – a motive referred to by the element *hind* 'hind'. The preferred habitats of *sigelhweorfa* 'common marigold' lit. 'sun-turn(ing plant)' and *sunfolgend* 'common marigold' lit. 'sun-following (plant)', namely, that they love sunny places, are expressed in the element *sigel* 'sun', respectively, *sun* 'sun'.

If the motive 'habitat' occurs in metonymic compounds, such as for instance in OE *eorþgealla* 'knapweed' lit. 'earth-gall', its function is not only to indicate the preferred habitat of the plant – although it certainly does – but also to disambiguate the second element: in the example *eorþgealla*, obviously, *gealla* 'gall' alone is so ambiguous a name for a plant – especially when it comes to preparing remedies – that the element *eorþe* 'earth' was, clearly, used to indicate the fact that the concept referred to was not the human organ, but a plant (see also chapter 6.3. above, and Catalogue, s.v. *eorþgealla*).[220]

'Habitat' is a motive that is often found in the Old English complex plant name corpus (c. 99 plant names). Many compounds conceptualising 'habitat' are formed with the elements *feld* 'field', *tūn* 'town, enclosure, garden' or *wudu* 'wood' as, for instance, *feldbisceopwyrt* 'mallow', *tūncærse* 'garden-cress' or *wudumerce* 'celery, etc.'. Here, the elements *feld*, *tūn* and *wudu* not only denote the habitats of the plant, but, implicitly also the information of whether these are wild-growing plants or whether they are cultivated (cf. chapter 6.4.). These plant names are good examples for the fact that motivation categories are not rigid, but blurred around the edges.

---

220 Cf. also the disambiguating function of the first elements in G *Wüstenschiff* 'camel', lit. 'desert-ship', or in OE *sǣhengest* 'ship' lit. 'see-stallion'.

(3) part of plant
Very different parts of the plants are conceptualised in the elements of the Old English complex plant names. The tendrils are referred to in the second element of, e.g., *ælfþone* 'dogwood' lit. 'elf tendril'; the boughs in the second element of *misteltān* 'mistletoe' lit. 'mistletoe-twig'; the husk, e.g., in the first element of *coddæppel* 'quince' lit. 'husk-apple' or 'husk-fruit'; the fruit of the plant in the first elements of, e.g., *æppeltrēow* 'apple-tree' lit. 'apple-tree', *begbēam* 'mulberry tree' lit. 'berry-tree' and *hnutbēam* 'walnut' lit. 'nut-tree'. The leaves found their way into the plant names *fīflēafe* 'fiveleaf' lit. 'five-leaf' and *hoclēaf* 'common mallow' lit. 'mallow leaf'; the pips are verbalised in, e.g., *cornæppel* 'pomegranate' lit. 'pip-, corn-apple' or 'pip-, corn-fruit'; the crops are found in, e.g., *croplēac* 'common leek' lit. 'umbel-leek', and *heorotcrop* 'huckleberry, etc.' lit. 'hart's cluster (of berries)'; the blossoms in, e.g., *goldblōma* 'marigold' lit. '? gold-blossom'; the thorns in, e.g., *heorotber(i)(g)e* 'bramble' lit. 'hart's berry' and *heorotbrēmel* 'bramble' lit. 'hart's bramble'; the tubers in, e.g., *moldcorn* 'saxifrage' lit. 'sand-corn', *sundcorn* 'saxifrage' lit. 'sea-corn'; the sap in, e.g., *wæterbōg* 'sucker' lit. 'water-bough'; the roots in, e.g., *wenclufu* 'pilewort' lit. 'tumour-clove'; and the seeds in, e.g., *fuglesbēan* '(tufted) vetch' lit. 'bird's bean' and in *sunnancorn* 'common gromwell' lit. 'sun's-corn'.

When the parts of plants are verbalised in the second elements, the complex plant names are exocentric, metonymic formations, based on a *pars pro toto* (or meronymic) relation.[221]

(4) shape
A number of plant names encode information on the plant morphology, especially on the shapes of the plants. The first element in *candelwyrt* 'great mullein' lit. 'candle-wort', for instance, relates to the shape of the flowering stems of the plant; the second element of *fornētes folm* 'cuckoo flower' lit. 'Fornēot's palm' to the shape of the root of the plant, which resembles a hand. Most of the plant names encoding information on the plants' shapes do it figuratively, i.e. with the help of a metaphor. Thus, the plant *dæges ēage* 'daisy, etc.' lit. 'day's eye' encodes the motive 'a ring of petals that open in the morning and close in the evening' metaphorically, alluding to the similarity of the eyes of humans and the 'habit' of this plant to open its petals at daybreak.

In other plant names encoding the motive 'shape', the associative relation is not metaphorical similarity but conceptual contiguity – when the attribute alluded to is named in a straightforward way: *hollēac* 'shallot' lit. 'hollow leek', for instance, encodes the motive 'has tubular, hollow leaves' directly, using the adjective *hol* 'hollow'. Further examples encoding the motive 'shape' are *clufþung* 'celery-leaved crowfoot' lit. 'clove(d) poisonous plant', *ēalifer* 'waterhemp, etc.' lit. 'water-liver' and *eorþnafela* 'ground-ivy' lit. 'earth-navel'.

(5) colour
Colour relates in many of the 43 plant names to the colours of the flowers, but in some instances, the stems or the fruits are alluded to. While in *goldwyrt* 'marigold' lit. 'gold-

221 For the importance of fruits in the naming of fructiferous trees, see Koch (1999b).

wort', the colour of the flowers is referred to metaphorically via the element *gold* 'gold', the black colour of the fruits of the ivy is verbalised directly in *blæc īfig* 'ivy' lit. 'black ivy'. In *hārhūne* 'hoarhound' lit. 'hoary hoarhound' the stems or the leaves have a special, noteworthy hue and texture, which is verbalised in the element *hār* 'hoary'. Further examples for the motive 'colour' are, e.g. *hwītingtrēow* 'guelder rose' lit. 'whiting tree', *sunnancorn* 'common gromwell' lit. 'sun's corn', and *rēade fille* 'wild thyme' lit. 'red thyme'.

(6) size
'Size' is a motivation category that can refer to various parts of the plants (c. 34 plant names). 'Size' is referred to either directly, as, e.g., in *grēate bānwyrt* 'common comfrey' lit. 'tall bone-wort', or figuratively as, e.g., in *cnēowholen* 'kneeholly' lit. 'knee-holly', where the fact that the plant kneeholly is not very high is verbalised metaphorically with the element *cnēow* 'knee'.

The second element of *wegbrǣde* 'waybread' lit. 'waybroad (plant)', for instance, holds information on the broadness of the plant's leaves, which is in contrast to the narrow leaves of its co-taxon ribwort (OE *lǣcewyrt* or *ribbe*; cf. G *Breitwegerich* lit. 'broad *Wegerich*' vs. *Spitzwegerich* lit. 'narrow, pointed *Wegerich*', where the size of the leaves distinguishes the two taxa in German folk taxonomy). Further examples verbalising the motive 'size' are *dweorged(w)os(t)(l)e* 'pennyroyal' lit. 'dwarf's *dost*', *curmealle sēo māre* 'knapweed, etc.' lit. 'bigger centaury' and *curmealle sēo læsse* 'common centaury' lit. 'lesser centaury'.

(7) animal
References to animals (in c. 30 plant names) can be of different kinds: *bēowyrt* 'common balm, etc.' lit. 'bee-wort' and *mucgwyrt* 'wormwood' lit. 'midge-wort', for instance, encode the fact that bees, respectively, flies are attracted by the plants' scents, while *cicena mete* 'chicken-weed' lit. 'chicken's food' conceptualises the fact that the plant is the preferred fodder for all kinds of birds. *Henn* 'hen' in *hennebelle* 'henbane' lit. 'hen's bell' refers to hens, but not in a very positive way: this plant is noxious to all kinds of fowl.

(8) action
Although plants are not especially famous for performing actions, as many as c. 23 plant names encode this motive. The actions encoded refer to the plants' clinging (e.g. *clifwyrt* 'great burr' lit. 'cleav(ing) plant') and creeping qualities (e.g. *grundeswelige* '(common) groundsel' lit. 'ground devouring (plant)'). The plants named *sigelhweorfa* 'common marigold, etc.' lit. 'sun-turning (plant)' and *sunfolgend* 'common marigold, etc.' lit. 'sun-following (plant)' turn their flowers towards the sun. Further examples are *pricþorn* 'holly' lit. 'pricking thorn' and *springwyrt* 'caper-spurge' lit. 'jump(ing) plant'.

(9) illness
A number of plant names conceptualise the parts of the body that can be affected by illnesses (c. 23 plant names), such as the lungs, the bones, the eyes, etc. (*bānwyrt*

'daisy, etc.' lit. 'bone-wort', or *ēagwyrt* 'eyebright, etc.' lit. 'eye-wort') or the illnesses themselves, such as *sticwyrt* 'greater stitchwort' lit. 'stitch-wort'.

(10) origin
Quite a number of Old English complex plant names refer to their original places of growth (c. 20 plant names). The 'origin' can be indicated by the conceptualisation of directions, such as in *sūþerne wudu* 'southernwood' lit. 'southernwood', of the element 'foreign', e.g. in *wealhmore* 'common parsnip' lit. 'foreigner carrot' and *wealhnutu* 'walnut' lit. 'foreigner-nut' or by naming the country (region, continent) of origin directly, such as can be observed in *englisc cost* 'tansy' lit. 'English costmary' or in *affricanisc æppel* 'pomegranate' lit. 'African apple'.

(11) texture
In contrast to the category 'effect' (no. 18), the salient attribute profiled in 'texture' is not the effect the plant or part of it has on humans, but the texture itself, even if it provokes, of course, a special effect when touched. 'Texture' can be encoded both in the first and in the second element (c. 16 plant names). The plant name *furhwudu* 'pine' lit. 'furrow-wood', for instance, relates to the fact that pine-wood is a heavily furrowed wood in contrast to the wood of other trees as that of the beech or the birch. As all other motives, 'texture', too, can be alluded to directly, as in *rūwe wegbrǣde* 'hoary plantain' lit. 'hairy plantain', where the hairy leaves of plantains are verbalised as *rūw* 'hairy', or indirectly, as is the case with the resin of the plant Dracaena, which is called metaphorically *dracanblōd*, i.e. lit. 'dragon's blood'.

(12) wild
The fact that some plants were cultivated while others grew wildly is expressed directly by the adjective *wild* 'wild' in a number of plant names (c. 14 plant names), e.g., *wilde cyrfet* 'gourd, etc.' lit. 'wild gourd', *wilde mealwe* 'mallow' lit. 'wild mallow', or *wilde minte* 'mint' lit. 'wild mint' (for further examples, see Catalogue, s.v. *wild*). The fact that plants grow wildly presupposes that these plants are not cultivated in the vicinity of and by men and, in this respect, there is a conceptual intersection with the motive 'habitat', where the fact that plants grow in woods or in the field indicates, simultaneously, that they are not grown on purpose. This is, of course, not the case with typical crops such as wheat, which cannot be grown in gardens, but are grown, as a rule, in fields (in contrast to, e.g., mint, or mallow).

(13) taste
Taste is a category that is important in terms of nutrition and medicine (c. 10 plant names). The second elements in OE *eorþgealla* 'knapweed' lit. 'earth-gall' and in *meargealla* 'lungflower' lit. '? horse-gall', for instance, relate metaphorically to the fact that the plants denoted are of a startingly bitter, gall-like taste. The first element in *meduwyrt* 'meadow-sweet, etc.' lit. 'mead-wort', in contrast, alludes to the sweet taste of the flowers of the plant(s). Since *medu* 'mead' is an extremely sweet beverage made from honey, the use of *medu* for 'sweet' is metaphorical. In contrast, *milscapuldor*

'apple-tree' lit. 'sweet apple-tree' and *sūrmilscapuldor* 'apple-tree' (or 'crab tree') lit. 'sour-sweet apple-tree' verbalise the fruit's tastes directly.

(14) number

'Number' is encoded in six plant names, but never as the second element. In most cases, the number refers to the number of leaves of the plant such as in *seofonlēafe* 'common tormentil' lit. 'seven-leaf', or to a numerically characteristic shape of the leaves as, e.g., in *fīffingre* 'five-leaf' lit. 'five-finger', where the leaves are divided into five parts, which look, however, as if they were individual leaves.

(15) tool

Plants can be be used for different purposes (c. 6 plant names): for example, as tinder – a motive expressed, e.g., in the first element of *dȳþhamor* 'reedmace, etc.' lit. 'tinder-hammer' – or as a means to form hedges, as expressed in the first element of *haguþorn* 'common hawthorn' lit. 'hedge-thorn'; other plants or parts of plants are used to make baskets, e.g., *windelstrēaw* 'windlestraw' lit. 'basket-straw'.

(16) food

Certainly one of the most important functions of plants is their use as food plants for humans and animals, but this is seldom expressed in the plant names. Examples are *mæstentrēow* 'cork oak' lit. 'mast-tree', whose fruit is eaten by pigs, or in the second element of *cicena mete* 'chicken-weed' lit. 'chicken's food', a plant loved by all kinds of birds; in total, c. 6 plant names encoded this motive.

(17) person

In a small number of plant names (c. 6 plant names), people are mentioned – either by name, such as in *fornētes folm* 'cuckoo flowers' lit. 'Forneot's hand', by the common noun *man*, *mon* 'man' in *menlufigende* 'cleavers' lit. 'men-loving (plant)' or by pronouns, e.g., in *selfǣte* 'common groundsel' lit. 'self-eating (plant)'.

(18) effect

The plants encoding this motive produce a certain effect when touched (c. 5 plant names). For instance, *blind(e)net(e)l(e)* 'dead nettle, etc.' lit. 'blind nettle' does not burn, in contrast to the *net(e)l(e)* 'nettle', whose leaves burn when they are touched.

The first element of *feltwyrt* 'great mullein' lit. 'felt wort', *felt* 'felt, wool' refers to the fact that the plant's leaves create the sensation of wool. In this plant name, the 'effect' is not verbalised directly by, e.g. an adjective, but figuratively, as it is named the 'felt- or wool-wort'.

(19) smell

Few names encode the motive 'smell' (c. 5 plant names). These are, for instance, *lēaccærse* 'garlic mustard' lit. 'leek-cress', *hundesmicge* 'henbane, etc.' lit. 'dog's urine' or *fūla bēam* 'alder buckthorn' lit. 'foul tree'. The element *lēac* 'leek, plant' in *lēaccærse* and *hundesmicge* 'dog's urine' refer to the onion-like smell of garlic mustard, respectively the smelliness of henbane metaphorically, while the element *fūl* 'foul' in *fūla bēam* encodes the mouldy smell of the bark of the alder buckthorn directly.

(20) healing virtue

Five plant names, refer to the healing power of the plants such as, for instance, *dolgrūne* 'pellitory of the wall' lit. 'wound-mystery' and *lǣcewyrt* 'rib-wort, etc.' lit. 'leech-wort': *lǣcewyrt* is a plant with a high healing virtue so that it is compared to a doctor; *dolgrūne* is obviously mysteriously powerful against wounds.

(21) pathogen

Five plant names refer to illness-inducing elements such as poison and elves; e.g. *āt(t)orlāþe* 'betony' lit. 'poison-harm', or *ælfþone* 'dog-wood' lit. 'elf-tendril'. The low number of plant names conceptualising this motive illustrates, probably, the fact that, apart from, e.g., strokes received in battles which were obvious, not many causes for illnesses were known in early England and also, that where no rational explanation could be found, supernatural beings or poison were made responsible for illnesses.

(22) inferiority

'Inferiority' (c. 5 plant names) is often expressed metaphorically, when an animal, such as the crow in *crāw(an)lēac* 'crow garlic' lit. 'crow's leek' stands for the inferior quality of the plant denoted by the complex formation in contrast to the plant, which is denoted by the second element. It is, however, important to note that 'inferiority' in this context does not refer to the inferiority of the animal named: While horses, for instance, were highly valued in the early and late Middle Ages (as they are, still, today), the element *hors* 'horse' in *horsminte* 'horse-mint' lit. 'horse-mint' is meant to indicate that this kind of mint is inferior in comparison with 'simple' *minte* 'mint'. According to Hoops, this is a common phenomenon in English and German plant names:[222]

> Pflanzennamen mit *horse*, bez. *ross*, *pferd* als erstem element dienen im engl. wie im deutschen mit vorliebe zur bezeichnung unechter, besonders gröberer, oft auch wildwachsender und ungeniessbarer arten gegenüber den echten, feinen, cultivierten. Schon ags. *horsminte* als bezeichnung der wilden minzen und minzähnlichen pflanzen gegenüber den zarteren garten-species; [...] Auch andere tiernamen werden manchmal in der gleichen function verwant (Hoops 1898: 565f.).
>
> [Plant names with *horse*, respectively *ross*, *pferd* as a first element serve both in English and in German to denote false, coarser, often wild growing and unpalatable species in contrast to the real, dainty, cultivated ones. Old English *horsminte* already served to denote the wild mints and

222 See also Boretzky (1987: 65).

plants similar to mint in contrast to the more delicate garden-species; other animal names are also used in the same function.]

(23) product

The category 'product' refers to plants or parts of plants that were subjected to some kind of lengthy processing in order to gain the product (c. 5 plant names). The element *ele-* 'oil' in the synonmys *elebēam* and *eletrēow* 'olive-tree' lit. 'oil-tree or olive-tree', for instance, refers to the product of the fruits of the olive-tree. The fact that parts of the plant *leoþowyrt* 'soap-wort' lit. 'soap-wort' can be manufactured into soap is conceptualised in the first element *leoþo* 'soap'. Other examples are *smeoruwyrt* 'smearwort, etc.' lit. 'grease-wort', a plant whose sap was used as an ointment, applied to wounds since antiquity and *hunigsūge* 'honeysuckle', which is full of sweet sap that is sucked by children.

(24) overall appearance

The motive 'overall appearance' indicates that more than one feature of two concepts are similar (c. 4 plant names). This is encoded, for instance, in the comparison of the carrot with the parsnip in the Old English element *more*: The similarity of shape and colour of the carrot and the parsnip led to the polysemy of OE *more*, because, in contrast to the plants as we know them today – one dark orange, the other white – originally, the carrot was rather whitish and much thinner than today and looked almost like the parsnip so that the two plants were easily confused (see also 6.3.). In *cirisæppel* 'Carian fig' lit. 'cherry-fruit', the shape, colour and size of the figs are compared to cherries. The second element of *snīþstrēo* 'carline thistle, etc.' lit. 'cut-straw', refers to the dry, straw-like petals of the plant.

(25) time

The features of four plants are connected to time specifications. The plant *dæges ēage* 'daisy' lit. 'day's eye', for instance, opens its rays of petals at the beginning of the day and closes them at the end of the day. This feature is conceptualised in the first element of the plant name, as the time specification is in, e.g., *nihtscada* 'morel' lit. 'night-shadow' and *ēfenlāste* 'mercury' lit. 'evening-lasting (plant)'.

(26) threat

Plants cannot only fend off illnesses, but were also believed to fend off threats such as thunder and lightening, expressed in the first element of *þunorwyrt* 'houseleek' lit. 'thunder-wort' and *þunorclǣfre* 'common bugle, etc.' lit. 'thunder-clover'.

(27) sex

The reproduction systems in the plant world are very diverse and represent the greatest diversity of reproduction of all biological systems. Two systems often found are plants with both female and male reproductive units on the same plant and plants that are either functionally male or functionally female (for a brief description of these and other sexual reproduction systems of plants, cf. Bayrhuber et al. 1989: 323ff.). Hops, denoted in Old English by *ēowuhymele*, is a dioecious plant. This means that each individual has reproductive units that are merely female or merely male. This motive is possibly encoded in the first element of the Old English plant name *ēowuhymele* 'hops, etc.' lit. 'ewe-hops', where the sex of the plant is represented metaphorically in the first element, *ēowu* 'ewe'. OE *ēowuhymele* is the only plant in the corpus of the present survey to sex.

(28) sound
One plant name – *hrætelwyrt* 'rattle' lit. 'rattle-wort' – encodes the motive 'sound': the seeds produce a rattling sound in the capsule when the plant is ripe.

(29) ?
A number of plant names cannot be explained in terms of motivation. This is either due to the fact that the elements could not be identified, such as the second element of *hǣwenh(n)ydele* 'scurvy-grass', to the fact that the plant names or parts of the plant names were the product of folk etymology without secondary semantic motivation, such as in *wermōd* 'wormwood', which means that they owe their shape and form to a phonological similarity between the original plant name and Old English lexemes, but that the denotations of the elements of the plant names do not have any relation to the plants denoted by the plant names. When the original form is known, such as it is for *lufestice*, the motive is 'sound pattern' (see (30) below); where it is not, it is marked as '?' and commented upon.

Sometimes, both the denotations of the complex plant names and the denotations of their elements are clear, but a connection between the plant and the denotation of the elements can, still, not be made. This is the case, for instance, in *æppellēaf* 'violet' lit. 'apple leaf'.

(30) sound pattern
When two different words (or parts of words) sound similar, and when one of the words is felt to be dissociated in the language system, the phonological similarity can trigger the change of one of the words in analogy to the other (c. 19 plant names). This process, better known as *folk etymology* (Knappe 2004: 492), is, first of all, purely formal, but it can be supported by secondary semantic motivation. For instance, the change of the Latin loan word *asarum* into OE *hæselwyrt* on the basis of the phonological similarity of *asarum* and *hæsel* 'hazel' is reinforced by the semantic aspect that the hazelwort grows preferably under hazel bushes (see Catalogue, s.v. *hæselwyrt*). For further examples with discussion, see chapter 7.5.

As was mentioned briefly before, the boundaries of motivation categories are not rigid, and often, overlapping motives can be found. The example of two motivation categories, 'habitat' and 'wild', – verbalised by the Old English lexemes *tūn*, *feld*, *wudu* and *wild* – may illustrate the motivation categories' flexible boundaries.

In a number of plant names, *tūn* 'town, garden, enclosure' is the first element: *tūncærse* 'garden-cress', *tūnhōfe* 'ground-ivy', *tūnmelde* 'orache', *tūnminte* 'spear-mint', *tūnnǣp* 'rape'. With the exception of ground-ivy,[223] all these plants were grown early in Europe, from antiquity onwards, and, as this was done by humans, naturally, they were grown in enclosed places, in some kind of *tūn*. This motive – 'is grown in enclosures' (as opposed to, e.g., grows 'in the woods', or 'in the field') – can be seen as part of the category 'habitat'. Yet, the fact that humans grow plants implies also cultivation – people sowing the plants, protecting them from, e.g., late frosts in spring or from wind, weeding out, watering them and choosing the best seeds for further cultivation – as opposed to plants growing wildly, where nobody takes care of them and the crop is, accordingly, not particularly good. Thus, the motive 'grown in enclosures' also alludes to the fact that these plants are 'cultivated' – as opposed to 'growing wild'. Consequently, the motive 'habitat' overlaps with a possible motive 'cultivated' and, in this respect, the motivation category 'habitat' is not rigid, but flexible.

Complementary cases yield the elements *feld* 'field' and *wudu* 'wood, forest'. The element *feld* 'field' in the plant names *feldbisceopwyrt* 'mallow' lit. 'field-bishopwort', *feldmædere* 'rosemary, etc.' lit. 'field-madder', *feldminte* 'horse-mint' lit. 'field-mint', *feldmore* 'wild carrot, etc.' lit. 'field-carrot or field-parsnip', *feldrūde* 'common meadow rue' lit. 'field rue', *feldswamm* 'mushroom' lit. 'field-mushroom', *feldwōp* 'hops' lit. 'field-hops' and *feldwyrt* 'felwort' lit. 'field-wort' describes the plants' habitats and, at the same time, the fact that they are growing wild (and are not cultivated by humans).

Wudu is an even more intricate element. As the second element in the plant names *furhwudu* 'pine' lit. 'furrow-wood', *healfwudu* 'calamint' lit. 'half-wood', *sūþerne wudu* 'southernwood' lit. 'southern-wood' and *wīelisc* [*wudu*] 'southernwood' lit. 'foreign wood', and as the first element in *wuduæppel* 'crab apple' lit. 'wood apple' and *wuduweax* 'dyer's broom' lit. 'wood-sinew' it refers to the texture of the plants and not to their habitats. Crab-apple, for instance, – OE *wuduæppel* – has small, woody fruits, hence the use of the element *wudu* 'wood'; the element *wudu* in *sūþerne wudu* refers to the fact that Artemisia abrotanum is, though as small as a herb, a shrub with woody stalks and twigs. In the plant names *wudubinde* 'woodbine' lit. 'wood-binding plant', *wuducerfille* 'cow parsley' lit. 'wood chervil', *wuduclǣfre* 'clover' lit. 'wood-clover', *wuduclāte* 'birthwort, etc.' lit. 'wood-cleavers', *wuducunille* 'wild thyme' lit. 'wood thyme', *wududocce* 'sorrel, etc.' lit. 'wood dock', *wudufille* 'wild thyme' lit. 'wood-thyme', *wudu(h)rōfe* 'woodruff-aperule' lit. 'wood-strong (plant)', *wudulēac* 'ramsons' lit. 'wood-leek', *wuduleahtric* 'lettuce' lit. 'wood-lettuce', *wudumerce* 'celery' lit. 'wood-celery', *wudusūræppel* 'crab-tree' lit. 'wood crab-apple' and *wuduþistel* 'sow-thistle, etc.' lit. 'wood-thistle', on the other hand, *wudu* refers to the habitat of these plants: they all grow in woods and, it is implicitly understood, these plants are not cultivated by humans.

The example of *wudu* and the complex plant names it is part of demonstrates not only that some of the motives overlap with other motives but also that one linguistic

223 OE *tūnhōfe* 'ground-ivy' is the exception because, due to the fact that it is poisonous, it is not a particularly cultivated, although ground-ivy is a beautiful creeper that produces many flowers. The reason ground-ivy is found in the vicinity of humans is that it prefers soils rich in nitrate, and this is the case with soils lived on by animals and humans, as human and animal urine contains nitrates.

sign can represent different motives in different plant names, and this does not depend on the position of the linguistic sign denoting the motive, as the following example will illustrate. When one linguistic sign can represent two or more different motives, the sign is either polysemous, such as *wudu* (i) 'wood', (ii) 'forest' (as shown above), or the sign is not polysemous, but different attributes of the concept are highlighted. To give an example: the element *āc* 'oak' in *āctrēow* 'oak' lit. 'oak tree' denotes the plant species Quercus. In the complex plant name *āctrēow*, the element *trēow* 'tree' is, semantically speaking, the ground, against which the figure, the genus *āc* 'oak' is highlighted: this kind of tree is an oak and the prominent information, the motive, is a botanical one. In contrast, in *ācmistel* 'oak mistletoe' lit. 'oak mistletoe', the predominant motive represented by *āc* 'oak' is not the botanical information, but the habitat of the plant of the genus *mistel* 'mistletoe': this kind of *mistel* is one that lives as a parasite preferably on oak trees.

When the elements of the plant names encode motives in a metaphorical way, it is obvious that these elements are not polysemous, but that they are used as a vehicle for transporting information on the attribute – the motive – that is expressed on the basis of the similarity between the tenor (the plant) and the vehicle. To give an example: The animal names in OE *crāwanlēac* 'crow garlic' lit. 'crow's leek' and *horselene* 'elecampane' lit. 'horse-elecampane', do not indicate that there is a contiguous relationship between these plants and the animals – instead, the motive is the 'inferiority' of the plants, and the animal names serve as a vehicle to indicate that the plants denoted by the complex plant names are perceived as inferior, as coarser in nature in comparison to the plants denoted by the second elements. The motive 'inferiority' is, thus, encoded metaphorically in these plant names (see also motive (22) above).

6.5. MOTIVATION CATEGORIES AND ASSOCIATIVE RELATIONS: SYNERGY EFFECTS

The focus of the present chapter was to fathom the semantics of the Old English complex plant names in terms of their motives, which are seen in specific associative relations to the denotations of the plant names as a whole. Analysing this relationship in detail is extremely useful as it helps to explain parallels between seemingly disparate compounds on the one hand and differences between seemingly similar compounds on the other, as the following examples will show.

Semasiological perspective: starting from the Old English complex plant names

Taking a semasiological perspective, it can be observed that the plant names *smælþistel* 'prickly cnicus' lit. 'narrow thistle' and *dweorged(w)os(t)(l)e* 'pennyroyal' lit. 'dwarf's *dost*' encode the same motive in their first elements, namely, the size of the plant, as, apparently, both plants are rather small. There is, still, a clear difference between the use of *smæl* 'small' and *dweorg* 'dwarf'. This semantic difference lies not in the motive – which is in both cases 'small in size' – but in the associative relations used to verbalise the motive in the two plant names: Whereas *smælþistel* verbalises the motive 'size' directly, via a relation of conceptual contiguity (the feature of being small is contiguous to the plant), *dweorged(w)os(t)(l)e* verbalises the motive via a metaphor or, more precisely, via a relation of metaphorical similarity: because of its smallness, the plant is associated with a dwarf (see Catalogue, s.vv. *dweorged(w)os(t)(l)e, smælþistel*).

The next example demonstrates that surface shape and structure do not automatically reflect the motivational structure of the plant names, and that this is revealed only when botanical and encyclopaedic evidence is consulted: The plant names *heorotclǣfer* 'common melliot' lit. 'hart's clover' and *heorotber(i)(g)e* 'bramble' lit. 'hart's berry' both feature the element *heorot* 'hart', and, thus, seem to be quite similar from a semantic point of view. However, they are not, as the element *heorot* relates to different motives in these two plant names. In *heorotber(i)(g)e* lit. 'hart's berry', the first element conceptualises the thorn-armed boughs of the plant (motive: part of plant), which are compared to the antlers of bucks (relation: 'metaphorical similarity'), while in *heorotclǣfer* lit. 'hart's clover', the first element conceptualises the motive that deer like to eat the plant common melliot (motive: (food for) animal; associative relation: conceptual contiguity) (see Catalogue, s.vv. *heorotber(i)(g)e, heorotclǣfer*). In these two plant names, the same lexeme, used in the same position, namely, as the first element, the modifier – encodes two different motives that are, in addition, associated differently.

Onomasiological perspective: starting from the plants

The analysis of the motives and the associative relations further allows a number of insights into the folk taxonomic structures of the Old English botanical terminology. Before the Swedish botanist Carl von Linné laid the basis for the modern scientific nomenclature, all botanical classifications in vernacular languages, including Latin, were rather unstable and unsystematic folk taxonomies, characterised by the fact that one plant name referred to more than one genus or species and one genus or species was often referred to by more than one vernacular plant name. In addition to this feature, discussed briefly in chapter 2.1., plant names which are synonymous from the point of view of denotation were not necessarily synonymous from the point of view of signification: a number of plant names encode different motives while denoting the same plant. To give an example: Approaching the plant cinnamon (Cinnamomum cassia Blume) from an onomasiological perspective, we find that three Old English names denote this plant: *ofersǣwisc rind* lit. 'overseas bark', *sūþerne rind* lit. 'southern bark', and *swēte wyrt* lit. 'sweet wort'. These three plant names differ both in the motives encoded in the elements of the plant names and in the associative relations that form the link between the concepts denoted by the elements and the plant cinnamon. The first element of the plant name *ofersǣwisc rind* conceptualises the fact that Cinnamomum cassia is not a plant native to England: this kind of *rind* 'bark' is of foreign origins, from 'over the sea'. The second element profiles the part of the plant used most often – the aromatic bark – in a metonymic, *pars pro toto* relationship (associative relation: conceptual contiguity). Both these motives are found in the second specimen denoting the cinnamon, in *sūþerne rind*, too, but here, the 'origin' is referred to by the indication of a direction: the plant name expresses that this kind of *rind* grows in countries south of England. Completely different motives are found in the third specimen denoting the cinnamon, in *swēte wyrt*: here, the second element provides botanical information, related by taxonomic sub-ordination to the plant denoted by the complex formation, as the second element is *wyrt* 'wort, plant'; the first element, in contrast, refers to the aromatic feature of the plant, namely, to its sweet scent (see Catalogue, s.vv. *ofersǣwisc rind, sūþerne rind, swēte wyrt*).

Onomasiological perspective: starting from the motives

Two different plant names denote the prickly cnicus (Cnicus ferox L.), but the motive and the associative relations are identical: OE *unbrāda þistel* lit. 'not broad thistle' and *smælþistel* lit. 'narrow thistle' both conceptualise the narrow leaves of the plant in the first element, but they use different linguistic material: in the first example, the adjective *smæl* 'narrow', in the second, the adjective *unbrād* 'not broad'; *unbrād* emphasises the difference to the species of thistle with *brād* 'broad' leaves slightly better than *smælþistel*, as the contrast is expressed overtly in the adjective by the prefix *un-* (see Catalogue, s.vv. *smælþistel*, *unbrāda þistel*).

The same motives relating to the same plant species are found in the plant names *croplēac* and *hēafdeht por(r)*, both denoting the 'common leek', but, in contrast to the pair of compounds denoting the prickly cnicus discussed above, this pair of compounds uses morphologically different lexemes: *croplēac* lit. 'umbel-leek' or 'umbel-plant' uses the noun *crop* 'umbel, crop', while *hēafdeht por(r)* lit. 'leek having a head' makes use of the adjective *hēafdeht* to verbalise that the common leek has flashy, round and big umbels (see Catalogue, s.vv. *smælþistel*, *hēafdeht por(r)*).

Just as concepts that are of no interest for a speech community never ever get a name, motives that are of no interest for the speech community do not acquire a name, either, while others are so omnipresent that they are encoded in a variety of ways. In the next example, not only the denotations, but also the motives and the way to verbalise these motives are identical. The water-cress (Nasturtium officinale L.) can be named by four complex and one simple term in Old English: by *cærse*, *brōccærse* lit. 'brook-cress', *ēacærse* lit. 'water-cress', *fencærse* lit. 'fen-cress'and *wyllecærse* lit. 'well-cress'. The morphological analysis of the complex plant names reveals identical structures (all are endocentric, rectional compounds), and the semantic analysis reveals identical structures, too: all four complex plant names encode the motive 'habitat' – the plant needs moist environments – in the first element (related by conceptual contiguity to the concept of the plant), and 'botanical information' (related by taxonomic sub-ordination to the concept of the plant) in the second element; however, these plant names are not totally synonymous from the point of view of signification, as the elements *brōc* 'brook', *ēa* 'river, large body of water', *fen* 'fen' and *wylle* 'well' are not synonyms. All the same, the semantic structure of these plant names tells us that the by far most salient feature of water-cress is its growing in the vicinity of water (see Catalogue, s.vv. *cærse*, *brōccærse*, *ēacærse*, *fencærse*, *wyllecærse*).

The plant marigold (Calendula officinalis L.) is represented with four plant names in the corpus: as *goldwyrt* lit. 'gold-wort', as *sigelhweorfa* lit. 'sun-turn(ing plant)', as *sōlsēce* lit. 'sun-seek(ing plant)' and as *sunfolgend* lit. 'sun-following (plant)', and, possibly, also as *goldblōma* lit. 'gold-blossom'. In these five names, basically two motives are verbalised: in the first elements of *goldblōma* and *goldwyrt*, the bright, yellow colour of the flowers are highlighted in a metaphorical way, while in *sigelhweorfa*, in *sōlsēce* and in *sunfolgend*, it is the plant's habit of facing the sun: this plant turns its blossoms following the movement of the sun. This example shows that the same plant concept can produce different associations, depending on the motives that are experienced as being most salient (see Catalogue, s.vv. *goldblōma, goldwyrt, sigelhweorfa, sōlsēce, sunfolgend*).

The pomegranate (Punica granatum L.) is represented with six plant names in the corpus, of which two refer to the fact that the fruits of the plant are full of seeds – *cornæppel* lit. 'pip-apple, pip-fruit' and *gecyrnlod æppel* lit. 'apple / fruit full of seeds' –, one refers to the colour of the fruit – *rēad æppel* lit. 'red apple / fruit'–, one to the plant's origin – *affricanisc æppel* lit. 'African apple / fruit' –, and one plant name to the 'heavenly' virtues of the plant, as a healing plant, as a plant that symbolises fertility and as a plant of importance in Greek mythology, in the Koran and in the Bible likewise: *swegles æppel* lit. 'heaven's apple' (see Catalogue, s.vv. *cornæppel, gecyrnlod æppel, rēad æppel, affricanisc æppel, swegles æppel*).

These examples, and many others (see, e.g., the Old English plant names denoting Solanum nigrum L., or Arctium lappa L.; *Index of botanical names*, s.vv.~) illustrate, once again, that folk taxonomies are not as clear-cut as scientific ones and that different speech communities can verbalise the plant concepts and their salient attributes in many different ways. There are a number of reasons for this variety of names. In the first place, every plant has different features that can all be encoded in the plant name. Then, regional and dialectal differences, or diachronic variation can account for the multiple verbalisation of reality: as mentioned earlier, the period of Old English covers c. six centuries and several generations of speakers. Some of the plant names that are recorded for example in one of the first glossaries, the *Épinal-Erfurt Glossary*, were probably not used towards the end of the Old English period and other plant names were used instead. While assigning the complex plant names to a dialectal area in Anglo-Saxon England or to a point in time on a diachronic axis relating to Old English is a challenging task in view of the textual situation, a major process of displacement of Old English plant names can be traced at the end of the period, when, in the wake of the Norman Conquest, Norman (French) plant names replaced the Old English ones. For instance, OE *bisceopwyrt* 'mallow', *cederbēam* 'cedar of Lebanon', *æscwyrt* 'vervain, etc.' and *candelwyrt* 'great mullein' were replaced by the French loans *betony* (< F *bétoine*), *cedar* (< OF *cedre*), *vervain* (< OF *verveine*) and (*great*) *mullein* (< OF *moleine*), but some of the Old English plant names have survived into Modern English, such as *daisy* < OE *dæges ēage* lit. 'day's eye', *garlic* < OE *gārlēac* lit. 'spear-leek', *groundsel* < OE *grundeswelige* lit. 'ground-devour(ing plant)' and *waybread* < OE *wegbrǣde* lit. 'way-broad plant'.[224]

224 The botanical vocabulary of Middle English was assembled and commented upon by Hunt (1989). An account of the linguistic structure of Middle English plant names is provided by Sauer (1995).

Finally, the importance of the plants for the community of speakers must be considered, as well. Plants with a high economic or medical value are probably named first or even get more than one name, while those without any value for the people are most probably ignored.

6.6. FOLK TAXONOMY, TAXONOMIC CENTRALITY AND SEMANTIC CHANGE

As mentioned earlier, research in folk taxonomy revealed that there are, basically, five taxonomic levels of class inclusion in folk taxonomies: the level of unique beginners (0), the level of life forms (1), the generic level (3), the specific level (4), and the varietal level (5) (see chapter 2.1. and 6.4. (1)). The studies into the botanical lexicon of the Tzeltal, an indigenous people in Mexico have revealed that of these five levels, the generic level displays a numerical superiority which is significant in comparison to the number of terms located on the other levels, and which is supported by the linguistic and cultural significance of the plants denoted by the generic terms (see Berlin Breedlove and Raven 1973; 1974).[225] Categories on the generic level, termed *basic level categories* in cognitive linguistics (Ungerer and Schmid 2006: 70), are characterised by the fact that they provide a large amount of relevant information on the class of objects accumulated in the prototype, and by a reliable *gestalt* perception (Ungerer and Schmid 2006: 75), which explains why these are the categories that come to mind first in naming situations. In most cases, names for generic categories are morphologically simple (Ungerer and Schmid 2006: 69). The taxonomic centrality of generic terms can result in what Ungerer and Schmid (2006: 77) have termed *parasitic categorisation*, which they described as follows:

> You would probably draw an orange, a banana, etc. to illustrate FRUIT, or a chair, a table and a bed for FURNITURE […]. In other words, you would 'borrow' the gestalt properties of the superordinate category from the basic level categories involved.

As far as the Old English plant name corpus is concerned, parasitic categorisation, as a result of the taxonomic centrality of terms on the generic level, can account for the meaning extension of a number of terms, of which the four most frequently used – *æppel* 'apple', *brēr* 'briar, etc.' and *brēmel* 'bramble, etc.', and *lēac* 'leek, etc.', are discussed, exemplarily, in the following. Namely, the meaning extension of the generic terms *æppel* 'apple', *lēac* 'leek', *brēr* 'briar' and *brēmel* 'bramble' to *æppel* 'fruit', *lēac* 'garden-plant, herb, plant', *brēr* 'thorn-bush' and *brēmel* 'thorn-bush', is a consequence of "the salience of the generic level [which] seems above all to be due to its cultural significance; sometimes it is even rooted in basic biological needs" (Ungerer and Schmid 2006: 69), and rests on parasitic categorisation.

225 A brief summary can be found in Krischke (2011: 85-9).

The generic term for 'apple' in Old English is *æppel* (see Catalogue, s.v. *æppel*). It occurs as a modifier in the compounds *æppellēaf* 'violet',[226] *æppeltrēow* 'apple-tree', and *æppelþorn* 'crab-tree' and as the head in *affricanisc æppel* 'pomegranate', *cirisæppel* 'Carian fig', *coddæppel* 'quince', *cornæppel* 'pomegranate', *eorþæppel* 'mandrake, etc.', *fingeræppel* 'date', *gecyrnlod æppel* 'pomegranate', *gōdæppel* 'quince', *rēad æppel* 'pomegranate', *swegles æppel* 'pomegranate' and *wuduæppel* 'crab apple'. It is not a surprise that *æppel* in *æppeltrēow* 'apple-tree', and *æppelþorn* 'crab-tree' denotes the fruit 'apple' (albeit not necessarily a particularly tasty species). We can also assume that *æppel* in *affricanisc æppel* 'pomegranate', *coddæppel* 'quince', *cornæppel* 'pomegranate', *gecyrnlod æppel* 'pomegranate', *gōdæppel* 'quince', *rēad æppel* 'pomegranate', *swegles æppel* 'pomegranate', and *wuduæppel* 'crab apple' denotes the 'apple', too, which is, in this interpretation, referred to in a figurative way in the plant names denoting the quince and the pomegranate (the pomegranate and the quince, it can be argued, can be compared to apples due to their shape, taste and colours), but this becomes more difficult with the plant names *cirisæppel* 'fig' and *fingeræppel* 'date', and almost impossible with *eorþæppel*, which denotes three different vegetables and no fruit at all: the mandrake, the cucumber and the sowbread. It is, therefore, much more likely that the *æppel* was, at some point, considered the prototypical fruit (based on parasitic categorisation), and that *æppel* had previously been subjected to the meaning extension from 'apple' to 'fruit'.

The cultural significance of species of Alliums is not only expressed in the high number of terms denoting Alliums (17 terms, see chapter 2.1.2.), but also in the apparent taxonomic centrality of *lēac*, whose meaning was extended to denote all kinds of garden-herbs and plants.

> Leeks and onions were an important part of the Anglo-Saxon diet: they not only provided flavour and have a relatively high vitamin C content, but could be kept for quite a long period. They were obviously ubiquitous and even feature in literary texts (Lee 2011: 151).

Besides non-botanical compounds such as *lēactūn* 'herb-garden', *lēactūnweard* and *lēacweard*, both 'gardener' – where it seems rather unlikely to assume that they denoted enclosures in which exclusively leeks were grown, or that the gardeners took care of the leeks only –, a number of plant names point to the denotation 'plant, herb' for OE *lēac*. Old English plant names featuring the element *lēac* are *lēaccærse* 'garlic mustard, etc.', *brādelēac* 'ramsons', *cīpelēac* 'onion, etc.', *crāwanlēac* 'crow garlic', *crāw(e)lēac* 'crow garlic', *croplēac* 'common leek', *fugleslēac* 'tufted vetch', *gārlēac* 'garlic', *hollēac* 'shallot', *hræfnes lēac* 'crowleek, etc.', *hwītlēac* 'leek', *por(r)lēac* 'common leek', *risclēac* 'rush garlic', *secglēac* 'rush garlic', *toscanlēac* 'horsetail', *wudulēac* 'ramsons', and *ynnelēac* 'onion'.

It is reasonable to assume the meaning 'leek' for the head *lēac* in the plant names that denote different Alliums – namely, in *brādlēac* 'ramsons', *crāwanlēac* 'crow

226 As the denotation of this plant name is not known, the plant name must be ignored in the following discussion.

garlic', *crāw(e)lēac* 'crow garlic', *croplēac* 'common leek', *gārlēac* 'garlic', *hollēac* 'shallot', *hræfnes lēac* 'crowleek, etc.', *hwītlēac* 'leek', *risclēac* 'rush garlic', *secglēac* 'rush garlic', *wudulēac* 'ramsons'. The case with *lēaccærse* 'garlic mustard, etc.', where *lēac* is the modifier, is quite easily explainable, as garlic mustard is a plant with a leek-like smell – a motive that makes more sense when we assume the meaning 'leek' instead of 'plant' for the modifier *lēac*.

The case with *fuglesl ēac* 'tufted vetch' and *toscanlēac* 'horsetail', on the one hand, and with *cīpelēac* 'onion, etc.', *por(r)lēac* 'common leek' and *ynnelēac* 'onion', on the other, is different, as here, the denotation 'leek' is excluded in the first two plant names and not very reasonable to assume for the last three plant names. Neither the plant tufted vetch nor the plant horsetail can be brought into a conceptual relationship with leeks, and it is much more likely that in these plant names the head *lēac* denotes 'plant' in general, as it does in its function as a modifier in the compounds *lēactūn* and *lēactūnweard* mentioned earlier. *Cīpelēac* 'onion, etc.', *por(r)lēac* 'common leek', and *ynnelēac* 'onion' are of a more special nature, as they are exact synonyms of their first elements, *cīpe* 'onion, etc.', *por(r)* 'common leek', and *ynne-* 'onion' (of which *cīpe* and *por(r)* are also used independently), but also because their first elements are loan words from Latin. These three plant names are, thus, in line with other (pleonastic) plant names (cf. chapters 5.3.2.2.1.1. (b) and 6.4. (1)), where the first element is a loan from Latin, whose meaning is identical with the meaning of the complex formation. It seems, therefore, more reasonable to assume that the second element *lēac* is not identical in denotation with the first elements *cīpe*, *por(r)*, and *ynne-*, respectively, and thus, also not identical with the denotation of the complex formations (which would produce tautological compounds), but that *lēac*, here, serves as an explanatory element, to indicate that the loans *cīpe*, *por(r)*, and *ynne-*, which were, probably, perceived as foreign within the system of Old English, denoted kinds of *lēac*, i.e. garden-plants. The meaning extension of *lēac* from 'leek' to 'plant' is based on a relation of taxonomic super-ordination between the original and the new meaning.

The simplex *brēmel* forms the compounds *brēmelbrēr* 'bramble', *brēmelþorn* 'bramble', *brēmelþȳfel* 'bramble', *brēmelþyrne* 'bramble', where *brēmel* is the modifier and *hēopbrēmel* 'dog rose', *heorotbrēmel* 'bramble', where *brēmel* is the head of the complex formations. The simplex *brēr* is a modifier in *brērþyrne* (i) 'bramble', (ii) 'thorn-bush' and head of the formations *brēmelbrēr* 'bramble', *heorotbrēr* 'bramble', and *hindbrēr* 'raspberry'. As simplexes, both *brēmel* and *brēr* can be used to denote the plants bramble (Rubus fruticosus L.) and dog rose (Rosa canina L.). This specific meaning can be derived from the compounds *brēmelþorn* lit. 'bramble thorn-bush', *brēmelþȳfel* 'bramble-thicket', *brēmelþyrne* 'bramble-bush', and *brērþyrne* 'bramble', where the second element denotes a life form and the first element, the modifier serves as *differentia specifica*, i.e. as a specifying element, which indicates that, e.g., *brēmelþorn* is a thorny bush that is a bramble or a dog rose (subsumptive, pleonastic compounds; cf. chapter 5.3.2.2.1.1. (b)). In contrast, in the plant names *hēopbrēmel* lit. 'hip bramble' or 'hip thorn-bush', *heorotbrēmel* 'hart's bramble' or 'hart's thorn-bush', *heorotbrēr* 'hart's bramble' or 'hart's thorn-bush' it can denote both the life form 'thorn-bush' or the generic taxon 'bramble', while in *brēmelbrēr* and *hindbrēr* the interpretation of *brēr* as 'bramble' or 'dog rose' is unlikely, as *brēmelbrēr* would, then, be tautologic, and as there would be a logical contradiction in *hindbrēr*: the raspberry cannot be a bramble and a dog rose at the same time. The meaning extension of *brēmel* and *brēr* from the specific to the generic meaning 'thorn-bush' (a case of taxonomic super-ordination) was probably caused by the taxonomic centrality of the terms, which is also reflected in the number of occurrences of these terms: while in the *DOEPN* (s.v. *brēmel*), which draws, basically, on the medico-botanical texts and the glosses and glossaries recording botanical terminology, only 27 occurrences of *brēmel* are recorded, the *DOE* (s.v. *brēmel*) notes 40 occurrences. The additional records are from the charters and from religious writings, and demonstrate the cultural significance of brambles (or dog roses) as the prototypical boundary markers. *Brēr* is clearly outnumbered by *brēmel*, as it is noted to be extant only 7 times in the Old English texts (*DOE*, s.v. *brēr*, *brǣr*; *DOEPN*, s.v. *brēr*).

It was claimed above that taxa on the generic level are taxonomically central and numerically superior to those on all other levels in folk taxonomies. These claims, developed by Berlin, Breedlove and Raven (1973; 1974) on the basis of taxonomic structures in the language of the Tzeltal were confirmed by data from Ælfric's *Glossary*, where 9 life form taxa, 10 taxa on the specific and 1 taxon on the varietal level are found to face 77 taxa on the generic level (Krischke 2011: 92). The analysis further revealed that the terms denoting taxa on the generic level can be morphologically simple or morphologically complex, with the complex plant names being either endocentric compounds with life form terms in the function of the head of the compounds (e.g. *hylwyrt* 'pennyroyal' lit. 'hill-wort'), or exocentric (metaphorical and metonymic) compounds (e.g. *cicena mete* 'chicken-weed' lit. 'chicken's food'). In contrast, complex plant names with the heads denoting specific or varietal taxa are plant names of the generic or of the specific level (Krischke 2011: 89-93), as, for instance, *sūþerne popig* 'opium poppy' and *wilde popig* 'common red poppy', which are taxonomically sub-ordinated to *popig* 'poppy'. When it is assumed that these findings, drawn from a rather small amount of data – the 97 botanical Old English terms in Ælfric's *Glossary* – echo the Old English botanical lexicon as a whole, the analysis of the complex plant names is expected to produce the conclusion that the sum of the c. 420 simplexes not included in the present survey, together with the endocentric complex plant names, where the second element is a life form term and with the exocentric formations, clearly outnumbers the endocentric plant names where the second element is a plant name itself – i.e., specific and varietal taxa.

Indeed, the numbers acknowledge this assumption. The life form terms used as heads of endocentric plant names are (in alphabetical order): *bēam* 'tree' with 19 complex formations, polysemous *brēmel* (i) 'bramble, etc.', (ii) 'thorn-bush' with 2 complex formations, *brēr* (i) 'bramble, etc.', (ii) 'thorn-bush' with 3 complex formations, *cærse* (i) 'cress', (ii) 'plant' with one formation and *lēac* (i) 'leek', (ii) 'garden herb, plant' with 5 formations; then, *gærs* 'grass' with one formation, *plante* 'plant' with one formation, *strēaw* 'straw' and *swamm* 'mushroom' with two formations each, *trēow* 'tree' with 30 formations, *þistel* 'thistle' with 7 formations, *þorn* 'thorn-bush' with 7 formations, polysemous *þung* (i) 'poisonous plant' (ii) 'monkshood' with 4 complex formations, *þȳfel* 'thicket' and *þyrne* 'bush, thorn-bush' with 2 formations each, *wyrt* 'wort, plant' with 72 complex formations, and *wyrtel* 'wort, plant' with one formation. In total, this makes c. 158 complex plant names denoting generic taxa. Thus, the sum of c. 660 generic plant names, i.e. c. 73 percent of the plant name corpus (c. 410 simplexes,[227] 158 generic endocentric and 90 generic exocentric terms) as opposed to c. 230 terms denoting specific and varietal taxa (i.e. c. 27 percent; e.g., *wilde popig* 'common wild poppy' and *micel worþignet(e)l(e)* 'great nettle' (albeit not from a scientific botanical point of view)) echoes the numbers from Tzeltal folk taxonomic structures, where c. 65 percent of the plant names were assigned to the generic level and c. 35 percent to the specific and varietal levels (Krischke 2011: 88) and from the numbers in Ælfric's *Glossary*, with c. 79 percent of the terms denoting generic taxa (cf. Krischke 2011: 92).

This numerical proportion is also the explanation of the significantly high number of plant names formed with the elements *bēam* (19 complex plant names), *trēow* (30 complex plant names) and *wyrt* (72 complex plant names) as these terms, in their function as super-ordinate life forms, are perfect to form plant names denoting generic taxa – taxonomically central, transparent and easy to retain complex formations.

6.7. SUMMARY

In taking the insight that paraphrasing the Old English complex plant names on the language-immanent level of knowledge is a tool not sufficiently elaborate enough to describe the semantics of all plant names as a starting point, the present chapter introduced a methodological approach to the semantics of simple and complex lexemes that was developed by Blank (1997) in his work on the processes and mechanisms involved in semantic change. This approach is based on the assumption that the semantics of any lexical item can be described best when cognitive and structural ways of analysis are combined, and presents a comprehensive semiotic model of the linguistic sign, where extra-linguistic, abstract knowledge (describable via prototype and frame theories) is classed along with language-dependent, abstract knowledge (e.g. the hierarchic structures of Anglo-Saxon folk taxonomy) and with language-dependent, concrete knowledge (the phonetic structure) on the lexical item (in addition to the concrete referent, which proved, however, to be irrelevant for the

227 The number results from the following calculations: c. 900 Old English plant names – the number is not given as absolute, as the corpus compiled by Bierbaumer includes not only terms denoting taxa, but also the names of parts of plants (simple and complex) – minus 467 complex plant names (see 1.3.), minus c. 20 life form terms.

These numbers disregard ambiguities concerning the (possible) denotations of some of the plant names, as well as ambiguities concerning the semantic and the morphological analysis of plant names not yet identified, or of plant names resulting from folk etymology, but it does take into account taxonomic differences between complex formations based on polysemous heads (parasitic categorisation and vertical polysemy). *Hollēac* 'shallot' and *por(r)lēac* 'common leek', for instance, are situated on different taxonomic levels, as *hollēac* lit. 'hollow leek' is sub-ordinated to the generic term *lēac* 'leek', while *por(r)lēac* lit. 'leek-plant' is sub-ordinated to the life form term *lēac* 'garden herb, plant'; thus, *lēac* is vertically polysemous, as the lexical unit *lēac* 'leek' is sub-ordinated to the lexical unit *lēac* 'garden-herb, plant'.

semantic description of the Old English plant name corpus). The lexical, semantic and phonological information located in one of the four parts of the quadripartite semiotic model is related via a set of ten associative relations, which go back, ultimately to the three processes Aristotle assumed to be at work in the process of remembering, namely, similarity, contiguity and contrast.

The semantic analysis of the complex Old English plant names, thus, starts from the plant name and asks first for the denotata of the elements of the plant names. These denotata, in turn, echo certain motives pertaining to the plant concepts in the extra-linguistic world, so that a relationship can be established between the concept denoted by the plant name – the plant – and the concepts denoted by the elements of the complex plant names. The motives are varied and can, theoretically, be used in both the first and the second element, but clear preferences, such as 'botanical information' being encoded predominantly in the second elements, or 'colour' being encoded predominantly in the first element, can be discerned. Among the most often used motives are, besides 'botanical information' and 'colour', also 'habitat', 'part of plant', 'shape' and 'size'.

The motives are alluded to differently. They can be used directly, i.e. related by conceptual contiguity to the plant concept. This is the case, for instance, with plant names verbalising their preferred habitats, as in *fencærse* lit. 'fen-cress', which refers to the moist ground water-cress preferably grows in, or with plant names verbalising the colour like the plant name *blæc īfig* 'ivy' lit. 'black ivy'. A different relationship exists between the motive 'colour' verbalised in the first element of *goldwyrt* 'marigold' lit. 'gold-plant' and the element *gold* 'gold': here, the colour is alluded to metaphorically, the relationship being termed 'metaphorical similarity'. Of the ten associative relations at work in semantic change, six were observed in the corpus of complex plant names.

The analysis of the semantics of the Old English complex plant names has proven to be useful to describe similarities between *prima facie* disparate compounds and discrepancies between *prima facie* similar compounds, as the example of *smælþistel* 'prickly cnicus' and *dweorged(w)os(t)(l)e* 'pennyroyal' illustrated, and to approach Old English plant names not only semasiologically, but also onomasiologically, starting either with the concepts denoted by the complex plant names asking for the different ways these concepts were verbalised, or with the concepts denoted by the elements of the plant names, asking for the motives encoded in these elements and their relation to the entire plant concept.

Further, the detailed analysis of a number of compounds using the same elements – *æppel*, *brēmel*, *brēr* and *lēac* – demonstrated how the information gained from the analysis of the motives and the associative relations can be used to determine semantic change, or to determine elements of special taxonomic centrality: For some of the compounds in question, the elements could be shown to have developed a secondary meaning via taxonomic super-ordination, resulting in vertical polysemy.

7. DIACHRONIC STRATIFICATION & THE EFFECTS OF LANGUAGE CONTACT

For much of its history, the English language was in a state of forced or deliberate co-existence with other languages – languages that have changed and shaped English to a considerable degree. It is a well-known fact that Modern English owes more than 50 percent of its vocabulary to the influence of foreign languages (Scheler 1977: 10-1; Townend 2008: 73), of which French and Latin are certainly the most important.[228] Old English was, however more fond of creating new lexemes by word-formation out of native resources than of borrowing words from other languages. Yet, as Thomason (2001: 31) insists, "the idea that monolingualism is the human norm is a myth", and this is certainly also true for Anglo-Saxon England.

Language contact occurred both before and after the Germanic invasion of Britain. Before, the Germanic tribes met Romans and Celts on the continent through trade and common military service (in the Roman army), and afterwards, at the beginning of the Old English period, their descendants, the Anglo-Saxons, had contact with the Celtic people they had conquered.[229] Beginning with the 8th century, contact with North Germanic tribes was established. Although they had initially come to plunder England, they settled from the 9th century onwards, mainly in the region known as the Danelaw, which was north and east of the border to Wessex, an imaginary line running roughly from Chester to London.[230] Regular cultural contact persisted further, throughout the period, between Latin, the medieval *lingua franca* of learning and education, and the vernacular Old English.[231]

The regular, though rather infrequent trade contact of the Anglo-Saxons with peoples from the Near and Far East and Africa left few traces in the Old English lexicon, except for some names for objects of trade or culture (Serjeantson 1935: 6, 313ff.). As is the case with all these loans, the plant names never entered directly, but took the route via Greek and Latin, as the following examples illustrate: OE *pipor* 'pepper' (> ModE *pepper*) < L *piper* < Gk *πέπερι* < OInd *pippali* 'pepper corn' –

228 The figures differ according to what part of the vocabulary is investigated and according to whether one counts types (in the dictionaries) or tokens (in texts); cf. Scheler (1977: 9-12).

229 The intensity of the contact between the Celts and the Anglo-Saxons is a matter of much discussion; cf. 7.3.

230 Scandinavian settlement in Anglo-Saxon England can be observed most obviously in the place names ending in *-by* 'settlement', *-þorp* 'hamlet' (among other topographical terms, which were, however, not so frequent (Clark 1992: 482-5; cf. also map 10 in Smith 1956, and the *VEPN*).

231 A comprehensive overview on the influence of Latin on English is provided, for instance, by Scheler (1977), Kastovsky (1992) or Leisi and Mair (1999). Valuable new search methods are offered by the third edition of the *OED*, where the data is searchable according to timelines, by language of origin, etc.

ultimately from Old Indic (see Catalogue, s.v. *pipor*); OE *cost* (ModE *costmary*) < L *costus, costum* < Gk *κόστος* < OInd *kústa(h)*, i.e., ultimately from Old Indic (see Catalogue, s.v. *cost*); OE *nǣp* 'turnip' (ModE *-nip* in *turnip* and *parsnip*) < L *nāpus* < Gk *σίναπι*, itself a loan from Egyptian (cf. Catalogue, s.v. *nǣp*; cf. Serjeantson 1935: 246); OE *cymen* < L *cumīnum* < Gk *κύμῑνον*, a loan from Semitic (Feulner 2000: 181-3; see also Catalogue, s.v. *cymen*).

The immense influx of direct and indirect loans[232] English experienced in the later Middle Ages and in the Modern Era does not pertain to Old English. Yet, while the ratio of loan words in Old English is, with less than 5 percent (according to Townend 2008: 73)[233], very much lower than that in Modern English, the fact that 36 percent[234] of the Old English complex plant names are affected by Latin loan influence (of which c. 19.5 percent are loan words attested in hybrid plant names) is outstanding. Medico-botanical discourse, thus, betrays a highly intensive language contact with Latin and the circumstances and the outcome of this contact in terms of numbers and types of loan influence merits a close inspection. The linguistic traces left by speakers of Old Norse and Brittonic (or Insular Celtic)[235], on the other hand, are so minimal that their 'non-occurrence' in the Old English botanical lexicon stands in a surprising contrast to the rather intimate language contact situation and the number of contact-induced changes in Old English. The focus of chapter 7.3. is, therefore, not only on discussing the question of how Latin left its traces in the botanical lexicon, but also of how the conspicuous discrepancy between its influence on the botanical lexicon and Old English in general, and also, between its impact on the Old English botanical lexicon and that of Old Norse and Celtic, can be explained.

Before the loan influence is examined in detail, however, the diachronic stratification of the Old English complex plant names will be investigated. Namely, the plant names going back to language contact are embedded in the same diachronic matrix as those of native Germanic origin. Both native and non-native plant names echo different time periods in the development of the language from Indo-European through Germanic and West Germanic into Old English. The Indo-European layer is, on the one hand, all-pervasive as most elements of the complex plant names, including the Greek and Latin loan words, are of Indo-European origin. On the other hand, as regards the complex plant names in Old English, no complex plant name can be traced back to its Indo-European roots completely (in contrast to their bases, which can), while the complex plant names that can be tracked to Germanic and West Germanic

232 For the distinction between direct and indirect loans, see 7.5.

233 Townend's number apparently does not include indirect loan influence, but it is higher than could be expected from the absolute numbers of loan words as provided, e.g. by Serjeantson (1935) and by Scheler (1977). According to Scheler (1977: 35), Old English borrowed c. 600 words from Latin; taking into account his estimated number of Old English words attested (c. 23,000-24,000; cf. Scheler 1977: 14), the ratio of loan words is c. 2.5 percent. If the estimated number of words attested from Old English followed is that of Healey (1997: 47, fn. 4), namely c. 35,000 words, the ratio drops to c. 1.7 percent.

234 This number includes 91 hybrid formations, 55 loan translations, 23 loan renditions and 1 loan creation. See, further, 7.4.2.1., 7.4.3. below.

235 I.e., the insular variety of the Celtic family of languages spoken in Britain; cf. Crystal (1995: 302-3; White 2003: 29).

origins are all the more numerous. In the following section, the diachronic stratification of the Old English complex plant names will be described, examples will be given and interesting results and correlations pointed out. The section on the origins and diachronic developments of the Old English complex plant names is concluded with an account of one of the most interesting diachronic processes, namely, the processes that are generally subsumed under the cover term 'folk etymology' (chapter 7.5.). As folk etymological processes affect native and borrowed terms in equal measure and transcend diachronic stratification, they are not subsumed to the two major foci of interest in chapter 7, diachronic stratification and loan influence, but are treated separately (in chapter 7.5.).

7.1. DIACHRONIC STRATA

It is, as Carr (1939: xxx) points out,[236] not easy to distinguish between formations that were formed as original compounds in Germanic or West Germanic and those that were formed independently in the Germanic and West Germanic languages and are, thus, coincidental parallel formations. While the (relative) age of particular formations can often be reconstructed through the comparison of form and meaning, through evidence of sound changes[237] or through syntactic developments,[238] the question of whether a formation is an original Germanic or West Germanic compound, i.e. formed before the peoples dispersed, or a coincidental parallel formation in more than one language can, in many cases, not be decided with certainty. Carr's (1939) classification of his corpus of compounds in the Germanic languages into 'Germanic compounds', 'parallel and independent formations in Germanic', 'West Germanic compounds' and 'parallel and independent formations in the West Germanic languages' is followed in the present work; it is based on a number of phonological, semantic and syntactic criteria of which some are more apt to identify the younger formations (see Carr 1939: xxx-xxxi), while others are more apt to fix the (relative) date of coinage of compounds found in more than one of the Germanic languages (see Carr 1939: xxxii-iv).

236 Though Carr's (1939) investigation into the *Nominal compounds in Germanic* is three quarters of a century old, it is still the most recent work dealing with the diachronic stratification of the Germanic compounds and will be, therefore, the major reference work in the present survey.

237 Verner's Law, for instance, affected all Germanic compounds, while the doubling of consonants triggered by *j* in the following syllable provides evidence of West Germanic origin; cf. Carr (1939: xxxii-iii); Campbell (1959; §§ 398 (2), (4), 407).

238 For instance, the coalescence of syntactic phrases into compounds is, according to Carr (1939: xxx), an indicator that the compound cannot be of old origin.

7.1.1. THE GERMANIC STRATUM
(c. 12 plant names)

Old English plant names that go back to an originally Germanic stratum, that is, formations that arose, according to Carr (1939: 40-68), in Proto-Germanic and developed individually in the Germanic languages are the following: *eorþber(i)(g)e* 'strawberry', *eorþhnutu* 'sowbread, etc.', *haguþorn* 'common hawthorn', *liferwyrt* 'liverwort', *meduwyrt* 'meadow-sweet, etc.', *misteltān* 'mistletoe', *wiþowinde* 'small bindweed, etc.', and *wudubinde* 'woodbine, honeysuckle'. Not recorded by Carr (1939) is OE *hwītlēac* 'leek', which is extant also in Old Norse as *hvítr laukr* (*Heizmann*, 27); both the Old English and the Old Norse terms are probably loan formations[239] formed after L *alba cēpa* lit. 'white leek' (see Catalogue, s.v. *hwītlēac*).

According to Carr (1939: 69-95), three plant names were formed independently in the variety of Germanic that was to become Old English: *fīctrēow* 'fig-tree', *goldblōma* '? marigold' and *wīntrēow* 'vine' are classified as 'parallel and independent formations in the Germanic languages'. The plant names *fīctrēow* and *wīntrēow* are hybrid formations (see 7.4.2.1.) with their first elements *fīc* 'fig, fig-tree' and *wīn* 'vine, wine' borrowed from Latin (L *fīcus* and *vīnea* 'vine' or *vīnum* 'wine', respectively).

7.1.2. THE WEST GERMANIC STRATUM
(c. 79 plant names)

The plant names that were created in Proto-West Germanic are classified as belonging to the West Germanic stratum of the Old English botanical lexicon. Their status is ascertained on the basis of several criteria, such as attestations in more than one West Germanic language, phonetic changes (especially Verner's Law and consonant gemination before *j*) and stem variation between simplex and compound element (see Carr 1939: xxxii-xxxiv).

The criterion 'parallel formations' in other West Germanic languages is somewhat problematic, especially when the only other parallel attestations come from Old High German. There are close similarities between the Anglo-Saxon and the Old High German glossaries, which may be due to the fact that copies of an assumed 'Canterbury glossary', a seventh-century, Anglo-Saxon glossary based on Isidore (see chapter 4.4.), were taken to north and central Europe by the Anglo-Saxon missionaries when they left England; these copies were probably consulted when the Old High German glossaries[240] were compiled (Porter 2011b: 167-8). The similarities are highly conspicuous especially between the *Antwerp-London Glossaries* and the Old High German glossaries, which seems to indicate that the German glossators were probably strongly influenced by the Old English interpretations of the Latin texts they were adapting, due to the fact that the Old English and the Old High German lexemes were

239 For a definition of loan formations, see chapter 7.4.3.
240 The Old High German glossaries were edited by Steinmeyer and Sievers (1879-1922).

phonetically very similar (Porter 2011b: 168). This being the case, one of the decisive criteria for classifying Old English plant names as, originally, West Germanic is, at least, weakened. To understand fully the exact relationship between the Old English and Old High German texts and the glosses, however, a detailed analysis has to be carried out.

According to Carr (1939: 96-127), the following plant names are of West Germanic origin: *bānwyrt* 'comfrey, etc.', *dȳþhamor* 'reed-mace', *eorþæppel* 'cucumber, etc.', *gelodwyrt* 'silverweed', *goldwyrt* 'marigold', *hǣþber(i)(g)e* 'whortleberry', *hind-ber(i)(g)e* 'raspberry', *hollēac* 'shallot', *īsenhearde* 'knapweed', *meledēaw* 'mildew', *nihtscada* 'garden nightshade', *openærs* 'medlar', *por(r)lēac* 'common leek', *slāhþorn* 'blackthorn', *smeoruwyrt* 'smearwort, etc.', *springwyrt* 'caper-spurge', *þēfanþorn* 'thevethorn', *þūþistel* 'sow-thistle', *wæterwyrt* 'water-hemp, etc.', *wealhhnutu* 'walnut', *wegbrǣde* 'waybread', *wōde(h)wistle* 'cowbane', *wyrmwyrt* 'wormgrass', and *ynnelēac* 'onion'.

The plant names *bēowyrt* 'common balm, etc.', *cederbēam* 'cedar of Lebanon', *cirisbēam* 'cherry-tree', *cistbēam* 'chestnut-tree', *cistelbēam* 'chestnut-tree', *cisten-bēam* 'chestnut-tree', *eorþgealla* 'knapweed', *feldminte* 'horse-mint', *fīcbēam* 'fig-tree', *fīcwyrt* 'lesser celandine, etc.', *fīffingre* 'fiveleaf', *hāmwyrt* 'house-leek', *hnutbēam* 'walnut', *hundeshēafod* 'small snapdragon', *hundestunge* 'dog's tongue', *lāwerbēam* 'laurel', *lēonfōt* 'lady's mantle', *lungenwyrt* 'golden lungwort, etc.', *mōrbēam* 'mulberry tree', *nǣdrewinde* 'bistort', *nǣdrewyrt* 'bistort', *pīnbēam* 'pine-tree', *sunnancorn* 'common gromwell' and *wīngeard* 'vine' are plant names formed independently in the West Germanic variety Old English (Carr 1939: 128-57).

A number of Old English plant names, which were not noted by Carr (1939), can be added to this list, because related plant names in other West Germanic languages are attested: *brādþistel* (cf. Björkman 1901: 207, 1902: 283), *brūn(e)wyrt* 'brownwort' (cf. Björkman 1902: 284), *dracanblōd* (but cf. OHG and OLG *drachenbluot*, Carr 1939: 313), *hæselwyrt* 'hazelwort' (cf. Björkman 1902: 289), *wermōd* 'wormwood' (*Kluge*, s.v. *wermōd*), *wilde elebēam* (cf. Björkman 1904: 191), *wilde minte* (cf. Björkman 1904: 190), *wilde rūde* (cf. Björkman 1904: 194) and *wulfestǣsel* (Björkman 1901: 231). To these we must add the plant names *fūla bēam* 'alder buckthorn', *oxantunge* 'alkanet', *selfhǣle* 'wood march', *wilde cyrfet* 'gourd' and various kinds of 'apples': *affricanisc æppel* 'pomegranate', *gecyrnlod æppel* 'pomegranate', *rēad æppel* 'pome-granate', *sūr æppeltrēow* 'crab apple(-tree)', *sūrapuldor* 'crab apple-tree', *swēte apuldor* 'apple-tree' and *wuduæppel* 'crab apple(-tree)'. They are not listed by Carr (1939) because they are syntactic groups (with the exception of *oxantunge* and *selfhǣle*), and were therefore not the scope of his study.[241] While parallel formations in Old High German (see below) suggest a common origin of these plant names at an earlier stage than Old English and Old High German, respectively, the fact that they are syntactic groups in Old English (and also in Old High German) makes them more likely to be independent formations (cf. Carr 1939: xxx), albeit many of them as loan formations modelled after the same Latin plant names:

241 Even the statuses of *oxantunge* (N_{gen}+N) and *selfhǣle* (synthetic) are not clearly unambiguous and different linguists assign them different statuses. For criteria that help to distinguish syntactic groups from compounds, e.g. inflection of the modifying element, see chapter 5.3.1.

- Old English *fūla bēam* lit. 'foul tree' is paralleled by OHG *fuelboum* and OLG *fulbom* (Björkman 1902: 305). As stated above, since the Old English plant name is a syntactic group, a common origin of the plant names is less likely.
- OE *oxantunge* and OHG *ohsenzunga* (Björkman 1901: 209; *OED*, s.v. *ox-tongue*) are loan translations of L *lingua bovis* (or *lingua būbula*), i.e. the Latin pattern was reproduced in both West Germanic languages; both have coalesced into compounds from syntactic phrases.
- OE *selfhǣle* lit. 'self-heal(ing plant)' is a loan rendition of L *sānicula* and denotes a plant that was renowned for its healing properties (Marzell, s.v. *Sanicula europaea* L.). Like its parallel formation OHG *selbheila* (Björkman 1902: 298), the Old English plant name consists of the personal pronoun OE *self* 'self' and the deverbal element *-hǣle*, a nominalisation of the verb *hǣlan* '(to) heal' not attested independently in Old English (see Catalogue, s.v. *selfhǣle*). The Old English formation is, thus, a synthetic compound and, as such, very probably not an original formation in West Germanic.
- OE *wilde cyrfet* lit. 'wild gourd' finds it West Germanic correspondents in OHG *wildekvirbs*, *wildekurbebez*, etc. (Björkman 1904: 186); both the Old English and the Old High German formations are independent loan renditions of L *cucurbita agrestis* (Bierbaumer 3, s.v. *cyrfet*).

The plant names denoting different species of apple, OE *affricanisc æppel*, *gecyrnlod æppel*, *rēad æppel*, *sūr æppeltrēow*, *sūrapuldor*, *swēte apuldor* and *wuduæppel* render, just as their counterparts in Old High German, *affricana ephili*, *gicurnit ephele*, *rotes apholes*, *surephele*, *suozephel* and *holzephele*, the same Latin formations (Björkman 1901: 210):

- OE *affricanisc æppel* and OHG *affricana ephili*, both lit. 'African apple', stand for L *mālum pūnicum*, as do also OE *rēad æppel* and OHG *rotes apholes*, both lit. 'red apple'.
- L *mālum grānātum* is translated by OE *gecyrnlod æppel* and by OHG *gicurnit ephele*, both lit. 'apple full of seeds'.
- OHG *holzephele* lit. 'wood apple' glosses L *māla agrestia* (Björkman 1901: 210) and corresponds in form and meaning to the OE hapax legomenon *wuduæppel*, also lit. 'wood apple'.
- OE *sūr æppeltrēow* and *sūrapuldor*, both lit. 'sour apple-tree' and *swēte apuldor* lit. 'sweet apple-tree' match OHG *surephele* lit. 'sweet apple(-tree)' (glossing L *māla acida*; Björkman 1901: 210) and OHG *suozephel* lit. 'sweet apple(-tree)' (glossing *malomelum*; Björkman 1901: 210) semantically, but the morphological patterns differ slightly in that the heads of the Old English complex plant names are complex themselves: *æppeltrēow* is an endocentric compound (see Catalogue, s.v. *æppeltrēow*) and *apuldor* is a suffix formation (see Catalogue, s.v. *apuldor*), while the head of the Old High German plant names is a simplex (*ephel* 'apple, apple-tree').

Loan influence makes its presence felt in the form of loan translations and loan renditions, as some of the examples mentioned above illustrate, but also in the form of all other possible shapes that loan influence can take. While more details are provided in chapter 7.5., the following examples will give an overview. The plant names *por(r)lēac* 'common leek' and *ynnelēac* 'onion' are hybrid formations with the elements *por(r)* and *ynne-* borrowed into Proto-West Germanic from Latin (L *porrus* and L *uniō*; see Catalogue, s.vv. *por(r)lēac, ynnelēac*); the loan elements in the hybrid formations *lēonfōt* 'lady's mantle' (< L *leōnem*, acc.sg. of *leo* 'lion'), and *mōrbēam* 'mulberry tree' (< L *mōrum*) were borrowed after 650 AD (see chapter 7.4.1.2.).

Not all of the 'early Latin loans', i.e. those borrowed before c. 650 AD (see chapter 7.4.1.1.) produced 'early' formations, i.e. formations in West Germanic. The early Latin loan *minte* 'mint' (< L *ment(h)a*), for instance, is part of a number of Old English plant names: *brōcminte* 'water-mint', *feldminte* 'horse-mint', *fenminte* 'water-mint, etc.', *hāreminte* 'wild mint', *horsminte* 'horse-mint', *sǣminte* 'catmint, etc.', *swearte minte* 'crisped mint', *tūnminte* 'spear-mint' and *wilde minte* 'mint', but only *feldminte* and *wilde minte* are present in other West Germanic languages: in Old High German as *feldminza*, etc. (Björkman 1904: 189) and as *wildeminza*, etc. (Björkman 1904: 190), and in Middle Low German as *veltminte* (Carr 1939: 131). The compounds *brōcminte, fenminte, hāreminte, horsminte, sǣminte, tūnminte* and the syntactic group *swearte minte*, in turn, are Old English formations (see Catalogue, s.vv. *brōcminte, fenminte, hāreminte, horsminte, sǣminte, tūnminte, swearte minte*).

7.1.3. THE OLD ENGLISH STRATUM (c. 376 plant names)

The overwhelming majority of the Old English complex plant names were formed in Old English. Beside the fact that they do not feature the properties inherent in the Germanic and West Germanic plant names, a number of distinct features point to the fact that some of these formations were most likely formed during the Old English period and not earlier. These features are presented in the following.

Age of the loan word

An indicator for the age of a hybrid plant name is the age of its loan element: the plant name is relatively young[242] when the loan element was borrowed after the Germanic Conquest of England, that is to say, when it belongs to the stock of 'later Latin loans' (after 650 AD; see chapter 7.4.1.2.).[243] Examples are, for instance, *magdalatrēow* 'almond-tree' (*magdala-* 'almond' < L *amygdala, amygdalum*), *mōrbēam* 'mulberry tree' (*mōr-* 'mulberry' < L *mōrum*), *palmtrēow* 'palm-tree' (*palm* 'palm-tree' < L *palma*), *wuducunille* 'wild thyme' (*cunille* 'thyme' < L *cunīla*) and *wilde lactuce* 'lettuce' (*lactuce* 'lettuce' < L *lactūca*). With OE *lēo* 'lion' as a 'later Latin loan' (< L

242 The qualification 'relatively' refers to the fact that the age of the plant names is not referred to as an absolute age, but in relation to other complex plant names in Old English.

243 With pre-650 AD loans it is not only often difficult but even in many cases impossible to decide whether they were borrowed on the continent (the so-called 'continental loans') or shortly after the conquest, i.e. between c. 450 and c. 650 AD. For this reason Wollmann (1990); following Campbell (1959) abdicates this distinction made, for instance, by Serjeantson (1935), Kastovsky (1992), and Baugh and Cable (2002): Campbell (1959) and Wollmann (1990) distinguish between 'early Latin loans' and 'later Latin loans'; for further details see chapter 7.4.1.

leōnem, acc. sg. of *leo*; cf. Serjeantson 1935: 285), the compound *lēonfōt* 'lady's mantle' is an Old English formation, while OHG *lewenfuoz* must be a coincidental parallel formation. Both are based on L *pēs leōnis* or *leontopodium* (for further details, see Catalogue, s.v. *lēonfōt*) and demonstrate how the double classification of Old English plant names as Old English and (independent formation in) West Germanic can be explained (see also chapter 7.1.3. below).

Morphological make-up

Another clear index for the relative age of a lexeme is its morphological make-up. Substantive bahuvrihis such as *hundesmicge* 'dog's tongue' and *foxes glōfa* 'foxglove' (N_{gen}+N) are relatively young formations, as this type of composition did not exist as early as Proto-Germanic (Carr 1939: 161ff.). It is, however, not easy to decide whether these plant names belong to the group of plant names formed after the settlement of the Germanic tribes in England or before. The plant name *hundeshēafod* 'snapdragon', for instance, is listed as an 'independent formation in West Germanic' by Carr (1939), but this does not necessarily mean that it was formed before c. 450 AD; it can just as well have been formed in early Old English with a parallel formation in another variety of West Germanic, created on the same Latin model, L *canis caput* (itself ultimately a loan translation from Greek; see Catalogue, s.v. *hundeshēafod*; compare also *lēonfōt* 'lady's mantle', mentioned above). Where the second element still shows inflection, such as exocentric N_{gen}+N formations (e.g. *hundesmicge*, see above) or with plant names that arose as syntactic groups and coalesced into compounds (e.g., *crāwanlēac* 'crow garlic, etc.'; see Catalogue, s.vv. *crāwanlēac*, *crāw(e)lēac*), we cannot assume formation in Germanic or West Germanic (Carr 1939: xxviii, xxx). Following the same reasoning, syntactic groups that are still recognizable as syntactic groups in Old English are most likely to be Old English formations (for the criteria, see chapter 5.3.1.). Examples of this kind of formation are, e.g. *basu popig* 'common red poppy', *brād cāwel* 'cabbage turnip', *brādbisceopwyrt* 'unidentified plant', *wilde caul* 'wild cabbage' and *wilde lactuce* 'lettuce'.

Semantic aspects

Semantic aspects can also serve as indicators of the relative age of the plant names. The attribute *englisc* 'English' in a complex name, for instance, points to the fact that the plant was either seen in contrast to another plant of the same genus or species *not* native to England, or that it was generally associated with England. Plant names conceptualising the attribute 'English' are, for instance, *englisc cost* 'tansy', *englisc more* 'wild carrot' and *englisc nǣp* 'turnip'.

Plants flagged as *affricanisc* 'African', as *ēastern* 'eastern', as *ofersǣwisc* 'from overseas, adj.', as *sūþern* 'southern', as *frencisc* 'French' or, more generally, as *wealh* 'foreign' – all attributes implying an awareness of plants being non-native to England – are semantically complementary to the formations with *englisc* 'English'. Plant names of this kind are, for instance, *affricanisc æppel* 'pomegranate', *ēasterne hnutbēam* 'almond' and *frencisc hnutu* 'walnut', *ofersǣwisc rind* 'cinnamon', *sūþerne bisceopwyrt* 'unidentified plant' and *sūþerne cymen* 'cumin', or *wealhhnutu* 'walnut' and *wealhmore* 'common parsnip, etc.' (see also chapter 6.4. (10)). Technically, the perception of plants as being 'African', 'foreign', or 'southern' does not have to be a post-Conquest one and can just as well have been an insight of the Pre-Old English tribes. This is not an option for the attribute *frencisc* 'French', but for *affricanisc æppel* and for *wealhhnutu*. The conquest of central Gaul by the Franks, from whom the country received the name of France, and whence the adjective OE *frencisc* derived, started in 486 AD, that is, the subsequent settlement took place from the 6th century onwards with the Franks being bilingual until the 9th century (Klare 1998: 42-3). The first texts attesting to the awareness that the language of the people in Gaul had departed from Latin to the extent that (classical) Latin was not understood any more at the court and in the churches date from the end of the 9th century (Klare 1998: 51ff.), from which it follows that (West) Germanic tribes on the continent referring to '*frencisc*' plants would have been an anachronism. Accordingly, the *DOE* defines the adjective *frencisc* as "French, of or belonging to France and its inhabitants; 'French, of France', referring to a person of French birth or descent, without an explicit distinction drawn between French and Norman; frencisc mann 'a Frenchman'" (*DOE*, s.v. *frencisc*, 1, 1a), and states that it appears "mainly in historical texts of s.xii and later".

The case is different with *affricanisc æppel*, which is extant not only in Old English, but also in Old High German, as *affricana ephili* (cf. chapter 7.1.2.), and with OE *wealhhnutu*, which is recorded in Middle Low German as *wallnut* and in Middle Dutch as *walnote*, etc. (see Catalogue, s.v. *wealhhnutu*; *Kluge*, s.v. *Walnuß*). The concept of 'opposed to what is a native *æppel* 'apple'' or 'a native *hnutu* 'nut'' is, for these plant names, certainly pre-Old English.

Diachronic processes: sound changes, loan blends, folk etymology

When the plant names are further subjected to diachronic processes such as sound changes, morphological processes (e.g. suffix substitutions; see below and chapter 7.4.2.1.1.) or processes of folk etymology (chapter 7.5.) it can be safely assumed that they are Old English formations. Cases in question are, for instance, *sūþerne bisceopwyrt* 'unidentified plant' and *sūþerne wermōd* 'southernwood' with *bisceop(wyrt)* and *wermōd* affected by folk etymology (see Catalogue, s.vv. *bisceop(wyrt)*, *wermōd*); *sūþerne cymen* 'cumin' and *sūþerne rǣdic* 'radish' with *cymen* changed by *i*-mutation from L *cumīnum* (Campbell 1959: §§ 190-204), and *rǣdic* from L *rādīx* (Gmc *ǣ* > WGmc *ā*, which appears as WS *ǣ*, nWS *ē* when not before nasal consonant; cf. Campbell 1959: §§ 128, 132) (see Catalogue, s.vv. *cymen*, *rǣdic*); or *sūþerne popig* 'opium poppy' with *popig* changed through suffix substitution from L *papāver* (see Catalogue, s.v. *popig*; cf. Campbell 1959: § 518).

As *popig* 'poppy', for instance, is a loan blend with a specific Old English suffix, the new lexemes it forms are Old English, all the more as these formations are also syntactic groups: *basu popig* 'common red poppy', *hwīt popig* 'opium poppy', or *wilde popig* 'common red poppy'. The same applies to formations built upon the head *bisceopwyrt* 'marsh mallow', for instance, which is a combination of the loan L *hibiscum* (or *hibiscus*) changed by folk etymology into OE *bisceop* and the Old English lexeme *wyrt* 'plant': the newly formed lexemes *brādbisceopwyrt* (*brāde bisceopwyrt*) 'unidentified plant name', *feldbisceopwyrt* 'mallow' and *læsse bisceopwyrt* 'betony' are Old English formations, accordingly.

Plant names affected by folk etymology often transcend the borderlines between two diachronic strata, as these plant names are, by definition, subjected to diachronic developments.[244] Plant names affected by folk etymological processes were either Old English or pre-Old English formations that had become isolated (dissociated[245]) elements (through various processes of phonological and morphological attrition), or loan elements which, as loans, were semantically and morphologically isolated in the system of Old English, as well (cf. chapter 7.5.). As the purpose of folk etymological processes is to "secure relative linguistic motivation", striving to integrate isolated words, "at least partially, with a family of 'consociated' words in a language" (Knappe 2004: 492), the isolated, unmotivated forms are changed in analogy to already existing, phonologically similar Old English terms. Phonological motivation is often accompanied by semantic motivation, i.e. the semantic association of the motivated (original) to the motivating (model) lexeme. As stated above, folk etymology affects native and foreign terms in equal measure. One example is the already mentioned Latin plant name *hibiscum* 'mallow', where the element *-bisc-* was changed in analogy to OE *bisceop* and formed with OE *wyrt* 'plant' the compound *bisceopwyrt* 'common mallow' (see Catalogue, s.v. *bisceopwyrt*); an example of native origins is the groundsel, OE *grundeswelige* lit. 'ground-devour(ing plant)', possibly a folk etymlogical change from *gundeswelige* 'pus-devour(ing plant)'. Sometimes, we assume folk etymological change as the semantic analysis of the plant name makes no sense, but the probable original etymon of the plant name can, still, not be deduced. Thus, OE *wermōd* lit. 'man-mood' cannot be brought into a semantically meaningful relation to the plant Artemisia absinthium, so that folk etymological change is a plausible explanation. From parallel formations in cognate languages, e.g. OHG *wer(i)muot(a)*, G *Wermut*, etc. (see Catalogue, s.v. *wermōd*; *Kluge*, s.v. *Wermut*) we know that *wermōd* is a West Germanic compound, but its formal assimilation to *wer* 'man' and *mōd* 'mood' is Old English and can, still, not be explained (just as the reflexes of this plant name in other West Germanic languages cannot be explained either).

Multiple marking

As has already been mentioned occasionally, many of the plant names are influenced in multiple ways by processes marking them as specifically Old English. As the processes have all been described above, a few examples shall give an idea of multiple marking affecting different linguistic levels, that is, where more than one characteristic identifies the plant name as Old English. The plant names *sūþerne cymen* 'cumin' and *sūþerne bisceopwyrt* 'unidentified plant', for instance, have already been described as being marked by specific Old English characteristics such as *i*-mutation (in *cymen*; level of phonologoy) and folk etymology (in *bisceop-*; phonological and morphological level), by being both syntactic groups (morphological level) and by being characterised as 'not English' by the (semantic) attribute 'southern'. *Brād cāwel*, for instance, is a syntactic group (morpho-syntactic level); *cāwel* is an early Latin loan

244 For a full account of folk etymological processes, see chapter 7.5.

245 For the distinction between dissociated and consociated lexical elements and definitions of these, see Leisi and Mair (1999: 51-9).

word going back to L *caulis*; it shows *w*-insertion (phonological level; see Catalogue, s.v. *cāwel*). Combinations with *pipor* such as *blæc pipor* and *lang pipor* are, again, syntactic groups (morpho-syntactic level) showing suffix substitution in *pipor* (morphological level). The plant name *rēade fille* '? wild thyme' is a syntactic group (morpho-syntactic level), where OE *fille* 'thyme' is clipped from OE *cerfille* 'thyme' (morphological level), itself being a loan going back to L *chaerephylla* (see Catalogue, s.vv. *fille*, *cerfille*).

The evidence of the manuscripts

There is the broad consensus among scholars investigating the stemmatic relationship between the Anglo-Saxon glosses and glossaries that not only the manuscripts of the *Épinal-Erfurt* and the *Corpus* glossaries are earlier than those of the other glossaries, but also that the glosses belong to the earliest period of written Old English (see chapter 4.4.). Both glossaries seem to have been among the first to have been copied from a common source, now lost. This was possibly a Latin-Old English glossary compiled in the seventh century on the basis of the Greek-Latin *Hermeneumata* (Rusche 2008) or a Latin-Old English glossary, compiled on the basis of Isidore's *Etymologiae* early in the seventh century (Porter 2011b: 158).[246] It follows that the plant names recorded in the *Épinal-Erfurt* or in the *Corpus* glossaries belong to the earliest stratum of botanical terminology within the Old English period. This, of course, does not preclude that the plant names not recorded in *Épinal-Erfurt* or in *Corpus* are definitely younger: it may very well be the case that a number of plant names were in use early in the history of Old English, without leaving written footprints, or where the early footprints in writing did not survive to our days. The plant names *fēferfuge* 'feverfew' and *lufestice* 'lovage', for instance, were borrowed from Latin before 650 AD (see chapter 7.4.1.1.), but they are not attested in the early glossaries. It means, though, that the 63 complex plant names marked with a superscript 'e' in the Catalogue (for 'early'; i.e. as occurring in the *Épinal-Erfurt Glossary*, in the *Corpus Glossary*, or in both, have definitely been formed in the seventh and eight centuries, at the latest.

The 'early' plant names are: *āt(t)orlāþe* 'betony', *æscþrote* 'goutweed', *bēowyrt* 'common balm', *bisceopwyrt* 'marshmallow', *blindenet(e)l(e)* 'dead nettle', *brādelēac* 'ramsons', *cirisbēam* 'cherry tree', *cistenbēam* 'chestnut tree', *cnēowholen* 'knee-holly', *cwicbēam* 'quickbeam', *dweorged(w)os(t)(l)e* 'pennyroyal', *ēarisc* 'rush', *eoforþrote* 'carline thistle', *eolhsecg* 'sedge', *feldwōp* 'hops', *fingerdocca* 'foxglove', *fuglesbēan* '(tufted) vetch', *fūle trēow* 'alder buckthorn', *furhwudu* 'pine', *gārlēac* 'garlic', *gēaces sūre* 'wood-sorrel', *gelodwyrt* 'silverweed', *geormanlēaf* 'common mallow', *gescādwyrt* 'great mullein', *gōdæppel* 'quince', *grundeswylige* 'groundsel', *haguþorn* 'common hawthorn', *heorotber(i)(g)e* 'bramble', *hindber(i)(g)e* 'raspberry plant', *hindber(i)(g)e* 'raspberry plant', *hnutbēam* 'walnut', *hræfnesfōt* 'crowfoot', *hunigsūge* 'honeysuckle', *hwītcwudu* 'mastic', *hwīteclǣfre* 'white clover', *lǣcewyrt* 'rib-wort', *lēaccærse* 'garlic mustard', *leloþre* 'dock', *lybcorn* 'caper-spurge', *plum-*

246 According to Porter (2011b: 155), the texts of the *Antwerp-London* glossaries are even older than those of *Épinal-Erfurt* and *Corpus*, despite the fact that the manuscripts are younger.

trēow 'plum-tree', *rēadclǣfre* 'red clover', *scaldhūlas* 'reed, sedge', *scaldþȳfel* 'reed, sedge', *slāhþorn* 'blackthorn', *smeoruwyrt* 'smearwort', *snīþstrēo* 'some kind of sedge', *sperewyrt* 'spearwort', *sunfolgend* 'common marigold', *tūncærse* 'garden-cress', *þēfanþorn* 'common buckthorn', *þūþistel* 'sow-thistle', *wananbēam* 'prick-wood', *wealhmore* 'common parsnip', *wealhwyrt* 'elecampane', *wealwyrt* 'dwarf elder', *wēdeber(i)(g)e* 'white hellebore', *wegbrǣde* 'greater plantain', *wermōd* 'wormwood', *wīngeard* 'vine', *wōde(h)wistle* 'cowbane', *wudubinde* 'woodbine', *wulfes-camb* 'wild teasel', *ynnelēac* 'onion'.

7.2. LANGUAGE CONTACT

We learn from contact linguistics that language contact is not only all-pervasive, but also so varied in type, socio-historical settings and linguistic outcome that almost any 'rule' can be invalidated by clear counterexamples. (Thomason 2001: 63).[247] Although 'robust generalisations' regarding the predictability of contact-induced language change, can be drawn from the data at hand, "any borrowing scale is a matter of probabilities, not possibilities" (Thomason 2001: 71). Yet, the linguistic outcome of any language contact can be correlated with certain socio-historical and linguistic factors: social predictors such as the kind of contact, its intensity (influenced by the duration of the contact, by the socio-economic effect of the contact situation or by the numbers of speakers) and the stability of the contact situation, its intensity in terms of attitudes of the speakers to each other (e.g. overt hostility vs. network construction and intermarriage) and towards the languages involved (prestige status of the respective languages), besides linguistic factors such as the typological distance of the languages, which is one of the most important factors in predictions of contact-induced change (Thomason 2001: 71). As Townend (2008: 69) emphasises, what is generally termed *language contact* is, in fact, not the contact between two or more languages, but the contact between the users of different languages. The contact situations are varied, reaching from forced contact (e.g., when speakers of one language conquer the territory of another language community or when one group of people enslaves another group) to voluntary contact (e.g., when two language communities are linked by trade or by cultural and scientific exchange).[248]

Research in contact linguistics has shown that socio-historical circumstances influence the outcome of language contact far more significantly than linguistic factors: Typological distance between the languages in contact, markedness of the interference features[249] or the degree of integration of the features in the source

247 For systematic investigations into language contact and its outcome see, e.g., Thomason and Kaufmann (1988), Thomason (2001), Winford (2003), and Riehl (2004).

248 For a detailed account of different ways languages may come into contact see Thomason (2001: 17-21).

249 Interference can be understood as "contact-induced change that involves the importation of material and/or structures from one language into another language, whether by borrowing or by shift-induced interference. All interference is contact-induced change, but not all contact-induced change is interference." (Thomason 2001: 267). In order to avoid the negative ring of the term

language can always "be overridden by social factors pushing in the opposite direction" (Thomason 2001: 77). Forced contact, for example, when one language group migrates into the territory of the other without the consent of the indigenous people, most likely leads to the shifting of the speakers of one language to the language of the other group. The original language of the shifting group disappears, in general, within a few generations of speakers, while contact situations that persist without any dramatic change for more than three or four generations can be considered stable (Thomason 2001: 23).

Following strata theory, the language 'surviving' the contact situation, the *stratum*, is influenced by the succumbing language(s) in mainly two different ways: the succumbing language leaves its traces in the stratum either as a *substrate* (or *substratum*) or as a *superstrate* (or *superstratum*).[250] When the succumbing language is the language of the socio-politically dominant people, e.g. Old Norse as the language of the victorious Vikings settling in the Danelaw from the 9th century onwards, or Old French as the language of the victorious Normans, mainly in the 13th century, it is called the *superstrate* and linguistic interference is due to *superstrate* influence (Krefeld 2003: 556). When the succumbing language is the language of the conquered people, e.g. Brittonic as the language of the Celts conquered by the Germanic tribes in 5th and 6th century Britain, its influence on the stratum is called *substrate* influence and the respective language is dubbed a *substrate* (Krefeld 2003: 556).

It is held by some linguists that there is a difference between the outcome of substrate interference and the outcome of superstrate interference. According to Vennemann (2004: 50), a substrate influences, above all, the structure of the receiving language (syntax and phonological structures), while a superstrate influences, above all, the lexicon of the receiving language. Thomason (2001: 75), in contrast, argues that the linguistic outcome depends on the question of whether speakers shift to another language or not. Accordingly, in her argumentation, there is a difference between *shift-induced interference*, i.e. when one group of speakers shift to the language of the other group of speakers and when imperfect learning plays a role (cf. Thomason 2001: 66-7) and *borrowing*, i.e. when speakers of one language introduce foreign terms into their native language:

> The crucial prediction about shift-induced interference is that, unlike borrowing, it does not start with the lexicon. Instead, it starts with phonology and syntax; […]. A possible exception to this generalization might occur if the shifting group is a superstrate, a socioeconomically dominant group rather than a substrate (Thomason 2001: 75).

'interference', Clyne (1991: 160) proposed the usage of the term 'transference'; see also Riehl (2004: 28).

250 Strata theory was initiated by the Romance linguist Graziadio Ascoli (1864), who created the term 'substrate' for the influence of Celtic languages on Latin; Hugo Schuchardt (1890) added the term G *Überschichtung*, which was, eventually, received in the disguise of W. von Wartburgs (1936) 'superstrate' and, finally, Marius Valkhoff's (1932) term 'adstrate' completed the schema. For an overview of strata theory, with critical remarks on its origins, on advantages and limits as a model for describing language contact see Krefeld (2003).

When the language contact is not the result of conquest but rather of trade relations and cultural exchange, the contact language is called the *adstrate* (or *adstratum*; Krefeld 2003: 557-8). However, any language contact is, initially, an adstrate influence, as the question of whether a contact language is a sub- or a superstrate can only be answered in retrospection, i.e. after the language shift; therefore, adstrate influence is, on the one hand, an alternative to sub- and superstrate influence and their super-ordinated concept at the same time.

Intense, daily contact between two (or more) peoples who inhabit the same territory and are linked by intermarriage, trade or religious practices, leads to intense language contact and to *intimate borrowing* (Bloomfield 1933: 461-75). The more intense the contact between the two (or more) peoples is, the more the structural and lexical cores of the recipient language are affected. Less intense contact, such as is the case with learned, cultural contact, affects, in general, only the non-basic vocabulary of the receiving language, i.e. involves the borrowing mainly of content words (*cultural borrowing*; cf. Bloomfield 1933: 444-69; Thomason 2001: 70-1).[251]

Adstrate influence (in its narrow sense, as language contact *in actu*) is triggered by casual or regular contact between two languages, but the borrowers need not necessarily be fluent in the source language. Normally, the two language communities do not inhabit the same territory; they can be immediate neighbours but do not have to be. In many cases – as with Latin during the Middle Ages up to the Renaissance – the source language is a language of high prestige and, in general, the *lingua franca* of two (or more) language communities.

7.3. CONTACT SITUATIONS AND TYPES OF CONTACT, WITH A FOCUS ON THE OLD ENGLISH COMPLEX PLANT NAMES

Old English was influenced mainly by four languages: Brittonic, the variety of Celtic spoken by the people inhabiting Britain before the arrival of the Germanic tribes; Latin, the language spoken by the Romans and the language of learning, science and diplomacy throughout Medieval Europe; Old Norse, the language spoken by the Danes and Norwegians who settled in the East Midlands and the North and founded the Danelaw from the end of the 9th century onwards; and Greek, as a an ancient language of learning and science, which, however, came into contact with English largely through the mediation of Latin (Feulner 2000: 4; see chapter 7.4.1.3.). Following the restoration of Edward the Confessor, who had grown up in exile in Normandy, to the English throne in 1042, a small number of French words entered Old English (Weinstock 2006: 1835) but these terms do not affect the plant name corpus.

251 The effects of borrowing can be arranged along a borrowing scale (Thomason 2001: 70-1) and depend on the intensity of contact: casual contact (content words); slightly more intense contact (content words, function words, some structural borrowing); more intense contact (function words, basic vocabulary, structural borrowing); intense contact (heavy lexical borrowing, heavy structural borrowing).

Old Norse

According to Kastovsky (1992: 300-1), it is generally accepted that Latin introduced the largest number of direct and indirect loans into Old English, followed by Old Norse (the bulk of which appeared as late as in Middle English) and Celtic. In terms of syntax and phonology, the ratio is probably the reverse, although the exact amount and impact of linguistic interference features that go beyond the lexicon is a matter of much dispute. Especially the linguistic influences exerted by the Scandinavians are considered by some linguists to be comparable to the outcome of creolisation (e.g. Poussa 1982)[252] or, at least, clear indices for the development of a supra-regional *koiné* based on Old English and Old Norse (cf. McWorther 2002), which ended (in either case) in the typological shift of English from a synthetic to an analytic language and in the incidence of a number of singular linguistic features in English not present in her Germanic sisters (according to McWorther 2002, for instance, the periphrastic progressive, the disappearance of verb-second word order, *have*-perfect, *be*-passive, etc.).[253] The intense contact between the Anglo-Saxons and the Scandinavian tribes as the result of settlement in the East Midlands and the North of Britain beginning in the second half of the 9th century led to intimate linguistic contact; the contact-induced changes on the part of Old English, however, did not fully surface before the Middle English period (Kastovsky 1992: 319-20).[254] Old Norse and Old English are closely related genealogically and typologically very similar, so that the question came up of whether the two can be regarded as dialects (e.g., McWorther 2002). As Kastovsky (1992: 327-30) points out, the question of mutual intelligibility between Anglo-Saxons and Scandinavians is hard to answer, mainly due to the fact that textual evidence from the dialect area in question is missing. It can be assumed, however, that in 10th and 11th century Danelaw

> there was a certain amount of bilingualism, notably with the offspring of mixed marriages or second- and third generation settlers; that the tendency towards bilingualism, or at least towards the acquisition of a working knowledge of the other language was probably more stronger among the Anglo-Saxons than among the Danes (Kastovksy 1992: 330).

252 This theory is, however, rejected in general; see fn. 253 (below).

253 To be complete, another hypothesis claims that French – English language contact led to the development of Middle English as a creole language, on the basis of mainly the same linguistic features (e.g. Bailey and Maroldt 1977, Domingue 1977). All of them have found their opponents; c.f., e.g., Görlach (1990), Thomason and Kaufmann (1988: 302-4, 306-15). More in line with the koinéisation approach (McWorther 2002) is the hypothesis that English developed as a learner interlanguage (Danchev 1997).

254 The intensity of contact can be deduced, e.g. from the far-reaching linguistic influence the superstrate Old Norse had on the surviving stratum English: e.g. the borrowing of so important a part of the (morpho-)lexicon as the plural forms of the 3rd person personal pronoun (*they, their, them*) or from the borrowing of basic-vocabulary words like *sky* (non-palatalised ON *sc-* /sk/ vs. palatised OE *sc-* /ʃ/), *husband* (< ON *bondi, buandi* 'husbandman') or *take* (< ON *taka*); for further information see, e.g., Kastovsky (1992: 320-36).

This bilingualism with the eventual shift of Norse people to the Midlands dialect of Old English or to an Old English-Old Norse koiné was facilitated by the typological similarity of the two languages.[255] Thus, for the influence of Old Norse, borrowing is a more likely scenario, while for the linguistic influence as it emerged in Middle English shift-induced interference as the result of imperfect learning is more probable. The loans from Old Norse in the Old English lexicon corroborate this assumption, as, at least for borrowings from Old Norse that show up early in the manuscripts, there are many technical terms, in contrast to the loans found later, in early Middle English manuscripts from the North and the East, such as, for instance, the *Orrmulum* (Kastovsky 1992: 321, 324).

In the corpus of Old English plant names, however, Old Norse influence is practically non-existent: The only loan possibly related to Old Norse is the complex plant name OE *hundescwelcan* (Hoops 1889: 76; see Catalogue, s.v. *hundescwelcan*), which, however, does not appear in the *Altwestnordische Pflanzennamen*, Heizmann's dictionary of plant names from Old West Norse (1993). One explanation for this lack of loans can be the diglossic situation[256] in the Midlands up to the end of the Old English period, where Late West Saxon was used as high variety and written standard, while the Midland dialect (whether a koiné or a modified form of Anglian) was used as the language of everyday spoken communication, i.e. was the language of communicative proximity (Kastovsky 1992: 330). In this scenario, even if Old Norse botanical terms were taken over into the spoken Midland variety, the likelihood for them being recorded in Late West Saxon texts is practically zero.

Yet another explanation for the absence of interferences from Old Norse can be that the two peoples had not only shared genetically and structurally related languages with at least partial intelligibility and a common stock of words, but also topographically and climatically similar territories on the continent and in England. The similar flora and fauna, then, would reduce the need to coin new words or borrow foreign terms (as happened with plants from the Mediterranean region such as *palmtrēow* 'palm-tree' , or *affricanisc æppel* 'pomegranate'). However, the complex botanical terms the Anglo-Saxons shared with their Viking neighbours are, still, few: they are *eorð-*

255 Typological similarity is exactly the factor Thomason and Kaufmann (1988: 307) spearhead *against* the assumption that Old Norse is responsible for the radical typological change of English, while Thomason (2001: 77) correlates structural borrowing with typological similarity in the following way: "even features that are highly marked or highly integrated into an interlocking structure are readily exchanged between typologically similar systems. The most obvious instance of this is in dialect borrowing, where borrowing is common even in the inflectional morphology."

256 According to Ferguson (1959: 336), "[diglossia is] a relatively stable language situation in which, in addition to the primary dialects of a language (which may include a standard or regional standards), there is a very divergent, highly codified (often grammatically more complex) superposed variety, the vehicle of a large and respectable body of written literature, either of an earlier period or in another speech community, which is learned largely by formal education and is used for most written and formal spoken purposes but is not used by any sector of the community for ordinary conversation." While Ferguson's high and low varieties are varieties of one language, in Fishman's (1967) concept of *extended diglossia* the high and the low varieties used do not have to be genetically related; an examples is, for instance, the use of Spanish (Indo-European language, high variety) and Guaraní (an indigeneous language of South America, low variety) in Paraguay.

ber(i)(g)e 'strawberry', *goldblōma* '? marigold', *fictrēow* 'fig-tree', *hwītlēac* 'leek', *wīntrēow* 'vine', *hwǣtecorn* 'corn of wheat' and *piporcorn* 'pepper-corn'. As the latter two do not denote species but parts of plants, they are not part of the corpus presently analysed.

Celtic

Recently, a number of researchers in Celtic studies (e.g., Vennemann 2004, Viereck 2000, White 2003)[257] promoted the view that it is the Celtic substrate that exerted the deepest and most far-reaching influence on English, accounting for a number of features such as the elimination of external possessors, the development of the periphrastic progressive or preposition stranding. Adherents of this hypothesis assume a diglossic situation in Anglo-Saxon England with a low variety, 'Celticised Anglo-Saxon' ("Keltisiertes Angelsächsisch"; Vennemann 2004: 47), used by the middle and lower classes and a high variety, 'Anglo-Saxon', used in written discourse by the intellectual elite. Accordingly, Celticised Anglo-Saxon could make its way to Middle English only when the Anglo-Saxon intellectual elite was replaced by the Normans in the wake of the Norman Conquest (1066).[258] Despite evident lexical borrowing from Brittonic (albeit small in the basic vocabulary), the language did not leave any noticeable traces in the Old English plant name lexicon. Of the two simplexes that can be brought back to Celtic, one term, OE *cearricge* '? rock parsley' can be associated with OIr *carrac* 'rock' (*DOE*, s.v. *cearricge*) while the other one, the tree name OE *sæppe* 'spruce-fir' is a (Continental) Celtic loan word that entered Old English indirectly via L *sappīnus* 'various species of Abies L. or Pinus L.' (Klare 1998: 39).[259]

This scarcity of loan influence is somewhat surprising with regard to Celtic-English language contact since in contact situations that involve the migration of one language group into the territory of another language group, as was the case with the West Germanic tribes invading Britain in the 5th and 6th centuries, among the first concepts that are taken over by the newcomers are, beside toponyms, terms related to the flora and fauna of the new territory.[260] With regard to toponymic borrowing, the substrate

257 The reader may find a number of essays dedicated to various aspects of 'Celtic Englishes', for instance, in the series of four volumes named *The Celtic Englishes* (*I*, *II*, *III*, *IV*), ed. Tristram.

258 Rather sceptical with regard to the impact of Celtic influence on English are Isaac (2003) and Sand (2003).

259 In contrast to Hoops's (1889: 76) assumption, the diminutive OE suffix *-oc* (ModE *-ock*) is Germanic and not Celtic (*ODEE*, s.v. *-ock*; *OED*, s.v. *-ock*), and the the plant names *cammoc* 'hog's fennel' (*DOEPN*, s.v. *cammoc*), *cassuc* 'a kind of grass or sedge' (*DOEPN*, s.v. *cassuc*), *cottuc* 'common mallow' (*DOEPN*, s.v. *cottuc*) and *coccel* 'corn-cockle' (*DOEPN*, s.v. *coccel*) are, thus, not hybrid formations.

260 Schneider's (2007: 35-40, 56) model of the developmental phases in (colonial) language contact situations describes that in the first and the second phases of settlement and in the process of stabilisation of the contact situation, it is the toponyms and the vocabulary of the flora and fauna that the conquerors take over from the indigenous people. As Schneider (2007) conceived his model primarily for language contact in colonial situations, it is not exactly applicable to language contact in early medieval England. On the other hand, parts of the events can be compared as both contexts involve the conquering and subjection of one group of speakers – the indigenous – by the conquering and settling people, with very unequal conditions in terms of military, administrative and economic power.

Brittonic left, indeed, indelible traces with c. 450 names from Romano-British times (Clark 1992: 479), but the number of loans in other semantic areas is minimal (Kastovsky 1992: 318). The quasi absence of Celtic botanical loan words thus ties in with the general picture, although, so far, no satisfactory explanation has been found. As a lack of textual evidence for 'Celticised Anglo-Saxon' hinders us from fully approving the existence of this variety and its impact on the English language, "we can only state that Celtic, contrary to all expectations, has not really left its mark on the English language" (Kastovksy 1992: 320).

So far, as regards considerable lexical borrowing in the botanical vocabulary of the Anglo-Saxons, we are left with Latin loan influence and, as mentioned earlier, the influence in this part of the Old English lexicon is not only considerable but disproportionately high in comparison to the overall loan influence of Latin on Old English. The amount and nature of Latin influence will be analysed in more detail in the following.

7.4. LATIN INFLUENCE AND THE BOTANICAL LEXICON

The language that exercised the largest influence on Old English was, as mentioned previously, Latin. The history of Latin-English contact before the Norman Conquest is one of spoken day-to-day language as well as one of the written word. Although the Germanic tribes on the continent had contact with the speakers of (vulgar) Latin, as they were part of the Roman army and met Romans through trade (Kastovsky 1992: 301-2), the larger number of Latin loan words came into Old English through the written texts in the centuries after the Germanic Conquest, when the educated Anglo-Saxons – mainly priests and monks – used Latin as their *lingua franca* in learning and education. Since throughout the Middle Ages learning and education meant primarily learning Latin and using Latin as a means of scientific and ecclesiastical communication, Latin was learned as a second language by those who had access to education, i.e. by the monks and their pupils. As a tool for communication it enjoyed a high prestige and was used predominantly in learned communicative situations (Kastovsky 1992: 306-7).

Loan influence can be of a direct and of an indirect kind. *Direct* loan influence refers to the borrowing of free or bound lexical and grammatical morphemes from a source (or donor) language, while *indirect* loan influence is associated with loan formations and semantic loans.[261] For the present survey, which is concerned with compounds and syntactic groups (cf. chapter 1.3.), the borrowing of bound lexical and

261 Kastovsky (1992: 300) makes a distinction between loan words ('direct') on the one hand, and semantic loans and loan translations ('indirect') on the other hand. Betz (1949) differentiates between 'external loans' (*Äußeres Lehngut*; including loan words, foreign words and hybrid formations), and 'internal loans' (*Inneres Lehngut*, including loan translations, loan formations and semantic loans), while Durkin's typology of lexical borrowing (2009: 132-4) is quadripartite and singles out loanwords, loan translations, semantic loans and loan blends. The sub-classification of loan influence employed here is that of Gneuss (1955); cf. chapter 7.4.3.

grammatical morphemes is not relevant and the term 'loan words' is employed to describe the direct loan influence from Latin on the Old English complex plant names.

When words pass on from one language to another, they may be adapted to the linguistic system of the receiving language and are, then, called *loan words* (Durkin 2009: 134). Loan words can adapt on the phonological level, i.e. when the loan word is subjected to sound changes operating in the receiving language (the plant names are, for instance, affected by *i*-mutation as in L *cumīnum* to OE *cymen* 'cumin' or by palatalisation of word-initial pre-OE /k/ > OE /tʃ/, as in L *cēpa* /k/ > OE *cīpe* with /tʃ/;[262] loan words may also be adapted to the inflectional system of the borrowing language, e.g. when they take on the plural morphemes of the receiving language.

In contrast, a *foreign word* is a word that "retains (broadly) its foreign-language pronunciation and may show non-native morphology (especially plural inflections which are not found in native words) and does not give rise to new derivatives within the borrowing language" (Durkin 2009: 139).[263] The distinction between loan words and foreign words is, however, hard to maintain unfailingly, especially when the plant names occur without inflection and out of context in the glossaries. Unambiguous cases like the following occur, but not as a rule: In the *Lǣcebōc* (68/22-3), *terebintina* is marked explicitly as foreign: "Nim ða wyrt þe hatte on suþerne terebintina [...]" ['Take the plant that is called in the language of the Southerners *terebintina*'].

Loans that are recorded with Old English and with Latin inflectional endings show different stages of accommodation. Thus, the morphemes *cist-*, *cistel* and *cisten/cysten* (in *cistbēam*, *cistelbēam*, *cistenbēam*, *stāncysten*), which all go back to L *castanea* 'chestnut', show different degrees of morphological accommodation: while *cisten/cysten* retains the putative suffix *-en* from L *castanea*, *cistel-* replaces *-en* with the native suffix *-el* (Campbell 1959: § 518) and *cist-* clips it altogether (see also 7.4.2.; see Catalogue, s.vv. *cistel*, *cisten*). For other plant names, the grade of accommodation to Old English cannot be ascertained since the attestations seem to allow both the classification as loan words and as foreign words. *Lactuca*, for instance, attested as "Herba lactuca leporina þæt is lactuca" ['The plant *lactuca leporina* that is *lactuca*'] (*Herbarium*, no. 114) and "þeos wyrt þe man [...] lactucam nemneþ" ['This plant which is called *lactucam*'] is regarded as Latin by de Vriend (1984: 314, no. 114; this is indicated by a blank space in the slot for the Old English name) and as Old English by Bierbaumer (2, s.v. *lactuca*).[264]

7.4.1. DIRECT LOAN INFLUENCE

The *direct loan influence* of Latin on Old English is, traditionally, sectioned into three (sometimes four) periods (cf., e.g., Serjeantson 1935: 11-6; Brunner 1960-1962, II:

262 For a detailed investigation into the early loans from Latin (up to c. 600 AD) see Wollmann (1990). A list of all Latin loans from the continental period can be found in Wollmann (1990: 152-80). Campbell (1959: chapter X) may be consulted for all phonological changes that the loan words could be subjected to in the course of Old English, depending on the date of borrowing.

263 See also Serjeantson (1935: 6-9), Gneuss (1955: 16-9) and Scheler (1977: 90-2).

264 For further examples with discussion, see Sauer (1992b: 387-8).

149-50; Scheler 1977: 35-9; Baugh and Cable 2002: 77ff.)[265], which are distinct with regard to the contact situation and can be singled out through specifics in the character of the loans, their integration into the sound system of Old English and their semantic content, i.e., their assignment to certain *Sachgruppen*, i.e. lexical fields:

(1) Latin – pre-Old English contact before the Anglo-Saxons migrated to Britain; this is the so-called *continental borrowing*, with Latin as a cultural adstrate influencing the West Germanic varieties; up to c. 400 AD;

(2) Borrowing of Latin terms that had entered Brittonic as superstrate influence (i.e. "Latin through Celtic transmission"; cf. Kastovsky 1992: 301) in the first period of settlement in Britain, i.e. after the Germanic Invasion of Britain – the so-called *insular loans* (c. 450-650 AD);

(3) The third period of Latin-Old English contact (c. 650-1100) is sometimes subdivided (a) into the time around the Christianisation of England beginning at the end of the 6th century AD (Irish mission around 600 AD; Roman mission beginning 597 AD), and (b) into the time following the Benedictine Reform in the second half of the 10th century, when the religious life in England was revitalised and the monastic life reformed by the bishops Dunstan, Oswald and Æþelwold with the support of king Edgar.

Following Wollmann (1990: 29-30) and Campbell (1959: § 493), who argue on the basis of phonological criteria (chronological order of pre-Old English, Old English and Latin sound changes and their occurrence, respectively, absence in loans from Latin),[266] the present work distinguishes in the main between (i) 'Early Latin loan words (up to c. 650 AD)' including continental and insular loans, i.e. "loan-words which show modifications owing to native and Vulgar Latin sound-changes" (Campbell 1959: § 545) and (ii) 'Later Latin loan words (c. 650-c. 1100)', i.e. words "which are taken not from Vulgar Latin but from the learned Latin of the Middle Ages, in the pronunciation of which the phonological changes of Vulgar Latin were largely eliminated, but in which the loan-words show considerable modifications of the classical quantities to have prevailed" (Campbell 1959: § 545). As Serjeantson's (1935) list of 542 Latin loan words passes for the most comprehensive, albeit not complete, corpus of Latin loans in Old English (Wollmann 1990: 63), and Wollmann's list of early Latin loan words (1990: 153-80) is based on Serjeantson's (1935) and is the most up-to-date list for the pre-650 loan words, the present survey follows Wollmann (1990) and Serjeantson (1935), and includes Campbell's (1959) reasonings,

---

265 The first linguist to propose this tripartite periodisation of the Latin loan influence was Pogatscher (1888). A different periodisation system is propagated, for istance, by Campbell (1959: 199-219), who distinguishes between "Early Latin Loan–words" (§§ 493-544) and "Later Latin Loan-words (§§ 545-64), the dividing line being Christianisation at the end of the 6th, beginning of the 7th century AD; for an overview of research on the subject since Pogatscher (1888) see (Wollmann 1990: 12-90).

266 The controversy on the question of when and how Latin words were received into Old English is summarised in Gneuss (1993: 112-8).

especially when he has an alternative explanation or classification.[267] Sometimes Carr's (1939) classification of compounds (which had not yet appeared when, in 1935, Serjeantson published her work), suggests early borrowing so that his assignment of plant names to certain diachronic layers is considered as well. It is, however, as mentioned before, not always easy to be sure about the relative ages of the lexemes, as many plant names may have been formed independently in the Germanic or West Germanic languages, while the similarities in their structures imply a common origin.

Following a short account of the pre-650 AD and the post-650 AD borrowings from Latin (chapters 7.4.1.1. and 7.4.1.2.) and the influence of Greek (chapter 7.4.1.3.), some specific characteristics of Old English complex plant names that go back to direct loan influence shall be described, notably the nature of *hybrid formations* and *loan blends* (chapter 7.4.2.1.) and different kinds of *doublets* (chapter 7.4.2.1.2.) resulting from loan influence. Plant names affected by folk etymology, though also part of 'direct' influence, are treated in chapter 7.5. because processes of folk etymology do not confine themselves to non-native, i.e. borrowed plant names, but can also affect native formations and span, at the same time, all diachronic strata.

In this context it is important to pay attention to the fact that, as the corpus of Old English is limited in many respects, and the corpus of the medico-botanical texts is even more limited (2.3.), all numbers and ratios regarding the kind and the extent of process of loan influence can only indicate tendencies. The analysis of the loan influence in the following, therefore, will concentrate on qualitative aspects rather than on the quantitative interpretation of the data. Instead of questioning whether one or the other type of loan influence is more prominent in terms of number than the other, the following sections concentrate on conspicuous features of the plant names and on general patterns that emerged during the analysis of the loan influence.

7.4.1.1. EARLY LATIN LOANS (UP TO C. 650 AD)

Before the Germanic conquest of Britain, speakers of Latin and speakers of West and North Germanic varieties met on the continent first as enemies, but the contacts gradually became peaceful with the peoples interacting as neighbours, as mercenary soldiers in the Roman army and as traders (Serjeantson 1935: 11-3). The very early loans reflect this casual adstrate contact situation, relating to superior goods of the Romans, such as, for instance, terms for household vessels, jewelry, plants and their products, and building terms (Kastovsky 1992: 302). In the 5th and the 6th centuries, when the Germanic tribes in England came into contact with the Celts, who had been under Roman rule for more than four centuries and had retained, as a result of this contact, not only Roman roads, towns and baths (Baugh and Cable 2002: 46), but also

267 OE *nǣp* 'turnip' from L *nāpus* and OE *rǣdic* 'radish' from L *rādīx* (Campbell 1959: § 504), for instance, are missing from Wollmann's (1990) list of early loans, although the representation of L *ā* as OE (WS) *ǣ* (sound change: WGmc *ā* > (OE) West Saxon *ǣ*, non-West Saxon *ē*; Campbell 1959: § 132) is second in his chronology of sound changes in Early Old English (Wollmann 1990: 131). Serjeantson (1935: 275-6), in contrast, includes them in her list of 'pre-conquest loan words from Latin'.

a number of superstrate loans, further borrowing of Latin terms occurred. As to the extent of the influence of Latin through Celtic transmission on Old English, scholars who think that with the withdrawal of the Romans Latin ceased to be spoken in England, assume that this kind of influence was rather small (e.g., Baugh and Cable 2002: 81-2), while those who claim that for another century or two Latin remained the official language in England assume borrowing on a larger scale (e.g., Serjeantson 1935: 277-81, who counts about 112 loan words from this period).

The botanical lexicon of the Anglo-Saxons shows a number of borrowings from Latin from the period up to c. 650 AD, most of them used independently *and* as one element of complex plant names. Examples are: *fēferfuge* 'feverfew' (< L *febrifugia*), *lufestice* 'lovage' (MedL *levisticum, luvesticum* < CL *ligusticum*), *box* 'box' (< L *buxus*; as part of *boxtrēow*), *candel* 'candle' (< L *candēla*; as part of *candelwyrt*), *ceder* 'cedar of Lebanon' (< L *cedrus*; as part of *cederbēam* and *cedertrēow*), and so on (see Catalogue, s.vv. *fēferfuge, lufestice, box, boxtrēow, candelwyrt, ceder, cederbēam, cedertrēow*). Only three of the Latin elements borrowed pre-650 AD are not attested as simplexes, which, however, does not necessarily entail that they were not used independently (cf. 5.4.): *corn-* 'cornel' (< L *cornus*; in *corntrēow*), *mōr-* 'mulberry' (possibly also a later Latin loan; < L *mōrum, mōrus*; in *mōrbēam* 'mulberry tree') and *ynne-* 'onion' (< L *uniō*; in *ynnelēac*) (see Catalogue, s.vv. *corn-, corntrēow, mōr-, mōrbēam, ynne-, ynnelēac*).

7.4.1.2. LATER LATIN LOANS (C. 650 AD TO C. 1100 AD)

The next influx of Latin loan words England saw was during and in the wake of the Christianisation of Anglo-Saxon England, which began at the end of the 6th century. The majority of these loan words is, however, not relevant to the present survey as they belong to the sphere of religion and refer to church organisation, ranks and functions (Kastovsky 1992: 305-8). A second influx of Latin terms can be observed from c. 955 onwards, when the Benedictine monastic reform movement triggered, among other things, the re-establishment of learning and scholarship in Anglo-Saxon England, which, in turn, led to the borrowing of about 150 Latin words (Strang 1970: 314).[268] These loans differ significantly from the loans of the previous periods as these loan words did not come in via the spoken language but via the written word. Few of them display characteristics of Vulgar Latin; most of them are drawn from Classical Latin and reflect the scholarly interests of the time. They are mostly technical terms, since they are not part of the basic vocabulary of the period and, very often, are not integrated into the Old English language system (Kastovsky 1992: 307). Examples of later Latin loans that are part of Old English complex plant names are *calcatreppe* 'caltrop' (< L *calcatrippe*), *cunille* 'thyme' (< L *cunīla*; in *wuducunille* 'wild thyme'), *lēo* 'lion' (< L *leōnem*, acc. sg. of *leo*; in *lēonfōt* 'lady's mantle') and *lactuce* 'lettuce' (< L *lactūca*; in *wilde lactuce*). Some of the plant names borrowed in this later period

268 For the intellectual foundations of the Benedictine reform movement in England and its influence on language in terms of standardisation, development of the lexicon, etc., see Gneuss (1972), Gretsch (e.g. 1973, 1999a, 1999b) and Hofstetter (1987).

form doublets with plant names borrowed earlier (see chapter 7.4.2.1.2.) (see Catalogue, s.vv. *calcatreppe*, *cunille*, *wuducunille*, *lēo*, *lēonfōt*, *lactuce*, *wilde lactuce*).

7.4.1.3. INFLUENCE OF GREEK THROUGH LATIN

With the exception of Theodore of Tarsus, archbishop in Canterbury from 669 onwards, and his companion Hadrian, both of whom taught in Canterbury in the 7th century, the Anglo-Saxons had no direct contact with Greek-speaking communities in the early Middle Ages, nor was the knowledge of Greek very widespread. One could even argue that Greek was not spoken or even known at all: "Greek was more honored than studied in the Middle Ages" (Berschin 1988: 19; cf. Bierbaumer 1989). It seems though, as if in Canterbury, during the time of Theodore and Hadrian, at least one glossary was compiled, which, then, served as the archetype for the glossaries that have survived to our days. This was possibly based a Greek-Latin one they had brought with them (possibly related to Dioscorides' *De materia medica*; cf. Rusche 2003: 191), or a copy of the Greek-Latin *Hermeneumata Pseudo-Dositheana* (Rusche 2008). Porter (2011), in constrast suggests that they initiated the compilation of a glossary at Canterbury, based on Isidore's *Etymologiae* (see chapter 4.4.).

It is clear, however, that Greek words seldom entered Old English directly, but mainly via the mediation of Latin, occasionally also via Celtic or Old French (Feulner 2000: 4). Greek elements in the lexicon of Old English complex plant names appear in the disguise of early or late Latin loan words, but also as loan formations based on Latin, which, themselves, are loan formations on Greek models.[269] Originally Greek elements in the corpus of the present study are, for instance, OE *rūde* 'rue' < L *rūta* < Gk *ρυτή* (in *feldrūde* 'common meadow rue', *wilde rūde* 'fen rue'); OE *persoc* 'peach' < MedL *persica* < L *mālum Persicum* < Gk *περσαία* < Gk *περσικός* (in *persoctrēow* 'peach tree') and OE *magdala-* 'almond' < L *amygdala*, *amygdalum* < Gk *ἀμυγδάλη*, *ἀμύγδαλον*, *ἀμύγδαλος* (in *magdalatrēow* 'almond tree') (see Catalogue, s.vv. *rūde*, *feldrūde*, *wilde rūde*, *persoc*, *persoctrēow*, *magdala-*, *magdalatrēow*).

Among the loan formations based, ultimately, on Greek, are, e.g., OE *hundeshēafod*, a loan translation of L *canis caput* ← Gk *κυνοκέφαλος*, OE *fīflēafe* loan translation of L *quinquefolium* ← Gk *πεντάφυλλον* and OE *eorþnafela*, a loan translation of L *umbilicus terrae* ← Gk *ὀμφαλόκαρπος*. One complex plant name seems to have been modelled directly on a Greek term, namely, OE *menlufigende* 'cleavers', apparently a translation of Gk *φιλάνθρωπος* as there is no Latin equivalent **amans hominum* (or a similar formation; Sauer 1992b: 390).

---

269 Patterns of loan influence with special reference to the influence of Greek are discussed in Sauer (2011).

7.4.2. PROCESSES OF ACCOMMODATION

As mentioned above (chapter 7.5.; Durkin 2009: 139-40), loan words differ from foreign words in that they are adapted to the system of the receiving language. The adaptation can occur on different levels of language, but especially on the phonological level, where the loans may be subjected to the same sound changes as the native vocabulary, and on the morphological level where they can give rise to new derivatives or are integrated into the morphological system by, e.g., taking over the plural inflection of the borrowing language.

In the following, some processes of accommodation in the complex plant name corpus will be discussed, notably the formation of (i) *hybrid* plant names, including hybrid plant names containing *loan blends* and (ii) the existence of *doublets* resulting from, e.g. multiple borrowing or from folk etymological changes.

7.4.2.1. HYBRID FORMATIONS

Complex plant names made up of a native and a borrowed word are called *hybrid formations* (c. 91 plant names). In the present survey, loan blends (cf. chapter 7.4.2.1.1.) are seen as a sub-group to hybrid formations although the dividing line between the two phenomena is not always easy to define.

Loan words are monomorphemic within the morphological system of the borrowing language even if their morphology is complex in the donor language (Marchand 1969: 5-8). Thus, initially, they appear as unanalysable items in the borrowing language and can be used to form new lexemes by word-formation.[270] As the corpus of the present survey is limited to complex plant names, naturally, the majority of both 'early' and 'later' Latin loans appear as monomorphemic elements of the complex Old English terms, that is, these Old English complex plant names are *hybrid*: they contain a native and a non-native element.

When the loan element takes the function of the head in the endocentric hybrid formations, it can be assumed that the loan word is fully institutionalised, as it is the basis for the new lexeme and can, in theory, replace the whole combination in a sentence (see chapter 5.3.2.2.1.). The complex plant names *swearte minte* 'crisped mint' lit. 'black mint', *wilde cyrfet* 'gourd' lit. 'wild gourd', or *sūþerne rǣdic* 'radish' lit. 'southern radish', for instance, can be replaced by their second elements *minte* 'mint' (< L *ment(h)a*), *cyrfet* 'gourd' (< L *cucurbita*) or *rǣdic* 'radish' (< L *rādīx*), respectively.

When the loan element acts as a modifier, the second, native element often seems to have been added as a kind of explanation to the first (see chapter 5.3.2.2.1.1. (b)). Thus, for instance, the modifiers *syrfe* (< L *sorbus*), *ceder* (< L *cedrus*), *box* (< L *buxus*), and *por(r)* (< L *porrum*) in the plant names *syrftrēow* 'service-tree', *cederbēam* 'cedar of Lebanon', *cedertrēow* 'cedar of Lebanon', *boxtrēow* 'box-tree'

---

270 This is, of course, unless they have become analysable lexemes because of reversal processes (see 7.5.).

and *por(r)lēac* are attested independently and are lexicalised with the same, identical meaning as their complex counterparts: they are pleonastic. This means that the first element is both attested independently and taxonomically sub-ordinated to the second element: a *ceder* is a kind of *bēam* just as a *box* is a kind of *trēow*, a *syrfe* a kind of *trēow* and a *por(r)* a kind of *lēac*.

Loan words are often subjected to changes on the phonological and/or the morphological level, although this is not a rule. Phonological changes in Latin loan words in Old English can be, for instance, *i*-mutation or palatalisation of /k/ and /g/ (for a full account with examples, see Wollmann 1990 and for some examples from the plant name corpus, see chapter 7.5.). Here, however, I would like to concentrate only on some conspicuous features on the morphological level, namely, to *clipping* and to *back-formation*; however, the examples are few.

Where *clipping* occurs, a submorphemic unit (often just one sound) is taken away (Schmid 2005: 218-20). This can be at the beginning of the word, as in L *amygdala, amygdalum* (ultimately from Greek; see Catalogue, s.v. *magdala-*), which was shortened to OE *magdala-* 'almond'; or in L *chaerephylla*, which was borrowed as *cerfille* and, then, reduced to OE *fille* 'tyhme', but clipping can occur also in the middle, or at the end of the word: L *cucurbita*, for instance, was borrowed into Old English as *cyrfette*, that is, as a clipped form and with the *i* triggering *i*-mutation of the preceding *u*; L *githago*, was clipped to OE *giþ-* (probably already in Latin; see Catalogue, s.vv. *giþ-*, *giþcorn*). When a shortened and an unshortened form exist side by side, as with OE *cerfille* 'thyme' and *fille* 'thyme' (both from L *chaer(e)phylla*), we speak of doublets (see chapter 7.4.2.1.2.). *Back-formation*, the process in which a suffix or a putative suffix is cut off at the end of a word (Schmid 2005: 217) is extant, for instance, in L *mōrum* 'mulberry tree', which was reduced to *mōr-* and in L *prūnum* 'plum, plum-tree', which was reduced to *plūm*, etc.[271]

Complex plant names affected by clipping or back-formation build a bridge between 'pure' hybrid formations (i.e. with formally largely unchanged loan word) and loan blends: In loan blends a morpheme or a putative morpheme of the loan word is replaced by an Old English affix (see chapter 7.4.2.1.2. below). OE *popig*, for instance, is classified as a loan blend because one morph of the loan was replaced by the native suffix *-ig* (< L *papāver*; with suffix substitution: native *-ig* for L *-āver*; cf. Campbell 1959: § 518). The complex formations based on the head *popig*, notably *basu popig* 'common red poppy', *hwīt popig* 'opium poppy', *sūþerne popig* 'opium poppy' and *wilde popig* 'common red poppy', are not exactly hybrid any more, as the head itself is not a 'pure' loan, but has been integrated into the Old English language system by the blending process. In contrast, plant names made up of two free lexical morphemes, one of which is a loan word that has not been affected by suffix substitution, e.g., *feldminte* 'horse-mint' (OE *feld* + *minte* < L *ment(h)a*), *cederbēam* 'cedar of Lebanon' (*ceder* < L *cedrus* + OE *bēam*) and *syrftrēow* 'service tree' (*syrfe* < L *sorbus* + OE *trēow*) are 'pure' hybrid formations.

---

271 For a detailed account of possible developments and processes of accomodation that gives far more examples as it draws from the whole corpus of Old English, see Campbell (1959: §§ 493-564).

In terms of morphology and semantics, the full range of variation is present in the group of hybrid plant names; hybridity is not an indicator of morphological and semantic constraints whatsoever. To give a few examples: Besides the already mentioned endocentric, subsumptive compounds such as *feldminte*, we can find, for instance, hybrid exocentric, metaphorical formations: e.g. N_{gen}+N: *lēonfōt* 'lady's mantle', lit. 'lion's foot'; *lēo* 'lion' < L *leo*, acc. sg. L *leōnem*) as well as hybrid exocentric, metonymic, formations: e.g. *giþcorn* 'caper-spurge, etc.'; *giþ-* < L *githago* and endocentric, rectional formations such as *cirisbēam* 'cherry-tree', lit. 'cherry-tree' and *ciristrēow* 'cherry-tree', lit. 'cherry-tree' with the first element *ciris* derived from MedL **ceresia, ceresea* (see Catalogue, s.vv. *feldminte, lēonfōt, giþcorn, ciris, cirisbēam, ciristrēow*).

7.4.2.1.1. LOAN BLENDS

Loan blends show "borrowing of a complex word with substitution of one or more native morphs for morphs in the borrowed word" (Durkin 2009: 138). In contrast to hybrid formations, where a new lexeme is created with a meaning different from that of its parts, loan blends show *suffix substitution* in the borrowed words while the denotational meaning stays the same. In loan blends a real or putative suffix of the Latin word is replaced by an Old English suffix.[272] Thus, for instance, L *papāver* was changed to OE *popig* (OE suffix *–ig*); L *castanea* was changed to OE *cistel* (OE suffix *-el*); L *humulus* changed to OE *hymele* (OE suffix *-ele*); L *piper* was changed to OE *pipor* (OE suffix *-or*) and L *persicum* was changed to OE *persoc* (suffix *-oc*). Since the original element of the loan word, i.e. the element not substituted is not an Old English morpheme (e.g., *pers-* in *persoc* 'peach', *cist-* in *cistel*, etc.), loan blends are rather to be classified as situated at one end on the cline between loan words which are hardly integrated into the Old English language system at the one end of the cline[273] and loan words that are fully integrated by a range of morphological and phonological processes at the other. As suffix substitution is an indicator for the integration of a loan word into the borrowing language's system, it can be argued that formations with loan blends should be treated as clearly different from hybrid formation, namely, as native formations with two Old English bases. In contrast to plant names like OE *bisceopwyrt* 'common mallow' and its derivatives, where *bisceopwyrt* is counted as a hybrid, but *brādbisceopwyrt* lit. 'broad mallow' is not, I have counted composites with loan blends as hybrid plant names.

7.4.2.1.2. DOUBLETS

After 650 AD, a number of words that had already been borrowed before, i.e. early Latin loans, were re-borrowed and, thus, reduplicated the 'older' loans with newer

272 For an overview and further examples, see Campbell (1959: § 518).
273 For the problems of distinguishing foreign words from loan words see 7.5. above.

ones. These new loans either lack some of the properties of the early loans, such as specific sound changes that the early Latin loans underwent, and/or show specific morphological features. There are three kind of doublets: *etymological*, *morphological* and *folk etymological doublets*, but there is much overlap between the three categories.

When "two borrowings occur which are ultimately from the same source" (Durkin 2009: 169), we can speak of *etymological doublets*. Etymological doublets in the present corpus are, for instance (the early borrowing is named first): *leahtric*, *-lectere* (with intrusion of *r*; cf. Campbell 1959: § 541 and breaking of L *a* > *ea*; cf. Campbell 1959: §§ 145, 594) and *lactuce* 'lettuce' from L *lactūca* (attested in *fenlectere* 'unidentified plant', *wilde lactuce* 'lettuce', *wuduleahtric* 'lettuce'; see Catalogue, s.vv. *fenlectere, wilde lactuce, wuduleahtric*); *cāwel*, *lāwer* (with parasiting *w* and L *au* > OE *ā*; cf. Campbell 1959: § 509) and *caul*, *laur* 'cabbage', 'laurel' from L *caulis*, *laurus* (attested in *brād cāwel* 'cabbage turnip' and *wilde caul* 'wild cabbage'; see Catalogue, s.vv. *brād cāwel, wilde caul, lāwerbēam, lāwertrēow*).

Morphological doublets, those doublets that go back to 'deliberate' modifications of the loan words by way of word-formation, are, for instance, *cerfille* 'chervil' and *fille* 'thyme' from L *chaer(e)phylla* (by way of clipping; in *wudufille* 'wild thyme', *wuducerfille* 'cow parsley', *rēade fille* 'wild thyme', *rēade wudufille* 'asparagus'; see Catalogue, s.vv. *wudufille, wuducerfille, rēade fille, rēade wudufille*); *cist-*, *cisten* and *cistel* 'chestnut (tree)' from L *castanea* (clipping; suffix substitution; in *cistbēam* 'chestnut tree', *cistelbēam* 'chestnut tree' with suffix substitution, *cistenbēam* 'chestnut tree' and in *stāncysten* 'chestnut tree'; see Catalogue, s.vv. *cistbēam, cistelbēam, cistenbēam, stāncysten*).

OE *cyrfet* 'gourd' and *-hwette* 'whetting plant' are not morphological but *folk etymological doublets* going back to L *cucurbita* (attested in *wilde cyrfet* 'wild gourd' and in *hwerhwette* 'cucumber'; see Catalogue, s.vv. *cyrfet, wilde cyrfet, hwerhwette*). These doublets result from changes of form of (parts of) loan words in analogy to (parts of) Old English words that sound phonologically similar and are consociated, i.e. integrated with a family of words in Old English (see chapter 7.5.). Further examples are, for instance, the first elements in *coddæppel* 'quince' lit. 'husk-apple' (folk etymology from L *cudōnia*) and *gōdæppel* 'quince' lit. 'good apple' (*gōd* 'good' being a folk etymological change of *codd* 'husk'), the second element in *colloncroh* 'white water-lily', lit. 'water-can', a folk etymology from the loan OE *crōg* 'crocus' (< L *crōcus*) in analogy to OE *crōg* 'can' and the second element in *giþ(h)rōfe* 'corn-cockle', changed from *giþrīfe* owing to the phonological similarity to OE *wudu(h)rōfe* 'sweet woodruff, etc.' (see Catalogue, s.vv. *coddæppel, gōdæppel, colloncroh, giþrīfe*).

7.4.3. INDIRECT LOAN INFLUENCE

Apart from donating words, languages can exercise *indirect loan influence* in the form of loan formations and semantic loans.[274] When new concepts have to be named, the processes subsumed under the heading *loan formation* (or calquing) represent good alternatives to borrowing words: loan formations use native, near-synonymous words in order to render the unmotivated foreign expression in the native language. The sub-classification of loan formations into loan translations, loan renditions and loan creations follows Gneuss (1955) and is governed by the morphological closeness of the replication to its Latin original.[275]

While it is, in general, no problem to recognise loan words and hybrid formations, it is not always easy to prove that a particular lexical item was modelled after a foreign term and is not a coincidental parallel. Semantic loans are the most difficult to prove, and even with loan formations, the circumstances are not always clear. When the Old English plant name happens to gloss its model, as, for instance, *hwītlēac*, which glosses L *alba cēpa* in the *Laud Herbal Glossary* (22), the evidence is at hand. For Old English *dæges ēage*, on the other hand, we cannot be sure whether it is a loan translation of L *oculus diēī* 'eye of the day' or a rendition of L *sōlis oculus* 'eye of the sun', as there is no explicit historical evidence: the Latin equivalents in the manuscripts are *anagallicum*, *consolida*, and *consolida mediana*, but not *oculus diēī* or *sōlis oculus* (*DOE*, s.v. *dæges ēage*).[276] In order to avoid misinterpretations, Gneuss (1955: 40) recommends the comparison of different translations of the Latin and Greek:

> Gegen die vielen Schwierigkeiten, die uns trotz der oben angeführten Kriterien noch bleiben, vor allem gegen die Gefahr, unabhängig-parallele Bedeutungsentwicklung als Lehnbedeutung oder Bauentsprechung als Lehnübersetzung anzusehen, bietet sich schließlich eine Untersuchungsmethode an, die im Teil III verwendet werden soll: der Vergleich mehrer Übersetzungsversionen für ein und dieselbe Vorlage.
>
> [As a solution for the many difficulties that are still present, despite the criteria mentioned above, especially against the danger of classifying independent and parallel semantic change as loan semantics or parallel constructions as loan translations, there is one method of analysis, which

274 When a word is passed on from one language to another through an intervening language it can also be called 'indirect loan influence' (Serjeantson 1935: 5); e.g., words from the Near or Far East, which have passed, for instance, through Greek and Latin before reaching Old English (e.g. OE *pipor* or OE *nǣp*).

275 On loan influence in the botanical lexicon of the Anglo-Saxons, see also, e.g., d'Aronco (1988); Sauer (1992b: 387-91; 1995: 303-8; 1996: 139-43, 2003: 170-1), Sauer and Krischke (2004: 181-3). On loan formations (calques) and their explanatory potentials in general, see, further, Kornexl (2003) and Johnson (2011).

276 Cf. Durkin (2009: 135) for this kind of problem and for further examples.

shall be employed in part III: the comparison of several translations of one and the same model text.]

This, however, is not easy, as most of the medico-botanical texts have come down to us in one copy only (see chapter 4.3.).

Loan translations
Loan translations are morpheme-by-morpheme translations of the foreign model (Gneuss 1955: 31-3).[277] It is the most often encountered form of loan formation and also the easiest one to recognise. In the plant name corpus of the present study, it is represented with c. 57 plant names.[278] Some examples are *candelwyrt* lit. 'candle-wort' ← L *herba lucernaria*, *coltnægl* lit. 'horse-nail' ← L *ungula caballi*, *eorþæppel* lit. 'earth-apple' ← L *mālum terrae*, *eorþīfig* lit. 'earth-ivy' ← L *hedera terrestris* and *þrilēfe* lit. 'three-leaf' ← L *trifolium*.

OE *candelwyrt* is not only a loan translation, but, as its first element is a loan from Latin (L *candēla*), it is also a hybrid lexeme. Other examples of hybrid loan translations are, for instance, *lēonfōt* 'lady's mantle' lit. 'lion's foot' ← L *pēs leōnis*, or *sōlsēce* 'common marygold, etc.' lit. 'sun-seek(ing plant)' ← L *sōlsequium*. The plant names *hierdewyrt sēo læsse*, *curmealle sēo læsse* and *curmealle sēo māre* are loan translations that display a morphological pattern native to English but not used very often, as the word order is reversed (see also chapter 5.3.3.3.).

Loan renditions
Loan renditions re-create the foreign word as loan translations do, but, in contrast to loan translations, they are not one-to-one representations of their model. One element of loan renditions always differs semantically and/or morphologically from the element(s) of the foreign model (Gneuss 1955: 33). Some examples from the corpus (in total, c. 23 plant names) are *brādþistel* 'milk thistle, etc.' lit. 'broad thistle' ← L *carduus folia habet lata* 'thistle that has broad leaves'; *wilde rūde* 'fen rue' lit. 'wild rue' ← L *rūta agrestis* lit. 'field rue', or *wilde wīngeard* 'black bryony' lit. 'wild vine' ← L *vitis agrestis* lit. 'field vine'. As *rūde* and *wīngeard* are Latin loan words, *wilde rūde* and *wilde wīngeard* are hybrid loan renditions.

Loan creations
Loan creations are creations of new lexemes triggered by foreign words; however, they do not translate the model as none of the elements of the new coin corresponds to the elements of the model (Gneuss 1955: 33-4). To my knowledge, only one of the complex plant names is assumed to have developed by way of loan creation, namely,

277 "Formations where the order of the elements is reversed [...], or where a Latin genitive is represented by an Old English nominative [...] or a Latin adjective [...] by an Old English noun [...] should probably still be regarded as loan translations" (Sauer 1992b: 391, fn. 34).

278 As for some of the plant names it could not be decided whether they are loan renditions or loan translations, there is some (but minor) overlap between the two categories.

OE *wēdeber(i)(g)e* lit. 'madness-berry' ← L *helleborum album* (see Catalogue, s.v. *wēdeber(i)(g)e*), but here – as with semantic loans (see below) – this is quite difficult to verify (Kastovsky 1992: 309).

Semantic loans

Semantic loans, i.e. cases where an existing native lexeme adopts the meaning of a foreign original, are most difficult to prove (Kastovsky 1992: 309; Gneuss 1955: 35-7); in the corpus of Old English complex plant names this kind of loan influence occurs, apparently, only once, namely, in the form of the hybrid formation OE *wīngeard* lit. 'vine-yard', which adopted the additional meaning 'vine' from the polysemous Latin lexeme *vīnea* 'vine, vineyard' (see Catalogue, s.v. *wīngeard*).[279]

7.5. FOLK ETYMOLOGY

Folk etymology (or popular etymology) is the process whereby a morphologically and semantically unmotivated word is changed in analogy to one (or more) existing word(s) and is thus given a new (secondary) motivation.[280] The aim of folk etymology is, according to Knappe (2004: 492-3),[281]

> [...] to secure relative linguistic motivation. Hearers achieve this aim through unconsciously establishing a group relationship between a dissociated element (motivated form) and an element in linguistic consociation in that language (motivating form) through a process of lexical reanalysis.

As Knappe (2004: 493) points out, phonological motivation is obligatory, while semantic motivation and formal or conceptual change are optional. Thus, a plant name like OE *lufestice* 'lovage', lit. 'love-stich', which has arisen from MedL *levisticum, luvesticum* < L *ligusticum* by folk etymological change on the basis of the phonological similarity of the Latin term and the Old English lexemes *lufu* 'love' and *stice* 'stitch' shows no sign of semantic motivation; the formal analysis of the Old English plant name into *lufe* and *stice* is possible, but to explain the semantics of this compound on the basis of the denotational meanings of its parts is not possible (see Catalogue, s.v. *lufestice*). In contrast, there is a semantic motivation in *codd* 'husk', part of OE *coddæppel* 'quince' lit. 'husk-apple', which was changed from L *cudōnia (mālum)* 'quince' by folk etymology and the concept denoted by the compound: quinces have noticeably hard skins (see Catalogue, s.v. *coddæppel*).

The association of two words (or parts of words) on the basis of phonological and/or semantic similarities is a mental process (Knappe 2004: 491) and it figures as

279 For cases of semantic loan among the simplexes of the Old English *Herbarium*, see d'Aronco (1988).

280 For secondary motivation, see further chapter 6.2.

281 For an overview of mechanisms and processes of folk etymology see, apart from Mayer (1962) and Olschansky (1996), Knappe (2004), whose model is followed in the present survey.

'formal similarity' in the set of associative relations (cf. chapter 6.3.). It can affect loan words and native lexemes in equal measure. Loan words seem to be more likely to be subjected to folk etymological change as they are unanalysable, dissociated lexemes in the receiving language: that means that they are not integrated into a family of formally, i.e. etymologically and semantically related words in the language (Leisi and Mair 1999: 51).

In contrast to foreign words changed by folk etymology, words of native origin changed by folk etymological processes are sometimes not easy to detect. For plant names such as *wermōd* 'wormwood' lit. 'man-mood', the assumption that folk etymological processes possibly accounted for the otherwise unexplainable combination of the lexemes *wer* 'man' and *mōd* 'mood' almost suggests itself (see Catalogue, s.v. *wermōd*). The case with *grundeswelige* 'groundsel' lit. 'ground-devour(ing plant)' on the other hand, is not that clear-cut. As the groundsel is a creeping plant that grows quite fast, the semantic analysis as 'plant devouring the ground' is obvious. This motivation is, however, only a secondary one, as, originally, the plant name was *gundeswelige,* i.e. literally 'pus-devour(ing) plant' (Sauer 2000: 30; see Catalogue, s.v. *grundeswelige*). As with the plant names of foreign origin, the native terms affected by folk etymology were perceived as unmotivated (dissociated) items (probably due to previous processes of morphological and/or phonological attrition), that needed to be re-integrated into the language system.

Integrating the thus isolated word or parts of it with a family of consociated words[282] in the receiving language is, apparently, a strong need of speakers of a language, and

> speakers tend to search for semantic motivation once phonological motivation has been established. [Mayer (1962: 86, 92)] notes the speakers' tendency to transgress the borderline to semantic motivation wherever possible so that the motivating form may start exerting its influence (Knappe 2004: 498).

Knappe (2004: 497-9) distinguishes between the (secondary, relative) motivation triggering the changes, on the one hand, and the resulting forms of change (form and denotation) on the other hand. Thus, four features can be distinguished with folk etymological change: phonological motivation, change of form, semantic motivation and change of denotational meaning. Only one of these, as mentioned above, is obligatory, namely, the phonological (formal) similarity between the motivated and the motivating elements. The cross-classification of these features leads to 6 types of folk etymological change: (Knappe 2004: 497-8): (1) phonological motivation; (2) phonological motivation and change of form; (3) phonological motivation and semantic motivation; (4) phonological motivation and change of denotational meaning; (5) phonological motivation and semantic motivation and change of form;

282 Consociated words are words that are integrated into a family of etymologically and semantically related words by procedures such as word-formation processes, *i*-mutation or others. Phonological similarity is not sufficient in order to speak of consociation (Leisi and Mair 1999: 51-9).

(6) phonological motivation and change of form and change of denotational meaning. Of these 6 types, two are present in the corpus of Old English complex plant names:

- Type (2), with the plant names:[283] *bisceop(wyrt)* 'marsh-mallow' lit. 'bishop's wort', *ceaster(æsc)* 'Christmas rose, etc.' lit. '*ceaster*-ash', *ceaster(wyrt)* 'Christmas rose, etc.' lit. '*ceaster*-wort', *(giþ)(h)rōfe* 'corn-cockle' lit. '*gith*-strong plant', *hamor(wyrt)* 'Christ-hellebore' lit. 'hammer-plant', *lufe-stice* 'lovage' lit. 'love-stitch', *(sumer)lida* lit. 'summer-sailor, shoot' and *wermōd* 'wormwood' lit. 'man-mood'.
- Type (5), with the plant names *codd(æppel)* 'quince' lit. 'husk-apple', *(collon)crōg* 'white waterlily' lit. 'water-can', *colte-træppe* 'caltrop, etc.' lit. 'colt's trap', *fēfer(-fuge)* 'feverfew' lit. 'fever [drive-away]', *feld(wyrt)* 'felwort' lit. 'field-plant', *gōd(æppel)* 'quince' lit. 'good apple', *grunde(swelige)* 'groundsel' lit. 'ground-devouring (plant)', *hæsel(wyrt)* 'hazelwort' lit. 'hazel-wort', *hwer-hwette* 'cucumber' lit. 'pot-whetting (plant)', *sǣ-minte* 'catmint, etc.' lit. 'sea-mint', *(sōl)sēce* 'common marygold' lit. 'sun-seeking (plant)', *stǣþ(wyrt)* 'sea-lavender' lit. 'shore-wort' and *(wulfes)teors* 'wolf's teasel' lit. 'wolf's penis'.[284]

As the numbers indicate, many of the plant names affected by folk etymology show no semantic or cognitive connection or association between the concept denoted by the plant name and the denotational meanings of the elements of the plant name, i.e. the concepts in analogy to which the original word is changed, but a number do. To expand on some examples: OE *fēferfuge* 'feverfew' goes back to L *febrifugia*. *Fēferfuge* shows phonological motivation as there is a strong phonological similarity between OE *fēfer* 'fever' and L *febris* 'fever' (owing not least to the fact that OE *fēfer* is a loan word going back to L *febris*). Apart from the phonological similarity of the two (Latin and Old English) words, semantic motivation supported the folk etymological formal change from *febrifugia* to OE *fēferfuge*: the plant feverfew is said to be helpful with fever (see Catalogue, s.v. *fēferfuge*). We can also construe a semantic connection between the concepts denoted by the native plant name OE *gundeswelige* 'groundsel' lit. 'pus-devourer' (*Épinal-Erfurt Glossary*, no. 976) and OE *grundeswelige* lit. 'ground devourer' (e.g. in *Herbarium*, no. 77): the plant is a creeper and seems to devour the ground it grows on.

The Old English ancestor of ModE *lovage* goes back to L *levisticum*, which was changed to OE *lufestice* lit. 'love-stitch' in analogy to OE *lufu* 'love' and *stice* 'stitch' without any plausible semantic connection between the senses of these two elements

---

283 The elements *not* changed by folk etymology are in brackets; where the complete plant name was changed in analogy to two Old English words, the plant name is written with a hyphen to indicate the two constituting elements. For more details for each plant name, see the entries in the Catalogue.

284 Mayer's findings (1962: 348-9), which suggest that type 5 represents the most frequent change, are somewhat corroborated by the present data with eight plant names of type (2) and thirteen plant names of type (5) (and none of the other types), but as Knappe (2004: 501) points out, the figures (based on the *OED*) have to be recalculated and checked against the results of an electronic search of the *MED*; in addition, our corpus is not representative as it covers only a part of the Old English vocabulary.

and the plant, whatsoever. In contrast, for instance, the first element of *hwer-hwette* 'cucumber', for instance, was changed from *cucur-* in L *cucurbita* 'gourd' in analogy to OE *hwer* 'pot' because the dried fruits of the plant develop a firm shell which can be used as water-bottles and pots; the second element, however, resulted from the phonological similarity of *-bita* and OE *hwettan* 'to whet' (see Catalogue, s.v. *hwerhwette*).

Sometimes, the part of the plant name changed by folk etymology is supported by a native, second, element, which was probably regarded necessary to disambiguate the new lexeme and to mark it as a plant name. L *hibiscum* 'mallow', for instance, was changed to OE *bisceop* 'bishop' on the basis of phonological motivation and forms, together with the native element *wyrt* 'plant', the plant name OE *bisceopwyrt* 'marsh mallow'; *coddæppel* 'quince', lit. 'husk-apple' goes back to the combination of L *cudōnium* and the native element OE *æppel*.[285]

7.6. SUMMARY

The amount of loan influence in the corpus of Old English complex plant names is remarkable, with c. 36 percent of the plant names affected.[286] Some of the plant names are affected in more than one way and cross-classifications are appropriate. The hybrid loan translations *candelwyrt* and *sōlsēce* and the hybrid loan rendition *wilde rūde*, for instance, attest to this, and demonstrate that the people engaged in reading, translating or interpreting Latin texts took pleasure in playing with their language, rendering the foreign terms in their vernacular in a wide variety of ways, in addition to taking them over as loan words and integrating them into the Old English language system.

From the nature of the Old English complex plant names, however, we also learn that the uneducated Anglo-Saxons were equally involved in the formation and change of native plant names and plant names of foreign origin, namely, when folk etymology comes into play. OE *gōdæppel* lit. 'good apple' is the result of the folk etymological change of OE *coddæppel* lit. 'husk apple'; and OE *coddæppel*, in turn, is the product of folk etymological change of L *cotōnium, cydōnium*, which, on the other hand, is a (learned) loan from Latin (see Catalogue, s.vv. *gōdæppel*, *coddæppel*).

In plant names affected by folk etymology, the original loan or the non-transparent native term is either partially or completely replaced by one (or more) Old English lexeme(s) with which it is associated formally, i.e. phonetically and/or semantically (cf. also chapter 6.4. (30)). As folk etymology reflects the "desire on the part of language users to establish iconic relationships with other words in the language" (Durkin 2009: 56), it results in at least morphologically transparent Old English lexemes, but, occasionally, semantic secondary motivation supports the morphological motivation.

285 Pleonastic plant names function along the same lines (see chapter 5.3.2.2.1.1. (b)).

286 As some of the plant names are affected twice by language contact – once directly, when one of the elements of the plant name is a loan word and indirectly, when the plant name is a loan translation, rendition or creation, they also appear twice in this number; compare, for instance, Catalogue (s.v. *lēonfōt*).

As pointed out in the preceding sections, some of the loans were taken over fairly early, i.e. in the West Germanic period (pre-450 AD), and some fairly late, that is towards the end of the Old English period (between c. 950 and 1100). This produces not only different chronological layers among the loan words, but also doublets when some of the terms were borrowed twice, but under different linguistic circumstances and with different linguistic consequences.

The exact period of borrowing cannot, however, be proved easily. When, for example, a loan word appears in more than one Germanic language, including Old English, this can either be the result of polygenetic borrowing, i.e. when the word was borrowed into each of the languages independently, or the result of a chain process, where the Latin word was borrowed into just one Germanic language, first, and then into all the others. This process is termed 'inter-Germanic borrowing' by Carr (1939: 3). Two cases of inter-Germanic borrowing in the botanical vocabulary are, according to Carr (1939: 32, 36), ON *geirlaukr* from OE *gārlēac* 'garlic' and ON *palmtrê* from OE *palmtrēow* 'palm tree'. In contrast, MedL *levisticum*, *luvesticum* < L *ligusticum* (Catalogue, s.v. *lufestice*), recorded in OE as *lufestice* lit. 'love-stitch', in OHG as *lubistecho*, MHG *liebstockel*, G *Liebstöckel* (remotivated after G *Liebe* 'love' and G *Stock* 'stick'), and in the North Germanic languages as Norw *lovstikke* and Sw *libsticka* was probably borrowed independently into the Germanic languages.

8. *BĒAM & TRĒOW*

One of the most salient paradigmatic relationships between two lexical items is the relation of *synonymy*. Two or more lexemes are synonymous if they possess "very closely related meanings [...]. [Synonyms] can often, though not always, be substituted for each other in sentences" (Yule 2006: 104). While two synonyms may be able to replace each other in a number of sentences, in certain contexts the use of one synonym may be appropriate while the use of the other is not: There is no such thing as *total* synonymy. One of the discriminating factors can be a different collocational range of the *partial* synonyms, while diatopic, diastratic or functional variation can also tip the scales in favour of one or the other lexeme. In natural languages, exact synonyms rarely occur, which is purely a question of economy: storing two synonyms and their identical meanings in one's mental lexicon unnecessarily locks up mental capacities that could be used elsewhere. It is highly un-economical for processes of speech production, as it is also for speech reception processing.

It is all the more surprising that two Old English lexemes, namely, *bēam* and *trēow*, at first sight, seem to have been used interchangeably in Anglo-Saxon times: Their most common and also most general sense is 'tree', both have Indo-European roots and both are used to form Old English complex plant names; as mentioned earlier (see chapter 6.6.), *bēam* produced 19 plant names, *trēow* 30 plant names, and – strikingly often, they form twin names as, e.g., *cederbēam* 'cedar of Lebanon' and *cedertrēow* 'cedar of Lebanon', *cirisbēam* and *ciristrēow* 'cherry-tree' or *cwicbēam* and *cwictrēow* 'quickbeam, etc.' (for a complete list, see further below). The question of why there are two synonyms denoting 'tree' in Old English, and why these form, again, synonymous Old English tree names will be addressed in the following, taking into account their denotation, their etymologies and further semantic developments, and the tree names of which they are constituents.

***Denotations of* bēam *and* trēow**

The *DOE* (s.v. *bēam*) notes four senses for *bēam*: (i) 'tree', (ii) 'timber', (iii) 'object made out of wood' and (iv) 'gallows'. According to the *ClH* (s.v. *trēow*), OE *trēow* has basically two senses: (i) 'tree' and related concepts, such as (in the order provided by *ClH*) 'wood, timber; beam, log, stake, stick; wood, grove; tree of the cross, cross, Rood'; and (ii) 'truth' and related concepts. As with *bēam*, the meaning 'tree' for *trēow* was, obviously, extended via conceptual contiguity to 'timber, etc.', 'wood, etc.' and to the 'Christian cross'.[287] Thus, the two terms were, apparently, almost completely synonymous.

287 The senses 'truth' and related concepts such as 'fidelity, faith, trust, belief, pledge, promise, agreement, treaty, favour, grace, kindness' can be explained as metaphorical extensions and are

Both were used with reference to 'tree, perennial plant with woody stem' in a rather interchangeable way, as can be seen in the following two examples: Here, the two terms are used in identical contexts serving to refer to the identical concepts:

> [...] ic seah on bearwe beam hlifian, tanum torhtne. þæt treow wæs on wynne, wudu weaxende (Riddle 53 in the *Exeter Book*, 207).
>
> [In the grove, I saw a tree rising high, with bright branches. The tree was in joy, a growing wood.]

> [...] nim þonne ele and hunig and beorman [...] and ælces treowcynnes dæl þe on þæm lande sy gewexen, butan heardan beaman […]. (Charm 'For unfruitful land' in *Anglo-Saxon Minor Poems*, 116).
>
> [take, then, oil and honey and yeast and a part of every kind of tree that has grown in this land, except for those of hard trees.]

In addition, both terms were used to denote metonymically the wood of trees as well as products made of wood, such as timber and logs, and both were used to denote the Christian cross. The sole difference, apparently, lay in the distinction of *bēam* as denoting, additionally, '(processed) objects made of wood' (e.g., stave-runes, gallows), while *trēow*, in contrast to *bēam*, apparently also served to denote collectivities of trees, i.e. 'woods' and 'groves'.

Diaphasic variation

In terms of text type and register, the simplexes *bēam* and *trēow* differ considerably. According to Hooke (2009: 75), OE *bēam* was "very common in literature, especially in ecclesiastical literature, but not common in place names and almost confined in Anglo-Saxon charters to southern England", while OE *trēow*, in contrast, was "particularly common in place names and charter boundaries, which are most likely to represent the everyday language of the period" (Hooke 2009: 77). The same applies, according to Hooke, to *bēam* and *trēow* used as elements in complex lexemes, both when denoting particular species and when referring to individual trees picked out as landmarks: While "the *elebēam* remains the only *bēam* tree species in charters and early place-names, at least in southern and central England […]" and only *æppeltrēow*, *mapultrēow* and *plūmtrēow* appear in the charters and as place names (Hooke 2009: 78), the use of *trēow*, especially when associated with a personal name, is particularly frequent.

---

presumably based on the fact that trees are considered to be strong and steadfast; cf. also G *Treue*. The *DOE*, which has been published only as far as letter G up to now, obviously, contains no entry for OE *trēow*.

Etymology of the terms

According to *Pokorny*, the Old English noun *bēam* goes back to the Indo-European stratum and is cognate with various East and West Germanic names for 'tree': with OFris *bām*, OS *bām, bōm*, MDu and Du *boom*, OHG, MHG *boum* and G *Baum* 'tree'. The relation of the West Germanic etymon **bauma* to its East Germanic cognates Goth *bagms* and ON *baðmr* 'tree' is not entirely clear according to *Pokorny* (s.v. *bheu-*, **bheuə-*, etc.), but the *American Heritage Dictionary* (appendix, s.v. *bheuə-*) posits two Germanic etymons: **bauma-z* and **bagma-z*. All stem from the Indo-European root **bheu-*, **bheuə-*, **bhou-*, **bhu-* 'to grow, to swell', which is extant, among others, also as the Old English copula *bēon* 'be'. It is, therefore, quite safe to say that, semantically, the sense of 'tree' as a 'growing, living organism' developed first, and that the sense 'wooden stem, log' probably evolved secondarily as a metonymic meaning extension, on the basis of a relation of contiguity between 'tree' and 'log, beam' since, economically, trees were (and still are) valued as timber.[288]

OE *trēow* is a Germanic word attested, e.g., as ME *trēo*, OFris *trē*, OS *trio, treo, trew-*, ON *trē*, Dan *træ*, Sw *trä* 'timber', *träd* 'tree' and Goth *triu*. It is not attested at all in Old High German texts[289] and in Low German and Dutch it is now obsolete (*ODEE*, s.v. *tree*; *OED*, s.v. *tree*). Its etymology can be traced back via Gmc **trewo-* to IE **deru-*, **doru-* (according to the *ODEE*, s.v. *tree*) or to IE **-tro-* (according to *Kluge*, s.v. *Holunder*); it is cognate with Skr *dru* 'tree, wood', and OSlav *drievo* (from *dervo*) 'tree, wood', etc. According to *Pokorny* (s.v. *deru-*, etc.), the sense of 'tree' was there first, and, probably, goes back to a substantivised adjective **doru* 'the hard one', from which the sense 'tree' evolved (cf. also L *rōbur* 'oak', which, etymologically and semantically, is cognate to L *rōbustus* 'hard, oaken'; *Lewis and Short*, s.vv. *rōbur, rōbustus*).

***Modern English* beam *and* tree**

While in Old English *bēam* and *trēow* were largely synonymous, as described above, their meanings changed during the Middle English period. OE *trēow* survived with its principle meaning as ModE *tree* 'tree', while OE *bēam* ceased to be used referring to 'tree' towards the end of the Anglo-Saxon period (*OED*, s.v. *beam*, I (1)). An allusion to the beam's original meaning is still felt in the compounds ModE *hornbeam, whitebeam, quickbeam*, yet these compounds are now labelled "unanalyzed" by the *OED* (s.v. *beam*, I (1)), which presumably reflects the fact that the original meaning, 'tree', is not present any more to speakers of Modern English and that the names are not fully transparent any more.

The principal meaning of ModE *beam* is "a large piece of squared timber, long in proportion to its breadth and thickness, such as is used in house- or ship-building,

288 This is in contrast to the more careful argumentation given by the *ODEE* (s.v. *beam*): "[...] but the uncertainty whether the orig. sense was 'wooden stem or block' or 'growing tree' makes the ult. origin doubtful."

289 A number of German and Old English plant names, e.g. G *Flieder*, *Holunder*, *Maßholder, Reckholder*, *Wachholder*, OE *apuldur*, *mapuldur*, *spaldur*, etc., are derivations with the tree name suffix *-dur*, *-dor*, *-dr*, which probably can be related to Gmc **trewo*, IE **-tro-* (cf. *Kluge*, s.v. *Holunder*).

where beams form important parts of the structure: originally, the squared timber of a whole tree, but now used without any such restriction [...]" (*OED*, s.v. *beam*, I (3a)), from which other, related meanings derived through metaphorical and metonymic meaning extensions. The senses of 'cross' and 'gallows' were completely lost for ModE *beam* while *tree* acquired them about the time *beam* was losing them. According to the *MED* (s.v. *trē*, 3), *tree* was first recorded c. 1150 as referring to the cross on which Christ was crucified;[290] a1225 it was recorded first as referring to gallows and, according to the *OED* (s.v. *tree*, 4 (a)), in 1366 it was recorded as referring to "the wooden shaft of a spear, handle of an implement, etc.; hence, a spear, lance".[291]

Bēam *and* trēow *as elements of complex tree names*

As mentioned in the introduction to this chapter, *bēam* and *trēow* are constituent elements in complex tree names, which are, from the point of view of denotation as well as from the point of view of signification and motivation of morphological shape and structure, identical. These are: *cederbēam* 'cedar of Lebanon' and *cedertrēow* 'cedar of Lebanon', *cirisbēam* and *ciristrēow* 'cherry-tree', *cwicbēam* and *cwictrēow* 'quick-beam, etc.', *elebēam* and *eletrēow* 'olive-tree', *fīcbēam* and *fīctrēow* 'fig-tree', *fūl bēam* and *fūletrēow* 'alder buckthorn', *lāwerbēam* and *lāwertrēow* 'laurel', and *pīnbēam* and *pīntrēow* 'pine-tree'. With the exception of *cwicbēam* and *cwictrēow*, and of *fūl bēam* and *fūletrēow*, the first elements of these compounds are all loans from Latin, which denote, when used as simplexes, the same trees as the complex tree names. These formations, called, from a morpho-semantic point of view, 'pleonastic', indicate that even loan words that are fully integrated into the system of Old English, apparently, were not always perceived as entirely part of the Old English language system; in these cases, they were equipped with an Old English, disambiguating term which indicated that the concepts they denoted belonged to the domain of plants (cf. chapter 5.3.2.2.1.1. (b)).

The question of why there are two different forms of these plant terms is, still, not evident. Apparently, they differed, if at all, only from the point of view of diaphasic variation (see above).

A clue to the puzzle?

As has been demonstrated so far, the differences in the meanings of both *bēam* and *trēow* cannot account for the synonymous use of the terms in Old English as exemplified by their use in one and the same text (see examples above). Neither do these differences find expression in the twin complex plant names.

---

290 An earlier reference to the cross can be found c. 1000, in the *Dream of the Rood*, line 25: "[…] ic.beheold hreowceari hælendes treow" ['I beheld sorrowful the tree of the Saviour'] (quoted from the *OED*, s.v. *tree*, 4 (a)), but here it seems as if *trēow* had not yet been lexicalised as 'cross', as *trēow* is used figuratively with the element *hælendes* '(the) Saviour's' serving as a disambiguating element.

291 The dates given by the *OED* (s.v. *tree*, 4 (a)) are: c. 1275 first record in the sense of 'cross' and 1425 first record of *tree* used for 'gallows'.

The semantic development of the terms in Middle and Modern English points, however, to the fact that, towards the end of the Old English period, the use of *bēam* with the metonymic extension of its original meaning, denoting, now, largely 'log, beam' became more prominent than its use as 'tree', eventually leading to a complete shift from 'tree' to 'log, beam' in Modern English. This seems to confirm Hooke's (2009) observation that, in literary texts, *bēam* was used more often than *trēow*; and that *trēow* occurs frequently in the charters: As Old English literary texts were seldom concerned with the natural world as a concrete object of observation (Neville 2003: 291, 297; see also fn. 26), the reference to 'concrete' trees is not to be expected, in contrast to the reference to the plant world in a rather figurative way.

What is the solution, then? The clue to this puzzle is offered by the analysis of the plant name corpus in terms of diachronic strata. Namely, it is striking that, for the West Germanic stratum of the botanical lexicon, 12 compounds with *bēam* face 2 compounds with *trēow*, while in Old English, 30 compounds with *trēow* stand against 19 compounds with *bēam* (see chapters 7.1.2. and 7.1.3.).

The compounds with *bēam* that go back to West Germanic are *cederbēam* 'cedar of Lebanon'; *cirisbēam* 'cherry-tree', *cistbēam* 'chestnut tree', *cistelbēam* 'chestnut tree', *cistenbēam* 'chestnut tree', *fīcbēam* 'fig-tree', *hnutbēam* 'walnut', *lāwerbēam* 'laurel', *mōrbēam* 'mulberry tree', *pīnbēam* 'pine-tree' and *fūla bēam* 'alder buckthorn'. The two compounds with *trēow*, namely, *fīctrēow* 'fig-tree' and *wīntrēow* 'vine', are considered to have been formed in the earlier, Germanic stage, as only Old Norse has the compounds *fīgtrê* and *vîntrê*.

From the practically stagnant and eventually falling numbers for the use of *bēam* and the steeply rising numbers for *trēow* as an element of complex tree names – 12 West Germanic vs. 19 Old English complex plant names with *bēam* as the second element, and 2 Germanic vs. 30 Old English complex plant names with *trēow* as second element – we can deduce that in contrast to previous stages of language, in Old English new tree names were formed with the element *trēow* rather than with the element *bēam*. This is strongly supported by the evidence from the charters, where for trees used as landmarks the (pre- or postmodified) term *trēow* is, clearly, preferred over *bēam*. The term *trēow* 'tree' was constantly gaining ground, while the use of *bēam* as 'tree' was, at first, stagnant, respectively, was in retreat towards the end of the period, then continuing as 'beam, log' into the Middle and Modern English periods.

As a result, we can conclude that while in West Germanic *bēam* was the 'default' term for 'tree', during the Old English period, the complex tree names with *bēam* as a second element were flanked by twin tree names with *trēow* as a second element because of the gradual retreat of the sense of *bēam* denoting 'tree'. The tree name pairs displaying morphological and semantic synonymy in Old English are, thus, explainable as reflecting this period of transition, when *bēam* still could be used to denote trees, and when *trēow* was a modern, perhaps a fashionable, means of expressing the concept TREE.

9. SUMMARY

> [...] it is like following the various intertwining threads in a tapestry until one finally ends up with a glimpse of what these intertwining threads produce: a representation of a major pattern, a tapestry from a synchronic point of view, which still lets us recognize its genesis.
>
> (Kastovsky 2006: 161)

Language is far more than a means of communication: it offers the possibility of conceptualising and organising extra-linguistic reality with the help of linguistic signs and linguistic categories. By encoding what we see, what we feel and what we experience into new linguistic units, we not only reshape our conceptual categorisation of the world, but also enlarge the lexicon of the language we use. Thus, processes of lexicogenesis, such as word-formation, enhance our means of expression and contribute decisively to the development of the language. A language equipped in this way with various means of expression is able to meet the demands of different communication situations, from oral, everyday verbal exchanges to highly elaborate text production.

Very early, the Anglo-Saxons made extensive use of their vernacular, Old English, and, thus contributed decisively to its internal and external elaboration (see chapter 2.2.). Internally, the language was refined with regard to its syntax and enlarged with regard to its lexicon; externally, Old English was expanded in its range of uses into all discourse traditions – from private letters to sermons, from Bible exegesis to scientific treatises – and also into medico-botanical literature. According to Mitchell and Robinson (2012: 93),

> At the time of the Norman Conquest, England – although she no longer led Western Europe in monastic learning, as she had in the eighth century – was fruitful ground for new forms of devotion, was famous for her craftsmen, and had a well-developed economy and the most advanced administration north of the Alps. It is of special interest here that her language was far more developed for the expression of both prose and poetry than any other contemporary European vernacular and that authors using it sometimes rose to very great heights.

One of the most impressive linguistic heritages of the Anglo-Saxons is their botanical lexicon. More than nine hundred Old English plant names testify not only to the Anglo-Saxons' various techniques for introducing new concepts and new lexemes, but also to their specific ways of observing the natural world and coining it into words.

The aim of the present study was to look at the 467 complex plant names of the Anglo-Saxons from different linguistic angles: from the point of view of word-formation, with regard to the semantics and by investigating the diachronic stratification, including loan influences from contact languages.

The specific nature of the Anglo-Saxon texts, in which the botanical lexicon was recorded and which were the products of a mainly adult, male, monastic elite trained in reading and speaking Latin, and writing texts in the language of communicative distance was explored in chapters 2 and 3 of the present work. In combination with the limited number of texts, due to unfit, sometimes even harmful storage conditions over the centuries, in consequence of the manuscript destruction by ravaging Vikings, or following the dissolution of the monasteries in the 16th century, the textual situation had to be mentioned repeatedly as a decisive factor that limits possible inferences from the linguistic analysis, and especially with regard to the quantitative evaluation of the findings. While numerical differences between, e.g., different morphological types, or between different motivation categories within the corpus investigated were pointed out, the numbers can do no more than indicate tendencies and preferences, and thus shift the focus of the analysis to the qualitative interpretation of the data.

In regard to the theoretical framework, the analysis began with a morphological description of the plant names, focusing especially on their modifier and head structure. The majority of the studies on the morphology of complex lexemes base their findings on intra-linguistic criteria, and semantic relationships between the elements of the complex formations are either described as abstract relations identified on the basis of syntactic-semantic paraphrases of the complex lexemes, with the help of so-called mnemonic devices, or via the notion of recoverably deletable predicates. While such approaches certainly provide extremely valuable data, it has been suggested in the present survey that the semantic structure of complex lexemes has to be analysed in a much wider framework, where structuralist and the cognitive approaches to language, the notions of associative relations taken over from *gestalt psychology*, and extra-linguistic, botanical information on the plants have to be integrated.

The formulation of paraphrases having been ruled out as a decisive means for the description of semantic relations between the elements of complex lexies, other ways of describing the meanings and the motivations of the elements had to be found. On the basis of Blank's (1997) theoretical framework, which he developed for the description of semantic change in Romance languages, it was shown that the formation of complex lexies is based on motives in extra-linguistic reality and, with the exception of plant names affected by processes of folk etymology, does not depend solely on intra-linguistic factors. The motives (grouped in motivation categories) – provided in the case of the Old English plant names by the plants' morphologies, typical habitats, and the like – are coded into language with the help of six different associative relations, the most frequent ones being taxonomic sub-ordination, conceptual contiguity and metaphorical similarity.

The different kinds of semantic information possibly encoded in each of the Old English complex plant names is summarised exemplarily in the following. Each plant name is equipped with a denotational meaning, i.e. "the relationship that holds between that lexeme and persons, things, places, properties, processes and activities external to the language-system" (Lyons 1977: 1, 207). The denotational meaning is represented on the language-internal level as the sememic meaning – the sum of all semes (the sememe) –, such as 'has green leaves', 'has flowers', or 'prefers moist habitats', etc., but is not identical with it. Sememic meaning is related to the language system, identifying, for instance, synonyms and antonyms, but does not necessarily include additional, nondistinctive attributes of the denotata; these are best described within the frameworks of the prototype and frame theories.

Plant names, further, have a 'meaning' related to style or register, and one determined by diachronic or diatopic variation. This kind of meaning is not easy to establish for the Old English complex plant names due to the limited number and the very specific nature of the medico-botanical texts, but it must, undoubtedly, have existed and, for some plant names, tentative interpretations can be made. It is, for instance, highly likely that plant names of the Germanic and West Germanic strata were used both in formal and informal contexts because they belong to the stock of the lexicon common to both the uneducated and the educated, formed long before the Christianity and the Benedictine Reform introduced learned Latin terminology to Anglo-Saxon England. In contrast, loans from Latin that were not integrated into the system of Old English (i.e. 'foreign words') were probably confined to a learned, monastic environment and a specific textual and stylistic situation.

Complex plant names have, in addition, a meaning that relates to the denotations of their elements in relation to the denotations of the compound plant names – a relation which is not necessarily reflected in the sememe and can also not be captured in the morpho-semantic analysis. This is evident especially in highly idiomatised plant names such as *hræfnes fōt* 'crowfoot, five-leaf', where the denotations of the individual elements – *hræfn* 'raven' and *fōt* 'foot', as well as their semes, e.g. 'bird', 'black', etc. for *hræfn* 'raven', and 'part of body', etc. for *fōt* 'foot' – are not necessarily part of the sememe of the plant name, just as the literal meaning of the combination, lit. 'raven's foot', is not a part of it.

Taking *hræfnes fōt* 'crowfoot' as an example to be analysed further, we can sketch the following description: crowfoot is a plant sprouting in meadows and lawns, often near brooks and fens; it has yellow flowers and an acrid taste and is, because of the latter, avoided by animals (Marzell, s.v. *Ranunculus (acer) acris* L.). Of all the attributes of the plant, the leaves of *hræfnes fōt* are so noticeably formed as to be compared to crows' feet; and this is of all attributes the one which, eventually, was coded in the plant's name. It is situated on the language-external level of knowledge (in contrast to meaning relating, for instance, to diatopic and diastratic marking, or the meaning established on the basis of distinctive semantic features, which are language-dependent) and was verbalised metaphorically. This occurred although the Old English language did provide means of expressing the fact that leaves of plants are physically divided (via conceptual contiguity) as can be seen from the example of *fīflēafe* 'fiveleaf', which denotes, just as *hræfnes fōt*, Potentilla reptans, and which also refers to the plant's divided leaves.

Plant names that make use of metaphorical similarity in either element (partial and total metaphors), and plant names that make use of conceptual contiguity in their second element (metonymic formations), testify to the linguistic creativity of the Anglo-Saxon. Expecially N_{gen}+N formations evoke in various ways the image of animals: besides *hræfnes fōt* alluding to ravens, there are also plant names alluding to, e.g. wolves (*wulfescamb* 'wild teasel' lit. 'wolf's comb' and *wulfestǣsel* 'wild teasel' lit. 'wolf's teasel'), to dogs (*hundeshēafod* 'small snapdragon' lit. 'dog's head'), to fowl (*cicena mete* 'chicken-weed' lit. 'chicken's food', *hennebelle* 'henbane' lit. 'hen's bell' and *hennewōl* 'henbane' lit. 'hen's disease') or to foxes (*foxes fōt* 'unbranched burr-reed, etc.' lit. 'fox's foot' and *foxes glōfa* 'foxglove, etc.' lit. 'fox's glove'). Notwithstanding the fact that the status of N_{gen}+N formations is disputed, there is no doubt that these plant names are highly lexicalised and should be recognised as Old English complex words in their own right.

The semantic analysis that distinguishes between the denotation (the plant) and the signification (the meaning of the plant name on the surface level of language, i.e. the semantic relationship between the contents of the compositional elements; cf. fn. 216), between the motivation and the associative relations, allows more fine-grained analyses of the semantics of the plant names than traditional approaches would have permitted, and, what is more, it helps in pinpointing differences between apparently similar plant names, and similarities between apparently very different plant names. This is the case when, for instance, plant names with antonymous signification have synonymous denotations (and, at times, different morphological shapes), as is the case, for instance, with those denoting Galium aparine L. 'cleavers', namely, *lange clite* lit. 'long cleavers', *smale clāte* lit. 'small cleavers', *smale clīfe* lit. 'small cleavers' and *smale cliþwyrt* 'small cleavers'. The adjectives *smale*, *smæl* 'small, narrow' and *lang* 'long', which both refer to the motive 'size', are antonymous while the second elements are identical, and, still, the complex plant names denote one and the same plant. The clue to these plant names is the motivational structure: both first elements encode the motive 'size', but the size refers to different parts of the plant: in *lange clite* the first element refers to the long stems of Galium aparine, while the first elements of *smale clāte*, *smale clīfe* and *smale cliþwyrt* refer to the cleaving fruits, which are much smaller than those of the plant denoted by the second elements *clāte*, *clīfe*, *clite* and *cliþwyrt*, respectively, namely, Arctium lappa L., cleavers.

In taking an onomasiological point of view, we gain the insight that motives might be expressed in quite diverse ways. Smallness in terms of 'size', for instance, is not only encoded as *smale* as in the example mentioned above, but also as *lȳtel* (in *lȳtle culmille* 'common centaury' lit. 'small centaury') and as *læsse* (the comparative of *lȳtel*; cf., e.g., *læsse bisceopwyrt* 'betony' lit. 'lesser betony'), as *dweorg* 'dwarf' (e.g. in *dweorged(w)os(t)(l)e* 'pennyroyal' lit. 'dwarf's *dost*') or via the element *cnēow* 'knee' (in *cnēowholen* 'knee-holly, etc.' lit. 'knee-holly'). These semantically and morphologically disparate elements – the adjectives *smale* 'small' and *lȳtel* 'small' (including *læsse* 'less'), and the nouns *dweorg* 'dwarf' and *cnēow* 'knee' – can be related to the plants and their concepts via their associative relations to the plants: these are, in the first two cases, conceptual contiguity, in the last two, metaphorical similarity. As the metaphorical similarity in the last two plant names mentioned pertains, however, to the relationship between the first element and the complex plant name, all these plant names are, morphologically, endocentric.

When taking an onomasiological perspective, but starting with the plant concept as a whole, the Old English botanical lexicon further attests to the fact that the same attributes of the same plants do not have to be rendered in the same way as Old English plant names. To give two examples: Cyclamen europaeum L., the plant sowbread, can be denoted by *eorþæppel* lit. 'earth-fruit', or 'earth-apple', by *eorþ-hnutu* lit. 'earth-nut' and by *eorþnafela* lit. 'earth-navel'. All three plant names are metaphorical formations, with the bulbous roots of the plant being compared to fruit (or, more specifically, to apples), to nuts or to navels.[292] In all plant names, however, the first element *eorþe* 'earth' is probably used to indicate that these elements do refer to plants and not to parts of the body (*nafela* 'navel'), respectively that these kinds of fruit (apples) and nuts grow in the earth and not, as usual, on trees.

Potentilla reptans is present in the Old English botanical lexicon with three plant names: *fīffingre* lit. 'five fingers', *fīflēafe* lit. 'five leaves' and *hræfnes fōt* lit. 'raven's foot': all of them refer to the shape of the leaves, which are so divided as to be compared, metaphorically, to five fingers (*fīffingre*) or to a raven's foot (*hræfnes fōt*); *fīflēafe*, in contrast, verbalises the contiguous feature that the plant's leaves are made up of five parts directly; since the second element denotes the plant's part, the relation is a meronymical one.

By contrast, the plant Melissa officinalis, the (common) balm, is present with three plant names, but all of them encode different motives. The first element of *bēowyrt* lit. 'bee-wort' verbalises the fact that the plant attracts bees with its sweet scent; *blind(e)net(e)l(e)* lit. 'blind nettle' focuses on the plant's similarity to nettles, stressing, however, that *blind(e)net(e)l(e)* lacks the stinging properties of nettles, i.e. is 'blind'. The third Old English name of the plant, *meduwyrt* lit. 'mead-wort' refers in its first element to the fact that the plant can be used to prepare sweet beverages.

Besides uncovering the different layers of semantic information, the analysis offered also notable insights into folk taxonomic structures of Old English. Numerically, the ratio of life form, generic and specific terms confirms largely the taxonomic system found in the language of the Tzeltal, a vernacular spoken in Mexico (see chapter 6.6.), and the majority of the complex Old English (and Tzeltal) plant names echo scientific botanical terminology as they are endocentric formations, as well (see below). However, in contrast to scientific taxonomy, where the plant world is grouped according to one, consistent criterion, namely, the physical properties of the sexual organs of the plants, Anglo-Saxon folk taxonomy makes use of life form terms that do not necessarily have to be based on the same set of criteria.[293] While, e.g., *trēow* as contrasted to *wyrt* in the pair of compounds *ellentrēow* 'elder' lit. 'elder-tree' and *ellenwyrt* 'dwarf-elder' lit. 'elder-wort' clearly refers to the difference in height of the related plants Sambucus nigra and Sambuculus ebulus, *þorn*,[294] *þyrne* and *þistel*

292 An 11th or 12th century illustration of Cyclamen europaeum L. in Oxford, Bodleian Library, Ashmole 1431 (S.C. 7523), fol. 10r (Ohlgren 221(17)) shows nicely the bulbous roots of the plant.

293 Scientific taxonomy is strictly based on the morphology of the sexual organs of the plants (Groß 2002: 15), and never on their effect on humans or animals, their colour, shape or on other attributes.

294 The meaning extension of *þorn* 'thorn' to 'thorn-bush' is based on a relation of conceptual contiguity, or meronymy: as the plant names *æppelþorn* 'crab-tree', *brēmelþorn* 'bramble',

refer to the plants' pricking qualities; the term *þung* 'poisonous plant', which reveals its potential as a life form when in the function of the head of the complex plant names *brōcþung* 'cowbane' lit. 'brook poisonous plant', *clufþung* 'celery-leaved crowfoot' lit. 'clove(d) poisonous plant', *micel þung* 'wolf's bane' lit. 'large poisonous plant' and *þelneþung* 'celery-leaved crowfoot' lit. '? poisonous plant', points to a feature that is not related to the morphology of the plants, but to a criterion of greatest (cultural) significance: The information that a plant is poisonous or not is, obviously, crucial since it can decide over life and death. The life form terms occurring are those denoting trees – OE *bēam*, *trēow* both 'tree' – which are distinguished from those denoting thorny bushes – OE *brēmel*, *brēr*, *þorn*, *þyrne*, all 'thorn-bush' –, from mushrooms – OE *swamm* 'mushroom' –, and from low growing plants – OE *wyrt*, *wyrtel* both 'plant, wort' –, as well as from pricking plants – OE *þistel* 'thistle' – and from the poisonous ones – *þung* 'poisonous plant'.

In addition, it turned out that the taxonomic centrality of terms denoting taxa on the generic level resulted in semantic shift in those cases where the plants were of particularly high economic and cultural value: most notably, *lēac* 'leek', and *brēmel* 'bramble'. Thereby, these plant names were moved to the super-ordinate taxonomic level (associative relation: taxonomic super-ordination): *lēac* 'leek', as a generic term developed the additional meaning 'garden-herb, plant', now a life form term, and *brēmel* 'bramble' developed the additional meaning 'thorn-bush'. These super-ordinate, additional meanings can be observed, for instance, in the plant names *ynnelēac* 'onion', *fuglesleāc* 'tufted vetch' and *hēopbrēmel* 'dog rose', but also in the non-botanical term *lēactūn* '(herb-)garden' (see chapter 6.6.).

As the original meanings of the terms subjected to semantic shift did not necessarily cease to exist, parasitic categorisation and semantic shift created structures of vertical polysemy: thus, for instance, *lēac* has proved to be, at the same time, a hyperonym in relation to generic terms when used as the life form term denoting 'garden-herb, plant', and its own hyponym when it denotes the generic taxon 'leek, onion, etc.' in particular. Vertical polysemy holding between the life form level and the generic level is also extant, for instance, in *æppel* 'apple' and 'fruit' (see Catalogue, s.vv. *æppeltrēow*, *eorþæppel*) and in *brēmel* 'bramble' and 'thorn-bush'. A number of elements are vertically polysemous, with the polysemy holding between the generic and the specific level: *wilde minte* 'mint', for instance, denotes the same plant as *minte*, *smale singrēne* 'common periwinkle' as *singrēne*, *wealhmore* 'common parsnip, etc.' the same as *more*.

The examples just given indicate, further, that folk taxonomies function along basically the same principles as scientific taxonomy: both have hierarchic terminological structures with the super-ordinate term – the *genus proximum* – modified by an element referring to a specific feature of the plant – the *differentia specifica*. Here, again, the parallel to morphology must be drawn, as 'endocentric

---

haguþorn 'common hawthorn', *lūsþorn* 'spindle-tree', *pricþorn* 'holly', *slāhþorn* 'blackthorn', and *þēfanþorn* 'common buckthorn' all denote thorny bushes, it is fairly safe to assume that *þorn* was subjected to a meaning extension based on contiguous relatiohship between the plant and its part(s): the meaning of *þorn* 'thorn' as the most salient attribute of thorn-bushes was extended to denoted the life form 'thorn-bush'.

structure' is, by far, the most often attested one in the Old English plant name corpus (see below). On the other hand, the categorisation of the natural world including phenomena like parasitic categorisation (whether this results, eventually, in semantic shift or not) that produces vertical polysemy, along with frequent cases of (taxonomically horizontal) polysemy and synonymy, and the existence of exocentric (metaphorical and metonymic) formations, attests to the linguistic flexibility of the Anglo-Saxon language community, as these kinds of folk taxonomic structures are far from resembling the unequivocal and clear-cut ones of scientific taxonomy.

As regards the morphological shapes and structures of the 467 Old English complex plant names, the analysis was able to determine an impressive variety of types: N+N compounds, combinations of N_{gen}+N, combinations of Adj+N, Num+N compounds, V_{stem}+N compounds, synthetic compounds, combinations of N+N_{deadj}, $PropN_{es/ing}$+N compounds, and a number of plant names where an exact analysis was not possible, as one or both elements of the plant names are not transparent. In terms of number, the majority of the Old English complex plant names are made up of two nouns (251 plant names), then, of Adj+N (128 plant names), with N_{gen}+N formations (43 plant names) forming the third-largest group of formations. These numerical results confirm expectations raised by findings in studies on the morphology of pre-Old English (Carr 1939: 162-3) and of Early Middle English (Sauer 1992a: 150, 166)[295], and are also in line with Schmid's (2005: 139) results recently gained from the analysis of the *BUMC*[296], where he notes:

> Erwartungsgemäß ist das Muster N + N mit großem Abstand am häufigsten belegt. Zu zahlenmäßig nennenswertem Auftreten bringen es daneben nur die beiden nominalen Muster Adj + N (*greenhouse*) und [V + ing] + N (*building-block*).
>
> [According to expectations, the pattern N + N is by far the most frequently attested one. Apart from this, only the nominal patterns Adj + N (*greenhouse*) and [V + ing] + N (*building-block*) make a noteworthy appearance.]

The numbers also confirm Kastovsky's (1992: 365) statements that, in Old English, "Noun+Noun compounds represent the most frequent pattern" and that the pattern Adj+Noun was fairly productive as well (1992: 370). This is in numerical contrast to morphological types that developed rather late in the history of the Germanic languages. Thus, the type V_{stem}+N, for instance, a recent development in Germanic languages according to Carr (1939: 162), is, with only 7 formations, only poorly represented in the Old English plant name lexicon.

295 While composition was a frequently used word-formation pattern in Old English, and still is one in Modern English, the Middle English period saw a decline in numbers, which can be explained by the fact that people, apparently, preferred borrowing from French (or Latin) than to forming new words out of the native stock of linguistic material (Sauer 1992a: 7).

296 The abbreviation stands for *Bayreuth UCL Morphology Corpus* (cf. Schmid 2005: 11).

With regard to their morphological structure, the endocentric plant names clearly outweigh (with c. 365 terms) the exocentric formations (c. 90 terms). This imbalance notwithstanding, the existence of 90 exocentric formations in the Old English plant name corpus is marked, particularly because this type was non-existent in Germanic and West Germanic (Carr 1939: 162-3), and it discloses a pronounced increase in exocentric formations in Old English. The ratio of endocentric and exocentric formations varies, however, with the morphological types: Whereas the majority of the N+N and of Adj+N combinations are endocentric, most N_{gen}+N combinations and all Num+N plant names are exocentric. This correlates with the finding that most of these formations are integral metaphors, respectively metonyms, which means that in these complex formations, the first element is related to the second element on the extra-linguistic level of knowledge and, thus, rather indirectly to the plant itself. In integral metaphors and metonyms, only one, 'integral' motive can be identified – in contrast to formations where both elements can be related individually to the plant; cf., for instance the difference between the N_{gen}+N *hundeshēafod* 'small snapdragon' lit. 'dog's head' and the Adj+N formation *rēad sealh* 'purple loosestrife' lit. 'red sallow'.

With regard to the motives, according to numbers, the motivation category 'botanical information' is represented with by far the largest group of specimens (352 plant names), followed by 'habitat' with 99 plant names, 'part of plant' with 63 plant names and 'shape' with 57 plant names. The high ratio of 'botanical information' echoes the results from the morphological analysis, where the overwhelming majority of plant names was analysed as endocentric: As 'botanical information' is encoded mainly in the second element, it fulfils the function of providing the *genus proximum* of the complex formation, which is modified by its first element, the *differentia specifica*, and, thus, provides the basis for the argument that complex formations follow the "natural human tendency to see a thing identical with another already existing and at the same time different from it" (Marchand 1969: 11). It is logically consistent that the associative relation holding between these plants denoted by these types of compounds and the concepts denoted by their second elements is 'taxonomic sub-ordination', which describes a partial similarity between two taxonomically related concepts in the extra-linguistic world.

'Botanical information' also fulfils, apparently, the function of setting the 'botanical frame' for some of the borrowed terms. As the pleonastic plant names such as *boxtrēow* 'box' lit. 'box-tree', *palmtrēow* 'palm tree' lit. 'palm-tree' and *ynnelēac* 'onion' lit. 'onion-plant' indicate, at some point in the history of Old English, the loans (the first elements) were equipped with a semantically supportive (second) element, which indicated that these terms denoted entities in the natural world, and this despite the fact that most of these loan simplexes were, indeed, integrated into the system of Old English, as can be inferred, for instance, from their shapes (e.g. *i*-mutation in *ynne-*), and/or from the number of their occurrences (*palm*, for instance, is attested 20 times).

It is not surprising that, as regards the diachronic stratification of the Old English complex plant name lexicon, the amount of complex plant names that were formed in the Old English period (c. 376 plant names), clearly outnumbers the plant names that can be traced back to West Germanic (c. 79 plant names) and to Germanic (c. 12 plant names). It is, on the other hand, a notable feature of the Old English botanical lexicon that about 36 percent of the complex plant names display loan influence of Latin. The number of c. 19.5 percent hybrid formations, i.e. complex plant names made up of a native and a loan word, is in clear contrast to the ratio of loan words of Latin on the complete Anglo-Saxon vocabulary, which amounts only to about 1.7-2.5 percent[297], and demonstrates the important role that Latin has played in the formation of the Old English complex plant name lexicon. From the impressive percentage of loan influence, it cannot, however, be inferred that the Anglo-Saxons were unable to produce their own categories and concepts. In fact, the large number of hybrid formations (c. 91) show that, as most of the borrowed elements are attested as simplexes, they were institutionalised, i.e. accepted and used as part of the Old English plant name lexicon and not necessarily felt to be foreign elements. This is especially true for the early Latin loans (pre-650), which can be said to belong to the common stock of Anglo-Saxon words. As a consequence, many hybrid formations can be regarded as based on the Anglo-Saxons' knowledge of the botanical world, on their associations and on their linguistic competence and creativity, while the Latin influence can, still, be felt in these formations. The ratio of loan translations, loan renditions and loan creations is, in comparison, with c. 79 plant names,[298] slightly lower.

The strong influence of Latin that can be felt in the corpus of Old English complex plant names is, thus, on the one hand, in line with the expectations raised by the sociolinguistic settings of Anglo-Saxon text production and the nature of the Anglo-Saxon medico-botanical texts as outlined in chapter 2. The learned environment, the situation of communicative distance, the fact that authors, compilers and scribes involved in the production of the texts were trained in receiving and producing Latin texts, to name but a few factors, are echoed in the high number of plant names influenced by Latin. On the other hand, the Anglo-Saxons being an agricultural society with most of their material culture based on the natural world, we can safely assume that their knowledge of the plant world was indeed more elaborated than ours is today and that the large number of plant names not affected by loan influence is a clear indicator of their intimate knowledge of the natural world they had to live in and cope with every day.

The etymological analysis of the botanical lexicon has, in addition, proved useful in solving the question of why there are a number of synonymous tree names in terms of denotation and of signification, such as *cederbēam* and *cedertrēow* 'cedar of Lebanon'; *cirisbēam* and *ciristrēow* 'cherry-tree'; *elebēam* and *eletrēow* 'olive-tree'; *fīcbēam* and *fīctrēow* 'fig-tree'. While no morphological and semantic criteria could account for the existence of these doublets, the analysis of the diachronic stratification of the lexicon revealed a shift from the simplex *bēam* being predominantly used in West Germanic to denote trees to *trēow* being predominantly used post-Old English,

297 Compare chapter 7, running text and fn. 233.

298 As regards a few plant names, the question as to whether they are renditions or translations could not be solved (cf., e.g., Catalogue, s.v. *ācmistel*) so that there is some, but a minor overlap in the numbers.

with Old English being a transition period, where *bēam* was still in use and where *trēow* was beginning to be used more often, eventually ousting *bēam*.

It is obvious that, still, much has to be done in order to fully understand the interplay of word-formation, semantics and diachronic stratification in the botanical lexicon of Old English. Aspects of semantic change, for instance, can be further investigated in the corpus of simple botanical terms, and the correlation of different morphological and semantic regularities, respectively anomalities, can be researched in more detail in the corpus of complex plant names on the basis of the Catalogue and the information provided there on individual plant names. One of the most interesting questions is perhaps the nature of the interrelation of motives and word-formation patterns. As mentioned above, it is striking, for example, that N_{gen}+N formations are exocentric to a much larger degree than, for example, N+N combinations. The obvious paucity of motives in the group of N_{gen}+N compounds is also eyecatching: the majority of N_{gen}+N compounds encode references to animals, whose body parts are compared to the shapes of parts of plants. This certainly calls for a discussion of conceptual metaphors and conceptual metonymies in the botanical lexicon and also raises the question of how often processes of analogy can account for the creation of individual plant names.

Research in a large lexical field like a botanical lexicon is rewarding in many respects. Besides uncovering some of the linguistic strategies of a people living more than one thousand years ago by 'following the individual threads' it was possible to produce a larger pattern covering the ways the Anglo-Saxons conceptualised and categorised extra-linguistic reality – from the recognition that folk taxonomic structures largely map structures in scientific taxonomy to the insight that among the most salient features of the plants encoded in the plant names are the habitats of the plants, culturally important parts and their shapes, with 'healing virtue', for instance, ranking among the least salient motives. The high percentage of loan influence confirmed the importance of Latin learning and Latin texts, but the numerous exocentric formations, partial metaphors and plant names changed by folk etymology demonstrated as well that the Anglo-Saxons were highly creative and innovative in their use of language. It is to be hoped that the information gathered here and in the Catalogue prove useful for future studies of the Old English lexicon, but also for studies of later stages of English. Just as it was helpful to apply insights gained from studies of modern English in this study of the past, it may be as helpful and promising to apply insights gained from studies of past stages of the language in order to explain some of its features in the present.

PART B

CATALOGUE OF THE OLD ENGLISH COMPLEX PLANT NAMES

INTRODUCTION AND STRUCTURE OF THE ENTRIES

The Catalogue contains all Old English complex plant names as specified in chapter 1.3., and all the elements of the complex plant names. To distinguish the complex plant names, which are the focus of the study, from the simple ones graphically, the complex plant names are marked in bold, while the simplexes are not. All entries are arranged in alphabetical order (*æ* follows *a*, and *þ* follows *t*). Assumed compound plant names are written with a hyphen between modifier and head while assumed syntactic groups are written as two separate words.

Structure of the entries: the complex formations

Each entry is headed by a lemma. The lemma is given in the form it appears in the *DOE* (for those beginning with the letters A-G), or in Bierbaumer (1, 2, 3), which is, generally, in the nominative singular, whether attested or reconstructed. For instance, *micel þung* is recorded in the accusative singular *þone micla þung*, but the suggested entry is *micel þung*. Spelling variants are not listed, as the *DOE* and the *DOEC* are easily accessible and searchable. However, when two variants of one plant name differ considerably, the spelling variant is given in brackets after the lemma and is, in addition, listed separately with a cross-reference to the headword (see, for instance, Catalogue, s.vv. *ilugsecg*, *eolhsecg*).

The identification of the plant is given next: first, the scientific name according to the International Code of Botanical Nomenclature (*ICBN*), then, the most common Modern English name(s) and the most common German name(s); further, a literal translation of the Old English plant name.[299]

The identification follows, unless otherwise indicated, Bierbaumer (1, 2, 3), respectively, the *DOEPN*[300]. The *DOE* was consulted as far as the respective entries have appeared, and especially, when the plant names were not recorded by Bierbaumer (1, 2, 3) or the *DOEPN*. As the editions of the medico-botanical texts and of the glosses and glossaries concentrate on individual texts and, therefore, do not take into account different possible identifications, Bierbaumer (1, 2, 3) and the *DOE* were considered to provide the most comprehensive identifications. The numbering does not

299 The conventions of botanical nomenclature observed by the *ICBN* were established by Carl von Linné (1707–1778), when he published his study *Species Plantarum* in 1753: A binomial name of, e.g., a species comprises a genus name an epithet; both are in Latin, respectively, obey Latin rules of morphology, but they may come from any other language. Genus and species names often describe a feature of the plant, or designate a person or a place. Binomials also include the name of an author or authority, denoting the person who originally described the taxon. To give two examples: Dipsacus silvestris Huds. denotes the plant wild teasel, with Huds. being derived from the name of the botanist William Hudson (1730–1793); Digitalis purpurea L. denoting the plant foxglove, with L. derived from the name of the botanist Carl von Linné (1707-1778).

300 For the *DOEPN* project, Bierbaumer's three volumes of *Der Botanische Wortschatz des Altenglischen* were digitised and the information on individual plant names, often dispersed on all three volumes, was merged into one entry. Work on the *DOEPN*, however, has not yet been finished.

express any preference for any identification proposed. It is also no indicator for the reliability of the identification. All question marks – indicators of doubt with regard to the identification proposed – follow Bierbaumer (1, 2, 3). Modern English names that are direct descendants of the Old English plant names are given in capitals. For instance, OE *dæges ēage* develops into DAISY; OE *haguþorn* develops into HAWTHORN. If only a part of the plant name is a direct descendant of a part of the Old English name, the Modern English plant name (or part of it) is not in capitals. Instead, the fact that this part of the plant name is still extant in Modern English is recorded in the entry of the simplex.

The next part provides the number of occurrences. As far as the lexemes beginning with the letters A to G are concerned, the numbers follow the numbers given in the *DOE*. For the rest and for plant names that are not lemmas in the *DOE*, such as, for instance, *æscapuldre* or *basu popig*, in the main, Bierbaumer (1, 2, 3) and the *DOEPN* were consulted, as they cover all major Old English texts where plant names occur, and are, therefore, likely to record most of the occurrences. The comparison of the attestations of some chance selections by the *DOE* and the *DOEPN* indicates, in general, agreement. The plant name *āt(t)orlāþe*, for instance, is recorded as occurring 45 times in the Old English texts by the *DOE* (s.v. *attor-lāþe*) and 44 times by the *DOEPN* (s.v. *āttor-lāþe*); the plant name *ælfþone* is in both dictionaries marked as occurring 10 times (*DOE*, s.v. *ælf-þone*; *DOEPN*, s.v. *ælfþone*) and for *æscþrote*, the numbers are: 31 occurrences according to the *DOEPN* (s.v. *æsc-þrote*) and 30 occurrences according to the *DOE* (s.v. *æsc-þrote*). In contrast, there is a marked discrepancy in terms of numbers for the simplex *brēmel* 'bramble, briar, etc.': the *DOE* (s.v. *brēmel*, *brembel*, *brember*) records 40 occurrences, while the *DOEPN* (s.v. *brēmel*) notes 27 occurrences, and even more marked is the difference, for instance, between the numbers given by the *DOEPN* for *þorn* and those that the *DOEC* produces (see →*þorn*). These discrepancies result from the fact that thorn-bushes are often used to grow hedges serving as boundary markers and these, in turn, appear very often as landmarks in the charters. As the charters were not covered by Bierbaumer (1, 2, 3), these attestations are missing.

When the term occurs only three times or less, the place(s) of occurrence are given. When a plant name does not occur in Bierbaumer (1, 2, 3) or in the *DOE* this fact is noted as well, and the source of identification of the plant is given. For example, OE *æscapuldre* is identified as denoting 'Sorbus domestica L. / service-tree, sorb-tree / Speierling' following Hooke (2003: 29).

Three manuscripts provide c. 97 illustrations of plants: manuscripts London, British Library, Cotton Vitellius C. iii, fols. 11-85 (Gneuss *Handlist*, no. 402; Ker *Catalogue*, no. 219); Oxford, Bodleian Library, Ashmole 1431 (S.C. 7523), fols. 3-43 (Gneuss *Handlist*, no. 527; Ker *Catalogue*, no. 289); and Oxford, Bodleian Library, Bodley 130 (Gneuss *Handlist*, no. 549; Ker *Catalogue*, no. 302). When the illustrations accompany an Old English plant name, the name(s) of the manuscript(s), the folio number(s) and the number(s) of the illustration(s) in Ohlgren's *Iconographic Catalogue* (1986) are given.

Then, the Latin equivalents in the manuscripts are listed in alphabetical order for each of the Old English botanical terms. Following the *DOE* (sub-chapter 'Entry

Format'), I understand by 'Latin equivalents in the manuscripts' Latin words corresponding to the Old English headwords and present in the same context in the same manuscripts. The Latin equivalents as they appear in the *DOE* are adopted as far as it is complete (A-G), with the exception, however, of equivalents relating to non-botanical senses. Bierbaumer (1, 2, 3) was checked for the Latin equivalents relating to the Catalogue entries H-Y. The nouns are in the nominative singular and all entries follow the spellings of the *Dictionary of Medieval Latin from British Sources* (*DMLBS*) as far as it is complete (A-Sal). I then used the *Revised Medieval Latin Word-List* (Latham 1965) and the *Latin Dictionary* (*Lewis and Short*). When the Latin lemmas are not entirely clear, they are given in the form in which they appear in the manuscripts preceded by a question mark and / or followed by a comment in brackets.

The etymological information indicates the chronological stratum the plant name belongs to, i.e. whether it can be traced back to Germanic, West Germanic or Old English, or whether it is goes back to loan influence (cf. chapter 7). This is followed by information on the morphological type (that is, shape and structure of the plant names) as discussed in chapter 5, and information on the motivations of the plant names. The indication of the motives is accompanied by a brief comment and further references, if necessary. Following the section indicating the associative relations that exist between the plants and the concepts referred to by the elements of the plant names (based on the theoretical framework developed in chapter 6), cross-references to other complex plant names or to the elements of the plant names complete the entries. The cross-references are subdivided into references to the simplexes, references to other complex plant names where the lemma is the first element of the plant name and references where the lemma is the second element of the plant name (see, e.g., Catalogue, s.v. *bisceopwyrt*).

Structure of the entries: the simplexes

As the present study focuses on the complex plant names, both botanical and non-botanical simplex terms that appear as elements of the complex plant names are listed, and, as with the complex plant names, information on their denotations, the number of occurrences and on their etymologies are given. However, this information is not exhaustive insofar as it is, in many cases, limited to the information needed to explain the complex plant names. Thus, for instance, while many simplex terms are polysemous, in the main, the meanings relevant to the analysis of the complex plant names are given.

When the simplex is not attested independently, it is preceded or followed by a hyphen, depending on the position it has in the corresponding complex plant name.

The Latin equivalents in the manuscripts are given only for the botanical terms (following the practice described above). No information on Latin equivalents is given for non-botanical simplex nouns, for adjectives or for verbs. Neither do I indicate the number of occurrences for the lexemes just mentioned.

Cross-references to the complex plant names they are part of conclude the entries. Cross-references where the lemma in question is the first element are given first; those, where the lemma is the second element of the compound are given second (see, e.g., Catalogue, s.v. *ber(i)(g)e*).

A

āc, noun
(i) 'Quercus robur L. / OAK / Eiche'; (ii) name of the rune for 'a'; (iii) place-name element
DOE (s.v. *āc*): c. 150 occ., mainly in bounds. According to Hooke (2009: 80), the most frequent tree name used in pre-Conquest place names and often named as a boundary marker in the charters.
LATIN EQUIV. in the manuscripts: *arbor*, *ilex*, *quercus*, *robur*, *quernus*.
ETYMOLOGY: Gmc **aik-* 'oak' (*Kluge*, s.v. *Eiche*).
▫ →*ācmistel*, →*āctrēow*

āc-mistel, lit. 'oak mistletoe'
(i) 'Loranthus europaeus Jacqu. / oak mistletoe / Europäische Riemenblume, Eichenmistel' or (ii) 'Viscum album L. / mistletoe / Weiße Mistel'
DOE (s.v. *āc-mistel*): 1 occ. (*Lǣcebōc*, 27/29).
ETYMOLOGY: OE formation; probably a loan translation of L *viscum quercinum* or a loan rendition of L *viscum de quercu* (Bierbaumer 1, s.v. *ācmistel*). MORPHOLOGY: N+N; endocentric, rectional. MOTIVATION: habitat & botanical information; The plant name encodes the information that this kind of *mistel* 'mistletoe' grows preferably on oaks (OE *āc*): "Mit der Mistel (Viscum album) verwandter, auf Bäumen (besonders Eichen) schmarotzender Strauch [...]" ['shrub of the same species as the mistletoe (*Viscum album*), living on trees (notably on oaks)'] (Marzell, s.v. *Loranthus europaeus* L.). ASSOCIATIVE RELATIONS: conceptual contiguity & taxonomic sub-ordination.
▫ →*āc*, →*mistel*

āc-trēow, lit. 'oak tree'
'Quercus L. / oak / Eiche'
DOE (s.v. *āc-trēow*): 2 occ. (Leslie (ed.) 1961: 47/28, 48/36); not in Bierbaumer (1, 2, 3), but a lemma in the *DOEPN* (s.v. *āc-trēow*).
ETYMOLOGY: OE formation. MORPHOLOGY: N+N; endocentric, copulative: subsumptive (pleonastic). MOTIVATION: plant & botanical information; The *āc* 'oak' is a kind of *trēow* 'tree'. ASSOCIATIVE RELATIONS: conceptual identity & taxonomic sub-ordination.
▫ →*āc*, →*trēow*

affricanisc, adjective
'AFRICAN / afrikanisch'
▫ →*affricanisc æppel*

affricanisc æppel, lit. 'African apple' or 'African fruit'
'fruit of Punica granatum L. / pomegranate / Granatapfel'
2 occ. (*Cleopatra Glossary* (Stryker), 304, no. 209; 305, no. 9); *affricanisc æppel* is not a lemma on its own in the *DOE*, but appears subsumed to the adjective *affricanisc*, *affrianisc*.
LATIN EQUIV. in the manuscripts: *malum punicum*.
ETYMOLOGY: Independent formation in WGmc; OHG *affricana ephili* for 'mala punica' (Björkman 1901: 210); a loan rendition of L *mālum pūnicum*, based on the erroneous assumption that

pūnicum refers to Carthage and its civilisation, referred to as 'Punic' by the Romans (see MOTIVATION below). MORPHOLOGY: Adj+N; endocentric, copulative: attributive. MOTIVATION: origin & botanical information; The pomegranate is an import from the Mediterranean region, cultivated especially in the Near East and Northern Africa, hence the first element *affricanisc* (Marzell, s.v. *Punica granatum* L.). According to *Genaust* (s.v. *pūnica*), however, *mālum pūnicum* is associated erroneously with the Phoenicians (referred to also as 'Punic'), as L *pūnicus* 'purple, red' derives from Gk *φοινίκεος* 'purple, red', and this is cognate with Gk *φοῖνιξ* 'date', from which Gk *φοῖνιξ* 'Phoenicians' derived (*Genaust*, s.v. *Phoenicaulis*). The two Medieval Latin names *punicum malum* and *malum granatum* refer to colour of the pomegranate (L *pūnicus* 'purple, red'), respectively to its juicy, red kernels (L *grānātus* 'with kernels') (*Genaust,* s.vv. *grānātum*, *pūnicum*). The element *affricanisc* 'African' in *affricanisc æppel*, however, clearly indicates the assumed origin of the plant. The characteristics 'red' and 'with kernels', in return, are encoded in the Old English plant names *→rēad æppel* and *→gecyrnlod æppel*. ASSOCIATIVE RELATIONS: conceptual contiguity & taxonomic sub-ordination.

▫ *→affricanisc, →æppel*

ampre, *ompre*, noun
'Rumex L. / dock / Ampfer'
DOE (s.v. *ampre²*): 19 occ.
LATIN EQUIVALENTS in the manuscripts: *cocilus*, *rhododaphne*.
ETYMOLOGY: Gmc **ampra-* 'sour, acrid' (*Kluge*, s.v. *Ampfer*).

▫ *→fenampre, →sundampre*

ananbēam →wananbēam

apuldor, *apuldre*, noun
'Malus communis Lam. / apple-tree / Apfelbaum'
DOE (s.v. *apuldor, apuldre*): c. 100 occ., "mainly in bounds"; *DOEPN* (s.v. *apuldor*): 14 occ.
LATIN EQUIV. in the manuscripts: *malus*.
ETYMOLOGY: Gmc derivative; the first element is Gmc **aplu-*, the second element is the Indo-European tree name suffix **-tro-*, which can be found in OE *mapuldor* 'maple', in OHG *wehhalter* 'juniper', G *Affolter* 'apple-tree', G *Holunder* 'elder' and G *Maßholder* 'field maple', too. IE **-tro-* is possibly the root of Gmc **trewa-* 'tree' (*Kluge*, s.vv. *Affolter*, *Holunder*, *Maßholder*; *OED*, s.v. *apple tree*, *maple tree*).

▫ *→æscapuldre, →milscapuldor, →sūrapuldor, →sūrmilscapuldor, →swēte apuldor*

āt(t)or, noun
(i) 'poison, venom / Gift', (ii) 'bile / Galle'

▫ *→āt(t)orlāþe*

āt(t)or-lāþe (*attor-*), lit. 'poison-harm' or 'bile-harm'
(i) '? Stachys officinalis (L.) Trevisan / betony / Betonie'; (ii) '? Panicum crus-galli L. / prickly grass, cock's foot / Hühnerhirse'; (iii) '? Solanum nigrum L. / morel / Schwarzer Nachtschatten'

DOE (s.v. *attor-lāþe*): c. 45 occ.[e], "mainly in medical recipes."
ILLUSTRATION in manuscript London, British Library, Cotton Vitellius C. iii, fol. 35r (Ohlgren 168(31)).
LATIN EQUIV. and OE synonyms in the manuscripts: *cyclaminos*, *gallicrus*, *morella*, *solanum*, *venenifuga*, *vettonica*; *mixenplante*.
ETYMOLOGY: OE formation; According to Cockayne (*Leechdoms*, 2: 370) and Stracke (1974: 76, no. 199) MedL *venenifuga* lit. 'something which puts poison to flight' is a loan formation after OE *āt(t)orlāþe*. Förster (1917: 130), however, takes the Old English gloss to be a loan formation modelled on *venenifuga* and not the other way round; in either direction, a loan rendition. MORPHOLOGY: N+N; exocentric, metonymic compound. MOTIVATION: pathogen & healing virtue; The plant name expresses the belief that the plant is an annoyance to poison- or bile-induced sicknesses (→*āt(t)or* (i) 'poison, venom'; (ii) 'bile' + →*lāþ(e)* 'pain, harm, annoyance'), i.e. it heals sicknesses caused by poison or by an abundance of bile (Bierbaumer 2, s.v. *āttorlāþe*; Sauer 1992b: 389, fn. 28). ASSOCIATIVE RELATION: conceptual contiguity.
▫ →*smale āt(t)orlāþe*

Æ

ǣgwyrt →*ēagwyrt*

ælf, noun
'ELF / Elfe'
▫ →*ælfþone*

ælf-þone, lit. 'elf tendril'
'Solanum dulcamara L. / dogwood, bittersweet / Bittersüß'
DOE (s.v. *ælf-þone*): 10 occ., "in medical recipes".
ETYMOLOGY: OE formation; MORPHOLOGY: N+N; exocentric, metonymic compound; *-þone* 'tendril' is a morpheme attested only in this combination, but it exists independently in other Indo-European languages; cf. OHG *dona* 'branch, tendril', OS *thona* 'branch, tendril', G *Dohne* 'branch, tendril', L *tenus*, *-oris* 'string, cord' (*Pokorny*, s.v. *ten-*, *tend-*). MOTIVATION: pathogen & part of plant; The element *ælf-* 'elf' indicates that the plant is helpful with diseases caused by elves, i.e. "used to treat ailments thought to have been caused by supernatural agency, especially skin disease and mental illness" (*DOE*, s.v. *ælf-þone*), while *-þone* 'tendril' refers to the fact that the plant is a kind of vine, with winding tendrils. ASSOCIATIVE RELATIONS: conceptual contiguity & conceptual contiguity (meronymy).
▫ →*ælf*, →*-þone*

æppel, noun
(i) 'fruit of Malus communis L. / APPLE / Apfel'; (ii) 'fruit / Frucht'; (iii) 'something round / etwas Rundes'; (iv) 'apple of the eye / Augapfel'
OE *æppel* is polysemous; it referred, initially, to the fruits of Malus communis L.; then, its meaning was extended metaphorically to all kinds of apple-like fruits (cf. *Kluge*, s.v. *Apfel*; *DOE*, s.v. *æppel*; compare also L *mālum* (i) 'apple', (ii) '(any kind of) fruit'; cf. *Lewis and Short*, s.v. *mālum*); further metaphorical extension of meaning to 'something round' and 'apple of the eye'.
DOE (s.v. *æppel*): c. 150 occ., many of them in bounds; as a simplex with a botanical meaning c. 31 occ. (*DOEPN*, s.v. *æppel* (1)).
LATIN EQUIV. in the manuscripts: *malum*, *pomum*.
ETYMOLOGY: Gmc **aplu*- 'apple' (*Kluge*, s.v. *Apfel*); ultimate origins unknown (*OED*, s.v. *apple*).
▫ →*æppellēaf*, →*æppeltrēow*, →*æppelþorn*
▫ →*affricanisc æppel*, →*cirisæppel*, →*coddæppel*, →*cornæppel*, →*eorþæppel*, →*fingeræppel*, →*gecyrnlod æppel*, →*gōdæppel*, →*rēad æppel*, →*swegles æppel*, →*wuduæppel*

æppel-lēaf, lit. 'apple leaf' or 'fruit leaf'
'? Viola L. / violet / Veilchen'
DOE (s.v. *æppel-lēaf*): 1 occ.; cf. Gough (1974: 286).
LATIN EQUIV. in the manuscripts: *viola*.
ETYMOLOGY: OE formation. MORPHOLOGY: N+N; exocentric, metonymic compound. MOTIVATION: ? & part of plant; The motivation of this plant name is not transparent according to the *DOEPN* (s.v. *æppel-lēaf*), and the description of the plant provided by Marzell (s.v. *Viola* L.), does not give useful hints as to its motivation, either. ASSOCIATIVE RELATIONS: Because of the unclear motivation, the associative relations are not transparent.
▫ →*æppel*, →*lēaf*

æppel-trēow, lit. 'apple-tree'
'Malus communis Lam. / APPLE-TREE / Apfelbaum'
DOE (s.v. *æppel-trēow*): 11 occ., "mainly in bounds".
LATIN EQUIV. in the manuscripts: *malus*.
ETYMOLOGY: OE formation. MORPHOLOGY: N+N; endocentric, rectional. MOTIVATION: part of plant & botanical information; This kind of *trēow* 'tree' bears apples. According to Biggam (2011: 34-5), not Malus communis, but Malus sylvestris, the crab-tree – a wild sub-species of Malus – was the native apple-tree of England (see →*æppelþorn*, →*sūrmilscapuldor*). ASSOCIATIVE RELATIONS: conceptual contiguity (meronymy) & taxonomic sub-ordination.
▫ →*æppel*, →*trēow*
▫ →*sūr æppeltrēow*

æppel-þorn, lit. 'apple thorn-bush'
'Malus silvestris (L.) Mill. / crab-tree / Wilder Apfelbaum'
DOE (s.v. *æppel-þorn*) 1 occ. (*Codex* (Kemble), 3: 448, no. 460); not in Bierbaumer (1, 2, 3), but it is lemmatised in the *DOEPN* (s.v. *æppelþorn*).
ETYMOLOGY: OE formation. MORPHOLOGY: N+N; endocentric, rectional. MOTIVATION: part of plant & botanical information; In contrast to Malus communis (not the native species of England; see →*æppeltrēow*), the native

species 'silvestris' is characterised by very sour fruit and thorny twigs: "dornige Zweige, unterseits kahle Blätter, kleinere Blüten und holzige Früchte" ['thorny twigs, leaves bare on the underside, smaller flowers and woody fruits'] (Marzell, s.v. *Malus communis* L.), hence the use of *æppel* 'apple' as the first and of *þorn* 'thorn-bush' as the second element. ASSOCIATIVE RELATIONS: conceptual contiguity (meronymy) & taxonomic sub-ordination.
▫ *→æppel, →þorn*

ærs, noun
'ARSE / Arsch'
▫ *→openærs*

æsc, noun
(i) 'Fraxinus excelsior L. / ASH / Esche'; (ii) 'spear, lance / Speer, Lanze'; (iii) 'ship / Schiff'
The wood of the ash is apparently "best for tools that involve impact because of its toughness and resistance to splintering" (Biggam 2011: 45). The fact that the denotation of OE *æsc* was extended metonymically to 'spear, lance' and to 'ship' indicates that, probably, it was one of the preferred woods for this kind of weapons and constructions.
DOE (s.v. *æsc*): c. 100 occ.; as a simplex attested c. 14 times (*DOEPN*, s.v. *æsc*).
LATIN EQUIV. in the manuscripts: *fraxinus*.
ETYMOLOGY: Gmc **ask-* 'ash' (*OED*, s.v. *ash*).
▫ *→æscapuldre, →æscþrote, →æscwyrt*
▫ *→ceasteræsc*

æsc-apuldre, lit. 'ash apple-tree'
'Sorbus domestica L. / service-tree, sorb-tree / Speierling'
DOEC: 1 occ., attested in the charters: "hit cymð to þan byhte to æsc apaldre leage" ['it comes to the bending to the lea of service-trees'] (*Charters* (Birch), 2: 162, no. 542); not in the *DOE*; not in Bierbaumer (1, 2, 3); identification as Sorbus domestica following Hooke (2003: 29).
ETYMOLOGY: OE formation. MORPHOLOGY: N+N; endocentric, rectional, partial metaphor; The second element, *apuldor* 'apple-tree' – the head of the combination – is a derivative made up of the free lexical morpheme {*apul*} and the tree name suffix {*dor, dre, dur*}, possibly a form of Gmc **trewa-* 'tree' (*→apuldor*; *→mapuldor*; cf. also G *Affolter*, *Holunder*, *Maßholder*; *Kluge*, s.vv. *Affolter*, *Holunder*, *Maßholder*). MOTIVATION: shape & botanical information; The first element of the composite refers to the leaves of Sorbus domestica, which, according to Marzell (s.v. *Sorbus domestica* L.), resemble heavily those of Fraxinus excelsior L., OE *æsc* 'ash'. The fruits of Sorbus domestica, apple-like in shape, are not palatable when raw, but can be eaten after a period of storage. Sorbus domestica was well known in medieval Europe (Marzell, s.v. *Sorbus domestica* L.). ASSOCIATIVE RELATIONS: metaphorical similarity & taxonomic subordination.
▫ *→apuldre, →æsc*

æsc-þrote, æsc-þrotu, lit. 'ash-throat'
(i) 'Aegopodium podagraria L. / goutweed, ashweed / Giersch, Geißfuß'; (ii) 'Verbena officinalis L. / vervain / Eisenkraut'; (iii) 'Anchusa

officinalis L. / alkanet, bugloss / Gemeine Ochsenzunge'
DOE (s.v. *æsc-þrote, æsc-þrotu*): 30 occ.[e]
ILLUSTRATION in manuscript London, British Library, Cotton Vitellius C. iii, fol. 23r (Ohlgren 168(7)).
LATIN EQUIV. in the manuscripts: *anchusa*, *ferula*, *harundo*, *hierabotane*, *nomen holeris*, *verbenaca*, *virgula*.
ETYMOLOGY: OE formation. MORPHOLOGY: N+N; partial metaphor. As the relationship between the second element and the plant name is not transparent, a morphological analysis that goes beyond stating the morphological type, is not possible. MOTIVATION: shape & ?. None of the attempts at explaining this plant name are entirely convincing: the use of *æsc* 'ash' is probably based on the similarity between the leaves of the goatweed and the leaves of ashes (see also →*æscwyrt*; Bierbaumer 1, 2, 3, s.v. *æscþrote*; Marzell, s.v. *Aegopodium podagraria* L.; ModE *ashweed* and *ground ash* also refer to the similarity between the leaves of Aegopodium podagraria and those of the ash), yet the reason for using *þrote* 'throat' as a second element is completely opaque, and is possibly the result of folk etymological processes. ASSOCIATIVE RELATIONS: metaphorical similarity & ?.
▫ →*æsc*, →*þrote*

æsc-wyrt, lit. 'ash-wort'
(i) 'Verbena officinalis L. / vervain / Eisenkraut'; (ii) 'Anchusa officinalis L. / alkanet, bugloss / Gemeine Ochsenzunge'
DOE (s.v. *æsc-wyrt*): 3 occ. (Meritt 1945: 58, no. 67; *Durham Plant Name Glossary*, no. 60; *Laud Herbal Glossary*, no. 168).
LATIN EQUIV. in the manuscripts: *anchusa*, *lustrago*, *verbena*, *verbenaca*.
ETYMOLOGY: OE formation. MORPHOLOGY: N+N; endocentric, rectional; partial metaphor. MOTIVATION: shape & botanical information; The first element refers to the leaves of *æscwyrt*, which resemble in shape those of the *æsc* 'ash' (Bierbaumer 1, s.v. *æscþrote*; 3, s.v. *æscwyrt*; →*æscþrote*) while the second element indicates that, in contrast to the ash, *æscwyrt* is not a tree but a plant. ASSOCIATIVE RELATIONS: metaphorical similarity & taxonomic sub-ordination.
▫ →*æsc*, →*wyrt*

(æt)clīþan, verb
'adhere / ankleben'
▫ →*cliþwyrt*

Æþelferþ, proper noun
Æþelferþ is recorded several times as a proper noun (*DOEC*, s.v. *Æþelferþ*: 9 hits, s.v. *Æðelfrið*: 12 hits); e.g. *Charters* ((Birch), 3: 35, no. 880): "Æðelferðes londe [...]" ['The land of *Æðelferð*'] and in the *Chronicle* ((Bately), 60, [896]): "Þær wearð ofslægen Lucumon cynges gerefa & Wulfheard Friesa & Æbbe Friesa & Æðelhere Friesa & Æðelferð cynges geneat, & ealra monna fresiscra & engliscra LXII & þara deniscena CXX" ['There, Lucumon, the king's reeve, and Wulfheard the Frisian and Æbbe the Frisian and Æðelhere the Frisian and Æðelferð of the king's household and of the Frisian and English men sixty-two and one hundred and twenty of the Danes were killed'].
▫ →*æþelferþingwyrt*

æþelferþing-wyrt, lit. '*Æþelferþ's* wort'
(i) '? Stellaria holostea L. / greater stitchwort / Sternmiere, Waldsternmiere' (according to Bierbaumer 2, s.v. *æþelferþing-wyrt*); (ii) '? Stellaria media L. / chicken-weed / Hühnermiere, Vogelmiere' (according to M.L. Cameron 1993: 112-3).
DOE (s.v. *æþel-feorþing-wyrt*): 16 occ. "in medical recipes and glosses".
LATIN EQUIV. in the manuscripts: *agrimonia alpha*, *spincium*.
ETYMOLOGY: OE formation; MORPHOLOGY: PropN_{ing}+N; endocentric, rectional. The *DOE* has the lemma *æþel-feorþing-wyrt*, marking that it is not clear which element(s) represent the head and which the modifier of the combination (see below). MOTIVATION: person & botanical information. There have been different attempts at explaining this plant name, but none of them is absolutely convincing. Hoops (1889: 18) ties the plant name to OE *fēorþung* 'a quarter, one quarter of a penny, farthing'. *Holthausen* (s.v. *æþelferþingwyrt*) neglects the element *æþel* and gives 'camomile' as the denotation for OE *ferþingwyrt*; however, *-ferþingwyrt* is not attested elsewhere. Lindheim (*Durham Plant Name Glossary*, no. 17) and Bierbaumer (1, s.v. *æþelferþingwyrt*) favour the explanation that the plant name is made up of the proper name *Æþelferþ* + seignoral {*-ing*} + {*wyrt*} 'wort, plant'. Apart from its occurrence in the plant name, *Æþelferþ* is recorded several times as a proper noun, for instance in the *Charters* ((Birch), 3: 35, no. 880) and in the *Chronicle* ((Bately), 60, [896]; see →*Æþelferþ*). Why the two species of Stellaria were associated with a person called *Æþelferþ* is, however, unclear. ASSOCIATIVE RELATIONS: ? & taxonomic sub-ordination.
▫ →*Æþelferþ*, →*wyrt*

B

bān, noun
'BONE / Knochen'
▫ →*bānwyrt*

bān-wyrt, lit. 'bone-wort'
(i) 'Bellis perennis L. / daisy / Gemeines Gänseblümchen'; (ii) 'Centaurium umbellatum Gilibert / common centaury / Tausendgüldenkraut'; (iii) 'Symphytum officinale L. / comfrey / Große Beinwurz, Beinwell'; (iv) 'Viola L. / violet / Veilchen'
DOE (s.v. *bān-wyrt*): 23 occ., "in medical recipes and glossaries".
ILLUSTRATION in manuscript London, British Library, Cotton Vitellius C. iii, fol. 62r (Ohlgren 168(84)).
LATIN EQUIV. and OE synonyms in the manuscripts: *centauris* (with *minor*), *consolida*, *viola* (1x with *herba*), *viola aurosa*, *viola purpurea*; *clǣfre*, *hōfe*.
ETYMOLOGY: WGmc compound (Carr 1939: 96); OHG *beinwurz* (cf. Björkman 1901: 207). MORPHOLOGY: N+N; endocentric, rectional. MOTIVATION: illness & botanical information; Symphytum officinale L. is said to heal broken bones; cf. also OHG *beinuuella*, *beinwalla*, G *Beinheil*, *Beinbruchwurzel*, etc.; L *consolida minor* (Marzell, s.v. *Symphytum officinale* L.); see also →*brȳsewyrt*. ASSOCIATIVE RELATIONS:

conceptual contiguity & taxonomic sub-ordination.
▫ →*bān*, →*wyrt*
▫ →*grēate bānwyrt*

basu, adjective
'purple / rot'
▫ →*basu popig*

basu popig, lit. 'purple poppy'
'Papaver rhoeas L. / common red poppy / Klatschmohn'
DOEC: 3 occ.; cf. also *DOEPN* (s.v. *popig*, *basu*); not a lemma in the *DOE*, despite the remark that it is "also interpreted as compound", (s.v. *basu*, b.i. "*basu popig*, glossing lemmas for 'asphodel', but perhaps 'red poppy'.").
LATIN EQUIV. in the manuscripts: *apodillis* (for *asphodelus*), *hastula regia*.
ETYMOLOGY: OE formation; hybrid formation (*popig* < L *papāver* with suffix substitution; see →*popig*). MORPHOLOGY: Adj+N; endocentric, copulative: attributive. MOTIVATION: colour & botanical information; Papaver rhoeas is a kind of *popig* 'poppy' with *basu* 'scarlet, red' flowers (Marzell, s.v. *Papaver rhoeas* L.). ASSOCIATIVE RELATIONS: conceptual contiguity & taxonomic sub-ordination.
▫ →*basu*, →*popig*

bēam, noun
(i) 'tree / Baum'; (ii) 'BEAM / Balken'; (iii) 'cross, gallows / Kreuz, Galgen'; (iv) 'timber / Bauholz'; (v) 'sunbeam / Sonnenstrahl'
DOE (s.v. *bēam*): c. 150 occ.; as a plant name, c. 20 occ. (*DOEPN*, s.v. *bēam*).
LATIN EQUIV. in the manuscripts: *cedrus*, *robur*, *stipes*, *trabs*, *tuba*.
ETYMOLOGY: WGmc **bauma*- 'tree'; the ultimate origins are unclear (*Kluge*, s.v. *Baum*).
▫ →*begbēam*, →*cederbēam*, →*cirisbēam*, →*cistbēam*, →*cistelbēam*, →*cistenbēam*, →*crāwanbēam*, →*cwicbēam*, →*elebēam*, →*ellenbēam*, →*fīcbēam*, →*fūla bēam*, →*fyrgenbēam*, →*gorstbēam*, →*hnutbēam*, →*lāwerbēam*, →*mōrbēam*, →*pīnbēam*, →*wananbēam*

bēan, noun
(i) 'fruit of Vicia faba L. / BEAN, horsebean / Bohne, Saubohne'; (ii) place name element; (iii) an element of personal names
DOE (s.v. *bēan*): c. 40 occ.; as a plant name, c. 30 occ. (*DOEPN*, s.v. *bēan*).
LATIN EQUIV. in the manuscripts: *cicer*, *faba*, *legumen*.
ETYMOLOGY: Gmc **bauno*- 'bean' (*Kluge*, s.v. *Bohne*).
▫ →*fuglesbēan*, →*swearte bēanen*

bēce, noun
'Fagus L. / BEECH (tree) / Buche'
DOE (s.v. *bēce*): 21 occ.
LATIN EQUIV. in the manuscripts: *aesculus* (2x with *ob edendo*), *fagus*.
ETYMOLOGY: Gmc **bōkjōn*- < Gmc **bōkā*- < IE **b^{h}āgā*- (*Kluge*, s.v. *Buche*); homonym with the dative singular and the nominative and accusative plural forms of the etymologically and semantically unrelated OE *bōc* 'book'.
▫ →*bēctrēow*

bēc-trēow, lit. 'beech-tree'
'Fagus L. / beech, BEECH-TREE / Buche'
DOE (s.v. *bēc-trēow*): 1 occ. (Napier 1900: 192/30).
LATIN EQUIV. in the manuscripts: *fagus*.

ETYMOLOGY: OE formation; MORPHOLOGY: N+N; endocentric, copulative: subsumptive (pleonastic). MOTIVATION: plant & botanical information; The *bēc* 'beech' is a kind of *trēow* 'tree'. ASSOCIATIVE RELATIONS: conceptual identity & taxonomic sub-ordination.
▫ *→bēc, →trēow*

beg-, beger →ber(i)(g)e

beg-bēam, lit. 'berry-tree'
(i) 'Morus nigra L. / mulberry tree / Maulbeerbaum'; (ii) '? Rubus fruticosus L. / bramble / Brombeere'
DOE (s.v. *beg-bēam*): 2 occ.
LATIN EQUIV. in the manuscripts: *morarius*.
ETYMOLOGY: OE formation. MORPHOLOGY: N+N; endocentric, rectional; {*beg-*} is clipped from *→ber(i)(g)e* 'berry'. MOTIVATION: part of plant & botanical information. OE *begbēam* glosses *morarius* 'mulberry tree' in *Cleopatra Glossary* ((Stryker), 313, no. 348) and renders *rubus* 'bramble' in Luke (*Gospels, Luke* 20, 37). While both the mulberry and the bramble bear black berries, only the former is a tree; Rubus fruticosus, in contrast, is a bush. The misinterpretation of OE *begbēam* for L *rubus* is perhaps due to Isidore, *Etymologiae* (17: 7, 19): "morus a Graecis uocata, quam Latini rubum appellant" ['called *morus* by the Greeks, (a tree) which the Romans call *rubus*']. The editor and translator André (1981: 98, fn. 219) comments: "Les paragraphes 19-20 trahissent une complète confusion. L'arbre qui donne le fruit nommé μόρον est le mûrier noir (*Morus nigra* L.), et non la ronce (*rubus*), dont il sera question ensuite sous le nom de *mora siluestris*" ['Paragraphs 19-20 betray a complete confusion. The tree producing the fruit called μόρον is the mulberry (*Morus nigra* L.) and not the bramble (*rubus*), which is the one discussed afterwards under the label *mora siluestris*']. The first element of *begbēam* denoting the 'mulberry' refers, hence, to the berries, the second element indicates that this plant is a kind of tree. ASSOCIATIVE RELATIONS: conceptual contiguity (meronymy) & taxonomic sub-ordination.
▫ *→bēam, →ber(i)(g)e*

belle, noun
'BELL / Glocke'
▫ *→hennebelle*

bēo, noun
'BEE / Biene'
▫ *→bēowyrt*

bēo-wyrt, lit. 'bee-wort'
(i) 'Melissa officinalis L. / common balm / Zitronenmelisse'; (ii) 'Ballota nigra L. / black horehound / Schwarzer Andorn'; (iii) '? Acorus calamus L. / sweet flag / Kalmus'; (iv) '?? Onopordon acanthium L. / Scottish thistle / Eselsdistel'; (v) '?? Acanthus mollis L. / bear's breech / Pracht-Akanthus' or ?? any other thistle species
DOE (s.v. *bēo-wyrt*): 25 occ.[e]
ILLUSTRATIONS in manuscript London, British Library, Cotton Vitellius C. iii, fol. 24v (Ohlgren 168(10)), fol. 65r (Ohlgren 168(90)).
LATIN EQUIV. and OE synonyms in the manuscripts: *acanthus*, *apiago*,

apiastrum, *leucacantha*, *marrubium*, *veneria*; *hūne*.
ETYMOLOGY: Independent formation in WGmc (Carr 1939: 128), according to Björkman (1901: 202 ff.) rendering L (Gk) *melisphilla* 'bee-leaf' or 'honey-leaf' (cf. OHG *biniwurz*, OLG *bini-wurt*). As OE *bēowyrt* glosses, among others, L *apiago* (e.g. *Laud Herbal Glossary*, no. 161) and L *apiastrum* (in *Cleopatra Glossary* (Stryker), 59, no. 455), more probably, it renders these two terms: L *apiago* < {*api(s)*} 'bee' + {*-ago*} (suffix); L *apiastrum* < {*api(s)*} 'bee' + {*-astrum*} (suffix) (Bierbaumer 2, s.v. *bēowyrt*; d'Aronco 1988: 30; Sauer 1992b: 391). MORPHOLOGY: N+N; endocentric, rectional. MOTIVATION: animal & botanical information. Melissa officinalis L. is extremely attractive to bees because of its sweet scent, so that it is easily associated with bees sitting on it; Marzell (s.v. *Melissa officinalis* L., 1) reports that, since antiquity, the plant was used as a fodder plant for bees and, in order to prevent the bee colony from leaving its hive, the people used to rub the hives with common balm (i.e. Melissa officinalis). Ballota nigra and Melissa officinalis strongly resemble each other (a case of co-taxonomic similarity). ASSOCIATIVE RELATIONS: conceptual contiguity & taxonomic sub-ordination.
▫ *→bēo*, *→wyrt*

bere, noun
'Hordeum vulgare L. / barley / Gerste'
DOE (s.v. *bere*): 20 occ.; as a simplex, 13 occ. (*DOEPN*, s.v. *bere*).
According to Biggam (2011: 31), the denotation of *bere* was probably extended metonymically to 'cereals': "*bere-* compounds, [...] apparently unlike the *hwǣte-* compounds, [...] often seem to apply to cereals in general. Examples include *bereern* 'barn, granary', *bereflōr* 'threshing floor' and *berewīc* 'grange, outlying or demesne farm'. It seems highly unlikely that a barn or a threshing floor would be restricted to barley only, and it is much more likely that barley was considered the *usual* (prototypical) cereal at the time these compound terms were coined." The *DOE* (s.v. *bere*) does not confirm this assessment, as *bere* is not recorded as denoting, very general, 'cereals'.
LATIN EQUIV. in the manuscripts: *hexastichum*, *hordeaceus*, *hordeum*.
ETYMOLOGY: Gmc **baroz-*, *bariz-* 'barley' (*OED*, s.v. *barley*).
▫ *→berewinde*

bere-winde, lit. 'barley-winder'
'Convulvulus arvensis L. / lesser bindweed / Ackerwinde'
DOE (s.v. *bere-winde*): 1 occ.; attested spelling *berwinde*; cf. *Brussels Glossary* (300/19).
ILLUSTRATION of *umbilicum* in manuscript London, British Library, Cotton Vitellius C. iii, fol. 35r (Ohlgren 168 (31)); here it is identified as 'Cotyledon umbilicus L. / lady's navel, wall pennywort'.
LATIN EQUIV. in the manuscripts: *umbilicum*.
ETYMOLOGY: OE formation; OE *-winde* is not attested independently in Old English, but in other Germanic languages; cf. G *Winde* and the second element in the compound G *Ackerwinde*. MORPHOLOGY: synthetic, $\text{N+N}_{\text{deverbal}(-e)}$; endocentric, rectional. MOTIVATION: habitat & action. According to the *DOE* (s.v. *berewinde*),

it is "so called from its binding together stalks of barley". ASSOCIATIVE RELATIONS: conceptual contiguity & taxonomic sub-ordination.
▫ *→bere, →winde*

ber(i)(g)e, *beger*, *beg-*, noun
'BERRY / Beere'
DOE (s.v. *berie*, *berige*): c. 55 occ.; as a simplex, 28 occ. (*DOEPN*, s.v. *berie*).
LATIN EQUIV. in the manuscripts: *acinus*, *baca*, *corymbus*, *morum*, *morus*, *racemus*, *uva*, *vaccinium*.
ETYMOLOGY: Gmc **basjôn-*, **bazjôn-* 'berry' (*OED*, s.v. *berry*).
▫ *→begbēam*
▫ *→blæcber(i)(g)e, →eorþber(i)(g)e, →hǣþber(i)(g)e, →heorotber(i)(g)e, →hindber(i)(g)e, →hundesber(i)(g)e, →strēawber(i)(g)e, →wēdeber(i)(g)e*

bindan, verb
'BIND / binden'
▫ *→weoþobend, →wudubinde*

-bindel, noun
'binding plant / Winde'
ETYMOLOGY: Derived from the verb OE *bindan* 'to BIND' + *-el* (suffix); cf. Sauer (2012: 218).
▫ *→wudubindel*

bisceop, noun
'BISHOP / Bischof'
ETYMOLOGY: OE *bisceop* < VL *(e)biscopus*, L *epīscopus* < Gk *ἐπίσκοπος*. The element *bisceop-* in *bisceopwyrt* is, however, not the Latin loan *epīscopus*, but the early Latin loan *hibiscus* or *hibiscum* (Wollmann 1990: 153), changed by processes of folk etymology in analogy to *bisceop* (for details, see *→bisceopwyrt*).
▫ *→bisceopwyrt, →bisceopwyrtel*

bisceop-wyrt, lit. 'bishop's wort'
'Althaea officinalis L. / marshmallow / Eibisch'
DOE (s.v. *bisceop-wyrt*): c. 90 occ.[e]
LATIN EQUIV. in the manuscripts: *hibiscum*, *hierobotanum*, *struthion*, *vettonica*.
ETYMOLOGY: OE formation; *bisceop-* < L *hibiscum* changed by folk etymology in analogy to OE *bisceop* 'bishop' (see MOTIVATION); hybrid formation (OE *bisceop* < L *epīscopus*; *→bisceop*); MORPHOLOGY: N+N; endocentric, rectional. The plant name *bisceopwyrt*, once lexicalised, was the base for further complex plant names (see below). MOTIVATION: secondary, phonological motivation of the first element: sound pattern & 'botanical information' in the second element. The first element of this frequently attested Old English plant name goes back to L *hibiscum* or *hibiscus* of unclear origin (Foulner 2000: 105), from which the non-morphemic element *-bisc-* was extracted and associated phonologically with the Old English lexeme *bisc(e)op* 'bishop' (< L *epīscopus* 'bishop', Gk *ἐπίσκοπος*; cf. Wollmann 1990: 153). The process of folk etymology was triggered, thus, by the phonological similarity between the lexeme and the non-morphemic element and resulted in the change of the word form of the loan, without, however, secondary semantic motivation. The head *wyrt* was probably added in order to single out this kind of *bisceop* as a kind of plant, i.e. has mainly explanatory function. L *hibiscum* was taken over

into other Germanic languages, too, e.g. as OHG *ibisca*, MHG *ibesch*, G *Eibisch* (cf. *Kluge*, s.v. *Eibisch*). ASSOCIATIVE RELATIONS: phonological similarity & taxonomic sub-ordination.

▫ *→bisceop*, *→wyrt*
▫ *→bisceopwyrtel*
▫ *→brādbisceopwyrt*, *→feldbisceopwyrt*, *→læsse bisceopwyrt*, *→sūþerne bisceopwyrt*

bisceop-wyrtel, lit. 'bishop's wort'
'Althaea officinalis L. / marsh mallow / Eibisch'
DOE (s.v. *bisceop-wyrtel*): 1 occ. (*Antwerp-London Glossaries*, 109/16), glossing *gerobotana ł uerbena ł sagmen*; according to Bierbaumer (3, s.v. *bisceopwyrt*) an erroneous gloss; the *DOE* (s.v. *bisceop-wyrtel*) notes that it may be an equivalent to *→(sēo) læsse bisceopwyrt* 'betony'.
LATIN EQUIV. in the manuscripts: *hierobotanum*, *sagmen*, *verbena*.
ETYMOLOGY: OE formation. MORPHOLOGY: N+N; endocentric, rectional; *bisceopwyrtel* is taken to be a compound with the second element enlarged by the diminutive suffix *-el* (Feulner 2000: 105); see also *DOE* (s.v. *bisceop-wyrtel*), where the morphological interpretation is indicated by the hyphen between *bisceop* and *wyrtel*. The plant name can also be regarded as a derivative, with the compound *bisceopwyrt* as the modifier and the suffix *-el* as the head of the combination. Semantically, it can be seen as an equivalent to *sēo læsse bisceopwyrt* 'betony'; for an account of the suffix family *-el*, *-il*, etc., see Sauer (2001) and (2012: 218, 221). MOTIVATION: see *→bisceopwyrt*. ASSOCIATIVE RELATIONS: see *→bisceopwyrt*.

▫ *→bisceop*, *→wyrtel*

blæc, adjective
'BLACK / schwarz'

▫ *→blæcber(i)(g)e*, *→blæc īfig*, *→blæc pipor*, *→blæc wīngeard*

blæc-ber(i)(g)e, lit. 'black berry'
(i) 'fruit of Vaccinium myrtillus L. / whortle-berry / Heidelbeere'; (ii) 'fruit of Rubus fruticosus L. / BLACKBERRY / Brombeere', or of 'Morus nigra L. / mulberry / Maulbeere'
Not a lemma in the *DOE*; two of the attestations are listed s.v. *blæc* (*Durham Plant Name Glossary*, no. 52; *Antwerp-London Glossaries*, 129/6; *Cleopatra Glossary* (Stryker), 208, no. 400); according to Bierbaumer (3, s.v. *blæcberie*), also in manuscript London, British Library, Cotton Otho E.i (resp. in its copy in Junius 77); this is not confirmed, however, by Voss (1996).
LATIN EQUIV. in the manuscripts: *bacinium* (? = *vaccinium*), *flavus*, *morus*.
ETYMOLOGY: OE formation. MORPHOLOGY: Adj+N; endocentric, copulative: attributive. MOTIVATION: colour & botanical information. OE *blacebergan* and *blaceberian* gloss the plural nouns *bacinia* (*Durham Plant Name Glossary*, no. 52) and *flaui uel mori* (*Antwerp-London Glossaries*, 129/6), that is, they refer to the fruits and not to the plants as such. The first element, *blæc* encodes the fact that the berries – those of the whortleberry, of the blackberry and of the mulberry alike – are of a dark, almost black colour, while the second provides botanical information: a *blæcber(i)(g)e* is a kind of *ber(i)(g)e* 'berry'. ASSOCIATIVE RELATIONS: conceptual contiguity & taxonomic sub-ordination.

▫ *→ber(i)(g)e*, *→blæc*

blæc īfig, lit. 'black ivy'
'Hedera helix L. / ivy / Efeu'
DOEC: 2 occ. (*Lǣcebōc*, 75/3, 80/26); not a lemma in the *DOE*.
ETYMOLOGY: OE formation; loan translation of L *hedera nigra* (Bierbaumer 1, s.v. *īfig*). MORPHOLOGY: Adj+N; endocentric, copulative: attributive. MOTIVATION: colour & botanical information. The berries of this *īfig* 'Hedera helix' are of a purple-black colour (Marzell, s.v. *Hedera helix* L.). ASSOCIATIVE RELATIONS: conceptual contiguity & taxonomic sub-ordination.
▫ *→blæc, →īfig*

blæc pipor, lit. 'black pepper'
'Piper nigrum L. / BLACK PEPPER / Schwarzer Pfeffer'
1 occ. (*Lǣcebōc* (70/17); not a lemma in the *DOE*.
ETYMOLOGY: OE formation; hybrid formation (OE *pipor* < L *piper*; *→pipor*). MORPHOLOGY: Adj+N; endocentric, copulative: attributive. MOTIVATION: colour & botanical information. The adjective black is used to distinguish this kind of pepper from the white pepper, which was thought to be a distinct species (Birkhan 2012: 148-9). In reality, white pepper is nothing but the peeled fruits of Piper nigrum, the black pepper (Marzell, s.v. *Piper nigrum* L.). ASSOCIATIVE RELATIONS: conceptual contiguity & taxonomic sub-ordination.
▫ *→blæc, →pipor*

blæc wīngeard, lit. 'black vine'
'Tamus communis L. / black bryony / Schmerwurz'
Not a lemma in the *DOE*; 1 occ. (*Antwerp-London Glossaries*, 146/2); a gloss to *ampelos male* as "blac wingeard ł brabasca"); cf. *DOEPN* (s.v. *wīn-gēard, blæc*).
LATIN EQUIV. in the manuscripts: *ampelos male, verbascum*.
ETYMOLOGY: OE formation; probably a loan translation after Isidore, *Etymologiae* (17: 9, 91): "Ampelos melena, id est vitis negra, eademque labrusca; folia hederae similia <habet>, in omnibus maior quam uitis alba, baca similiter habens, quae in maturitate nigrescunt, unde uocabulum sumpsit" ['Ampelos melena, that is black vine, also called wild vine; it has leaves which resemble those of the white vine, [which are], in general, much larger than those of the white vine, having berries that turn black as well, when mature, whence its name']; *wīngeard* 'vine, vineyard' lit. 'vine-yard' is a polysemous, hybrid compound (*wīn* 'wine, vine' < L *vīnea*; *→wīn*, *→wīngeard*), which is possibly a semantic loan from the polysemous Latin lexeme *vīnea* 'vine, vineyard'. MORPHOLOGY: Adj+N; endocentric, copulative: attributive; MOTIVATION: colour & botanical information. As in one of the German names of Tamus communis, in G *Schwarzwurzel* lit. 'black root', the first element of the Old English name refers to the dark axis of the plant: "nach der außen dunklen Grundachse" ['with reference to the dark outside of the base rhachilla'] (Marzell, s.v. *Tamus communis* L.); the use of *wīngeard* as the head of the plant name is, as mentioned above, probably based on the polysemy of L *vīnea* and provides botanical information. ASSOCIATIVE RELATIONS: conceptual contiguity & taxonomic sub-ordination.
▫ *→blæc, →wīngeard*

blind, adjective
'BLIND / blind'
▫ →*blindenet(e)l(e)*

blind-net(e)l(e), blinde net(e)l(e), lit. 'blind nettle'
(i) 'Lamium L. / dead nettle / Taubnessel'; (ii) 'Melissa officinalis L. / common balm / Zitronenmelisse'
DOEPN (s.v. *netele, blinde*): 12 occ.[e]; not a lemma on its own in the *DOE*, but subsumed to the adjective *blind*, 7.
LATIN EQUIV. in the manuscripts: *acanthus, archangelicus*.
ETYMOLOGY: OE formation; MORPHOLOGY: Adj+N; endocentric, copulative: attributive; attested as a compound, without inflection of the first element, for instance, in Ælfric's *Glossary* (Zupitza 1880: 311) and as a syntactic group (with (rests of) inflection of the first element in *Antwerp-London Glossaries* (117/7). MOTIVATION: effect & botanical information. A kind of nettle whose leaves do not burn when touched (Bierbaumer 1, s.v. *netele*), "lacking its characteristic stinging property" (*DOE*, s.v. *blind*, 7). Because of the outward similarity of the plants Lamium L. and Melissa officinalis L., the name of the former was used to denote the latter plant, too (Bierbaumer 3, s.v. *netele*). The motivation for this transfer is the co-taxonomic similarity between the concepts of the two plants as a whole. ASSOCIATIVE RELATIONS: conceptual contiguity & taxonomic sub-ordination.
▫ →*blind*, →*net(e)l(e)*

blōd, noun
'BLOOD / Blut'
▫ →*dracanblōd*

blōma, -blōma, noun
(i) 'mass or lump of metal / Metallbrocken'; (ii) ? 'bloom, blossom / Blüte, Blume'
DOE (s.v. *blōma*): 6 occ. in the sense of 'mass of metal', glossing *massa, metallum*; 2 attestations as *-blōma*, probably in the sense of 'bloom'.
ETYMOLOGY: Gmc **blomon-* < **blo-* 'blow' + **-mon-* (suffix forming action nouns) (*OED*, s.v. *bloom*); ModE *bloom* is derived from ON *blóm* 'flower' and developed via ME *blom(e)*. According to the *DOE* (s.v. *gold-blōma*), "the sense of *blōma* is uncertain: either 'blossom', unattested in the simplex (but cf. OS *blômo*, OHG *bluomo*, ON *blóm*, Goth *bloma* in this sense and cf. *OED2 gold-bloom* 'marigold') or 'mass of gold' […]."
▫ →*goldblōma*

blōstm, noun
(i) 'BLOSSOMS / Blüte(n)'; (ii) fig. 'the best part of, the choiciest of / das Beste'
DOE (s.v. *blōstm, blōstma*): c. 160 occ.; as a simplex in the sense of 'blossoms', c. 88 occ. according to the *DOEPN* (s.v. *blōstma*).
LATIN EQUIV. in the manuscripts: *caducum, flos, flosculus, ligustrum, ornamentum, pappus, quintina*.
▫ →*hwīte blōstman*

bōc, noun
'Fagus L. / beech / Buche'
DOE (s.v. *bōc*[2]): 8 occ., "mainly in glosses and glossaries".
LATIN EQUIV. in the manuscripts: *aesculus, arbor, faginus, fagus*.
ETYMOLOGY: Gmc **bōkō-* < IE **b^{h}āgā-* 'Buche' (*Kluge*, s.v. *Buche*). Though

homonyms in Old English, OE *bōc* 'book' and OE *bōc* 'beech' are neither semantically, nor etymologically related (cf. *Kluge*, s.vv. *Buch*, *Buche*).
▫ →*bōctrēow*

bōc-trēow, lit. 'beech-tree'
'Fagus L. / beech / Buche'
DOE (s.v. *bōc-trēow*): 3 occ. (Zupitza 1880: 29/16, 312/4; Napier 1900: 229/355).
LATIN EQUIV. in the manuscripts: *fagus*.
ETYMOLOGY: OE formation; MORPHOLOGY: N+N; endocentric, copulative: subsumptive (pleonastic). MOTIVATION: plant & botanical information. The *bōc* 'beech' is a kind of *trēow* 'tree'. ASSOCIATIVE RELATIONS: conceptual identity & taxonomic sub-ordination.
▫ →*bōc*, →*trēow*

bōg, noun
(i) 'BOUGH, twig, shoot / Ast, Zweig, Spross, Trieb'; (ii) 'shoulder / Schulter' (with various sub-senses; cf. *DOE*, s.v. *bōg*)
DOE (s.v. *bōg*): c. 90 occ.; as a botanical term attested c. 41 times (*DOEPN*, s.v. *bōg*).
ETYMOLOGY: Gmc **bogu-* < IE *$b^h\bar{a}g^hu$- 'arm' (*Kluge*, s.v. *Bug*).
▫ →*wæterbōg*

box, noun
'Buxus sempervirens L. / BOX / Buchsbaum'
DOE (s.v. *box*[1]): 8 occ., "mainly in glossaries".
LATIN EQUIV. and OE synonyms in the manuscripts: *belsa*, *buxus*; *þȳfel*.
ETYMOLOGY: Early Latin loan (Wollmann 1990: 166); < L *buxus* < Gk *πύξος*, fem. (*OED*, s.v. *box*). Homonym of OE *box* 'BOX, receptacle', but not related semantically or etymologically.
▫ →*boxtrēow*

box-trēow, lit. 'box-tree'
'Buxus sempervirens L. / BOX-TREE / Buchs(baum)'
DOE (s.v. *box-trēow*): 1 occ. (Zupitza 1880: 20).
LATIN EQUIV. in the manuscript: *buxus*.
ETYMOLOGY: OE formation; hybrid compound: *box* is a loan from Latin *buxus*, which, ultimately, goes back to Gk *πύξος*, fem. (*OED*, s.v. *box*; see →*box*). MORPHOLOGY: N+N; endocentric, copulative: subsumptive (pleonastic). MOTIVATION: plant & botanical information. The *box* 'box-tree' is a kind of *trēow* 'tree'; cf. also G *Buchs*, *Buchsbaum*. ASSOCIATIVE RELATIONS: conceptual identity & taxonomic sub-ordination.
▫ →*box*, →*trēow*

brād, adjective
'BROAD / breit'
▫ →*brād cāwel*, →*brādbisceopwyrt*, →*brāde wulfescamb*, →*brāde wulfestǣsel*, →*brādlēac*, →*brādþistel*
▫ →*wegbrǣde*

brāde wulfescamb, lit. 'broad (wild) teasel'
'Dipsacus silvestris Huds. / wild teasel / Kardendistel, Wilde Karde'
Not an entry on its own in the *DOE*, but subsumed to *brād*, 1.a.: "of plants: presumably 'broad-leaved'"; 1 occ. (*Brussels Glossary*, 296/27).
LATIN EQUIV. in the manuscripts: *chamaeleon alba*.

ETYMOLOGY: OE formation. MORPHOLOGY: Adj+N; endocentric, copulative: attributive; the second element *wulfescamb* is a complex plant name itself (see →*wulfescamb*). MOTIVATION: size & botanical information. A kind of *wulfescamb* 'teasel' with broad leaves (Bierbaumer 3, s.v. *wulfescamb*). ASSOCIATIVE RELATIONS: conceptual contiguity & taxonomic sub-ordination.
▫ →*brād*, →*wulfescamb*

brād-bisceopwyrt, brāde bisceopwyrt, lit. 'broad marshmallow'
unidentified plant
DOE (s.v. *brād-bisceopwyrt*): 1 occ., but the plant name is attested twice: once as a compound (*brād-bisceopwyrt*; *Lācnunga*, 41) and once as a syntactic group (*brāde bisceopwyrt*, in *Lācnunga*, 698).
ETYMOLOGY: OE formation; see also →*bisceopwyrt*; MORPHOLOGY: Adj+N; endocentric, copulative: attributive. MOTIVATION: size & botanical information; This plant name seems to refer to a kind of *bisceopwyrt* with broad leaves (Bierbaumer 2, s.v. *brādbisceopwyrt*). ASSOCIATIVE RELATIONS: conceptual contiguity & taxonomic sub-ordination.
▫ →*bisceopwyrt*, →*brād*

brād cāwel, lit. 'broad cabbage'
'? Brassica oleracea L. var. gongylodes L. / cabbage turnip / Kohlrabi'
DOEC: 1 occ. (*Lǣcebōc*, 25/11); not a lemma on its own in the *DOE*, but it appears subsumed to the entry *cāwel*, 1.a. "in combinations: *brad cawel*, *read cawel*, *wilde cawel*"; cf. *DOEPN* (s.v. *cāwel, sē brāda*).
ETYMOLOGY: OE formation; hybrid formation: OE *cāwel* < L *caulis* (with *w*-insertion and L *au* > OE *ā*; cf. Campbell 1959: § 509; see also →*cāwel*). MORPHOLOGY: Adj+N; endocentric, copulative: attributive. MOTIVATION: size & botanical information. The first element of this plant name refers to the lower part of the turnip, which is thickened: "[...] mit Bezug auf den am Grunde (über dem Boden) kugelig angeschwollenen Stengel, worauf auch die Vorschrift, den unteren Teil zu nehmen, hinweist" (Bierbaumer 1, s.v. *cāwel*) ['referring to the stem, whose lower part is thickened, a fact which is also referred to by the instruction to take the lower part of the plant']; see also Marzell (s.v. *Brassica oleracea L.*): "bei der var. gongylodes ist der Stengelgrund verdickt" ['the lower part of the stem of the var. gonglyodes is thickened']. The second element indicates that the plant is a kind of cabbage. ASSOCIATIVE RELATIONS: conceptual contiguity & taxonomic sub-ordination.
▫ →*brād*, →*cāwel*

brāde-lēac, lit. 'broad leek'
(i) 'Allium ursinum L. / ramsons, broad-leaved garlic / Bärlauch' or (ii) 'Allium porrum L. / common leek / Porree, Breitlauch'
DOE (s.v. *brāde-lēac*): 7 occ.[e], "mainly in glossaries".
LATIN EQUIV. in the manuscripts: *serpyllum*.
ETYMOLOGY: OE formation; probably a loan translation of L *ampeloprasum* < Gk *ἀμπελόπρασον* (Gk *ἄμπελος* 'vine', cf. *Genaust*, s.v. *ampelóprasum*) which was probably misinterpreted as **amploprasum*, i.e. *amplo* 'broad' +

prasum 'leek' (Bierbaumer 3, s.v. *brādelēac*). MORPHOLOGY: Adj+N; endocentric, copulative: attributive. MOTIVATION: size & botanical information. The leaves of ramsons are broader than those of other Alliums, hence the use of the first element; the head indicates that the plant is a kind of *lēac* 'leek'. ASSOCIATIVE RELATIONS: conceptual contiguity & taxonomic sub-ordination.

▫ →*brād*, →*lēac*

(se) brāde wulfestǣsel, lit. 'broad (wild) teasel'

'Dipsacus silvestris Huds. / wild teasel / Kardendistel, Wilde Karde'

Not an entry on its own in the *DOE*, but subsumed to *brād*, 1.a. "of plants: presumably 'broad-leaved'"; 1 occ. (*Durham Plant Name Glossary*, no. 81).

LATIN EQUIV. in the manuscripts: *chamaeleon*, *chamaeleon alba*.

ETYMOLOGY: OE formation. MORPHOLOGY: Adj+N; endocentric, copulative: attributive; the second element, *wulfestǣsel*, is a complex plant name itself (see →*wulfestǣsel*). MOTIVATION: size & botanical information. A kind of *wulfestǣsel* 'teasel' with broad leaves (Bierbaumer 3, s.v. *wulfestæsel*). ASSOCIATIVE RELATIONS: conceptual contiguity & taxonomic sub-ordination.

▫ →*brād*, →*wulfestǣsel*

brād-þistel, lit. 'broad thistle'

(i) '? Silybum marianum Gärtner / milk thistle / Mariendistel' or (ii) '?? Dipsacus silvester Huds. / wild teasel / wilde Karde'

2 occ. (*DOEPN*, s.v. *brādþistel*); not a lemma on its own in the *DOE* (s.v. *brād*, 1.a.i.), but the comment says: "brad þistel 'broad thistle', a kind of thistle (the form may alternatively be read as a compound)"; attested in *Durham Plant Name Glossary* (no. 169), and Voss (1996: 186/58).

LATIN EQUIV. in the manuscripts: *eringius* (? for *eryngion*), *erithius* (? for *eryngion*).

ETYMOLOGY: WGmc formation; OHG *breitdistele* (Björkman 1902: 283; 1901: 207); loan rendition: Bierbaumer (3, s.v. *brādþistel*) suggests loan formation modelled on L *cardus folia habet lata* lit. 'thistle (which) has broad leaves'. MORPHOLOGY: Adj+N; endocentric, copulative: attributive. MOTIVATION: size & botanical information. The leaves of Silybum marianum are not only broad, as the first element indicates, but also eye-catching, since they are green and marked by white stripes (Marzell, s.v. *Silybum marianum* L.). ASSOCIATIVE RELATIONS: conceptual contiguity & taxonomic sub-ordination.

▫ →*brād*, →*þistel*

brēmel, noun

(i) 'Rubus fruticosus L. / BRAMBLE, briar, blackberry bush / Brombeere'; (ii) 'thorn-bush / Dornenstrauch'; (iii) 'Rosa canina L. / dogrose / Hundsrose'

DOE (s.v. *brēmel*, *brembel*, *brember*): c. 40 occ.

ILLUSTRATION in manuscript London, British Library, Cotton Vitellius C. iii, fol. 45v (Ohlgren 168(52)).

LATIN EQUIV. in the manuscripts: *anguens*, *eruscum*, *morus*, *senticosus sirculus* (with *spinosa virgula*), *sentix*, *vepres*.

ETYMOLOGY: Gmc **bræm*- 'broom' + *-*ilo*- (> OE -*el*; cf. Sauer 2001; Sauer

2012: 218, 220); ModE *bramble* with euphonic [*b*] (*OED*, s.v. *bramble*).

▫ →*brēmelbrēr*, →*brēmelþorn*, →*brēmelþȳfel*, →*brēmelþyrne*

▫ →*hēopbrēmel*, →*heorotbrēmel*

brēmel-brēr, lit. 'bramble-briar'
'Rubus fruticosus L. / bramble / Brombeere'
DOE (s.v. *brēmel-brēr*): 1 occ. (*Cleopatra Glossary*, (Quinn), 38/1). Bierbaumer (3, s.v. *brēmelbrēr*) argues that *brēmelbrēr* is not a compound but an erroneous gloss, because, in his opinion, the glossator glossed *tribulus* twice instead of once. OE *brēr* is polysemous, denoting the bramble but also all kinds of thorny bushes and roses (see →*brēr*; *OED*, s.v. *briar*). The three plant names *brēmelþorn* 'bramble' lit. 'bramble thorn-bush', *brēmelþȳfel* lit. 'bramble-thicket' and *brēmelþyrne* lit. 'bramble thorn-bush' (s.vv.), however, are not erroneous glosses and yet quite similar with *brēmelbrēr* from the point of view of their structure and the meaning of their constituents. *Brēmelbrēr* may be, therefore, considered a parallel formation consisting of the elements 'bramble' + 'thorn-bush' and not an 'erroneous' compound.
LATIN EQUIV. in the manuscripts: *tribulus*.
ETYMOLOGY: OE formation; MORPHOLOGY: N+N; endocentric, copulative: subsumptive (pleonastic). MOTIVATION: plant & botanical information. The *brēmel* 'bramble' is a kind of *brēr* 'briar, thorn-bush'. ASSOCIATIVE RELATIONS: conceptual identity & taxonomic sub-ordination.

▫ →*brēmel*, →*brēr*

brēmel-þorn, lit. 'bramble thorn-bush'
'Rubus fruticosus L. / bramble / Brombeere'
DOE (s.v. *brēmel-þorn*): 2 occ. (*Charters* (Birch), 3: 475, no. 1199; 3: 211, no. 648); not in Bierbaumer (1, 2, 3).
ETYMOLOGY: OE formation; MORPHOLOGY: N+N; endocentric, copulative: subsumptive (pleonastic). MOTIVATION: plant & botanical information. The *brēmel* 'bramble' is a kind of *þorn* 'thorn-bush'. ASSOCIATIVE RELATIONS: conceptual identity & taxonomic sub-ordination.

▫ →*brēmel*, →*þorn*

brēmel-þȳfel, lit. 'bramble-thicket'
'Rubus fruticosus L. / bramble / Brombeere'
DOE (s.v. *brēmel-þȳfel*): 1 occ. (*Charters* (Birch), 3: 62, no. 902); not in Bierbaumer (1, 2, 3).
ETYMOLOGY: OE formation; MORPHOLOGY: N+N; endocentric, copulative: subsumptive (pleonastic). MOTIVATION: plant & botanical information; The *brēmel* 'bramble' is a kind of *þȳfel* 'thicket', i.e. it is a thicket-forming plant. ASSOCIATIVE RELATIONS: conceptual identity & taxonomic sub-ordination.

▫ →*brēmel*, →*þȳfel*

brēmel-þyrne, lit. 'bramble-bush'
'Rubus fruticosus L. / bramble / Brombeere'
DOE (s.v. *brēmelþyrne*): 7 occ.; not in Bierbaumer (1, 2, 3).
ETYMOLOGY: OE formation. MORPHOLOGY: N+N; endocentric, copulative: subsumptive (pleonastic). MOTIVATION: plant & botanical information; The *brēmel* 'bramble' is a kind of *þyrne*

'bush'. ASSOCIATIVE RELATIONS: conceptual identity & taxonomic sub-ordination.
▫ →*brēmel*, →*þyrne*

brēr, noun
(i) 'Rubus fruticosus L. / BRIAR / Brombeere'; (ii) 'thorn-bush / Dornenstrauch'; (iii) '? Rosa canina L. / dog-rose / Hundsrose'
DOE (s.v. *brēr*, *brǣr*): 7 occ.
LATIN EQUIV. in the manuscripts: *anguens*, *morus*, *tribulus*.
ETYMOLOGY: Of unknown origin (*OED*, s.v. *briar*).
▫ →*brērþyrne*
▫ →*brēmelbrēr*, →*heorotbrēr*, →*hindbrēr*

brēr-þyrne, lit. 'briar thorn-bush'
(i) 'Rubus fruticosus L. / bramble / Brombeere'; (ii) 'thorn-bush / Dornenbusch'
DOE (s.v. *brēr-þyrne*): 1 occ. (*Codex* (Kemble), 6: 221, no. 1368); not in Bierbaumer (1, 2, 3), but a lemma in *DOEPN* (s.v. *brēr-þyrne*).
ETYMOLOGY: OE formation; MORPHOLOGY: N+N; endocentric, copulative: subsumptive (pleonastic). MOTIVATION: plant & botanical information. The *brēr* 'briar' is a kind of *þyrne* 'thorn-bush'. ASSOCIATIVE RELATIONS: conceptual identity & taxonomic sub-ordination.
▫ →*brēr*, →*þyrne*

brōc, noun
'BROOK, water / Bach, Wasser'
▫ →*brōccærse*, →*brōcminte*, →*brōcþung*

brōc-cærse, lit. 'brook-cress'
'Nasturtium officinale R. Br. / water-cress / Brunnenkresse'
DOE (s.v. *brōc-cærse*): 1 occ. (*Lǣcebōc*, 105/2).
ETYMOLOGY: OE formation; MORPHOLOGY: N+N; endocentric, rectional. MOTIVATION: habitat & botanical information; This kind of *cærse* 'cress' grows preferably in moist environments, e.g. along brooks (Marzell, s.v. *Nasturtium officinale* R. Br.; Bierbaumer 1, s.v. *cærse*). ASSOCIATIVE RELATIONS: conceptual contiguity & taxonomic sub-ordination.
▫ →*brōc*, →*cærse*

brōc-minte, lit. 'brook-mint'
(i) 'Mentha aquatica L. / water-mint / Bachminze'; or (ii) 'Mentha longifolia Huds. / horse-mint / Roßminze'
DOE (s.v. *brōc-minte*): 7 occ.
ILLUSTRATION in Oxford, Bodleian Library, Ashmole 1431 (S.C. 7523), fol. 24r (Ohlgren 221(81)).
LATIN EQUIV. in the manuscripts: *absinthium ponticum*, *balsamita*, *mentastrum*, *sisymbrium*.
ETYMOLOGY: OE formation; according to Sauer (1992b: 405), possibly a loan translation of L *mentha aquatica*; hybrid compound (OE *minte* < L *ment(h)a*; →*minte*). MORPHOLOGY: N+N; endocentric, rectional. MOTIVATION: habitat & botanical information; This kind of *minte* 'mint' grows preferably in moist environments, e.g. alongside brooks; cf. OHG *bachminza* lit. 'brook-mint' (Björkman 1901: 207). ASSOCIATIVE RELATIONS: conceptual contiguity & taxonomic sub-ordination.
▫ →*brōc*, →*cærse*

brōc-þung, lit. 'brook poisonous plant'
'Cicuta virosa L. / cowbane / Wasserschierling'

DOE (s.v. *brōc-þung*): 1 occ. (*Laud Herbal Glossary*, no. 1448).
LATIN EQUIV. in the manuscripts: *typhus*.
ETYMOLOGY: OE formation; MORPHOLOGY: N+N; endocentric, rectional. MOTIVATION: habitat & botanical information. This kind of *þung* 'poisonous plant' grows preferably in moist environments, e.g. along brooks (Marzell, s.v. *Cicuta virosa* L.). ASSOCIATIVE RELATIONS: conceptual contiguity & taxonomic sub-ordination.
▫ →*brōc*, →*þung*

brōþor, noun
'BROTHER / Bruder'
▫ →*brōþorwyrt*

brōþor-wyrt, lit. 'brother-wort'
'Mentha pulegium L. / pennyroyal / Polei, Flöhkraut'
DOE (s.v. *brōþor-wyrt*): 1 occ. (*Brussels Glossary*, 300/24).
LATIN EQUIV. and OE synonyms in the manuscripts: *puleium*; *dweorge-d(w)os(t)(l)e, hylwyrt*.
ETYMOLOGY: OE formation. MORPHOLOGY: N+N; endocentric, rectional. MOTIVATION: ? & botanical information. There is no obvious reason for the use of *brōþor* 'brother' as a modifier to *wyrt* with reference to Mentha pulegium (Marzell, s.v. *Mentha pulegium* L.; Bierbaumer 3, s.v. *brōþorwyrt*). ASSOCIATIVE RELATIONS: ? & taxonomic sub-ordination.
▫ →*brōþor*, →*wyrt*

brūn, adjective
'BROWN / braun'
▫ →*brūn(e)wyrt*

brūn(e)-wyrt, lit. 'brown wort'
(i) 'Scrophularia aquatica L. / water-betony, BROWNWORT / Wasser-braunwurz'; or (ii) 'Scrophularia nodosa L. / fig-wort / Knotige Braunwurz'; or (iii) 'Asplenium ceterach L. / brown back, finger-fern / Schriftfarn, Milzfarn'; or (iv) 'Phyllitis scolopendrium Newm. / hart's tongue / Hirschzunge'; (v) 'Prunella vulgaris L. / common self-heal / Kleine Braunelle'
DOE (s.v. *brūn-wyrt, brūne-wyrt*): 30 occ.
ILLUSTRATION in manuscript London, British Library, Cotton Vitellius C. iii, fol. 38r (Ohlgren 168(37)). According to M.L. Cameron (1993: 113f.), the illustration of the plant *brūn(e)wyrt* in the manuscript Cotton Vitellius C. iii points to the plant Prunella vulgaris, but, based on the recipes, species of Scrophularia cannot be excluded either as possible denotata.
LATIN EQUIV. in the manuscripts: *radiolus*, *splenium*, *teucrion*, *vaccinium*.
ETYMOLOGY: WGmc formation; OHG *brunwurz* (Björkman 1902: 284); MORPHOLOGY: Adj+N; endocentric, copulative: attributive. MOTIVATION: colour & botanical information. With regard to the first element, two motives are suggested by M.L. Cameron (1993: 113f.): either it refers to the colour 'brown' – a feature displayed by the leaves of many of the plants suggested above and the motive assumed here – or it is a reference to a disease called in OE *bruneþa*, which was probably diphtheria. ASSOCIATIVE RELATIONS: conceptual contiguity & taxonomic sub-ordination.
▫ →*brūn*, →*wyrt*

brȳsan, verb
(i) '(to) BRUISE, crush / (zer-)brechen'; (ii) '(to) season / würzen'
▫ →*brȳsewyrt*

brȳse-wyrt, lit. 'bruise-wort' or 'season(ing)-wort'
(i) 'Symphytum officinale L. / comfrey / Beinwell, Wallwurz'; or (ii) 'Bellis perennis L. / daisy / Gänseblümchen'
DOE (s.v. *brȳse-wyrt*): 1 occ. (*Leechdoms*, 1: 374); not in Bierbaumer (1, 2, 3); identification following the *DOE*.
ETYMOLOGY: OE formation. MORPHOLOGY: V_{stem}+N; endocentric, rectional. MOTIVATION: illness & botanical information. Symphytum officinale L. was used to cure broken bones; compare also OHG *beinuuella*, *beinwalla*, G *Beinbruchwurzel*, etc. (Marzell, s.v. *Symphytum officinale* L.). Less likely is the allusion to seasoning properties of *brȳsewyrt*, if interpreted according to the second meaning of the verb *brȳsan* (cf. *OED*, s.vv. *bruisewort*, *bonewort*; see also →*brȳsan*). ASSOCIATIVE RELATIONS: conceptual contiguity & taxonomic sub-ordination.
▫ →*brȳsan*, →*wyrt*

byrc, noun
(i) 'Betula L. / BIRCH / Birke'; (ii) ? 'Populus L. / poplar / Pappel'; (iii) 'name of the rune for 'b' / Name der Rune für b'
DOE (s.v. *byrc, byrce, beorc*): c. 35 occ.; according to the *DOEPN* (s.v. *byrc*), 22 attestations as 'birch'.
LATIN EQUIV. in the manuscripts: *beta*, *betulus*, *populus*.
ETYMOLOGY: Gmc **berk-jo*/**berko* < IE **b^h^erəgo-* '(to) shine' (*Kluge*, s.v. *Birke*).
▫ →*byrctrēow*

byrc-trēow, lit. 'birch-tree'
'Betula L. / birch / Birke'
DOE (s.v. *byrc-trēow*): 1 occ. (*Antwerp-London Glossaries*, 206/12).
LATIN EQUIV. in the manuscripts: *populus*.
ETYMOLOGY: OE formation. Bierbaumer (3, s.v. *byrctrēow*) assumes that the gloss is not a compound, but just the co-occurrence of the lexemes *byrc* 'birch' and *trēow* 'tree'; however, OE *byrctrēow* is listed as a compound in the *TOE* (1: 101) and in the *DOE* (s.v. *byrc-trēow*). MORPHOLOGY: N+N; endocentric, copulative: subsumptive (pleonastic). MOTIVATION: plant & botanical information. The birch is a kind of tree. ASSOCIATIVE RELATIONS: conceptual identity & taxonomic subordination.
▫ →*byrc*, →*trēow*

C

camb, noun
'COMB / Kamm'
▫ →*wulfescamb*

candel, noun
'CANDLE / Kerze'
ETYMOLOGY: Early Latin loan, < L *candēla* (Wollmann 1990: 174).
▫ →*candelwyrt*

candel-wyrt, lit. 'candel-wort'
'Verbascum thapsus L. / great mullein / Königskerze'
DOE (s.v. *candel-wyrt*): 1 occ. (*Antwerp-London Glossaries*, 145/8).
LATIN EQUIV. in the manuscripts: ? *insana*, *lucernaris* [*herba*], *lucubrum*, *phlomos*.
ETYMOLOGY: OE formation; hybrid loan translation into Old English (OE *candel* < L *candēla*) modelled on L *herba lucernaria* (Bierbaumer 3, s.v. *candelwyrt*; Sauer 1992b: 391); cf. Isidore, *Etymologiae* (17: 9, 73) "Flomos, quam Latini herbam lucernarem uocant ab eo quod ad licinia facit" ['*Flomos*, which the Romans call 'candel plant', since one can make candle wicks out of it']. MORPHOLOGY: N+N; endocentric, rectional; partial metaphor. MOTIVATION: shape & botanical information; The flowering stems of the plant – it can grow up to c. 6.6 ft. (2 m) high and c. 0.8 ft. (2.5 cm) in diameter – resemble candles. Also likely seems to be the explanation that the plant was used for making candle wicks, as suggested by the *Kluge* (s.v. *Königskerze*), an explanation based on Isidore's *Etymologiae* (see the quotation above). ASSOCIATIVE RELATIONS: metaphorical similarity & taxonomic sub-ordination.
▫ →*candel*, →*wyrt*

cāwel, *caul*, noun
'Brassica L. / cabbage / Kohl'
DOE (s.v. *cāwel*): 21 occ.
ILLUSTRATIONS in manuscripts London, British Library, Cotton Vitellius C. iii, fol. 56v (Ohlgren 168(73)), and Oxford, Bodleian Library, Ashmole 1431 (S.C. 7523), fol. 30r (Ohlgren 221(119)).
LATIN EQUIV. in the manuscripts: *arboracia* (for *armoracia*), *brassica*, *caulis*, *corvus*, *crambe*, *lapsana*, *magudaris*, *serpyllum*.
ETYMOLOGY: OE *cāwel* is an early Latin loan with parasiting *w*; it was borrowed a second time, where L *au* was retained; hence the doublets OE *caul* and *cāl* (Campbell 1959: § 509); < L *caulis* (*DOE*, s.v. *cāwel*).
▫ →*brād cāwel*, →*wilde caul*

cærse, noun
(i) 'Nasturtium officinale R.Br. / water-CRESS / Kresse' or (ii) 'Lepidium sativum L. / garden-cress / Garten-kresse'; (iii) 'plant / Pflanze' (see →*hole cærse*)
DOE (s.v. *cærse*): c. 45 occ.; as a simplex, 36 occ. (*DOEPN*, s.v. *cærse*).
ILLUSTRATIONS in manuscripts London, British Library, Cotton Vitellius C. iii, fol. 28v (Ohlgren 168(18)), and Oxford, Bodleian Library, Ashmole 1431 (S.C. 7523), fol. 11r (Ohlgren 221(20)).
LATIN EQUIV. in the manuscripts: *britia*, *cardamum*, *cresso*, *nasturtium*; *sinapio* (? = *sinapis*).
ETYMOLOGY: Gmc **krasjon-* (*OED*, s.v. *cress*).
▫ →*brōccærse*, →*ēacærse*, →*fencærse*, →*hole cærse*, →*lambescærse*, →*lēaccærse*, →*tūncærse*, →*worþigcærse*, →*wyllecærse*

ceaster-, noun
unidentified plant
Not attested independently.
ETYMOLOGY: < L *cestros*, *cestron*, *cestrum* < Gk *κέστρος*, *κέστρον* (*DOE*, s.v. *ceasterœsc*; Feulner 2000: 122-3); Gk *κέστρον* is a derivative of the Greek verb *κεντέω* '(to) prick' (Feulner 2000: 122-3). Probably, the loan word was

associated with OE *ceaster* 'castle, fort, town' (a Latin loan word, too, which derives from L *castra*).
▫ →*ceasteræsc*, →*ceasterwyrt*

ceaster-æsc, lit. '*ceaster*-ash'
(i) 'Helleborus niger L. / Christmas rose, Christ-hellebore / Schneerose, Nieswurz'; (ii) 'Veratrum album L. / neese-wort, white hellebore / Weißer Germer'; (iii) 'Daphne mezereum L. / spurge olive, dwarf-bay / Seidelbast, Kollerhals'
DOE (s.v. *ceaster-æsc*): 6 occ.
LATIN EQUIV. in the manuscripts: *helleborus*.
ETYMOLOGY: OE formation; hybrid formation (→*ceaster-*). MORPHOLOGY: N+N; exocentric, metaphorical compound. MOTIVATION: plant & shape. L *elleborus*, the gloss to *ceasteræsc* and *ceasterwyrt* in the Old English glosses and glossaries referred to Helleborus niger, to Stachys betonica or to Daphne gnidium.

The use of the second element, *æsc* 'ash', is probably based on the shape of the leaves of these plants, as the leaves of Daphne mezereum and of Veratrum album are all lancet-shaped, just like the leaves of the ash (Marzell, s.vv. *Daphne Mezereum* L., *Veratrum album* L.); according to Marzell (s.v. *Veratrum album* L.), Veratrum album and Helleborus niger were often denoted by the same plant names, although they are not related and also do not really resemble each other. As all possible denotata of *ceasteræsc* are plants and not trees, the explanation that *ceasteræsc* is a sub-group of *æsc* cannot be assumed; thus, the explanation that *ceasteræsc* is compared to the ash because of the similarity of the leaves, is more likely; in this case, the relation between the second element and the whole compound is one of metaphorical similarity. ASSOCIATIVE RELATIONS: conceptual identity & metaphorical similarity.
▫ →*æsc*, →*ceaster-*

ceaster-wyrt, lit. '*ceaster*-wort'
'Helleborus niger L. / Christmas rose, Christ-hellebore / Schneerose, Nieswurz'
DOE (s.v. *ceaster-wyrt*): 1 occ. (*Lǣcebōc*, 32/1).
ETYMOLOGY: OE formation; hybrid formation (→*ceaster-*). MORPHOLOGY: N+N; endocentric, copulative: subsumptive (pleonastic). MOTIVATION: plant & botanical information. The second element is probably a kind of explanatory element, indicating that this kind of *ceaster-* is a plant. ASSOCIATIVE RELATIONS: conceptual identity & taxonomic sub-ordination.
▫ →*ceaster-*, →*wyrt*

ceder, noun
'Cedrus libani Barrel. / cedar of Lebanon / Zeder'
DOE (s.v. *ceder*): 10 occ.
LATIN EQUIV. in the manuscripts: *cedrus*.
ETYMOLOGY: Probably an early Latin loan (though not listed in Wollmann 1990 and subgrouped by Serjeantson 1935: 284 to the category (C), i.e. to the later loans); the compound *cederbēam* is counted among the WGmc formations by Carr (1939: 129; OHG *cedirboum*, MLG *cederbôm*, MDu *cederboom*); < L *cedrus* (*OED*, s.v. *cedar*) < Gk *κέδρος* (Feulner 2000: 123). ModE *cedar* is not the descendant

of OE *ceder* but a more recent loan from Latin via OFr *cedre* into Middle English (ME *cedre*, etc.); cf. Feulner (2000: 123-4).
▫ →*cederbēam*, →*cedertrēow*

ceder-bēam, lit. 'cedar-tree'
'Cedrus libani Barrel. / cedar of Lebanon / Zeder'
DOE (s.v. *ceder-bēam*): c. 60 occ., "mainly in psalter glosses".
LATIN EQUIV. in the manuscripts: *cedrus*.
ETYMOLOGY: Independent formation in WGmc (Carr 1939: 129; OHG *cedirboum*, MLG *cederbôm*, MDu *cederboom*); hybrid formation (OE *ceder* < L *cedrus*; →*ceder*). MORPHOLOGY: N+N; endocentric, copulative: subsumptive (pleonastic). MOTIVATION: plant & botanical information; The *ceder* 'cedar' is a kind of *bēam* 'tree'. ASSOCIATIVE RELATIONS: conceptual identity & taxonomic sub-ordination.
▫ →*bēam*, →*ceder*

ceder-trēow, lit. 'cedar-tree'
'Cedrus libani Barrel. / cedar of Lebanon / Zeder'
DOE (s.v. *ceder-trēow*): c. 19 occ., "mainly in psalter glosses".
LATIN EQUIV. in the manuscripts: *cedrus*.
ETYMOLOGY: OE formation; hybrid formation (OE *ceder* < L *cedrus*; →*ceder*). MORPHOLOGY: N+N; endocentric, copulative: subsumptive (pleonastic). MOTIVATION: plant & botanical information; The *ceder* 'cedar' is a kind of *trēow* 'tree'. ASSOCIATIVE RELATIONS: conceptual identity & taxonomic sub-ordination.
▫ →*ceder*, →*trēow*

cerfille, noun
'Anthriscus cerefolium (L.) Hoffm. / garden chervil / Kerbel'
DOE (s.v. *cerfille*): 24 occ.
ILLUSTRATION in manuscript London, British Library, Cotton Vitellius C. iii, fol. 51v (Ohlgren 168(63)).
LATIN EQUIV. in the manuscripts: *chaerophyllum*.
ETYMOLOGY: Early Latin loan (Wollmann 1990: 158) < L *chaer(e)phylla*, pl. of *chaerephyllum* < Gk *χαιρέφυλλον* (cf. *Genaust*, s.v. *chaerophýllum*; *OED*, s.v. *chervil*; Feulner 2000: 132-4); see also →*fille*.
▫ →*wuducerfille*

cicen, noun
'CHICKEN / Huhn'
▫ →*cicena mete*

cicena mete, lit. 'chicken's food'
'Stellaria media L. / chicken-weed / Hühnermiere, Vogelmiere'
DOE (s.v. *cicena mete*): 14 occ., "in glosses and medical recipes".
LATIN EQUIV. in the manuscripts: *intubum*, *senecio*.
ETYMOLOGY: OE formation; possibly a loan translation of L *morsus gallinae* (see MOTIVATION). MORPHOLOGY: N_{gen}+N; exocentric, metonymic compound. MOTIVATION: animal & food. Stellaria media is the preferred fodder for all kinds of birds (Marzell, s.v. *Stellaria media* L.). Bierbaumer (3, s.v. *cicena mete*) suggests loan formation modelled on L *morsus gallinae*, but as this plant is a common weed and loved by all kinds of birds (cf. also the high number of corresponding German plant names, e.g., *Vogelmiere* and *Hühnermyrte* listed by Marzell, s.v. *Stellaria*

media L.), it cannot be excluded that it was formed independently in Old English. ASSOCIATIVE RELATIONS: conceptual contiguity.
▫ →*cicen*, →*mete*

cīpe, noun
(i) 'Allium cepa L. / onion / Zwiebel';
(ii) 'Allium ascalonium L. / shallot / Schalotte'
DOE (s.v. *cīpe*): 13 occ.
LATIN EQUIV. in the manuscripts: *ascalonius*, *caepa*.
ETYMOLOGY: Early Latin loan; < L *cēpa* (*DOE*, s.v. *cīpe*) < Gk **κήπια* (Wollmann 1990: 161; Feulner 2000: 137-8).
▫ →*cīpelēac*

cīpe-lēac, lit. 'onion-leek' or 'onion-plant'
(i) 'Allium cepa L. / onion / Zwiebel'; (ii) 'Allium porrum L. / leek, common leek / Porree, Winterlauch'; (iii) 'Allium ascalonium L. / shallot / Schalotte'
DOE (s.v. *cīpe lēac*): 3 occ.
LATIN EQUIV. in the manuscripts: *ascalonius*, *caepa*.
ETYMOLOGY: OE formation; hybrid formation (OE *cīpe* < L *cēpa* 'onion';[301] →*cīpe*). MORPHOLOGY: N+N; endocentric, copulative: subsumptive (pleonastic). MOTIVATION: plant & botanical information; As different kinds of Alliums were important food for the Anglo-Saxons (see →*lēac*), OE *lēac* became the prototypical food plant, which can be seen not only in compounds such as *cīpelēac*, but also in *lēactūn* 'herb-garden' lit. 'enclosure for plants', where *lēac* has extended its denotation to 'plant' in general. ASSOCIATIVE RELATIONS: conceptual identity & conceptual identity.
▫ →*cīpe*, →*lēac*

ciris, *cisir*, noun
'fruit of Prunus cerasus L. / cherry / Kirsche'
DOE (s.v. *ciris*, *cirse*): 2 occ.
LATIN EQUIV. in the manuscripts: *fructus*, *myrtus*.
ETYMOLOGY: Early Latin loan (Wollmann 1990: 160); OE *ciris* < ? WGmc **kirissa*: **keresjâ* < MedL **ceresia*, **ceresea* (see *OED*, s.v. *cherry*) CL *cerasus* 'cherry-tree', *cerasum* 'cherry' < Gk *κέρασος* (Feulner 2000: 141-3; see also Wollmann 1990: 160); with reciprocal metathesis of *r* and *s* and modification with Gmc suffix (Campbell 1959: § 541(6)). ModE *cherry* is a remotivated form of Agn *cherise* 'cherry'; the /s/ was reinterpreted as a plural marker and therefore dropped when the word was used in the singular.
▫ →*cirisæppel*, →*cirisbēam*, →*ciristrēow*

ciris-æppel (**cirs-**, **cisir-**, etc.), lit. 'cherry-fruit'
'fruit of Ficus carica L. / Carian fig / Feige'
DOE (s.v. *ciris-æppel*): 1 occ. (*Cleopatra Glossary* (Stryker), 96, no. 210); not in Bierbaumer (1, 2, 3); identification following the *DOE* (s.v. *cirisæppel*).
LATIN EQUIV. in the manuscripts: *carica*.
ETYMOLOGY: OE formation; hybrid formation; OE *ciris* is a loan word from

301 Carr (1939: 293) regards *cīpe* as a nominative or as a genitive form with dissimilated *n*.

Latin; see →*ciris*. MORPHOLOGY: N+N; endocentric, rectional; partial metaphor. MOTIVATION: overall appearance & botanical information. The use of the element *ciris-* seems to draw on a comparison between figs and cherries; the second element, *æppel*, is polysemous (see →*æppel*): here, it is rather the meaning 'fruit', which is activated. The fig-tree was probably introduced in middle Europe in the 10th century (Marzell, s.v. *Ficus carica* L). ASSOCIATIVE RELATIONS: metaphorical similarity & taxonomic sub-ordination.

▫ →*æppel*, →*ciris*

ciris-bēam (***cirs-***, ***cisir-***, etc), lit. 'cherry-tree'

'Prunus cerasus L. / cherry-tree / Kirschbaum'

DOE (s.v. *ciris-bēam*): 4 occ.[e], "in glossaries".

LATIN EQUIV. in the manuscripts: *cerasus*.

ETYMOLOGY: Independent formation in WGmc; OHG *kersboum*, OLG *kersbôm*, MDu *kerseboom* (Carr 1939: 129); hybrid formation; OE *ciris* is a loan word from Latin (see →*ciris*). MORPHOLOGY: N+N; endocentric, rectional. MOTIVATION: part of plant & botanical information: this kind of tree bears cherries. ASSOCIATIVE RELATIONS: conceptual contiguity (meronymy) & taxonomic sub-ordination.

▫ →*bēam*, →*ciris*

ciris-trēow (***cirs-***, ***cisir-***, etc.), lit. 'cherry-tree'

'Prunus cerasus L. / CHERRY-TREE / Kirschbaum'

DOE (s.v. *ciris-trēow*): 2 occ. (*Antwerp-London Glossaries*, 122/12; *Bodley 730*, no. 189).

LATIN EQUIV. in the manuscripts: *cerasus*.

ETYMOLOGY: OE formation; hybrid formation; OE *ciris* is an early loan word from Latin (see →*ciris*). MORPHOLOGY: N+N; endocentric, rectional. MOTIVATION: part of plant & botanical information: this kind of tree bears cherries. ASSOCIATIVE RELATIONS: conceptual contiguity (meronymy) & taxonomic sub-ordination.

▫ →*ciris*, →*trēow*

cist- →*cisten*

cist-bēam, lit. 'chestnut-tree'

'Castanea sativa Mill. / chestnut tree / Kastanienbaum'

DOE (s.v. *cist-bēam*): 1 occ.; attested spelling *cystbeam* (*Antwerp-London Glossaries*, 123/6).

LATIN EQUIV. and OE synonyms in the manuscripts: *castanea*; *cistel*, *cisten*.

ETYMOLOGY: Independent formation in WGmc (Carr 1939: 130); hybrid compound; The first elements of *cistbēam*, *cistelbēam* and *cistenbēam* go back to L *castanea*, which was probably borrowed into WGmc as **kastinja-* (Sauer 2000: 18) and further disseminated into the WGmc languages (G *Kastanie*, etc.): *cist(en, el)* < WGmc **kastinna*, **kastinja* < **kastanja* < L *castanea* (*OED*, s.v. *chesteine, chesten*; Feulner 2000: 145-7). According to Carr (1939: 130), however, the West Germanic formations are independent. MORPHOLOGY: N+N; endocentric, copulative: subsumptive (pleonastic). *Cist-* is clipped from *cisten-* and is attested only in this compound. OE *cist-* 'chestnut' must not be confused with OE *cist* 'chest, box, coffer', which is a Latin loan word, too (< L *cista* 'chest, box'; also extant in G *Kiste*, Du *kist*, etc.; *OED*, s.v. *chest* n[1]). MOTIVATION: plant & botanical information; see →*cistenbēam*. ASSOCIATIVE RELATIONS: conceptual identity & taxonomic sub-ordination.

▫ →*bēam*, →*cisten*

cistel, noun

'Castanea sativa Mill. / chestnut (tree) / Kastanie'

DOE (s.v. *cisten*): 2 occ. (*Herbarium*, no. 15; *Antwerp-London Glossaries*, 123/8).

LATIN EQUIV. and OE synonyms in the manuscripts: *castanea*; *cistbeam*, *cistin*. ETYMOLOGY: Early Latin loan (Wollmann 1990: 178); WGmc **kastinna*, **kastinja* + *-el* < **kastanja* + *-el* < L *castanea* + *-el* (suffix substitution; cf. Campbell 1959: § 518; Feulner 2000: 145-7).

▫ →*cistelbēam*

cistel-bēam, lit. 'chestnut-tree'

'Castanea sativa Mill. / chestnut tree / Kastanienbaum'

DOE (s.v. *cistel-bēam*): 1 occ. (*Durham Plant Name Glossary*, no. 103).

LATIN EQUIV. in the manuscript: *castanea*.

ETYMOLOGY: Independent formation in WGmc (Carr 1939: 130); hybrid compound (see →*cistbēam*). MORPHOLOGY: N+N; endocentric, copulative: subsumptive (pleonastic). Suffix substitution of native *-el* for putative suffix *-en* in *cisten-* (< L *castanea*); (Feulner 2000: 145-7; Campbell 1959: § 518); see also →*cistenbēam*. MOTIVATION: plant & botanical information. ASSOCIATIVE RELATIONS: conceptual identity & taxonomic sub-ordination.

▫ →*bēam*, →*cistel*

cisten (*cysten*), noun

'Castanea sativa Mill. / chestnut (tree) / Kastanie'

DOE (s.v. *cisten*): 2 occ. (*Herbarium*, no. 15 and *Antwerp-London Glossaries*, 123/8).

LATIN EQUIV. and OE synonyms in the manuscripts: *castanea*; *cistbeam*, *cistel*. ETYMOLOGY: Early Latin loan (Wollmann 1990: 178); WGmc **kastinna*, **kastinja* < **kastanja* < L *castanea* < Gk *κάστανον*, etc. (*OED*, s.v. *chesteine*, *chesten*; Feulner 2000: 145-7; *Genaust*, s.v. *castánea*).

▫ →*cistbēam*, →*cistenbēam*

▫ →*stāncysten*

cisten-bēam, lit. 'chestnut-tree'

'Castanea sativa Mill. / chestnut tree / Kastanienbaum'

DOE (s.v. *cisten-bēam*): 5 occ.[e]

LATIN EQUIV. in the manuscripts: *castanea*, *castaneus*.

ETYMOLOGY: Independent formation in WGmc (Carr 1939: 130); hybrid compound (see →*cisten*, →*cistbēam*). MORPHOLOGY: N+N; endocentric, copulative: subsumptive (pleonastic). L *castanea* is extant in OE *cisten* 'chestnut tree' (attested two times), in OE *cistel* 'chestnut tree', a hapax legomenon where the pseudo-suffix *-en* obviously was replaced by the suffix *-el* (see →*cistelbēam*), and in *cist-*, a clipped version of the Latin loan, which is not attested elsewhere (→*cist-*). OE *cistbēam* and *cistelbēam* are hapaxes

(*Antwerp-London Glossaries*, 123/6; resp. *Durham Plant Name Glossary*, no. 103), while *cistenbēam* is recorded five times. MOTIVATION: plant & botanical information. The *cisten* 'chestnut tree' is a kind of *bēam* 'tree'. ASSOCIATIVE RELATIONS: conceptual identity & taxonomic sub-ordination.
▫ →*bēam*, →*cisten*

clāte, noun
(i) 'Arctium lappa L. / CLOTE, greater burdock / Klette'; (ii) 'Galium aparine L. / cleavers / Kletten-Labkraut'
DOE (s.v. *clāte*): c. 50 occ.; as a simplex, 46 occ. (*DOEPN*, s.v. *clāte*).
LATIN EQUIV. in the manuscripts: *amorfolia* (1x with *gliconus* [for *glechon*]), *blitum*, *cliton*, *lappa*, *oxylapathum*, *personata*, *philanthropos*, *tuber*.
ETYMOLOGY: Gmc **klaiton-*, derived from Gmc **klei-* 'cleave, adhere' (*Kluge*, s.v. *Klette*; *OED*, s.v. *clote*).
▫ →*clātecrop*
▫ →*foxesclāte*, →*smale clāte*, →*wuduclāte*

clāte-crop, lit. 'burr-umbel'
'Arctium lappa L. / great burr / Große Klette'
DOE (s.v. *clāte-crop*): hapax legomenon (Steinmeyer 1889: 250).
LATIN EQUIV. in the manuscript: *personata*.
ETYMOLOGY: OE formation; MORPHOLOGY: N+N; exocentric, metonymic compound. MOTIVATION: plant & part of plant. OE *clāte* refers to the burr; the element *crop* 'umbel, cluster, head, bunch' is probably used because of the noticeably big heads of the burr (Bierbaumer 3, s.v. *clāte-crop*). ASSOCIATIVE RELATIONS: conceptual identity & conceptual contiguity (meronymy).
▫ →*clāte*, →*crop*

clǣfre, *clǣfer*, noun
'Trifolium L. / CLOVER / Klee'
DOE (s.v. *clǣfre*): 17 occ.
ILLUSTRATION in manuscript London, British Library, Cotton Vitellius C. iii, fol. 41r (Ohlgren 168(43)).
LATIN EQUIV. in the manuscripts: *calt(h)a* (1x with *trifolion*), *chamaedrys*, *crision* (= *cirsion*), ? *fecta*, *trifolium*, *viola*.
ETYMOLOGY: Gmc **klaibron-* (*OED*, s.v. *clover*).
▫ →*clǣferwyrt*
▫ →*heorotclǣfer*, →*hwīteclǣfre*, →*rēad clǣfre*, →*wuduclǣfre*

clǣfre-wyrt, lit. 'clover-wort'
'Trifolium L. / clover / Klee'
DOE (s.v. *clǣfre-wyrt*): 1 occ. (*Lǣcebōc*, 32/6) in the phrase *þā smālan clǣferwyrt,* where it refers probably to the smaller species 'Trifolium minus Relh. ex Sm. / slender yellow trefoil / Fadenklee, Kleiner Himmelhopfen (Bierbaumer 1, s.v. *clǣferwyrt*).
ETYMOLOGY: OE formation. MORPHOLOGY: N+N; endocentric, copulative: subsumptive (pleonastic). MOTIVATION: plant & botanical information. The clover is a kind of plant. ASSOCIATIVE RELATIONS: conceptual identity & taxonomic sub-ordination.
▫ →*clǣfre*, →*wyrt*

clīfan (str.) or *clifian* (wk.), verb
'(to) CLEAVE, adhere / kleben'
The strong verb *clīfan* is attested once in Gregory's *Pastoral Care* (ed. Sweet 1871-1887; cf. *DOE*, s.v. *clīfan*); the

weak verb *clifian* has 27 occ. (cf. *DOE*, s.v. *clifian*).
▫ →*clifwyrt*

clīfe, *clife*, noun
(i) 'Arctium lappa L. / greater burdock / Klette'; (ii) 'Galium aparine L. / CLIVERS, CLEAVERS / Kletten-Labkraut'
DOE (s.v. *clife*): 14 occ.; 11 occ. as a simplex (*DOEPN*, s.v. *clīfe*).
LATIN EQUIV. in the manuscripts: *aparine*, *cirros* (? = *cirrus*), *glitilia* (?), *personata*.
ETYMOLOGY: derived from the verb OE *clīfan* '(to) cleave' < Gmc **klei-* (*Kluge*, s.v. *Klette*; *OED*, s.v. *clote*); cf. OS *biklīban*, OHG *klīban*, ON *klífa* (*DOE*, s.v. *clīfan*) or from *clifian* '(to) cleave'.
▫ →*foxesclīfe*, →*gārclīfe*, →*hegeclīfe*, →*smale clīfe*

clif-wyrt, lit. 'cleav(ing) plant'
'Arctium lappa L. / great burr / Klette'
DOE (s.v. *clif-wyrt*): 6 occ.
LATIN EQUIV. in the manuscripts: *agrimonia*, *cliton*, *lappa* (1x with *blitum*).
ETYMOLOGY: OE formation. MORPHOLOGY: V_{stem}+N; endocentric, rectional. MOTIVATION: action & botanical information. Arctium lappa L. is a plant that cleaves to the garments of people and to the furs of animals (cf. OE *clīfan* 'to cleave'; Bierbaumer 1, 2, 3, s.v. *clif-wyrt*). ASSOCIATIVE RELATIONS: conceptual contiguity & taxonomic sub-ordination.
▫ →*clīfan*, →*wyrt*

clite, noun
(i) '? Arctium lappa L. / greater burdock, Große Klette'; or (ii) '? Galium aparine L. / cleavers, CLITE / Kletten-Labkraut'; or (iii) '? Tussilago farfara L. / colt's foot / Huflattich'
As a simplex, attested 6 times (*DOEPN*, s.v. *clite*; *DOE*, s.v. *clite*).
LATIN EQUIV. in the manuscripts: *lappa*.
ETYMOLOGY: Parallel to *clāte* and, apparently, based on the same root (*OED*, s.v. *clote, clete, clett*).
▫ →*lange clite*

cliþ-wyrt, lit. 'cleaving plant'
'Arctium lappa L. / great burr / Klette'
DOE (s.v. *cliþ-wyrt*): 1 occ. (*Lācnunga*, 798); the second occurrence is as *smale cliþwyrt* in *Lācnunga* (742), which is listed separately in the present survey (→*smale cliþwyrt*).
ETYMOLOGY: OE formation; MORPHOLOGY: V_{stem}+N; endocentric, rectional. MOTIVATION: action & botanical information. Arctium lappa L. cleaves to the garments of people and to the fur of animals (cf. →*(æt)clīðan* '(to) adhere'). ASSOCIATIVE RELATIONS: conceptual contiguity & taxonomic sub-ordination.
▫ →*(æt)clīðan*, →*wyrt*
▫ →*smale cliþwyrt*

clufiht, adjective
'cloved, bulbed / knollenartig, knollig'
DOE (s.v. *clufiht*): 11 occ. The adjective *clufiht* occurs only in combination with OE *wenwyrt* (cf. *DOE*, s.v. *clufiht*).
▫ →*clufihte wenwyrt*

clufihte wenwyrt, lit. 'bulbous wenwort'
'? Ranunculus ficaria L. / figwort, pilewort / Scharbockskraut, Feigwurz'
DOE (s.v. *clufiht*): 11 occ., "only in the collocation (seo) clufihte wenwyrt 'the cloved or bulbed lesser celandine'".
ETYMOLOGY: OE formation. MORPHOLOGY: Adj+N; endocentric, copulative: attributive. MOTIVATION: shape & botanical information. Ranunculus ficaria has small and bulbous roots, which can be associated with tumours (hence perhaps the OE name *wenwyrt*; see →*wenwyrt*), but which can also be compared to cloves – probably also the motivation for the first element, *clufiht(e)* 'bulbous, cloved'. ASSOCIATIVE RELATIONS: conceptual contiguity & taxonomic sub-ordination.
▫ →*clufiht*, →*wenwyrt*

cluf-þung, cluf-þunge, lit. 'clove(d) poisonous plant'
'Ranunculus sceleratus L. / celery-leaved crowfoot / Gift-Hahnenfuß'
DOE (s.v. *cluf-þung, cluf-þunge*): 22 occ.
ILLUSTRATIONS in manuscripts London, British Library, Cotton Vitellius C. iii, fol. 25r (Ohlgren 168(11)), and Oxford, Bodleian Library, Ashmole 1431 (S.C. 7523), fol. 7v (Ohlgren 221(8)).
LATIN EQUIV. and Old English synonyms in the manuscripts: *artemisia monoclonos*, *batrachion*, ? *pamfesus* (*-um*), *scelerata*; *þelneþung*, *þung*.
ETYMOLOGY: OE formation (*OED*, s.v. *cloffing*). MORPHOLOGY: N+N; endocentric, rectional. MOTIVATION: shape & botanical information. The Old English noun *clufu* 'clove (of garlic, etc.), bulb, tuber' is derived ultimately from the IE stem **gleub^h^*- 'to cleave, to cut' (*ODEE*, s.v. *cleave*[1]) and relates, according to Bierbaumer (1, s.v. *cluf-þung*), to the cleft leaves of the plant; the related Old English verb is *clēofan* 'to cleave'. But, since the species Ranunculus sceleratus L., Ranunculus acer L. and Ranunculus bulbosus L. (see →*clufwyrt*) were often not distinguished by the laymen, the element *clufu* more probably relates to the stems of Ranunculus bulbosus, which are thickened onion-like right above the earth (Marzell, s.v. *Ranunculus bulbosus* L.). The plant is poisonous, hence the use of *þung* 'poisonous plant' (cf. G *Gift-Hahnenfuß*). ASSOCIATIVE RELATIONS: conceptual contiguity & taxonomic sub-ordination.
▫ →*cluf*, →*þung*

clufu, noun
'CLOVE, bulbil / Knolle'
DOE (s.v. *clufu*): 5 occ.
LATIN EQUIV. in the manuscripts: *capiclum* (prob. for *capitulum*).
ETYMOLOGY: < Gmc **kluba*- (*OED*, s.v. *clove*). The Old English noun *clufu* 'clove (of garlic, etc.), bulb, tuber' is derived from the IE stem **gleub^h^*- 'to cleave, to cut' (Bierbaumer 1, s.v. *cluf-þung*).
▫ →*clufþung*, →*clufwyrt*
▫ →*wenclufu*

cluf-wyrt, lit. 'clove-wort'
(i) 'Ranunculus acer L. / buttercup, tall crowfoot / Scharfer Hahnenfuß'; (ii) 'Ranunculus bulbosus L. / bulbous buttercup / Knollen-Hahnenfuß'
DOE (s.v. *cluf-wyrt*): 9 occ.
ILLUSTRATIONS in manuscripts London, British Library, Cotton Vitellius C. iii, fol. 25r (Ohlgren 168), and Oxford,

Bodleian Library, Ashmole 1431 (S.C. 7523), fol. 8r (Ohlgren 221(9)).
LATIN EQUIV. in the manuscripts: *batrachion* (1x with *os munda*).
ETYMOLOGY: OE formation (*OED*, s.v. *clovewort*). MORPHOLOGY: N+N; endocentric, rectional. MOTIVATION: shape & botanical information. The stems of Ranunculus bulbosus are onion-like thickened right above the earth (Marzell, s.v. *Ranunculus bulbosus* L.); hence, the use of the elements *bulbosus* in Ranunculus bulbosus and *Knollen-* in G *Knollen-Hahnenfuß*. The species Ranunculus acer and Ranunculus bulbosus were often confused by the laymen, which explains the identification as Ranunculus acer by Bierbaumer (1, s.v. *clufwyrt*; see also Marzell, s.v. *Ranunculus (acer) acris*). ASSOCIATIVE RELATIONS: conceptual contiguity & taxonomic sub-ordination.
▫ →*cluf*, →*wyrt*

cnēow, noun
'KNEE / Knie'
▫ →*cnēowholen*

cnēow-holen, lit. 'knee-holly'
(i) 'Ruscus aculeatus L. / KNEE-HOLLY, butcher's broom / Stechender Mäusedorn, Dornmyrte'; (ii) '?? Daphne laureola L. / spurge laurel / Lorbeer-Seidelbast'
DOE (s.v. *cnēow-holen*): 28 occ.[e]
ILLUSTRATIONS in manuscripts London, British Library, Cotton Vitellius C. iii, fol. 38v (Ohlgren 168(38)), and Oxford, Bodleian Library, Ashmole 1431 (S.C. 7523), fol. 16v (Ohlgren 221(48)).
LATIN EQUIV. in the manuscripts: *chamaedaphne*, ? *mustelago*, *ruscus*, *symphytum*, *victoriola*.
ETYMOLOGY: OE formation (*OED*, s.v. *knee-holly*). MORPHOLOGY: N+N; endocentric, rectional; partial metaphor. MOTIVATION: size & botanical information. Ruscus aculeatus is only knee-high; its leaves are narrowed at the top and prick like the leaves of the holly (Ilex aquifolium L.), OE *holen* (Marzell, s.v. *Ilex aquifolium* L.). The height of the plant is not expressed directly, but metaphorically: 'this kind of holen is as high as a knee.' In Anglo-Saxon folk taxonomy, *cnēowholen* is taxonomically sub-ordinated to the *holen* 'holly' (based on the characteristic pricking leaves), but in modern scientific taxonomy, these two plants belong to totally unrelated species. ASSOCIATIVE RELATIONS: metaphorical similarity & taxonomic sub-ordination.
▫ →*cnēow*, →*holen*

cnēowiht, adjective
'knotty / knotig'
ETYMOLOGY: Derived from *cnēo(w)* 'KNEE'.
DOE (s.v. *cnēowiht*): 1 occ., in the plant name →*cnēowihte wenwyrt*.
▫ →*cnēowihte wenwyrt*

cnēowihte wenwyrt, lit. ? 'knotty wenwort'
unidentified plant; perhaps (i) '?? Scrophularia nodosa L. / figwort / Knoten-Braunwurz' or (ii) '?? Polygonum L. / knotweed / Knöterich'
DOE (s.v. *cnēowiht*): attested once, as "cneoehte wenwyrt" (*Lǣcebōc*, 42/33).
ETYMOLOGY: OE formation; MORPHOLOGY: Adj+N; endocentric, copulative: attributive. MOTIVATION: shape & botanical information. According to Bierbaumer (1, s.v. *wenwyrt*; 3, s.v.

wenclufu), probably a plant with a knotty, windy stem. ASSOCIATIVE RELATIONS: conceptual contiguity & taxonomic sub-ordination.
▫ →*cnēowiht*, →*wenwyrt*

codd, noun
(i) 'CODD, husk, peel / Schale, Hülse'; (ii) 'bag, sack / Tasche, Sack'; (iii) 'scrotum / Hodensack'
DOE (s.v. *codd*): 15 occ.; as a botanical term, 3 occ. (*DOEPN*, s.v. *codd*).
LATIN EQUIV. in the manuscripts: *folliculus*.
ETYMOLOGY: The element *codd-* in *coddæppel* is, according to Wollmann (1990: 166, 673), an Early Latin loan with VL /o/ instead of CL /u/ (L *cudōnia (mālum)*), but cf. *cotōnium, cotōneum* 'from Cydonia' (→*codd-æppel*); probably, the loan word was changed in analogy to OE *codd* 'husk' (< Gmc **kuddo-z*; *OED*, s.v. *codd*).
▫ →*coddæppel*

cod(d)-æppel, lit. 'husk-apple' or 'husk-fruit'
'fruit of Cydonia oblonga Mill. / quince / Quitte'
DOE (s.v. *codd-æppel*): 1 occ. (*Cleopatra Glossary*, (Stryker), 212, no 456).
LATIN EQUIV. in the manuscripts: *malum cotonium*.
ETYMOLOGY: OE formation; loan rendition of *mālum cotōnium*; hybrid formation, folk etymology (see MOTIVATION). MORPHOLOGY: N+N; endocentric, rectional. MOTIVATION: secondary, phonological and semantic motivation of the first element: part of plant & botanical information. The first element of this hapax legomenon was derived from L *cydōnium, cotōnium, cotōneum* lit. 'from Cydonia (a town on the island Crete)', a loan which was changed to *codd* because of the phonological similarity between the loan word and the Old English lexeme *codd* 'husk, hull, peel' (Feulner 2000: 150), probably supported by the denotational meaning of *codd* 'husk, peel' and the flashingly hard peel of the quince (semantic motivation). The whole plant name is a loan rendition of *malum cotonium* attested in the *Cleopatra Glossary* ((Stryker), 212, no 456); cf. also It *cotogno*, Prov *codonh*, etc. (*ODEE*, s.v. *quince*). Both L *mālum* (*Lewis and Short*, s.v. *mālum*) and OE *æppel* are polysemous, denoting the 'apple' on the one hand, and 'fruit' in general (see also →*æppel*). ASSOCIATIVE RELATIONS: phonological similarity & taxonomic sub-ordination.
▫ →*æppel*, →*codd*

collon-, noun
'? well / Quelle'
DOE -; not attested independently.
▫ →*colloncroh*

collon-croh, lit. 'water-crocus'
collon-crōg, lit. '? water-can'
'? Nymphaea alba L. / white waterlily / Weiße Seerose'
DOE (s.v. *collon-croh*): 6 occ.
LATIN EQUIV. in the manuscripts: *achilleos, nymphaea, typhus*.
ETYMOLOGY: OE formations; *collon-croh* is a hybrid formation; *colloncrōg* is affected by folk etymology (see MOTIVATION). MORPHOLOGY: *collon-croh*: N+N; endocentric, rectional; *colloncrōg*: N+N, exocentric, metaphorical compound; the element *collon-*

is not attested independently. MOTIVATION: *colloncroh*: habitat & botanical information; *colloncrōg*: habitat & secondary, phonological and semantic motivation: sound pattern (on the intralinguistic level) and shape (on the conceptual level). Several interpretations as to the exact etymology of the plant name have been proposed; see, for instance, Lehmann (1907) and Gardner (1968: 175). The second element is probably a loan word from Latin, OE *croc, croh* 'crocus' < L *crōcus* (Wollmann 1990: 168), which was changed to *crōg* 'can, jug, crock' because of the shape of the blossoms (phonological and semantic motivation); cf. Bierbaumer (3, s.v. *colloncroh*). As far as *collon-* is concerned, meaning and form are even more unclear than those of *crōg*; Bierbaumer (3, s.v. *colloncrōh*) connects it to OHG *quella* 'well'. ASSOCIATIVE RELATIONS: conceptual contiguity & metaphorical similarity.

▫ *→collon-, →crōg, →croh*

colt, noun
'COLT / Pferd'

▫ *→coltetræppe, →coltnægl*

colte-træppe, calca-treppe, lit. 'colt's trap'

(i) 'Centaurea calcitrappa L. / CALTROP, star-thistle / Sterndistel, Stern-Flockenblume'; (ii) 'Eryngium maritimum L. / sea-holly / Strand-Mannstreu'

DOE (s.v. *calca-treppe*): 2 occ.; as *calcatrippe* (*Brussels Glossary*, 298/30) and as *coltetræppe* (*Cleopatra Glossary* (Quinn), 37/4) "with *t* [mistakenly?] for *c*".

LATIN EQUIV. in the manuscripts: *heraclea*, *rhamnus*.

ETYMOLOGY: OE formation; later Latin loan word, changed by processes of folk etymology (see MOTIVATION). MORPHOLOGY: N+N; exocentric, metaphorical compound; MOTIVATION: secondary, phonological motivation: sound pattern & sound pattern, and secondary semantic motivation of both elements. The Latin loan *calcatreppe* (< L *calcatrippe*; Serjeantson 1935: 284), was changed by folk etymology in analogy to the OE terms *colt* 'colt, foal' and *træppe* 'trap, snare' based on the phonological similarity of the Latin elements and the Old English terms (Bierbaumer 3, s.v. *coltetræppe*), further supported by the semantic analogy of the plant's characteristic of entangling feet and the denotational meaning of OE *træppe* as 'trap' (cf. also *OED*, s.v. *caltrop*); ASSOCIATIVE RELATIONS: phonological similarity & phonological similarity; secondary, semantic motivation: metaphorical similarity.

▫ *→colt, →træppe*

colt-nægl (***colt-grǣg***), lit. 'colt-gray' or '? colt-nail'

'Tussilago farfara L. / coltsfoot / Huflattich'

DOE (s.v. *colt-grǣg*): 1 occ. (*Antwerp-London Glossaries*, 118/9).

LATIN EQUIV. in the manuscripts: *caballopedia*, *ungula caballi*.

ETYMOLOGY: OE formation; according to Bierbaumer (2003: 158), the second element does not represent the adjective *grǣg* but should be emended to *nægl* 'nail' as the Old English plant name is probably a literal translation of the Latin lemma *ungula caballi* 'horse's nail' (in "caballopodia uel ungula caballi colograeig", *Antwerp-London Glossaries*, 118/9); MORPHOLOGY: As *coltnægl* a N+N, exocentric, metaphorical compound. MOTIVATION: animal & shape. L *ungula caballi* and its translation, OE *coltnægl*, allude to the shape of the plant's leaves, which resemble the hooves of horses; cf. also F *ungle de cheval*, It *ungula cavalli*, G *Huflattich* (Marzell, s.v. *Tussilago farfara* L.) and ModE *coltsfoot*. ASSOCIATIVE RELATION: metaphorical similarity.

▫ →*colt*, →*nægl*

corn 1, noun

(i) 'CORN, pip, seed, berry / Korn, Kern, Samen, Frucht'; (ii) 'cereals / Getreide' (iii) fig. use for manna, hail (in poetry), etc.

DOE (s.v. *corn*[1]): c. 150 occ.; as a simplex denoting 'corn, pip' attested c. 72 times (*DOEPN*, s.v. *corn*).

LATIN EQUIV. in the manuscripts: *armentum* (? for *frumentum*), *frumentum*, *granum*, *granum frumenti*, *granum sinapis*, *scandala*, *ptisana*.

ETYMOLOGY: Gmc **korno-* < Gmc **kurnóm* < IE **grənó-* (*Kluge*, s.v. *Korn*).

▫ →*cornæppel*

▫ →*gipcorn*, →*lybcorn*, →*moldcorn*, →*sundcorn*, →*sunnancorn*

corn- 2, noun

'Cornus mas L. / cornel / Kornelkirschenbaum'

Not attested independently.

ETYMOLOGY: Early Latin loan (Wollmann 1990: 168); < L *cornus* (*Kluge*, s.v. *Kornelkirsche*); ModE *cornel* goes back to an assumed diminutive form VL **cornuculus* (> MedL *cornolium*) derived from CL *cornus* or *cornum* (the fruit of *cornus*; *OED*, s.v. *cornel*).

▫ →*corntrēow*

corn-æppel, lit. 'pip-, corn-apple' or 'pip-, corn-fruit'

'fruit of Punica granatum L. / pomegranate / Granatapfel'

DOE (s.v. *corn-æppel*): 4 occ.

LATIN EQUIV. in the manuscripts: *malum punicum*.

ETYMOLOGY: OE formation; MORPHOLOGY: N+N; endocentric, rectional. MOTIVATION: part of plant & botanical information. As the edible parts of the pomegranate are the juicy red arils – in contrast to the species of 'apple', where not the pips but the pulp is edible – *corn* 'pip, corn' is used as the first element of this compound. The second element, *æppel*, refers either to the plant genus Malum 'apple', or to 'fruit' in general (see →*æppel*): the *cornæppel* 'pomegranate' is a kind of *æppel* 'apple, fruit'. ASSOCIATIVE RELATIONS: conceptual contiguity (meronymy) & taxonomic sub-ordination.

▫ →*æppel*, →*corn* 1

corn-trēow, lit. 'cornel tree'

'Cornus mas L. / cornel/ Kornelkirsche'

DOE (s.v. *corn-trēow*): 5 occ.

LATIN EQUIV. in the manuscripts: *cornus*.

ETYMOLOGY: OE formation; hybrid compound (OE *corn-* < L *cornus*; →*corn-* 2). MORPHOLOGY: N+N; endocentric, copulative: subsumptive (pleonastic); *corn-* is not attested independently. MOTIVATION: plant & botanical information; The *corn-* 'cornel' is a kind of *trēow* 'tree'.

ASSOCIATIVE RELATIONS: conceptual identity & taxonomic sub-ordination.
▫ →*corn-* 2, →*trēow*

cost, noun
'Tanacetum balsamita L. / aleCOST, COSTmary / Balsamkraut'
DOE (s.v. *cost*, noun[1]; *DOEPN*, s.v. *cost* 1): 27 occ.
LATIN EQUIV. in the manuscripts: *costum*.
ETYMOLOGY: Early Latin loan (Wollmann 1990: 168); < L *costus, costum* < Gk *κόστος* < OInd *kústa(h)* (*Genaust*, s.v. *costus*; Feulner 2000: 159-60); homonym with OE *cost* 'way, manner, etc.'.
▫ →*englisc cost*

coþu, noun
'disease, illness, sickness / Krankheit'
DOE (s.v. *coþu*): c. 40 occ.
▫ →*coþuwyrt*

coþu-wyrt, lit. 'disease-plant'
unidentified plant
DOE (s.v. *coþu-wyrt*): 1 occ. (*Durham Plant Name Glossary*, no. 255).
LATIN EQUIV. and OE synonyms in the manuscripts: *heliotropium*; *ox(e)nalybb*.
ETYMOLOGY: OE formation; MORPHOLOGY: N+N; endocentric, rectional.
MOTIVATION: illness & botanical information. L *hēliotropium* 'heliotrope' is glossed by OE *ox(e)nalybb* 'setterwort, hellebore' (→*ox(e)nalybb*) and by *coþuwyrt*. The heliotropes (e.g. Heliotropium europaeum L.) are not particularly helpful against any kinds of diseases and the hellebores (e.g. Helleborus foetidus L.) are even poisonous. This does not fit with the literal meaning of *coþuwyrt* as 'disease-plant'. As the plant name is unidentified, the motivation of the plant name and the associative relations are based on the denotata of the two constituing elements of the plant name.
ASSOCIATIVE RELATIONS: conceptual contiguity & taxonomic sub-ordination.
▫ →*coþu*, →*wyrt*

crāwan-bēam, lit. 'crow's tree'
(i) '? Tragopon pratensis L. / goat's beard / Bocksbart'; (ii) '? Daphne gnidium L. / spurge laurel / Südlicher Seidelbast'
Not an entry in the *DOE*, but attestation s.v. *crāwe*, a.ii.: 1 occ. (*Durham Plant Name Glossary*, no. 4).
LATIN EQUIV. in the manuscripts: *ablacta*.
ETYMOLOGY: OE formation; MORPHOLOGY: N_{gen}+N; endocentric, rectional.
MOTIVATION: animal & botanical information. Bierbaumer (3, s.v. *crāwan-bēam*) favours the identification of *crāwanbēam* as Daphne gnidium L., because Daphne is a tree and fits the element *bēam* better (in contrast to Tragopon pratensis L., which is a herb). The motivation for *crāwan-* is unclear; perhaps crows like to sit on the tree.
ASSOCIATIVE RELATIONS: conceptual contiguity & taxonomic sub-ordination.
▫ →*bēam*, →*crāwe*

crāwan-lēac, lit. 'crow's leek'
(i) 'Allium vineale L. / crow garlic / Bärlauch'; (ii) 'Colchium autumnale L. / meadow saffron / Herbstzeitlose'; (iii) 'Crocus albiflorus Kit. / purple crocus / Frühlingskrokus'
DOEPN (s.v. *crāwan-lēac*): 5 occ., but two of them are *cravelec* (in the

Durham Plant Name Glossary, no. 279 and no. 322), which I consider to be a separate plant name; see →*crāw(e)lēac*; Not an entry on its own in the *DOE*, but examples subsumed s.v. *crāwe*, a.i.
LATIN EQUIV. in the manuscripts: *ballote* (manuscript: *polloten*), *hermodactylus*, ? *tidolosa*.
ETYMOLOGY: OE formation; MORPHOLOGY: N_{gen}+N; endocentric, rectional; partial metaphor; also attested as →*crāw(e)lēac*. MOTIVATION: inferiority & botanical information. *Tidolosa* is a latinised borrowing of OLG *tidelosa*, OHG *zītilōsa*, a name which originally denoted the crocus (Crocus albiflorus Kits.), a plant flowering in spring (cf. Marzell, s.v. *Colchium autumnale* L., 1). This name was later given to the plant Colchicum autumnale L. on the basis of the great outward similarity of the two plants. The Latin vernacular *hermodactula* denoted Colchicum autumnale L. Marzell (s.v. *Allium vineale* L.) notes that there are parallels between the shape of the plant's leaves and the bulbs to those of the garlic, Allium vineale L.; this resemblance and also the similarity in smell (all species of Alliums, denoted in Old English by generic *lēac* 'leek', including Allium vineale, have a strong smell) is also alluded to in OHG *huntlouch*, G *Wilder Knoblauch, Hundslauch*, etc. and here our OE *crāwanlēac* fits in. ModE *crow's leek* denotes Allium vineale L., whose ML name *allium agreste* could stand for Colchium autumnale L., too. The motivation is that the plant is a kind of leek, but it is inferior with regard to the leek; to mark this inferiority its name contains the element *crāwan-* (Marzell, s.v. *Allium vineale* L.).[302] ASSOCIATIVE RELATIONS: metaphorical similarity & taxonomic sub-ordination.
▫ →*crāwe*, →*lēac*

crāwe, noun
'CROW / Krähe'
▫ →*crāwanbēam*, →*crāwanlēac*, →*crāw(e)-lēac*

crāw(e)-lēac, lit. 'crow's leek'
(i) 'Allium vineale L. / crow garlic / Bärlauch'; (ii) 'Colchium autumnale L. / meadow saffron / Herbstzeitlose' or (iii) Crocus albiflorus Kit. / purple crocus / Frühlingskrokus'
DOE (s.v. *crāw-lēac*): 4 occ.
LATIN EQUIV. in the manuscripts: *ballote* (manuscript: *polloten*), *hermodactylus*, ? *tidolosa*.
ETYMOLOGY: OE formation; MORPHOLOGY: N+N; endocentric, rectional; partial metaphor. MOTIVATION: inferiority

302 Marzell (s.v. *Allium vineale* L.): "Aussehen und Geruch des Knoblauchs (Schnittlauchs, usw.), von dem die Art als minderwertig durch verächtliche Zusätze unterschieden wird" ['[It has the] outward appearance and smell of the garlic (chives, etc.), from which the species is distinguished by the information that it is inferior; this fact is expressed by epithets with a pejorative meaning']. In my corpus, *crāwanlēac* glosses only *hermodactula* and *tidolosa* (*polloten* seems to be based on a misunderstanding, see Bierbaumer 3, s.v. *crāwanlēac*). Instead of equating *crāwanlēac* with Allium vineale, it is possible to follow d'Aronco & Cameron (1998: 48) and to regard the Latin lemmas alone, neglecting similarities with plant names preserved in Middle and Modern English. Then, however, Allium vineale could not be accepted as an identification of OE *crāwanlēac*.

& botanical information. For the explanation, see →*crāwanlēac*. ASSOCIATIVE RELATIONS: metaphorical similarity & taxonomic sub-ordination.
▫ →*crāwe*, →*lēac*

crōg, noun
'CROCK, jug / Krug'
DOE (s.v. *crōg*): 13 occ.
▫ →*colloncroh*

croh, noun
'Crocus sativus L. / saffron / Safran'
DOE (s.v. *croh*): 7 occ. "in medical recipes and glossaries".
ETYMOLOGY: Early Latin loan, < L *crocus* (Wollmann 1990: 168).
▫ →*colloncroh, colloncrōg*

crop, noun
(i) 'umbel, rounded head or top of a herb, bunch, cluster / Dolde, Büschel, Bündel, Traube'; (ii) 'sprout, shoot, stalk / Spross, Halm'
The sense 'produce of the field, harvest', specified by the *OED* (s.v. *crop*, III. 8, a.) as "the annual produce of plants cultivated or preserved for food, esp. that of the cereals; the produce of the land, either while growing or when gathered; harvest" developed in the 13th century.
DOE (s.v. *crop*, *croppa*): 75 occ., which include the attestations with a non-botanical meaning such as 'crop of a bird' or 'kidney'; as a botanical term attested 65 times (*DOEPN*, s.v. *crop*).
LATIN EQUIV. in the manuscripts: *acidula*, *acucula* (= ? *acidula* or ? *acucia*), *baca*, *botrys*, *caulis*, *circullus*, *corymbus*, *cyma*, *racemus*, *spica*, *summitas*, *thyrsus*, *uva*.
ETYMOLOGY: Gmc **kruppa-* 'swelling, bump' (*Kluge*, s.v. *Kropf*).
▫ →*croplēac*
▫ →*clātecrop*, →*ēacrop*, →*eorþcrop*, →*heorotcrop*, →*stāncrop*

crop-lēac, lit. 'umbel-leek'
'Allium porrum L. / common leek / Porree'
DOE (s.v. *crop-lēac*): 26 occ.
LATIN EQUIV. in the manuscripts: *caepa*, *serpyllum*.
ETYMOLOGY: OE formation; MORPHOLOGY: N+N; endocentric, rectional. MOTIVATION: part of plant & botanical information. This kind of *lēac* 'leek', Allium porrum, has flashy, spherical umbels (Marzell, s.v. A*llium porrum* L.); cf. →*hēafdeht por(r)*. ASSOCIATIVE RELATIONS: conceptual contiguity (meronymy) & taxonomic sub-ordination.
▫ →*crop*, →*lēac*

cū, noun
'COW / Kuh'
DOE (s.v. *cū*): c. 80 occ.
▫ →*cūslyppe*

cudu →*cwidu*

culmille →*curmealle*

cunille, noun
(i) 'Thymus L. / thyme / Thymian'; (ii) Ruta L. / rue / Gartenraute'; (iii) 'Anthriscus cerefolium L. / chervil / Kerbel'
DOEPN (s.v. *cunelle*), *DOE* (s.v. *cunelle*): 6 occ.

LATIN EQUIV. in the manuscripts: *caerefolium*, *ruta*.
ETYMOLOGY: Later Latin loan (Serjeantson 1935: 284); < L *cunīla* (*DOE*, s.v. *cunelle*); ultimately from Greek *κονίλη*; cf. *Etymologiae* (17: 206, fn. 507; Feulner 2000: 180).
▫ →*wuducunille*

curmealle, *culmille*, noun
(i) 'Centaurea L. / knapweed / Flockenblume'; (ii) 'Blackstonia perfoliata (L.) Huds. / yellow wort / Durchwachsener Bitterling'; (iii) '? Centaurium erythraea Rafn. / common centaury / Echtes Tausendgüldenkraut'
DOE (s.v. *curmealle*): 29 occ.; as a simplex, 19 occ. (*DOEPN*, s.v. *curmealle*).
LATIN EQUIV. in the manuscripts: *centaurea*.
ETYMOLOGY: IE **geu-l-os* 'pot, ball' + **mel-* 'to crush, crash' (Bierbaumer 1, s.v. *curmealle*).
▫ →*curmealle sēo læsse*, →*curmealle sēo māre*
▫ →*lytle culmille*

curmealle sēo læsse, lit. 'lesser centaury'
'Centaurium umbellatum Gilibert / common centaury / Echtes Tausendgüldenkraut'
DOE (s.v. *curmealle*, b): 3 occ; the *DOE* subsumes *lȳtle culmille* to *curmealle sēo læsse*; however, the two plant names are structurally different: the first is a noun phrase postmodified by an adjective in the comparative, while the latter is a premodified noun (Adj+N) with the adjective in the positive; cf. →*lȳtle culmille*.
ILLUSTRATION in Oxford, Bodleian Library, Bodley 130, fol. 1v (Ohlgren 222(2)).
LATIN EQUIV. in the manuscripts: (*herba*) *centaurea minor*.
ETYMOLOGY: OE formation; possibly a loan translation of L *centauria minor* (Sauer 1992b: 394), visible in the structure of the Old English formation, where the noun is not premodified by the adjective, but postmodified, imitating, structurally, the Latin model. MORPHOLOGY: postmodified noun phrase: noun + demonstrative + adjective. MOTIVATION: botanical information & 'size'. *Curmealle sēo læsse* is a smaller species of *curmealle* and is distinct from →*curmealle sēo māre*. ASSOCIATIVE RELATIONS: taxonomic sub-ordination & conceptual contiguity.
▫ →*curmealle*, →*lȳtel*

curmealle sēo māre, lit. 'bigger centaury'
(i) 'Centaurea L. / knapweed / Flockenblume'; (ii) 'Blackstonia perfoliata (L.) Huds. / yellow centaury / Durchwachs-Bitterling'
DOE (s.v. *curmealle*, b): 2 occ. (*Herbarium*, no. 35; *Laud Herbal Glossary*, no. 335).
ILLUSTRATION in Oxford, Bodleian Library, Bodley 130, fol. 1r (Ohlgren 222(1)).
LATIN EQUIV. in the manuscripts: *centaurea maior*.
ETYMOLOGY: OE formation; loan translation of L *centauria maior* (Sauer 1992b: 394). MORPHOLOGY: post-modified noun phrase: noun + demonstrative + adjective. MOTIVATION: botanical information & size. *Curmealle sēo māre* is bigger than →*curmealle* and than →*curmealle sēo læsse*.

ASSOCIATIVE RELATIONS: taxonomic sub-ordination & conceptual contiguity.
▫ *→curmealle, →māre*

cū-slyppe (*-sloppe*), lit. 'cow-dung, cow-slime'
'Primula veris L. / COWSLIP, paigle / Frühlingsschlüsselblume'
DOE (s.v. *cū-slyppe*): 8 occ.
LATIN EQUIV. in the manuscripts: *britannica, hieribulbum.*
ETYMOLOGY: OE formation; MORPHOLOGY: N+N; exocentric, metaphorical compound. MOTIVATION: animal & shape. The plant's large leaves turn brown and slimy when they wither; then, they easily resemble a cow's pat. Not really convincing seems to me the suggestion to connect *slyppe* with *slip*, *slop* 'trousers, garment' on the basis of the inflated calyx of the cowslip reported in Rydén (1981: 308); according to him, 'cow-trousers' would be a humorous plant name, coined for children. ASSOCIATIVE RELATIONS: metaphorical similarity.
▫ *→cū, →slyppe*

-cwelcan (pl.), ?
unidentified
Not attested independently.
ETYMOLOGY: Origin unknown; perhaps cognate with G *Schwelke*, Dan *kvalkved*, etc. (Marzell, s.v. *Viburnum opulus* L.); perhaps a loan word from Old Norse (Hoops 1889: 76).
▫ *→hundescwelcan*

cwic, adjective
'living, alive / lebend, lebendig', with various related and figurative senses (cf. *DOE*, s.v. *cwic*)
DOE (s.v. *cwic*): c. 350 occ.
▫ *→cwicbēam, →cwictrēow*

cwic-bēam, lit. 'quick tree'
(i) 'Sorbus aucuparia L. / QUICK-BEAM, mountain-ash / Vogelbeerbaum, Eberesche'; (ii) 'Ulmus montana With. / wych elm, mountain elm / Berg-Ulme'; (iii) 'Juniperus communis L. / juniper / Wacholder'
DOE (s.v. *cwic-bēam*): 11 occ.[e]
LATIN EQUIV. in the manuscripts: *cariscus, juniperus, ? modica musca.*
ETYMOLOGY: OE formation; MORPHOLOGY: Adj+N; endocentric, copulative: attributive. MOTIVATION: effect & botanical information. The first element, *cwic* 'living, alive, lively, vivid, etc.' possibly refers to the revitalizing fruits (Bierbaumer 1, s.v. *cwicbēam*). ASSOCIATIVE RELATIONS: conceptual contiguity & taxonomic sub-ordination.
▫ *→bēam, →cwic*

cwic-trēow, lit. 'quick tree'
(i) 'Sorbus aucuparia L. / quick-beam, mountain-ash / Vogelbeerbaum, Eberesche'; (ii) 'Ulmus montana With. / wych elm, mountain elm / Berg-Ulme'; (iii) 'Juniperus communis L. / juniper / Wacholder'
DOE (s.v. *cwic-trēow*): 3 occ. (*Antwerp-London Glossaries*, 129/3; *Harley Glossary*, 113, no. 286; *Cleopatra Glossary* (Stryker), 255, no. 152).
LATIN EQUIV. in the manuscripts: ? *cresis*, ? *gresis*, possibly a corruption of *cariscus* (see *→cwicbēam*).
ETYMOLOGY: OE formation; MORPHOLOGY: Adj+N; endocentric, copulative: attributive. MOTIVATION: effect & botanical information. The first element,

cwic 'living, alive, lively, vivid, etc.' possibly refers to the revitalizing fruits of the tree (Bierbaumer 1, s.v. *cwic-bēam*). ASSOCIATIVE RELATIONS: conceptual contiguity & taxonomic sub-ordination.
▫ →*cwic*, →*trēow*

cwidu, noun
'mastic, resin / Mastixharz, Harz'
DOE (s.v. *cwudu, cudu*): c. 35 occ.
ETYMOLOGY: WGmc **kwedu-* 'Leim, Kitt' (*Kluge*, s.v. *Kitt*; *ODEE*, s.v. *cud*).
▫ →*hwītcwidu*

cymen, noun
(i) 'Cuminum cyminum L. / CUMIN / Kumin'; (ii) 'Cinnamomum zeylanicum Blume / ceylon cinnamon / Ceylon-Zimtbaum', perhaps by confusion of a Latin abbreviated form of *cuminum* (*DOE*, s.v. *cymen*).
35 occ. as 'cumin' (*DOEPN*, s.v. *cymen* (1)); 4 occ. as 'cinnamon' (*DOEPN*, s.v. *cymen* (2)).
ILLUSTRATION in manuscript London, British Library, Cotton Vitellius C. iii, fol. 65r (Ohlgren 168(90)).
LATIN EQUIV. in the manuscripts: *cuminum*, *cinnamomum* (perhaps by confusion with *cuminum*).
ETYMOLOGY: Early Latin loan (with *i*-mutation; Wollmann 1990: 165); < L *cumīnum* < Gk *κύμῖνον*; originally, a Semitic loan word (Feulner 2000: 181-3).
▫ →*sūþerne cymen*

cyning, noun
'KING, monarch / König, Monarch'
DOE (s.v. *cyning, cyng*): c. 8000 occ.
▫ →*cyningeswyrt*

cyninges-wyrt, lit. 'king's wort'
(i) '? Majorana hortensis Moench / marjoram / Majoran'; (ii) '? Sambucus nigra L. / elder / Schwarzer Holunder'
DOE (s.v. *cyninges-wyrt*): 2 occ. (*Brussels Glossary*, 301/17; *Durham Plant Name Glossary*, no. 316).
LATIN EQUIV. in the manuscripts: *sampsuchum* (? confused with *sambucus*).
ETYMOLOGY: OE formation; MORPHOLOGY: N_{gen}+N; endocentric, rectional; partial metaphor. MOTIVATION: ? & botanical information. According to Lipp (2002: 65) the elder is of major importance in various religions: people from the Baltico-Slavonic regions believe that the god of the earth lives in elder trees and, according to Northern mythology, it is the goddess Holde who lives in elder trees. In addition, people used to plant elder trees in cemeteries in Northern England and in Tyrol because elders were also associated with death. In view of these indices, the plant name *cyningeswyrt* seems to point towards an identification of the plant as Sambucus elder L. with *cyning* referring to Christ.

A different explanation of this plant name is based on the medical virtues and the usability of the elder: it is analgic, anti-bacterial, anti-inflammatory, anti-hypertensive and anti-convulsant, and its flowers and berries can be processed to beverages or to jam.[303] Hence, the motivation for the element *cyning* is that the plant is highly valued, i.e. it is the best of all plants, a plant worthy of kings. Cf. also *cynelic adl* lit. 'king's disease', prob-

303 As regards the medical values of elder as described in Hildegard von Bingen's *Physica* and Konrad von Megenburg's *Liber de natura rerum*, see Birkhan (2012: 127-8).

ably referring to jaundice, G *Gelbsucht*: in the Middle Ages, it was said to be curable only by a king or through the agency of a king, owing to the saintly character of the king: either by a king's personal touch or through the agency of water in which his hands had been washed. ASSOCIATIVE RELATIONS: metaphorical similarity & taxonomic sub-ordination.

▫ →*cyning*, →*wyrt*

cyrfet, noun
'Lagenaria Ser. / bottle gourd / Flaschenkürbis'
DOE (s.v. *cyrfæt*): 11 occ.
LATIN EQUIV. in the manuscripts: *bryonia*, *colocynthis*, *cucurbita* (1x with *agrestis*).
ETYMOLOGY: Early Latin loan (Wollmann 1990: 165); < L *cucurbita* (*DOE*, s.v. *cyrfæt*); see also →*-hwette*.

▫ →*wilde cyrfet*

D

dæg, noun
(i) 'DAY / Tag'; (ii) name of the rune for 'd'; (iii) an element in place names
DOE (s.v. *dæg*): c. 9100 occ.

▫ →*dæges ēage*

dæges ēage, lit. 'day's eye'
'Bellis perennis L. / DAISY / Gänseblümchen'
DOE (s.v. *dæges ēage*): 5 occ.
LATIN EQUIV. in the manuscripts: *anagallicum* (manuscript: *anagarica*), *consolida*, *consolida mediana*.
ETYMOLOGY: OE formation; according to Bierbaumer (3, s.v. *dæges ēage*) possibly a loan rendition of L *sōlis oculus* 'eye of the sun', or a loan translation of *oculus diēī* 'day's eye'; however, the plant name does not gloss *sōlis oculus* (nor *oculus diēī*) in the extant Old English texts and *sōlis oculus*, in fact, denotes either Matricaria camomilla L. or Chrysanthemum parthenium Berth., but not Bellis perennis L. MORPHOLOGY: N_{gen}+N; exocentric, metaphorical compound; MOTIVATION: time & shape. The fact that the plant opens its petals in the morning, i.e. at sunrise, and closes in the evening, i.e. at sunset, is verbalised metaphorically in the comparison of the plant's flower with the eye of a person opening in the mornings, when the person wakes up, and closing in the evenings, when the person goes to sleep. ASSOCIATIVE RELATION: metaphorical similarity.

▫ →*dæg*, →*ēage*

dēaw, noun
'DEW / Tau'
DOE (s.v. *dēaw*): c. 65 occ.

▫ →*meledēaw*, →*sundēaw*

-die, ?
unidentified
Not attested independently.

▫ →*gasdie*

docce, *docca*, noun
(i) 'Rumex obtusifolius L. / broad-leaved DOCK / Stumpfblättriger Ampfer'; (ii) 'Rumex crispus L. / curled DOCK / Krauser Ampfer'
DOE (s.v. *docce*, *docca*): c. 35 occ.

ILLUSTRATIONS in manuscripts London, British Library, Cotton Vitellius C. iii, fol. 26v (Ohlgren 168(14)), and Oxford, Bodleian Library, Ashmole 1431 (S.C. 7523), fol. 9r (Ohlgren 221(13)).

LATIN EQUIV. in the manuscripts: *acrocorium*, *dilla* (for *paradilla* or *paratella*), *lapathium*, *rodinaps* (= *rhododaphne*).

ETYMOLOGY: Gmc **dukkon* (*OED*, s.v. *dock*); OE *docce* (cf. also G *Docke*, Du *breede dokke*, *dokkewortel*, F *docque*) is a plant name based on a relation of contiguity: it refers to the children's habit of wrapping their 'doll' (OHG *tocka*, ON *dokka*, OS *dokka* 'doll' < Gmc **dukkōn*; *Kluge*, s.v. *Docke*) in broad leaves (of any kinds of plants with broad leaves, e.g. of Rumex L., of Malva silvestris L., etc.).[304] Thus, the name of the doll was transferred to plants having broad leaves (Marzell, s.v. *Rumex obtusifolius* L.).

▫ →*ēadocce*, →*fealwe docce*, →*fingerdocca*, →*rēad docce*, →*scearpe docce*, →*wudu-docce*

dolg, noun

'wound, sore / Wunde'

DOE (s.v. *dolg*): c. 85 occ.

▫ →*dolgrūne*

dolg-rūne, lit. 'wound-mystery, wound-secrecy'

'Parietaria officinalis L. / pellitory of the wall / Aufrechtes Glaskraut'

DOE (s.v. *dolg-rūne*): 19 occ.

ILLUSTRATION in manuscript London, British Library, Cotton Vitellius C. iii, fol. 44r (Ohlgren 168(49)).

LATIN EQUIV. in the manuscripts: *perdicalis*.

ETYMOLOGY: OE formation; MORPHOLOGY: N+N; exocentric, metaphorical compound. MOTIVATION: illness & healing virtue. According to Bierbaumer (1, s.v. *dolgrūne*), the plant *dolgrūne* is "die mit den Wunden Vertraute, Wundzauberin" ['the one which is familiar with wounds, the wound enchantress']. ASSOCIATIVE RELATION: metaphorical similarity.

▫ →*dolg*, →*rūne*

draca, noun

(i) 'dragon, snake, serpent, beast / Drache, Schlange, Bestie'; (ii) 'standard / Standarte'; (iii) place name element

DOE (s.v. *draca*): c. 225 occ.

ETYMOLOGY: Early Latin loan word, < L *draco*, *dracōnis* 'dragon, snake, etc.' (Wollmann 1990: 175); ModE *dragon* is not a direct descendant of OE *draca*, but was borrowed from OFr *dragon* (< accusative form, L *dracōnem*; *OED*, s.v. *dragon*).

▫ →*dracanblōd*

dracan-blōd, lit. 'dragon's blood'

'resin of Dracaena L. / dragon's blood tree / Drachenblutbaum'

Not an entry in the *DOE*; hapax legomenon (Meritt 1961: 210), this is not confirmed, however, by Voss (1996). *Dracanblōd* is attested in a copy of a manuscript (London, British Library, Cotton Otho E.i; cf. Voss 1996), made by Junius right before the fire in Ashburnham house destroyed large parts of the original manuscript in

304 The *OED* (s.v. *dock*, n²) notes: "Identical with mod.Icel. *dockr* short stumpy tail (Haldorssen). Ulterior etymology obscure. Cf. Fris *dok* bundle, bunch, ball (of twine, straw, etc.), LG. *dokke* bundle (of straw, thread), skein of yarn, mod.G. *docke* bundle, skein, plug, peg."

1731. Ker (*Catalogue*, 239) notes that "[a] comparison with the still legible fragments shows that this copy, which was made in old age, is incomplete and untrustworthy." Bierbaumer (3, s.v. *dracanblōd*) probably therefore lists this compound, though only in brackets.
ETYMOLOGY: WGmc formation; OHG *drachenbluot*, OLG *drachenbluot* (Carr 1939: 313); hybrid formation (OE *draca* 'dragon, devil' < L *draco, dracōnis*; →*draca*). MORPHOLOGY: N_{gen}+N; exocentric, metaphorical compound. MOTIVATION: animal & part of plant; OE *dracanblōd* has cognates in other Germanic languages, as for instance OHG *drachenbluot* from the 11th or 12th century (Carr 1939: 313). The use of *blōd* is based on the phenomenon that when the plant is cut, the resin runs off the plant as if it were blood. As to the first element, several motivations overlap. First of all, very probably, the element *dracan-*, an early loan from Latin (L *draco, dracōnis* < Gk *δράκων*, see Feulner 2000: 201-2) is used by analogy to the first part of the Old English loan *dracontse, dracentse, draconze* (Dracunculus vulgaris Schott., 'adderwort', G *Schlangenwurz*) from Latin *dracontea*.[305] The use of *draca* is further supported by the idea that the dragon's blood tree grows out of dragon's blood; the *Herbarium* notes with regard to the plant *dracentse*: "Ðeos wyrt þe man *dracontea* & oðrum naman dracentse nemneð ys sæd þæt heo of of dracan blode acenned beon sceolde" ['it is said that it grows out of dragon's blood'] (*Herbarium*, no. 15). It is noteworthy that OE *draca* (as well as its Latin precursor) was polysemous and referred not only to the mythical creature but also to snakes in general. This fits the Anglo-Saxon conception of snakes as one possible cause for ailments. According to Marzell (s.v. *Dracaena* L.), the resin was already known as 'dragon's blood' in antiquity and mentioned as *αἷμα δρακόντιον* in Dioscorides' *De materia medica*. Hence, OE *dracanblōd* may be a loan translation of the Greek term via an unattested **dracōnis sanguis*. This composite is exceptional in my list because it denotes the resin of Dracaena L. and not the tree itself. In contrast, OHG and OLG *drachenbluot* refer to the tree (Carr 1939: 313); ME *anglice dragans-blode* can refer both to the resin and to the tree (see Hunt 1989: 98, *demechia*). ASSOCIATIVE RELATION: metaphorical similarity.
▫ →*blōd*, →*draca*

dweorg(e), noun
(i) 'DWARF / Zwerg'; (ii) 'fever / Fieber'
DOE (s.v. *dweorg*): 15 occ.
▫ →*dweorged(w)os(t)(l)e*

dweorge-d(w)os(t)(l)e, lit. 'dwarf's *dost*'
'Mentha pulegium L. / pennyroyal / Poleiminze'
DOE (s.v. *dweorge-dwostle*): 35 occ.[e]
ILLUSTRATION in manuscript London, British Library, Cotton Vitellius C. iii, fol. 48r (Ohlgren 168(56)).
LATIN EQUIV. and OE synonyms in the manuscripts: *puleium*; *brōþorwyrt*, *hylwyrt*.

305 An illustration of *dracontse* can be found in manuscript London, British Library, Cotton Vitellius C. iii, fol. 26v (Ohlgren 168(14)).

ETYMOLOGY: OE formation; MORPHOLOGY: *-dwos(t)(l)e* is not attested independently in Old English, but in other WGmc languages, e.g. as OHG *dost(o)*, G *Dost*; N+N; endocentric, rectional; partial metaphor; Waxenberger (1996: 279-80) classifies it erroneously as a N_{gen}+N compound, arguing that there is but one attestation without linking, resp. genitive *-s*; the reverse, however, is the case (*DOE*, s.v. *dweorge-dwostle*): the only attestation with intermediate *-s-* is *Brussels Glossary* (300/24-5). MOTIVATION: size & botanical information. The motivation for this plant name is rather unclear and several suggestions have been put forward. OE *dweorg* can mean both 'dwarf' and 'cramp, fever'; the second element *-d(w)os(t)(l)e*, etc. is probably a plant name (cf. G *Dost* 'Origanum vulgare L., wild marjoram' or 'bunch', OHG *dost(o)*, MHG *doste*, *toste*; cf. *Kluge*, s.v. *Dost*) which is, however, not attested independently in Old English and totally extinct in Modern English. The plant name probably refers to the size of the plant, i.e. lit. 'dwarf *dost*' = 'small *dost*', but a connection to Germanic mythology is not excluded, either (Bierbaumer 1, s.v. *dweorgedwostle*). ASSOCIATIVE RELATIONS: metaphorical similarity & taxonomic sub-ordination.

▫ *→dweorg(e)*, *→d(w)os(t)(l)e*

-d(w)os(t)(l)e, noun
'Origanum majorana L. / marjoram / Majoran'
Not attested independently.
ETYMOLOGY: Derivative (*dost* + *-el*) of unclear origin (*Kluge,* s.v. *Dost*); cf. also Sauer (2012: 218, 221) not attested independently in Old English but as the plant name exists in OHG as *dost(o),* in MHG as *doste* (*Kluge*, s.v. *Dost*) the plant can be identified as marjoram.

▫ *→dweorged(w)os(t)(l)e*

dȳþ, noun
'tinder, dry twigs used for kindling / Zunder'
DOE (s.v. *dȳþ*): 2 occ.

▫ *→dȳþhamor*

dȳþ-hamor *(dūþ-, -homar)*, lit. 'tinder-hammer'
(i) 'Typha L. / reedmace, locks / Rohrkolben'; (ii) 'glossing papyrus'; (iii) 'dry twigs used for kindling / trockene Zweige, Span'
DOE (s.v. *dȳþ-hamor*): 6 occ.
LATIN EQUIV. in the manuscripts: *malleolus*, *papyrus*.
ETYMOLOGY: WGmc compound; LG *duthoamer* (Carr 1939: 109); MORPHOLOGY: N+N; exocentric, metaphorical compound. MOTIVATION: tool & shape. The stems of the plants were used as tinder, hence 'tool', while the spadices of reedmace are compared to hammers. ASSOCIATIVE RELATIONS: conceptual contiguity & metaphorical similarity.

▫ *→dȳþ*, *→hamor*

E

ēa, noun
'river, large body of running water / Fluss, Wasser'
DOE (s.v. *ēa*): c. 500 occ.

▫ *→ēacærse*, *→ēacrop*, *→ēadocce*, *→ēalifer*, *→ēarisc*, *→ēawyrt*

ēa-cœrse, lit. 'water-cress'
'Nasturtium officinale R.Br. / water-cress / Brunnenkresse'
DOE (s.v. *ēa-cœrse*): 1 occ. (*Lǣcebōc*, 29/21).
ETYMOLOGY: OE formation; MORPHOLOGY: N+N; endocentric, rectional. MOTIVATION: habitat & botanical information. This kind of *cœrse* 'cress' grows preferably in moist environments near brooks and in fens, hence OE *ēa* 'water' (Marzell, s.v. *Nasturtium officinale* R. Br.). ASSOCIATIVE RELATIONS: conceptual contiguity & taxonomic sub-ordination.
▫ *→cœrse, →ēa*

ēa-crop, lit. 'water-umbel'
unidentified plant
DOE (s.v. *ēa-crop*): 1 occ. (*Durham Plant Name Glossary*, no. 85).
The *DOE* (s.v. *ēa-crop*) notes that *ēa-crop* is a "plant of uncertain identification, perhaps (water) *germander*, *ground pine* or *hemp* (cf. *Durham Plant Name Glossary*, 88, and *Durham Plant Name Glossary*, 140, where *henep* 'hemp' glosses chamaepitys)" while Bierbaumer (3, s.v. *ēacrop*) believes it to be a miswriting of *eorþcrop* lit. 'earth-umbel'.
LATIN EQUIV. in the manuscripts: *chamaepitys*.
ETYMOLOGY: OE formation; MORPHOLOGY: N+N; exocentric, metonymic compound. MOTIVATION: habitat & part of plant. Since the identification of the plant is not clear, the analysis of the plant name is also more or less speculative. It seems, though, as if the first element could refer to the preferred habitat of the plant, while the second element refers to its umbels. ASSOCIATIVE RELATIONS: conceptual contiguity & conceptual contiguity (meronymy).
▫ *→crop, →ēa*

ēa-docce, lit. 'water-dock'
'Nuphar Smith / water-lily / Teichrose'
DOE (s.v. *ēa-docce*): 2 occ. (*Lācnunga*, 50; *Antwerp-London Glossaries*, 118/8).
LATIN EQUIV. in the manuscripts: *nymphaea*.
ETYMOLOGY: OE formation; MORPHOLOGY: N+N; endocentric, rectional. MOTIVATION: habitat & botanical information. The *ēadocce* grows in ponds. In Anglo-Saxon folk taxonomy *ēadocce* is a kind of *docce*, while in modern, scientific taxonomy the water-lily and the dock belong to two different species. ASSOCIATIVE RELATIONS: conceptual contiguity & taxonomic sub-ordination.
▫ *→docce, →ēa*

ēage, noun
(i) 'EYE / Auge'; (ii) 'eye-shaped hole'
DOE (s.v. *ēage*): c. 2100 occ.
▫ *→ēagwyrt*
▫ *→dæges ēage*

ēag-wyrt (*ǣg-wyrt*), lit. 'eye-wort' or 'egg-wort'
'? Euphrasia rostkoviana Hayne / eyebright / Gemeiner Augentrost' or '? Taraxacum officinale Wiggers / dandelion / Löwenzahn'
DOE (s.v. *ēag-wyrt*): 2 occ. (*Lācnunga*, 516; *Lǣcebōc*, 99/20).
Ǣgwyrt lit. 'egg-wort' in *Lācnunga* (516) appears as *ēagwyrt* lit. 'eye-wort' in the almost identical recipe in *Lǣcebōc* (99/20). As both lexemes are hapax legomena, they may denote two

different plants, but they may also be variants of one and the same plant name, namely, when OE *ǣgwyrt* is seen as a smoothed (Anglian) variant or a late West-Saxon form of OE *ēagwyrt*.[306]
ETYMOLOGY: OE formation; MORPHOLOGY: N+N; endocentric, rectional. MOTIVATION: illness & botanical information. The plant name lit. 'eye-wort' points to the plant's potential as a cure fore sore eyes; the plant name lit. 'egg-wort' would point to "the round form of the pappus" of the dandelion (Cockayne, *Leechdoms*, 2: 368). ASSOCIATIVE RELATIONS: conceptual contiguity & taxonomic sub-ordination.
▫ →*ēage*, →*wyrt*

ēa-lifer, lit. 'water-liver'
(i) 'Eupatorium cannabium L. / water-hemp, thoroughwort / Gemeiner Wasserdost'; (ii) 'Ranunculus aquatilis L. / water liverwort / Gemeiner Wasser-Hahnenfuß'
DOE (s.v. *ēa-lifer*): 3 occ. (*Lǣcebōc*, (20/35, 65/21; *Durham Plant Name Glossary*, no. 78).
LATIN EQUIV. and OE synonyms in the manuscripts: *callitrichum*; *wæterwyrt*.
ETYMOLOGY: OE formation; MORPHOLOGY: N+N; exocentric, metaphorical compound. MOTIVATION: habitat & shape. Both Eupatorium cannabium and Ranunculus aquatilis live in moist environments, hence the use of the first element *ēa* 'water, river'. Eupatorium helps with diseases of the liver (cf. Bierbaumer 1 and 3, s.v. *ēalifer*) and old names for Ranunculus are *hepatica aquatica* and *hepatica palustris*; cf. Marzell (s.v. *Ranunculus aquatilis* L.). But it is rather unusual for a sick organ to be used as a second element in a compound denoting a plant (compare, e.g. →*bānwyrt*, where the sick organ, i.e. the bone, is the modifying element of the compound). Therefore, I would rather suggest that the shape of the leaves of this plant motivated the plant name. The plants Agrimonia eupatoria, Eupatorium cannabium and Asperula odorata all share the old Latin names *trifolium hepaticum* and *hepatica terrestris* with the plant Anemone hepatica L. (Marzell, s.v. *Anemone hepatica* L.), and all refer to the leaves' shapes. This motivation may of course have supported the use of the plant for diseases of the liver, which corresponds to the practice of healing after the principle *similia similibus*. ASSOCIATIVE RELATIONS: conceptual contiguity & metaphorical similarity.
▫ →*ēa*, →*lifer*

eallhwīt, adjective
'ALL-WHITE, pure white / ganz weiß'
DOE (s.v. *eall-hwīt*): 3 occ.
▫ →*eallhwīte pise*

eallhwīte pise, lit. 'entirely white pea'
'? Lathyrus L. / vetchling / Platterbse'
Not an entry on its own in the *DOE*, but subsumed to the adjective *eall-hwīt*; hapax legomenon (*Cleopatra Glossary* (Stryker), 225, no. 153).
LATIN EQUIV. in the manuscripts: *gesce* (? for *gessia*).
ETYMOLOGY: OE formation; hybrid formation (OE *pise* < L *pīsum*; see →*pise*; Wollmann 1990: 154). MORPHOLOGY: Adj+N; endocentric, copulative: attributive. MOTIVATION: colour &

306 Anglian smoothing from *ēa* > *ǣ* took place in the 7th century, and *ǣ* was raised to *ē* in the 8th century (cf. Lehnert 1990, § 35).

botanical information. The fruits of Lathyrus resemble peas, lentils, chickpeas and sweet peas (Marzell, s.v. *Lathyrus* L.). The subspecies Lathyrus sativus L. has white flowers and is the only species of Lathyrus whose fruits can be eaten by humans and cattle; it has been known and grown in Europe since antiquity, eaten, in the main, by the poor (*Genaust*, s.v. *láthyris*; Marzell, s.v. *Lathyrus sativus* L.). ASSOCIATIVE RELATIONS: conceptual contiguity & taxonomic sub-ordination.

▫ *→eallhwīt, →pise*

ēa-risc, lit. 'water-rush'
'Juncus L. / rush / Binse'
DOE (s.v. *ēa-risc*, *ēa-risce*): 13 occ.[e]
LATIN EQUIV. in the manuscripts: *bremium*, *papyrus*.
ETYMOLOGY: OE formation; MORPHOLOGY: N+N; endocentric, rectional. MOTIVATION: habitat & botanical information. Rushes (OE *risc*) grow in moist environments, hence, the first element OE *ēa* 'water, river' (Marzell, s.v. *Juncus* L.). ASSOCIATIVE RELATIONS: conceptual contiguity & taxonomic sub-ordination.

▫ *→ēa, →risc*

ēastern, adjective
'EASTERN / östlich'
DOE (s.v. *ēasterne*): 14 occ.

▫ *→ēasterne hnutbēam*

ēasterne hnutbēam, lit. 'eastern / oriental nut-tree'
'Prunus amygdalus Batsch 1801 / almond / Mandelbaum'
DOE (s.v. *ēasterne*): not a lemma on its own in the *DOE*, but subsumed to the adjective *ēasterne* (sense 3.a.); hapax legomenon (*Durham Plant Name Glossary*, no. 26).
LATIN EQUIV. in the manuscripts: *amygdalus*.
ETYMOLOGY: OE formation; MORPHOLOGY: Adj+N; endocentric, copulative: attributive. MOTIVATION: origin & botanical information. The plant name stresses the difference between the nut-tree native to northern Europe, Juglans regia L., and the almond-tree, which is of oriental origin and does not produce almonds in the colder regions of middle and Northern Europe, although it can be grown there (Marzell, s.v. *Prunus amygdalus* Batsch). ASSOCIATIVE RELATIONS: conceptual contiguity & taxonomic sub-ordination.

▫ *→ēastern, →hnutbēam*

ēa-wyrt, lit. 'water-wort'
'Arctium lappa L. / great burr / Große Klette'
DOE (s.v. *ēa-wyrt*): 3 occ.
ETYMOLOGY: OE formation; MORPHOLOGY: N+N; endocentric, rectional. MOTIVATION: habitat & 'botanical information'. This kind of plant lives in moist environments. ASSOCIATIVE RELATIONS: conceptual contiguity & taxonomic sub-ordination.

▫ *→ēa, →wyrt*

ēfen, *ǣfen*, noun
'evening, EVEN / Abend'

▫ *→ēfenlāste*

ēfen-lāste, lit. 'evening-lasting (plant)'
'Mercurialis perennis L. / mercury / Wald-Bingelkraut, Ausdauerndes Bingelkraut'

DOE (s.v. *efenlāste*): 8 occ.; according to the *DOE* (s.v. *efenlāste*) and the *OED* (s.v. *evenlesten*), the plant is of uncertain identification. Bierbaumer's (1, s.v. *efelāste*) identification of the plant is probably based on the Middle English occurrence of the plant name where it glosses L *mercurialis*, the Mercurialis perennis L., mercury (*MED*, s.v. *evenlesten*).
ETYMOLOGY: OE formation; MORPHOLOGY: synthetic, $N+N_{deverbal(-e)}$; endocentric, rectional. MOTIVATION: ? time & action. In contrast to Mercurialis annua, Mercurialis perennis is a perennial, i.e. a 'lasting', plant (Marzell, s.v. *Mercurialis perennis* L.), hence the second element of the compound, *-lāste* < *lǣstan* '(to) last, follow, pursue, endure'. The first element, OE *ēfen, ǣfen* 'evening', probably points to a time specification (*DOE*, s.v. *ǣfen*), but it may also be the adverb 'evenly, equally, alike' (*DOE*, s.v. *efen*). ASSOCIATIVE RELATIONS: ? conceptual contiguity & taxonomic sub-ordination.
▫ →*ēfen*, →*lǣstan*

ehheolobe →*hēahheolobe*

ele, noun
'oil / Öl'
DOE (s.v. *ele*): c. 500 occ.
LATIN EQUIV. in the manuscripts: *oleum*.
ETYMOLOGY: Early Latin loan (Wollmann 1990: 170); < L *olea* 'olive, olive-tree' < Gk **ἐλαίβα* (*Genaust*, s.v. *olea*; Feulner 2000: 206-7). ModE *olive* goes back to Agn *olive* (*OED*, s.v. *olive*). Latin *olea* and *oliva* are polysemous, referring to 'olive' and to the 'olive-tree', *oleum* denotes 'oil, olive-oil' (*Lewis and Short*, s.vv. *olea*, *oleum*, *oliva*). The loan *ele* glosses L *oleum*, and never *olea* or *oliva*. The Old English compounds *elebearu* 'olive-grove' and *elelēaf* 'leaf of the olive-tree', however, suggest that *ele-* could also allude to the 'olive' and to the 'olive-tree'.
▫ →*elebēam*, →*eletrēow*

ele-bēam, lit. 'oil-/olive-tree'
'Olea europaea L. / olive-tree / Olivenbaum'
DOE (s.v. *ele-bēam*): 40 occ.
LATIN EQUIV. in the manuscripts: *olea*, *oliva*.
ETYMOLOGY: OE formation; hybrid compound (OE *ele* < L *olea*; see →*ele*). MORPHOLOGY: N+N; endocentric, rectional. MOTIVATION: product & botanical information. In regards to the polysemy of OE *ele*, see →*ele*. The interpretation here is that *elebēam* is a *bēam* 'tree' that yields olives and oil; hence, the first element refers to the product of the plant; all three denotata, 'oil', 'olive' and 'olive-tree' for *ele*, make sense, however, in this compound; cf. MHG *oelpaum*, G *Ölbaum* lit. 'oil-tree' (Birkhan 2012: 83-4). ASSOCIATIVE RELATIONS: conceptual contiguity & taxonomic sub-ordination.
▫ →*bēam*, →*ele*
▫ →*wilde elebēam*

elene, *eolone*, noun
'Inula helenium L. / elecampane / Alant'
DOE (s.v. *eolone, elene*): c. 80 occ.

LATIN EQUIV. in the manuscripts: *ibis*, *inula*, ? *orestion*, *symphoniaca*, *inula campana*.
ETYMOLOGY: Early Latin loan (Wollmann 1990: 154, 158); VL **iluna* < L *inula (helenium)* < Gk *ἑλένιον* (*DOE*, s.v. *eolone*; *OED*, s.v. *elecampane*; Gk *ἑλένιον*; cf. Feulner 2000: 217).
▫ →*horselene*

ele-trēow, lit. 'oil-/olive-tree'
'Olea europaea L. / olive-tree / Olivenbaum'
DOE (s.v. *ele-trēow*): 13 occ.
LATIN EQUIV. in the manuscripts: *oliva*.
ETYMOLOGY: OE formation; hybrid compound (OE *ele* < L *olea*; cf. →*ele*). MORPHOLOGY: N+N; endocentric, rectional. MOTIVATION: product & botanical information. For details see →*elebēam*. ASSOCIATIVE RELATIONS: conceptual contiguity & taxonomic sub-ordination.
▫ →*ele*, →*trēow*
▫ →*wilde eletrēow*

ellen 1, adjective
'of elderwood, elder- / Holunder-'
DOE (s.v. *ellen*, adj.): 5 occ., "mainly in bounds".
▫ →*ellenbēam*

ellen 2, noun
'Sambucus nigra L. / elder-tree / Holunder'
DOE (s.v. *ellen* noun²): c. 40 occ.
ILLUSTRATION in manuscript London, British Library, Cotton Vitellius C. iii, fol. 63v (Ohlgren 168(87)).
LATIN EQUIV. in the manuscripts: *acte*, *lignum*, *sambucus*, *sampsuchum* (by confusion with *sambucus*).
▫ →*ellentrēow*, →*ellenwyrt*

ellen-bēam, lit. 'elder-tree'
'Sambucus nigra L. / elder / Schwarzer Holunder'
Not an entry on its own in the *DOE*, but as an attestation for the adjective *ellen* 'of elder-wood, elder-' (s.v. *ellen*, adj.); 1 occ.; not in Bierbaumer (1, 2, 3); hapax legomenon (*Charters* (Birch), 2: 581, no. 817).
ETYMOLOGY: OE formation. MORPHOLOGY: Adj+N; endocentric, copulative: attributive; in contrast to →*ellentrēow* 'elder-tree', which is a N+N formation, *ellenbēam* is analysed as an Adj+N formation based on the form: the element *ellen* is inflected ("to þan ellenebem" ['to the elder tree']; *Charters* (Birch), 2: 581, no. 817); MOTIVATION: texture & botanical information. In contrast to Sambucus ebulus (→*ellenwyrt*), which is small, Sambucus nigra, the *ellenbēam*, can grow high as a tree. ASSOCIATIVE RELATIONS: conceptual contiguity & taxonomic sub-ordination.
▫ →*bēam*, →*ellen* 1

ellen-trēow, lit. 'elder-tree'
'Sambucus nigra L. / elder / Schwarzer Holunder'
DOE (s.v. *ellen-trēow*): 1 occ. (*Codex* (Kemble), 3: 397, no. 115); not in Bierbaumer (1, 2, 3).
ETYMOLOGY: OE formation. MORPHOLOGY: N+N; endocentric, copulative: subsumptive (pleonastic). MOTIVATION: plant & botanical information. In contrast to Sambucus ebulus (→*ellenwyrt*), which is small, Sambucus nigra, the *ellentrēow*, can grow high as a tree: This kind of tree is an elder. ASSOCIATIVE RELATIONS: conceptual identity & taxonomic sub-ordination.
▫ →*ellen* 2, →*trēow*

ellen-wyrt, lit. 'elder-wort'
'Sambucus ebulus L. / dwarf-elder / Zwerg-Holunder'
DOE (s.v. *ellen-wyrt*): 12 occ.
LATIN EQUIV. in the manuscripts: *chamaesyce* (for *chamaeacte*), *ebulum*, *ebulus*, *sampsuchum* (by confusion with *sambucus*).
ETYMOLOGY: OE formation; MORPHOLOGY: N+N; endocentric, copulative: subsumptive (pleonastic). MOTIVATION: plant & botanical information. This kind of *ellen* is a plant. In order to distinguish between the dwarf-elder (a smaller plant) and the elder (a bush, high like a tree), the elements *wyrt*, respectively *bēam* and *trēow* (see →*ellenbēam*, →*ellentrēow* above) were used. ASSOCIATIVE RELATIONS: conceptual identity & taxonomic sub-ordination.
▫ →*ellen* 2, →*wyrt*

englisc, adjective
'ENGLISH / englisch'
DOE (s.v. *englisc*): c. 550 occ.
▫ →*englisc cost*, →*englisc more*, →*englisc nǣp*

englisc cost, lit. 'English cost(mary)'
'Tanacetum vulgare L. / tansy / Rainfarn'
Not a lemma on its own in the *DOE*; it appears once, as an attestation for meaning A.1.d. "in names of various plants" of the adjective *englisc* (*DOE*, s.v. *englisc*); attested in *Lācnunga* (242).
ETYMOLOGY: OE formation; hybrid formation (OE *cost* < L *costus*, *costum*; →*cost*). MORPHOLOGY: Adj+N; endocentric, copulative: attributive. MOTIVATION: origin & botanical information. The *englisc cost* can be grown in England – in contrast to the *cost* (Tanacetum balsamita L.) native to the Caucasus and Southern Europe (Marzell, s.v. *Tanacetum balsamita* L.). ASSOCIATIVE RELATIONS: conceptual contiguity & taxonomic sub-ordination.
▫ →*cost*, →*englisc*

englisc more, lit. 'English carrot'
'Daucus carota L. / wild carrot / Gemeine Mohrrübe, Gelbe Rübe'
Not an entry in the *DOE*; attestation can be found s.v. *englisc*, A.1.d; 2 occ. (*Lǣcebōc*, 12/38, 96/12).
ETYMOLOGY: OE formation; MORPHOLOGY: Adj+N; endocentric, copulative: attributive. MOTIVATION: origin & botanical information. Daucus carota L., the carrot was native to England, while Pastinacia sativa L., the parsnip, a Mediterranean plant, was introduced later. The two could be easily confused since the carrot was not dark orange as it is today, but rather whitish (Banham 2003: 25). The shape and size of both plants are so similar that obviously there was a need to stress that by *more* the native – the *englisc more* – was meant; in contrast to the *wealhmore* and the *wīelisc moru*, which both denoted the parsnip. ASSOCIATIVE RELATIONS: conceptual contiguity & taxonomic sub-ordination.
▫ →*englisc*, →*more*

englisc nǣp, lit. 'English rape'
'? Brassica rapa L. var. silvestris / turnip / Rübenkohl'
Not an entry in the *DOE*; attestations can be found s.v. *englisc*, A.1.d; 2 occ. (*DOEPN*, s.v. *nǣp, englisc*); attested in

Cleopatra Glossary ((Stryker), 319, no. 36) and *Lācnunga* (117).
LATIN EQUIV. in the manuscripts: *napus*.
ETYMOLOGY: OE formation; hybrid formation (OE *nǣp* < L *nāpus*; →*nǣp*). MORPHOLOGY: Adj+N; endocentric, copulative: attributive. MOTIVATION: origin & botanical information. The *nǣp* 'rape' is a plant native to England, hence the use of the element *englisc* 'English'; there is, however, no obvious reason for the use of the element *englisc* since the rape was commonly grown in England and no *wīelisc nǣp suggests that it was important to distinguish the *englisc* species from the foreign one (compare, for instance, the pair →*englisc more* and →*wīelisc moru*). ASSOCIATIVE RELATIONS: conceptual contiguity & taxonomic subordination.
▫ →*englisc*, →*nǣp*

enne- →*ynne-*

eofor, noun
'wild boar / Eber'
DOE (s.v. *eofor*): c. 35 occ.
▫ →*eoforfearn*, →*eoforþrote*

eofor-fearn, lit. 'boar-fern'
'Polypodium vulgare L. / polypody / Engelsüß, Tüpfelfarn'
DOE (s.v. *eofor-fearn*): c. 35 occ.
ILLUSTRATION in manuscript London, British Library, Cotton Vitellius C. iii, fol. 44v (Ohlgren 168(50)).
LATIN EQUIV. in the manuscripts: *filicina*, *filicina arboratica*, *filix*, *filix minuta*, *filix arboratica felgera*, *polypodium*, *radiolus*.
ETYMOLOGY: OE formation; MORPHOLOGY: N+N; endocentric, rectional. MOTIVATION: animal & botanical information. The *eoforfearn* is a kind of *fearn* 'farn', the rhizomes of which are a preferred fodder for (wild) boars, OE *eofor* 'boar' (Bierbaumer 1, s.v. *eofor-fearn*). ASSOCIATIVE RELATIONS: conceptual contiguity & taxonomic subordination.
▫ →*eofor*, →*fearn*

eofor-þrote, lit. 'boar-throat'
'? Carlina acaulis L. / carline thistle, stemless carline / Stengellose Eberwurz, Silberdistel'
DOE (s.v. *eofor-þrote*): c. 40 occ.[e]
LATIN EQUIV. in the manuscripts: *acanthus*, *colicus*, ? *foragium*.
ETYMOLOGY: OE formation; MORPHOLOGY: N+N; exocentric, metaphorical compound. MOTIVATION: animal & shape. According to some of Marzell's lay informants (s.v. *Carlina acaulis*), the carline thistle was used as a remedy for various kinds of diseases of pigs, especially against the pigs's cough. Other informants stated, however, that the carline thistle is rather poisonous for pigs, and it seems, in any case, strange that the name of the sick organ is the head of the compound and not the modifying element (see also *ēalifer*). The Old English plant name may as well be a metaphorical formation: Carlina acaulis is called stemless carline thistle in Modern English because it has 'no throat' – similar to boars, where the throat is very short, thick and hairy (Marzell, s.v. *Carlina acaulis* L.). ASSOCIATIVE RELATIONS: conceptual contiguity & metaphorical similarity.
▫ →*eofor*, →*þrote*

eolh, eolha, noun
'elk / Elch'
The element *eolh* and *ilug-* in *eolhsecg* and *ilugsecg* are probably variants of *eolete* 'water' (cf. *DOE*, s.v. *? eolh-secg*; Pheifer 1974: 111, no. 781).
DOE (s.v. *eolh, eolha*): 9 occ.
▫ *→eolhsecg*

? *eolh-secg* (? *ilug-*), lit. ? 'water-sedge'
(i) '? Carex L. / sedge / Riedgras, Schilf'; (ii) ? 'Eryngium maritimum L. / sea-holly / Strand-Mannstreu'
DOE (s.v. ? *eolh*-secg): 6 occ.[e], "mainly in glossaries".
LATIN EQUIV. in the manuscripts: ? *papyrus* (manuscripts: *papillum, papillus, papiluus*).
ETYMOLOGY: OE formation; *eolh* and *ilug-* are probably variants of *eolete* 'water', but *eolh* may also be a variant of *helix, helice* 'winding plant, kind of willow' (*DOE*, s.v. ? *eolh-secg*; Pheifer 1974: 111, no. 781). MORPHOLOGY: N+N; endocentric, rectional. MOTIVATION: habitat & botanical information. The plant is a kind of *secg* 'sedge'. When the first element is connected to *eolete* 'water', it refers to the preferred habitat of the plant (cf. Bierbaumer 3, s.v. *eolhsecg*; Sauer 2000: 4, 7). ASSOCIATIVE RELATIONS: conceptual contiguity & taxonomic sub-ordination.
▫ *→eolh, →secg*

eormelēaf →geormanlēaf

eorþ-æppel, lit. 'earth-fruit' or 'earth-apple'
(i) 'Mandragora officinarium L. / mandrake / Alraune'; (ii) 'Cyclamen europaeum L. / sowbread / Alpenveilchen'; (iii) 'Cucumis sativus L. / cucumber / Salatgurke'
DOE (s.v. *eorþ-æppel*): 4 occ., "in multiple MSS"; the identification as 'cucumber' follows the *DOE*, where it is based on the equation of L *cucumis* with *eorþæppel* in the Old English *Heptateuch* (11, 5).
LATIN EQUIV. in the manuscripts: *cyclamen, mandragoras, terrae malum.*
ETYMOLOGY: WGmc formation (Carr 1939: 98); loan translation of L *mālum terrae* (in Meritt 1945: 62, no. 73, Dc 186/1; Bierbaumer 3, s.v. *eorþæppel*). MORPHOLOGY: N+N; exocentric, metonymic compound. MOTIVATION: habitat & part of plant. The motives for calling the different plants – the cucumber, the mandrake and the sowbread – *eorþæppel* lit. 'earth-fruit', or 'earth-apple' – are different. In all instances, the polysemy of OE *æppel* as 'apple' on the one hand, and, generically, as 'fruit', is not easy to solve as both denotata can be applied.
(i) The name describing mandrake as 'earth-fruit' is a loan translation of L *mālum terrae* lit. 'earth-fruit' and refers to the eye-catching, fleshy root of the mandrake. (ii) The edible, bulbous roots of Cyclamen europaeum resemble apples (Marzell, s.v. *Mandragora officinarium* L.), although they are smaller and of a dark brown colour; cf. also the German name of the plant, G *Erdapfel* lit. 'earth-apple'; cf. also *→eorþhnutu.* (iii) The cucumber is a kind of fruit that does not grow on trees, but right above the earth (OE *eorþe* 'earth'). In all three cases, the root, the edible part of the plant, stands

for the whole plant.[307] ASSOCIATIVE RELATIONS: conceptual contiguity & conceptual contiguity (meronymy).
▫ *→æppel, →eorþe*

eorþ-ber(i)(g)e, lit. 'earth-berry'
'fruit of Fragaria vesca L. / strawberry / Wald-Erdbeere'
DOE (s.v. *eorþ-berie, eorþ-berige*): 1 occ. (*Harley Glossary*, 195, no. 656).
LATIN EQUIV. and OE synonyms in the manuscripts: *fraga*; *merscmealwe*, *ofet* 'fruit, vegetable', *strēawber(i)(g)e*.
ETYMOLOGY: Gmc compound; OHG *erdberi*, OLG *erðberi*, MDu *erdbere*, Fris *ierdbei*, Dan *jordbær*, Sw *jordbär*, Icel *jarðarber* (Carr 1939: 47); MORPHOLOGY: N+N; endocentric, rectional. MOTIVATION: habitat & botanical information. This kind of *ber(i)(g)e* 'berry' grows right above the earth (OE *eorþ*); compare also G *Erdbeere* lit. 'earth-berry'. ASSOCIATIVE RELATIONS: conceptual contiguity & taxonomic subordination.
▫ *→beri(g)(e), →eorþe*

eorþ-crop, lit. 'earth-umbel'
unidentified plant
DOE (s.v. *eorþ-crop*): 1 occ. (*Durham Plant Name Glossary*, no. 83).
LATIN EQUIV. in the manuscripts: *chamaeleon*.
ETYMOLOGY: OE formation; MORPHOLOGY: N+N; exocentric, metonymic compound. MOTIVATION: habitat & part of plant. As the plant has not been identified, the analysis must rest on the denotata of the elements of the complex formation, which means that this formation is regarded as a completely transparent lexeme, with *eorþe* 'earth' presumably referring to the habitat, and *crop* 'umbel, cluster, head of the plant, bunch' to a part of the plant. ASSOCIATIVE RELATIONS: conceptual contiguity & conceptual contiguity (meronymy).
▫ *→crop, →eorþe*

eorþe, noun
'EARTH / Erde'
DOE (s.v. *eorþe*): c. 5200 occ.
▫ *→eorþæppel, →eorþber(i)(g)e, →eorþcrop, →eorþgealla, →eorþhnutu, →eorþīfig, →eorþmistel, →eorþnafela, →eorþrima*

eorþ-gealla, lit. 'earth-gall'
(ii) 'Centaurea L. / knapweed / Tausendgüldenkraut', or (ii) 'Centaurium umbellatum Gilibert / common centaury / Tausendgüldenkraut'
DOE (s.v. *eorþ-gealla*): c. 30 occ.
ILLUSTRATION in manuscript London, British Library, Cotton Vitellius C. iii, fol. 32v (Ohlgren 168(26)).
LATIN EQUIV. and OE synonyms in the manuscripts: *centaurea*, *limnestis*, *centauria maior*; *fel terrae*.
ETYMOLOGY: Independent formation in WGmc; OHG *erdgalla*, OLG *erthgalla*; loan translation of L *fel terrae* (Carr 1939: 130); MORPHOLOGY: N+N; exocentric, metaphorical compound. MOTIVATION: habitat & taste; The taste of the plant is quite bitter, hence the metaphorical use of OE *gealla* 'gall'. The reference to *eorþe* 'earth' is used to mark that the 'gall' grows in or near the earth, i.e. is a plant. Pliny the Elder (*Naturalis historia*, 25, 31) notes with regard to *centaurium lipton*: "hoc

307 For an account of Mandragora as a plant with medicinal and magic properties, see Birkhan (2012: 107-8), and van Arsdall, Klug and Blanz (2009: 285-346).

centaurium nostri fel terrae vocant propter amaritudinem summam" ['This *centaurium*, called by us *fel terrae*, because of its intense bitterness']. The Old English name is probably a loan translation of L *fel terrae* 'gall of the earth', that is to say, metaphorically 'bitterness of the earth'. According to Birkhan (2012: 167), the Old High German plant name – *erdgalla* lit. 'earth-gall' is seen by the German abbess Hildegard von Bingen as an allusion to the conception of the earth as a creature with all organs that humans and animals have. ASSOCIATIVE RELATIONS: conceptual contiguity & metaphorical similarity.
▫ *→eorþe, →gealla*

eorþ-hnutu, lit. 'earth-nut'
(i) 'Cyclamen europaeum L. / sowbread / Alpenveilchen'; (ii) 'Bunium bulbocastanum L. / pignut / Erdkastanie'
DOE (s.v. *eorþ-hnutu*): 2 occ. (*Charters* (Birch), 2: 161, no. 541; *Durham Plant Name Glossary*, no. 186).
LATIN EQUIV. in the manuscripts: mistakenly glossing *gentiana* (*DOE*, s.v. *eorþ-hnutu*).
ETYMOLOGY: Gmc compound; OHG *erdnuz*, MDu *erdnote*, Dan *jordnød*, Sw *jordnöt* (Carr 1939: 47); MORPHOLOGY: N+N; exocentric, metaphorical compound. MOTIVATION: habitat & shape. "Die ausgegrabenen Knollen werden da und dort (besonders von Kindern) wie Kastanien geröstet oder wie Kartoffeln gekocht gegessen" ['After being dug up, the tubers are eaten roasted like chestnuts (especially by children) or boiled like potatoes'] (Marzell, s.v. *Bunium bulbocastanum* L.). The bulbous roots are compared to nuts growing in the earth, because they are of a similar shape and a similar, dark brown colour. ASSOCIATIVE RELATIONS: conceptual contiguity & metaphorical similarity.
▫ *→eorþe, →hnutu*

eorþ-īfig, lit. 'earth-ivy'
'Glechoma hederaceum L. / ground-ivy / Gundelrebe'
DOE (s.v. *eorþ-īfig*): 14 occ.
ILLUSTRATIONS in manuscripts London, British Library, Cotton Vitellius C. iii, fol. 50r (Ohlgren 168(60)), and Oxford, Bodleian Library, Ashmole 1431 (S.C. 7523), fol. 25v (Ohlgren 221(89)).
LATIN EQUIV. in the manuscripts: *chamaemelon*, (*herba*) *hedera nigra* (1x with *chrysocanthos*).
ETYMOLOGY: OE formation; loan translation of L *hedera terrestris* (Bierbaumer 1, s.v. *eorþīfig*). MORPHOLOGY: N+N; endocentric, rectional. MOTIVATION: habitat & botanical information. Ground-ivy is a creeping plant with leaves similar to those of the ivy (Hedera helix). Although Glechoma hederaceum and Hedera helix are not related according to modern, scientific taxonomy, the Anglo-Saxons obviously saw them as related taxonomically, i.e. as kinds of ivies. ASSOCIATIVE RELATIONS: conceptual contiguity & taxonomic sub-ordination.
▫ *→eorþe, →īfig*

eorþ-mistel, lit. 'earth-mistletoe'
'?? Clinopodium vulgare L. / wild basil / Wirbeldost'
DOE (s.v. *eorþ-mistel*): 1 occ. (*Lǣcebōc*, 27/23).
ETYMOLOGY: OE formation; MORPHOLOGY: N+N; endocentric, rectional. MOTIVATION: habitat & botanical informa-

tion. "Da *eorþmistel* im selben Rezept wie *acmistel* [...] vorkommt, wird damit wohl – zum Zwecke der Unterscheidung – die andere, auf dem Boden wachsende „Mistel", eben S. vulgaris, bezeichnet sein" ['Since *eorþmistel* occurs in the same recipe as *acmistel*, it probably denotes – for a better distinction of the two plants – the other mistle, which grows right above the earth'] (Bierbaumer 1, s.v. *eorþmistel*). ASSOCIATIVE RELATIONS: conceptual contiguity & taxonomic sub-ordination.
▫ *→eorþe*, *→mistel*

eorþ-nafela, lit. 'earth-navel'
(i) 'Asparagus officinalis L. / asparagus / Spargel'; (ii) 'Cotyledon umbiculus L. / navelwort, wall pennywort / Nabelkraut'; (iii) 'Cyclamen europaeum L. / sowbread / Alpenveilchen'
DOE (s.v. *eorþ-nafela*): 9 occ.
LATIN EQUIV. in the manuscripts: *asparagus*, *asparagus agrestis*, *cotyledon*.
ETYMOLOGY: OE formation; loan translation of L *umbilicus terrae* < Gk *ὀμφαλόκαρπος* lit. 'earth fruit' (cf. Sauer 1992b: 390; Bierbaumer 2, s.v. *eorþnafela*). MORPHOLOGY: N+N; exocentric, metaphorical compound. MOTIVATION: habitat & shape. According to Bierbaumer (2 and 3, s.v. *eorþnafela*), the plant denoted by OE *eorþnafela* is most probably Cyclamen europaeum, which could also be referred to by *→eorþhnutu* and *→eorþæppel*. All three Old English names take their names from the bulbous roots, which grow in the earth and were associated with fruit (apples) and nuts likewise. Just as ModE *navelwort*, OE *eorþnafela* refers to plants with a feature likened to a navel (*OED*, s.v. *navelwort*). ASSOCIATIVE RELATIONS: conceptual contiguity & metaphorical similarity.
▫ *→eorþe*, *→nafela*

eorþ-rima, lit. '? earth-belt'
'? Polygonum convolvulus L. / black bindweed / Winden-Knöterich'
DOE (s.v. *eorþ-rima*): 1 occ. (*Lǣcebōc*, 102/14).
ETYMOLOGY: OE formation; MORPHOLOGY: N+N; exocentric, metaphorical compound. MOTIVATION: habitat & shape. This creeping plant may be seen as a kind of belt (OE *rima*) which entwines itself around the earth (Bierbaumer 1, s.v. *eorþrima*). ASSOCIATIVE RELATIONS: conceptual contiguity & metaphorical similarity.
▫ *→eorþe*, *→rima*

ēowu, noun
'EWE / Aue'
DOE (s.v. *ēowu*, *ēow*): 15 occ.
▫ *→ēowuhymele*

ēowu-hymele, lit. 'ewe-hops'
(i) 'Humulus lupulus L. (female) / hop / Hopfen'; or (ii) 'Bryonia L. / bryony / Zaunrübe'
DOE (s.v. *ēowu-hymele*): 1 occ. (*Lǣcebōc*, 105/11).
ETYMOLOGY: OE formation; hybrid compound: OE *hymele* < VL **humilus* ~ L *humulus* (Wollmann 1990: 164; see also *→hymele*; and Bierbaumer 1, s.v. *ēowohumele*; *Genaust*, s.v. *humulus*). MORPHOLOGY: N+N; endocentric, rectional; partial metaphor. MOTIVATION: sex & botanical information. The first element, *ēowu* 'ewe' is not entirely clear, but it may refer to the

fact that the hop (OE *hymele*) is a dioecious plant, i.e. female and male flowers do not grow on the same plant but on separate specimens. This fact may have been known to the Anglo-Saxons and the element *ēowu* would, then, point to the female part. The identification of OE *hymele* as 'hops' seems uncertain to Bierbaumer (1, *ēowohymele*), since *hymele* can also refer to other kinds of creepers. Yet, both the bryony and the hop are climbing plants which can grow rather high and resemble each other in their general appearance and with regard to the shape of the leaves (see also →*hegehymele* below). Since the bryony is a monocieous plant, there is no convincing way to explain the element *ēowu-*; that is why I favour the identification of *ēowohymele* as 'hop'. ASSOCIATIVE RELATIONS: metaphorical similarity & taxonomic sub-ordination.
▫ →*ēowu*, →*hymele*

etan, verb
'EAT / essen'
DOE (s.v. *ēowu*, *ēow*): c. 1200 occ.
▫ →*selfǣte*

F

fealu, adjective
'pale, yellowish, reddish / gelblich, rötlich'
The reflex *fallow* is now obsolete, except when used of the coats of animals (*OED*, s.v. *fallow*).
DOE (s.v. *fealu*): c. 40 occ.
▫ →*fealwe docce*

fealwe docce, lit. 'yellow dock'
? 'Rumex crispus L. / yellow dock / Krauser Ampfer'
Not an entry on its own in the *DOE*; attestation given s.v. *fealu*, adj.; hapax legomenon (*Lǣcebōc*, 38/4).
ETYMOLOGY: OE formation; MORPHOLOGY: Adj+N; endocentric, copulative: attributive. MOTIVATION: colour & botanical information. *Fealwe* 'pale, yellowish' probably refers to the pale, unspectacular colour of the flowers of Rumex crispus (Marzell, s.v. *Rumex crispus* L.). ASSOCIATIVE RELATIONS: conceptual contiguity & taxonomic sub-ordination.
▫ →*docce*, →*fealwe*

fearn, noun
'Filix L. / FERN / Farn'
DOE (s.v. *fearn*): 30 occ.
ILLUSTRATION in manuscript London, British Library, Cotton Vitellius C. iii, fol. 43r (Ohlgren 168(47)).
LATIN EQUIVALENTS and OE synonyms in the manuscripts: *felgera*, *filix*, *manus*; *fornētes folm*.
ETYMOLOGY: WGmc **farna-* < IE **pte/or-(no-)* 'fern' (*Kluge*, s.v. *Farn*).
▫ →*eoforfearn*, →*fenfearn*, →*micel fearn*

fēfer, noun
'FEVER / Fieber'
DOE (s.v. *fēfer*): c. 110 occ.
ETYMOLOGY: Early Latin loan, < L *febris* (Wollmann 1990: 158).
▫ →*fēferfuge*

fēfer-fuge (*-fūge*), lit. 'fever-?drive-away'
(i) 'Chrysanthemum parthenium (L.) Bernh. / FEVERFEW / Mutterkraut';

(ii) 'Centaurium umbellatum Gilibert / common centaury / Tausendgüldenkraut'
DOE (s.v. *fēfer-fuge, fēfer-fugie*): 25 occ. ("freq. in medical recipes").
ILLUSTRATION in manuscript London, British Library, Cotton Vitellius C. iii, fol. 32v (Ohlgren 168(26)).
LATIN EQUIV. in the manuscripts: *febrifugia.*
ETYMOLOGY: OE formation; early Latin loan, OE *fēferfuge* < L *febrifugia* (Wollmann 1990: 158). Serjeantson (1935: 284) and Campbell (1959: § 512), however, date it later than 650. Several processes overlap in OE *fēferfuge*: the first element, OE *fēfer* is a loan from Latin (< L *febris* 'fever'; Wollmann 1990: 158), so that the formal change of *febris* to *fēfer* is at the same time a loan translation process and a process of folk etymology (phonological motivation), in addition to the fact that the second element is not attested independently in Old English, and the formation can be regarded as a hybrid loan blend, too. The change of *febri-* to *fēfer* is supported by (secondary) semantic motivation: the plant is used with fever. MORPHOLOGY: unanalysable formation; second element not attested independently. MOTIVATION: first element: secondary, phonological and semantic motivation: sound pattern & illness; second element: ? action. L *febrifugia*, lit. 'who drives the fever away' (L *febris* 'fever, goddess who protected the people against fever and malaria' + *fugāre* 'drive away'; *Genaust*, s.v. *febrífuga*); from ancient times onwards, Chrysanthemum parthenium had been used as a medical plant; hence the motivation of the first element of the plant name. The element *-fuge*, which is not attested elsewhere, can be explained as 'drive-away'. With exception of the educated elite, however, the Anglo-Saxons probably did not recognise the OE reflex of the Latin lexeme *fugāre* 'drive away'. The plant name developed into ME *fewerfue*, ModE *feverfew*, i.e. the second element underwent further processes of folk etymology (see Sauer 1996: 146). ASSOCIATIVE RELATIONS: phonological similarity, conceptual contiguity & ?.

▫ *→fēfer, →-fuge*

feld, noun
'FIELD / Feld'
DOE (s.v. *feld*): c. 375 occ., "freq. in charters and bounds".

▫ *→feldbisceopwyrt, →feldmædere, →feldminte, →feldmore, →feldrūde, →feldswamm, →feldwōp, →feldwyrt*

feld-bisceopwyrt, lit. 'field-bishopwort'
'? Malva silvestris L. / mallow / Roßpappel'
DOE (s.v. *feld-bisceopwyrt*): 1 occ. (Napier 1890: 325-6).
ETYMOLOGY: OE formation. MORPHOLOGY: N+N; endocentric, rectional; MOTIVATION: habitat & botanical information. L *malva*, *ibiscum* and *althea* were used interchangeably to denote Malva silvestris and Althaea officinalis. Malva silvestris is rarely cultivated; hence its Old English name is *feldbisceopwyrt*: the element 'feld' is used in this context metaphorically to mark that the plant grew wildly, and not in the garden; it was probably meant to set it in contrast to the *bisceopwyrt*, which was cultivated in private gardens; compare also G *Feldmalve* lit. 'field-mallow' and *Wilde Malve* lit. 'wild mallow', which are in contrast to G

Malve 'mallow' (Marzell, s.v. *Malva silvestris*). ASSOCIATIVE RELATIONS: conceptual contiguity & taxonomic sub-ordination.
▫ *→bisceopwyrt, →feld*

feld-mædere, lit. 'field-madder'
(i) '?? Rosmarinus officinalis L. / rosemary / Rosmarin', or (ii) '? Galium mollugo L. / hedge bedstraw / Gemeines Labkraut'
DOE (s.v. *feld-mædere*): 2 occ. (*Durham Plant Name Glossary*, no. 290 and *Brussels Glossary*, 300/10).
LATIN EQUIV. and OE synonyms in the manuscripts: *ros marinus*; *boþen* 'rosemary, corn-marygold', *sundēaw*.
ETYMOLOGY: OE formation; MORPHOLOGY: N+N; endocentric, rectional; MOTIVATION: habitat & botanical information. OE *feldmædere* glosses L *ros marinus* 'rosemary'. The identification of *feldmædere* as Rosmarinus officinalis was questioned by Bierbaumer (3, s.v. *feldmædere*) on the grounds that the Old English name clearly refers to a wild colouring plant. This is, probably, Galium mollugo L., called in ModE *wild madder*, in German *Wilde Färberröte*; 'a kind of *mædere* 'madder' that grows in the field, i.e. wildly'. ASSOCIATIVE RELATIONS: conceptual contiguity & taxonomic sub-ordination.
▫ *→feld, →mædere*

feld-minte, lit. 'field-mint'
'Mentha longifolia Huds. / horse-mint / Roßminze'
DOE (s.v. *feld-minte*): 2 occ. (*Cleopatra Glossary* (Stryker), 316, no. 388; *Brussels Glossary*, 301/30).
LATIN EQUIV. in the manuscripts: *mentastrum*.
ETYMOLOGY: Independent formation in WGmc; OHG *feltminza*, *veltmīze* (Björkman 1904: 189); MLG *veltminte* (Carr 1939: 131); hybrid compound (OE *minte* < L *ment(h)a*; *→minte*). MORPHOLOGY: N+N; endocentric, rectional. MOTIVATION: habitat & botanical information; The *feldminte* is a kind of *minte* which is not cultivated but grows in the field, i.e. wildly. ASSOCIATIVE RELATIONS: conceptual contiguity & taxonomic sub-ordination.
▫ *→feld, →minte*

feld-more, feld-moru, lit. 'field-carrot, field-parsnip'
(i) 'Daucus carota L. / wild carrot / Gelbe Rübe, Gemeine Mohrrübe'; (ii) 'Pastinaca sativa L. / wild parsnip, common parsnip / Gemeiner Pastinak, Hirschmöhre'
DOE (s.v. *feld-more, feld-moru*): 17 occ.
ILLUSTRATION in manuscript London, British Library, Cotton Vitellius C. iii, fol. 44r (Ohlgren 168(49)).
LATIN EQUIV. in the manuscripts: *pastinaca, pastinaca silvatica*.
ETYMOLOGY: OE formation; MORPHOLOGY: N+N; endocentric, rectional; MOTIVATION: habitat & botanical information. The plant *feldmore* is a kind of *more* 'carrot, root' which is not cultivated, but grows in the field, i.e. wildly; cf. the Latin equivalent *pastinaca silvatica*, with the epithet *silvaticus* "transf[erred], of plants and animals, *growing* or *running wild*, wild" (*Lewis and Short*, s.v. *silvāticus*). ASSOCIATIVE RELATIONS: conceptual contiguity & taxonomic sub-ordination.
▫ *→feld, →more*

feld-rūde, lit. 'field-rue'
'Thalictrum flavum L. / common meadow rue / Gelbe Wiesenraute'
DOE (s.v. *feld-rūde*): 1 occ. (*Laud Herbal Glossary*, no. 1262).
LATIN EQUIV. in the manuscripts: *ruta montana*.
ETYMOLOGY: OE formation; OHG *felt-rute* (Björkman 1904: 181); hybrid formation (OE *rūde* < L *rūta*; →*rūde*); loan rendition of L *rūta montāna* (Stracke 1974: 104, no. 1261). MORPHOLOGY: N+N; endocentric, rectional. MOTIVATION: habitat & botanical information. The *feldrūde* is a kind of *rūde* which is not cultivated in gardens but grows in the fields, i.e. wildly. ASSOCIATIVE RELATIONS: conceptual contiguity & taxonomic sub-ordination.
▫ →*feld*, →*rūde*

feld-swamm, lit. 'field-mushroom'
'mushroom, toadstool / Schwamm, Pilz'
DOE (s.v. *feld-swamm*): 1 occ. (*Cleopatra Glossary* (Stryker), 196, no. 222).
LATIN EQUIV. and OE synonyms in the manuscripts: *fungus*; *swamm*.
ETYMOLOGY: OE formation; MORPHOLOGY: N+N; endocentric, rectional. MOTIVATION: habitat & botanical information. In contrast to →*meteswamm*, *feldswamm* is not edible; it is a kind of *swamm* 'mushroom' which grows wildly (expressed metaphorically by the element *feld* 'field'). ASSOCIATIVE RELATIONS: conceptual contiguity & taxonomic sub-ordination.
▫ →*feld*, →*swamm*

feld-wōp (= ? **feld-hopp*), lit. '? field-hops'
'Humulus lupulus L. / hops / Hopfen'
DOE (s.v. *feld-wōp*): 5 occ.[e]
OE *-wōp* is either a morpheme of uncertain meaning or a corruption of *hop(p)* 'hops'. According to the *DOE*, the identification of OE *feldwōp* is "a crux, where both lemma and gloss are of uncertain meaning. [It glosses] *bradigabo*; the lemma has been understood either as a form of LL *bardicatio* 'clamour, wailing' (cf. CL *baritus, barditus* 'battle-cry'); *feldwop*, literally 'field-cry', might then be interpreted as 'cry or clamour on the (battle-) field' (cf. *herewōp* 'cry of an army'); or, with similar sense, as a compound of LL *bradia* 'field' and Gmc **gab(b)-* (OF *gab*, It *gabbio* 'mockery'), which would correspond to the two elements of OE *feldwop* 'field-shouting / wailing'; alternatively, the lemma has been taken as a corruption of *bracii galio* < *bracium* 'malt, sprouted grain' ([...] [with second element ? cognate with OHG *hopfo* 'hops' or ? cognate with OE *-hoppa* 'hopper']); *feldwop* might then be a corruption of **feld-hopp(a)*, to be interpreted either as 'wild hops' or less probably as 'field-hopper', a synonym of *gærshoppa* 'grass-hopper'" (*DOE*, s.v. *feld-wōp*).
LATIN EQUIV. in the manuscripts: *bardicatio*, *bradigabo*, *bradigatio* (cf. *DMLBS* (fasc. vi, corrigenda A-L, s.v. *bardicatio*).
ETYMOLOGY: OE formation; MORPHOLOGY: N+N; endocentric, rectional. MOTIVATION: habitat & botanical information. The analysis of the motivation and the associative relations is based on the asumption that *-wōp* is a corrupt form

of *hopp* 'hops' (see above): then, the plant name denotes a kind of hop that grows in the field. ASSOCIATIVE RELATIONS: conceptual contiguity & taxonomic sub-ordination.
▫ →*feld*, →*-wōp*

feld-wyrt (*ferd-*), lit. 'field-wort'
'Gentiana amarella L. / felwort / Bitterer Enzian'
DOE (s.v. *feld-wyrt*): 9 occ.; *ferdwyrt* (*Lǣcebōc*, 46/26) is probably misspelled for *feldwyrt*.
ILLUSTRATIONS in manuscripts London, British Library, Cotton Vitellius C. iii, fol. 27r (Ohlgren 168(15)), and Oxford, Bodleian Library, Ashmole 1431 (S.C. 7523), fol. 9v (Ohlgren 221(16)).
ETYMOLOGY: OE formation; folk etymological change of L *fel* 'gall' in analogy to OE *feld* 'field' (see MOTIVATION below). MORPHOLOGY: N+N; endocentric, rectional. MOTIVATION: secondary, phonological and semantic motivation in the first element: sound pattern & habitat; second element: botanical information. Bierbaumer (1, s.v. *feldwyrt*) notes that *feldwyrt* is a remotivated form of an unattested **felwyrt* (made up of L *fel-* 'gall' + OE *wyrt* 'plant') based on the bitter taste of the plant (compare →*eorþgealla* lit. 'earth-gall' and L *fel terrae* lit. 'gall of the earth'). The change of L *fel* 'gall' to OE *feld* 'field' is motivated by the phonological similarity of the two nouns and by the semantic association of the plant with its habitat, the open field. ASSOCIATIVE RELATIONS: conceptual contiguity & taxonomic sub-ordination.
▫ →*feld*, →*wyrt*

felt, noun
'FELT /Fell'
DOE (s.v. *felt*): 3 occ.
▫ →*feltwurma*, →*feltwyrt*

felterre, noun
'Centaurium erythraea Rafn. / common centaury / Echtes Tausendgüldenkraut'
DOE (s.v. *feltere*): 9 occ.
LATIN EQUIV. and OE synonyms in the manuscripts: *centaurea*; *eorþgealla*, *hierdewyrt*.
ETYMOLOGY: < L *fel terrae* (*DOE*, s.v. *feltere*).
▫ →*smale felterre*

felt-wurma, lit. '? felt-marjoram', '? field-marjoram'
'Origanum vulgare L. / wild marjoram / Dost'
DOE (s.v. *felt-wurma*): 1 occ. (*Cleopatra Glossary* (Stryker), 349, no. 216).
LATIN EQUIV. in the manuscripts: *origanum*.
ETYMOLOGY: OE formation; MORPHOLOGY: N+N; endocentric, rectional; partial metaphor. MOTIVATION: habitat & botanical information. OE *wurma* is a colouring plant. That marjoram is a colouring plant, too, was suggested by Isidor (*Etymologiae*, 17: 9, 76): "Origanum, quod latine colena interpretatur propter quod infusa coloret uinum" ['The wild marjoram, whose Latin equivalent is *colena*, because, when put into wine, it colours the wine']. André (*Etymologiae* 17: 206, fn. 507) confirms that, according to Dioscorides and Pliny, "on préparait un vin médicinal à l'origan [...]" ['people used to prepare a medicinal wine on the basis of marjoram']. *Felt* is probably

misspelled for *feld* (*DOE*, s.v. *felt-wurma*) and indicates that, in contrast to *wurma*, it is a wild-growing plant (expressed metaphorically by the element *feld* 'field'). ASSOCIATIVE RELATIONS: conceptual contiguity & taxonomic sub-ordination.

▫ →*felt*, →*wurma*

felt-wyrt, lit. 'felt-wort'
'Verbascum thapsus L. / great mullein / Königskerze'
DOE (s.v. *felt-wyrt*): 11 occ., "mainly in glossaries".
ILLUSTRATION in manuscript London, British Library, Cotton Vitellius C. iii, fol. 41v (Ohlgren 168(44)).
LATIN EQUIV. in the manuscripts: *anadonia*, *verbascum*.
ETYMOLOGY: OE formation; MORPHOLOGY: N+N; endocentric, rectional; partial metaphor. MOTIVATION: effect & botanical information. The plant's leaves feel like felt (Cockayne, *Leechdoms*, 2: 383). ASSOCIATIVE RELATIONS: metaphorical similarity & taxonomic sub-ordination.

▫ →*felt*, →*wyrt*

fen(n), noun
(i) 'FEN / Moor, Sumpf'; (ii) 'dirt / Schmutz'
DOE (s.v. *fenn*): c. 120 occ.

▫ →*fenampre*, →*fencærse*, →*fenfearn*, →*fenlectere*, →*fenminte*

fen-ampre, lit. 'fen-dock'
'Rumex hydrolapathum Huds. / water-dock / Flussampfer'
DOE (s.v. *fen-ampre*): 1 occ. (*Lǣcebōc*, 31/24).
ETYMOLOGY: OE formation; MORPHOLOGY: N+N; endocentric, rectional. MOTIVATION: habitat & botanical information. Rumex hydrolapathum, a kind of *ampre* 'dock' grows preferably in the vicinity of marshes and fens (see Marzell, s.v. *Rumex obtusifolius* L.). ASSOCIATIVE RELATIONS: conceptual contiguity & taxonomic sub-ordination.

▫ →*ampre*, →*fen*

fen-cærse, lit. 'fen-cress'
'Nasturtium officinale R. Br. / water-cress / Brunnenkresse'
DOE (s.v. *fen-cærse*): 4 occ.
ETYMOLOGY: OE formation; MORPHOLOGY: N+N; endocentric, rectional. MOTIVATION: habitat & botanical information. This kind of *cærse* 'cress' grows preferably in moist environments, that is, alongside fens and brooks (Marzell, s.v. *Nasturtium officinale* R.Br.). ASSOCIATIVE RELATIONS: conceptual contiguity & taxonomic sub-ordination.

▫ →*cærse*, →*fen*

fen-fearn, lit. 'fen-fern'
'? Osmunda regalis L. / osmund-royal / Königsfarn'
DOE (s.v. *fen-fearn*): 1 occ. (*Antwerp-London Glossaries*, 110/12).
Although OE *fenfearn* glosses L *salvia* 'sage', it probably does not denote the sage but refers to a kind of fern which prefers marshy ground, such as Osmunda regalis (Bierbaumer 3, s.v. *fenfearn*). Old Latin names of Osmunda are *filix aquatica* lit. 'water-fern' and *filix palustris* 'marsh-fern', so that the OE plant name may be a loan formation based on one of them.
LATIN EQUIV. in the manuscripts: *salvia*.

ETYMOLOGY: OE formation; probably a loan translation of L *filix aquatica* or *filix palustris* (see comment above). MORPHOLOGY: N+N; endocentric, rectional. MOTIVATION: habitat & botanical information. Fenfearn is a kind of fern that prefers marshy grounds (Marzell, s.v. *Osmunda regalis* L.). ASSOCIATIVE RELATIONS: conceptual contiguity & taxonomic sub-ordination.
▫ →*fearn*, →*fen*

fen-lectere (= ? *-leahtric*), lit. '? fen-lettuce'
unidentified plant
DOE (s.v. *fen-lectere*): 1 occ. (*Laud Herbal Glossary*, no. 1161). The Latin lemma *peristeria* suggests 'vervain', but the *DOE* rejects this interpretation as the gloss indicates that it is a plant whose preferred habitat is moist ground (*DOE*, s.v. *fen-lectere*).
LATIN EQUIV. in the manuscripts: *peristeria*.
ETYMOLOGY: OE formation; hybrid formation: *-lectere*, a morph not attested independently in Old English is probably derived from the Latin loan *lactūca*, which was changed in analogy to OE *lēac* 'leek' (see →*leahtric*). MORPHOLOGY: N+N; endocentric, rectional. MOTIVATION: first element: habitat & second element: secondary, phonological and semantic motivation: sound pattern & botanical information. This kind of *-lectere* '? lettuce' prefers moist ground, hence the element *fen* 'fen'. ASSOCIATIVE RELATIONS: conceptual contiguity & phonological similarity, taxonomic sub-ordination.
▫ →*fen*, →*leahtric*

fen-minte, lit. 'fen-mint'
(i) 'Mentha aquatica L. / water-mint / Bachminze'; or (ii) 'Mentha longifolia Huds. / horse-mint / Roßminze'
DOE (s.v. *fen-minte*): 5 occ.
LATIN EQUIV. in the manuscripts: *mentastrum*.
ETYMOLOGY: OE formation; hybrid compound (OE *minte* < L *ment(h)a*; →*minte*). MORPHOLOGY: N+N; endocentric, rectional. MOTIVATION: habitat & botanical information. This kind of *minte* 'mint' prefers as its habitat moist ground, i.e. fens; cf. also ModE *water-mint*, G *Bachminze* lit. 'brook-mint' and the scientific botanical name, Mentha aquatica, all encoding this attribute. ASSOCIATIVE RELATIONS: conceptual contiguity & taxonomic sub-ordination.
▫ →*fen*, →*minte*

fīc, noun
(i) 'fruit of Ficus carica L. / fig / Feige'; (ii) '(diseases with) fig-like outgrowths / (Krankheiten mit) feigenartigen Auswüchsen' (iii) 'Ficus carica L. / fig-tree / Feigenbaum'
DOE (s.v. *fīc*): 19 occ.
LATIN EQUIV. in the manuscripts: *carica*, *ficulnea*, *ficus*.
ETYMOLOGY: Early Latin loan (Wollmann 1990: 156) < L *fīcus* (*OED*, s.v. *fig*). ModE *fig* goes back to the OFr loan word *figue*, *fige*.
▫ →*fīcbēam*, →*fīctrēow*, →*fīcwyrt*

fīc-bēam, lit. 'fig-tree'
'Ficus carica L. / fig-tree / Feigenbaum'
DOE (s.v. *fīc-bēam*): 45 occ.
LATIN EQUIV. in the manuscripts: *arbor fici*, *cariscus*, *ficulnea*, *ficus*.
ETYMOLOGY: Independent formation in WGmc (Carr 1939: 131); OHG *fīgboum*

(*DOE*, s.v. *fīc-bēam*); hybrid compound (OE *fīc* < L *ficus*; →*fīc*). MORPHOLOGY: N+N; endocentric, copulative: subsumptive (pleonastic). MOTIVATION: plant & botanical information. OE *fīc* is polysemous: it can refer to both the tree and to its fruits, the figs. Therefore two motivations seem plausible: a. the fig is a kind of tree; or b. this is a tree, which bears figs. In analogy to *pīntrēow* the first explanation is favoured: figs had been imported to middle and northern Europe since antiquity and were quite popular in (late) medieval cuisine (Birkhan 2012: 85). ASSOCIATIVE RELATIONS: conceptual identity & taxonomic sub-ordination.

▫ →*bēam*, →*fīc*

fīc-trēow, lit. 'fig-tree'

'Ficus carica L. / fig-tree / Feigenbaum'

DOE (s.v. *fīc-trēow*): 45 occ.

LATIN EQUIV. in the manuscripts: *arbor fici*, *ficulnea*, *ficus*.

ETYMOLOGY: Independent formation in Gmc; ON *fígtrê* (Carr 1939: 80); hybrid compound (OE *fīc* < L *ficus*; →*fīc*). MORPHOLOGY: N+N; endocentric, copulative: subsumptive (pleonastic). MOTIVATION: plant & botanical information; see →*fīcbēam*. ASSOCIATIVE RELATIONS: conceptual identity & taxonomic sub-ordination.

▫ →*fīc*, →*trēow*

fīc-wyrt, lit. 'fig-wort'

(i) '? Ranunculus ficaria L. / lesser celandine / Scharbockskraut'; or (ii) '? Potentilla erecta Räuschel / common tormentil / Blutwurz'

DOE (s.v. **fīc-wyrt*): 1 occ. (*Antwerp-London Glossaries*, 109/8).

LATIN EQUIV. in the manuscripts: *ficus*.

ETYMOLOGY: Independent formation in WGmc (Carr 1939: 138), cf. OHG *figwurz* (Björkman 1904: 182); hybrid compound (OE *fīc* < L *ficus*; →*fīc*). MORPHOLOGY: N+N; endocentric, rectional. MOTIVATION: illness & botanical information. The plant is of uncertain identification. *Fīcwyrt* can denote a plant used to treat hemorrhoids (*DOE*, s.v. **fīcwyrt*) after the principle *similia similibus*: hemorrhoids are the result of a disease characterised by fig-like outgrowths, so that the fig-like, bulbous roots of Ranunculus ficaria and Potentilla erecta were possibly thought to help; according to Birkhan (2012: 122), it was used to treat fevers. ASSOCIATIVE RELATIONS: metaphorical similarity & taxonomic sub-ordination.

▫ →*fīc*, →*wyrt*

fīf, numeral

'FIVE / five'

DOE (s.v. *fīf*): c. 1300 occ.

▫ →*fīffingre*, →*fīflēafe*

fīf-fingre, lit. 'five-finger'

'? Potentilla reptans L. / fiveleaf, cinquefoil / Fünffingerkraut'

DOE (s.v. *fīf-fingre*): 1 occ. (*Peri Didaxeon*, 35/5).

ETYMOLOGY: Independent formation in WGmc; MLG *vîffingere* (Carr 1939: 155); there is no attestation from Old High German, but compare the translations of L *quinquefolium* into OHG *finfblat* (Carr 1939: 155) and MHG *funffblat* (Birkhan 2012: 122). MORPHOLOGY: Num+N; exocentric, metaphorical compound. MOTIVATION: number & shape. Hoops (1889: 17) explains the motivation as follows: "Zur Kennzeichnung der Pflanze war ‚Fünfblatt' vollkommen ausreichend; der Vergleich mit den menschlichen Fingern bezeugt eine Anteilnahme der Phantasie: es zeigt sich hier, wenn auch noch in roher Form, bereits ein gewisses ästhetisches Interesse an der Pflanze ohne Rücksicht auf ihre praktische Verwendung." ['Fiveleaf' was perfectly sufficient to denote the plant; the comparison of the plant with human fingers is imaginative: a certain aesthetic interest in the plant manifests itself, albeit in a coarse manner, without any interest in the practical uses of the plant']. Bierbaumer (2, s.v. *fīffingre*), questions the identification of the plant given by Hoops (which was followed by *ClH*, s.v. *fīffingre*) since it is based on the meaning of the ModE plant name *five-finger* and neglects the Latin lemma it glosses (L *stafisagriam*). If we accept the identification as Potentilla reptans L., however, the motivation is the resemblance of the plant's leaves with the five fingers of a human hand. ASSOCIATIVE RELATION: metaphorical similarity.

▫ *→fīf, →fingre*

fīf-lēaf(e), lit. 'five-leaf'
'Potentilla reptans L. / FIVELEAF, cinquefoil / Fünffingerkraut'
DOE (s.v. *fīf-lēafe, fīf-lēaf*): 35 occ.
ILLUSTRATIONS in manuscripts London, British Library, Cotton Vitellius C. iii, fol. 22v (Ohlgren 168(6)), and Oxford, Bodleian Library, Ashmole 1431 (S.C. 7523), fol. 6r (Ohlgren 221(3)).
LATIN EQUIV. in the manuscripts: *pentaphyllon*, *quinquefolium*.
ETYMOLOGY: OE formation; The Old English plant name is probably a loan translation of L *quinquefolium*, itself a translation of Gk *πεντάφυλλον* (Bierbaumer 1, s.v. *fīflēafe*); cf. OHG *finfblat*, *fuinfbletter*, *vunfblat*, etc. for L *quinquefolium* (Björkman 1902: 287; Carr 1939: 155). MORPHOLOGY: Num+N; exocentric, metonymic compound. MOTIVATION: number & part of plant. The plant's leaves are divided into five parts, but as these have a special form, they seem to consist of five leaves (Marzell, s.v. *Potentilla reptans* L.). ASSOCIATIVE RELATIONS: conceptual contiguity (meronymy).

▫ *→fīf, →lēafe*

fille, noun
'Thymus serpyllum L. / breckland thyme / Sand-Thymian'
DOE (s.v. *fille*): 7 occ.
LATIN EQUIV. in the manuscripts: *chaerophyllum*, *serpyllum*.
ETYMOLOGY: OE *fille* is shortened from *→cerfille* < L *chaer(e)phylla*, pl. of *chaerephyllum* < Gk *χαιρέφυλλον* (cf. *Genaust*, s.v. *chaerophýllum*; *OED*, s.v. *chervil*, *OED*, s.v. *fille*).

▫ *→rēade fille, →wudufille*

finger, noun
(i) 'FINGER / Finger'; (ii) 'a unit of measure / Maßeinheit'; (iii) 'place name element'
DOE (s.v. *finger*): c. 250 occ.

▫ *→fingeræppel, →fingerdocca*

▫ *→fīffingre*

finger-æppel, lit. 'finger-fruit'
'fruits of Phoenix dactylifera L. / date / Dattel'
DOE (s.v. *finger-æppel*): 8 occ.
LATIN EQUIV. in the manuscripts: *dactylus*.

ETYMOLOGY: OE formation; possibly a loan rendition after L *dactylus* (< Gk *δάκτυλος*), which was polysemous and could denote both the 'finger' and the 'date' (*DOE*, s.v. *finger-æppel*); MORPHOLOGY: N+N; endocentric, rectional; partial metaphor. MOTIVATION: shape & botanical information. As OE *fingerappla* glosses L *dactylus* < polysemous Gk *δάκτυλος* '1. finger, 2. date', the first element of *fingeræppel* is probably based on the shape of the fruits of the date palm, while the second element indicates that this 'finger' is a fruit, thus disambiguating the first element and transposing it from the domain of body parts to the domain of plants and part of plants. ASSOCIATIVE RELATIONS: metaphorical similarity & taxonomic sub-ordination.

▫ *→æppel, →finger*

finger-docca, lit. '? finger-dock, or finge-sorrel'

'? Digitalis purpurea L. / fox-glove / Roter Fingerhut'

DOE (s.v. *finger-docce*): 3 occ.[e] (*Corpus Glossary*, no. 687; *Épinal-Erfurt Glossary*, no. 345a; Meritt 1945: 60, no. 70/22); according to the *DOE*, "the lemma is obscure: *digitalis* 'fitting the finger' may refer to the flowers of the foxglove whose modern vernacular names include 'finger-flower, finger-root, purple fingers' etc.; the second word of the lemma is recorded variously as *musculus* 'muscle', *munusculum* 'little gift' (cf. VERG. Ecl. 4.18) and *masculus* 'male'; a meaning 'finger-muscle' has also been suggested, but this is based on the Latin *digitalis musculus* and takes little account of the second element of the OE compound, which is always a plant name". For Bierbaumer (3, s.v. *fingerdocca*), the identification as 'finger-muscle' seems more plausible.

LATIN EQUIV. in the manuscripts: *digitalis masculus*, *digitalis munusculum*, *digitalis musculus*.

ETYMOLOGY: OE formation; possibly a loan rendition of L *digitalis musculus* (*DOE*, s.v. *finger-docce*); MORPHOLOGY: N+N; endocentric, rectional, partial metaphor. MOTIVATION: shape & botanical information. If we assume that OE *fingerdocca* denotes the foxglove, the first element refers to the fact that the flowers resemble small gloves, while the second element indicates the taxonomic affiliation of the plant. ASSOCIATIVE RELATIONS: metaphorical similarity & taxonomic sub-ordination.

▫ *→docca, →finger*

finu(g)le, noun

'Foeniculum vulgare Mill. / FENNEL / Fenchel'

DOE (s.v. *finol*, *finule*, *finugle*): c. 90 occ.

ILLUSTRATION in manuscript London, British Library, Cotton Vitellius C. iii, fol. 55v (Ohlgren 168(71)).

LATIN EQUIV. in the manuscripts: *faeniculum*, *marathrus*.

ETYMOLOGY: Early Latin loan (Wollmann 1990: 164); < MedL *fēnuclum* < CL *fēniculum* from *fēnum* 'hay' (*OED*, s.v. *fennel*; Wollmann 1990: 164).

▫ *→sūþerne finu(g)le*

flēaþe, noun

'Nymphaea L. / waterlily' / Seerose'

DOE (s.v. *flēaþe*): 2 occ.

▫ *→flēaþe wyrt*

flēaþe wyrt (*flēawyrt*), lit. '? waterlily plant' or '? floating plant'
(i) '? Nymphaea alba L. / white water-lily / Weiße Seerose'; (ii) '? Nuphar Smith L. / water-lily / Teichrose'
DOE (s.v. *flēaþe*): 2 occ., attested as *fleaþan wyrte* in *Lǣcebōc* (79/35) and as *fleoþanwyrte* in *Lǣcebōc* (80/21); *DOE* (s.v. *flēaþor-wyrt*): 1 occ. (*Durham Plant Name Glossary*, no. 252).
DOE (s.v. *flēawyrt*): 2 occ.
ILLUSTRATION in Oxford, Bodleian Library, Ashmole 1431 (S.C. 7523), fol. 19r (Ohlgren 221(58)).
LATIN EQUIV. in the manuscripts: *nymphaea, papyrus*.
ETYMOLOGY: OE formation; MORPHOLOGY: N+N; endocentric, copulative: subsumptive (pleonastic). This plant name is not easy to analyse morphologically. For *BT* and *ClH*, *flēaþe wyrt* (*Lǣcebōc*, 79/35f.) is a noun phrase, with premodifying *flēaþe* 'waterlily' and the head *wyrt*, while *flēaþor wyrt* (*Durham Plant Name Glossary*, no. 252) is seen as a compound. Bierbaumer (1, s.v. *flēaþe wyrt*), however, argues that *flēaþe* is a premodifying adjective, i.e. that the composite is an Adj+N formation. The *DOE*, whose interpretation is followed here, is more inclined to believe that *flēaþe* is the noun 'water-lily' , but does not reject completely the idea that it may also be an adjective, since it flags it with a question mark and notes "Noun, f. wk. (? or adjective)". MOTIVATION: plant & botanical information. This kind of *wyrt* is a water-lily. ASSOCIATIVE RELATIONS: conceptual identity & taxonomic sub-ordination.
▫ →*flēaþe*, →*wyrt*

flēawyrt →*flēaþe wyrt*

flēot, noun
'tidal inlet, stretch of water, river / Wasser, Flut, Fluss'
DOE (s.v. *flēot*[1]): c. 75 occ.
▫ →*flēotwyrt*

flēot-wyrt, lit. 'water-wort'
'sea-weed / Alge'
DOE (s.v. *flēot-wyrt*): 1 occ. (*Lǣcebōc*, 81/7).
ETYMOLOGY: OE formation; MORPHOLOGY: N+N; endocentric, rectional. MOTIVATION: habitat & botanical information. Algae, a kind of *wyrt* that lives in water (see →*flēot*). ASSOCIATIVE RELATIONS: conceptual contiguity & taxonomic sub-ordination.
▫ →*flēot*, →*wyrt*

folgian, verb
'FOLLOW / folgen'
DOE (s.v. *folgian*): c. 275 occ.
▫ →*sunfolgend*

folm, noun
'palm / Hand'
DOE (s.v. *folm, folme, ? folma*): c. 70 occ.
▫ →*fornētes folm*

Fornēot, proper name
Fornēot was possibly the mythological name of a god, derived from OIcel *Fornjótr*, the etymology of which is uncertain (*Holthausen*, s.v. *forn*, and Bierbaumer 1974: 175, with further references).
▫ →*fornētes folm*

fornētes folm, lit. 'Forneot's hand'
'Orchis L. / cuckoo flower, cullions / Knabenkraut'
Not a lemma on its own in the *DOE*, but recorded s.v. *folm*, 1.g.i.: 1 occ. (*Cleopatra Glossary* (Stryker), 316, no. 382; Voss 1988: 132).
LATIN EQUIV. and OE synonyms in the manuscripts: *manus*; *fearn*.
ETYMOLOGY: OE formation; loan translation of *manus færne*. MORPHOLOGY: $PropN_{es}$+N; exocentric, metaphorical compound. MOTIVATION: proper name & shape. According to Marzell (s.v. *Orchis*), the form of the root resembles a hand (a palm) (see also Bierbaumer 1974), hence the use of the second element *folm*. Why the first element, *fornēot* '? god *Fornjótr*', was used is not evident. ASSOCIATIVE RELATION: metaphorical similarity.
▫ →*folm*, →*Fornēot*

fōt, noun
'FOOT / Fuß'
DOE (s.v. *fōt*): c. 550 occ.
▫ →*foxes fōt*, →*hræfnes fōt*, →*lēonfōt*

fox, noun
'FOX / Fuchs'
DOE (s.v. *fox*): c. 80 occ.
▫ →*foxesclāte*, →*foxesclīfe*, →*foxesfōt*, →*foxesglōfa*

foxes-clāte, lit. 'fox's clivers'
'Arctium lappa L. / greater burdock / Große Klette'
Not a lemma on its own in the *DOE*, but attested s.v. *fox*, 3.a.: 1 occ. (*Lǣcebōc*, 44/3).
ETYMOLOGY: OE formation; MORPHOLOGY: N_{gen}+N; endocentric, rectional. MOTIVATION: animal & botanical information. According to Bierbaumer (1, s.v. *clāte*) the element *fox(es)* refers to the plant's natural environment, the wood: the plant – a kind of burdock – grows where foxes live. Compare also →*hindheolaþ* for a similar motivation. ASSOCIATIVE RELATIONS: conceptual contiguity & taxonomic sub-ordination.
▫ →*clāte*, →*fox*

foxes-clīfe, lit. 'fox's clivers'
'Arctium lappa L. / greater burdock / Große Klette'
Not a lemma on its own in the *DOE*, but attested s.v. *fox*, 3.a. (e.g., *Lācnunga*, 1007).
ETYMOLOGY: OE formation; MORPHOLOGY: N_{gen}+N; endocentric, rectional. MOTIVATION: animal & botanical information. See →*foxesclāte*. ASSOCIATIVE RELATIONS: conceptual contiguity & taxonomic sub-ordination.
▫ →*clīfe*, →*fox*

foxes fōt, lit. 'fox's foot'
(i) '? Sparganium emersum Rehman / unbranched burr-reed / Igelkolben'; (ii) 'Gladiolus segetum Gawler / gladiolus / Siegwurz'
Not a lemma on its own in the *DOE*, but attested s.v. *fox*, 3.b.: 4 occ.; It is not entirely clear whether *foxes fōt* is a real plant name, as suggested by Cockayne (*Leechdoms*, 2: 385), or whether the gloss is due to a misreading of *adipes vulpinus* 'fox's fat' as *pēs vulpinus* 'fox's foot' and hence a ghost plant name as suggested by Bierbaumer (3, s.v. *foxes fōt* and Bierbaumer 2003: 158). De Vriend (1984: 299, no. 47) does not support the idea that *foxes fōt* is a misinterpretation of *adipes vulpinus*. D'Aronco and Cameron

(1998: 52) favour interpretation (ii), which corresponds to the plant referred to by "herba xifion" (*Herbarium*, manuscript Ca, no. 47). OE *foxes fōt* survives as ME *voxesfot* (Hunt 1989: 266) and ModE *fox's foot*.
ILLUSTRATIONS in manuscripts London, British Library, Cotton Vitellius C. iii, fol. 36r (Ohlgren 168(33)), and Oxford, Bodleian Library, Ashmole 1431 (S.C. 7523), fol. 14r (Ohlgren 221(36)).
LATIN EQUIV. in the manuscripts: *xiphion*.
ETYMOLOGY: OE formation. MORPHOLOGY: N_{gen}+N; exocentric, metaphorical compound. MOTIVATION: Since its identification is unclear, the motivation of the plant name cannot be substantiated. Sparganium and Gladiolus cannot be brought into associative relations with foxes or their feet, and in other languages, these plants do not have names using the elements 'fox' or 'foot', either. ASSOCIATIVE RELATIONS: As the motivation cannot be ascertained, the associative relations are unclear, too.
▫ →*fōt*, →*fox*

foxes glōfa, lit. 'fox's glove'
(i) 'Digitalis purpurea L. / FOXGLOVE / Roter Fingerhut'; (ii) 'Datura stramonium L. / thorn apple / Stechapfel'; or (iii) 'Atropa belladonna L. / deadly nightshade / Tollkirsche'
Not a lemma on its own in the *DOE*, but attested s.v. *fox*, 3.c.: 5 occ.
ILLUSTRATIONS in manuscripts London, British Library, Cotton Vitellius C. iii, fol. 64r (Ohlgren 168(88)), and Oxford, Bodleian Library, Bodley 130, fol. 44r (Ohlgren 222(85)).
LATIN EQUIV. in the manuscripts: *buglossa*, *saliunca*, *strychnos manicos*.
ETYMOLOGY: OE formation; MORPHOLOGY: N_{gen}+N; exocentric, metaphorical compound. MOTIVATION: ? colour & shape. ad (i) Digitalis purpurea L. / foxglove', is a tall plant with tubular flowers in different colours – purple, pink, white or yellow – depending on the species. Obviously, the shape of the flowers – they can be easily fitted over a human fingertip like a glove – was salient enough to stand sponsor to the Old English plant name. The motivation for the first element, *fox*, is unclear to me. The *OED* (s.v. *foxglove*) mentions the Norwegian vernacular name of Digitalis purpurea, namely, *revbjelde* lit. 'fox bell', which uses the element 'fox' as well. The German name *Fingerhut* encodes the concept GLOVE, too. Obviously, the authors of the Old English plant name associated the shape of the flowers with a FOX'S GLOVE (relation of metaphorical similarity). Why it is exactly the fox's glove and not that of other animals, is not so clear. There is, perhaps, a relation of similarity between the plant and foxes, based on the reddish colour of both the plant and the foxes' furs. ad (ii) For Bierbaumer (3, s.v. *foxesglōfa*; 2003: 157-8), *foxes glōfa* in the *Herbarium* (no. 144) is the result of a misunderstanding of the Greek lemma on the part of the glossators: who seems to have interpreted Gk *μανικόν* 'frantic, manic' (< Gk *μανία* 'madness') as L *manica* 'glove' and therefore used *foxes glōfa* as gloss for *trychnos manicos*. The Greek plant name *στρύχνος* referred to a variety of plants from the order of the Solanaceae ('a kind of nightshade'; *Lewis and Short*, s.v. *strychnos*); Bierbaumer suggests here the Atropa belladonna L. De Vriend (1984: 320, no. 140) favors the identi-

fication as Datura stramonium L., 'thorn-apple', since Datura belongs to the order of the Solanaceae, but is not stimulating like Digitalis purpurea L. – it is rather a sedative, as is required by the recipes in the *Herbarium*. D'Aronco and Cameron (1998: 57) and van Arsdall (2002: 213) follow de Vriend; see also Hunt (1989: 254). Marzell (s.v. *Datura stramonium* L.), however, reports that the Datura was native to the American continent and was introduced to Europe as late as the 16th century; hence, the identification of *foxes glōfa* as Datura in the *Herbarium* should be rejected. ASSOCIATIVE RELATIONS: ? conceptual contiguity & metaphorical similarity.

▫ →*fox*, →*glōfa*

frencisc, adjective
'FRENCH / französisch'
DOE (s.v. *frencisc*): c. 70 occ., "mainly in historical texts of s.xii and later."

▫ →*frencisc hnutu*

frencisc hnutu, lit. 'French nut'
'fruit of Juglans regia L. / walnut / Walnuss'
Not a lemma on its own in the *DOE*, but attested s.v. *frencisc*, 1.a.iii.: 1 occ. (*Peri Didaxeon*, 35/29).
ETYMOLOGY: OE formation; MORPHOLOGY: Adj+N; endocentric, copulative: attributive. MOTIVATION: origin & botanical information. The walnut is not native to England, but was often grown in Gaul; see also →*wealhnutu*. ASSOCIATIVE RELATIONS: conceptual contiguity & taxonomic sub-ordination.

▫ →*frencisc*, →*hnutu*

-fuge, ?
'? to drive away / vertreiben'
This element of *fēferfuge* goes back to L *fugāre* '(to) drive away' (see →*fēferfuge*).

▫ →*fēferfuge*

fugel, *fugol*, noun
(i) 'bird, FOWL / Vogel'; (ii) element in plant names; (iii) personal name or element in personal names; (iv) an element in place names
DOE (s.v. *fugel*): c. 500 occ.

▫ →*fuglesbēan*, →*fugleslēac*, →*fugleswīse*

fugles-bēan, lit. 'bird's bean'
'Vicia cracca L. / (tufted) vetch / Vogelwicke'
Not a lemma on its own in the *DOE*, but attested s.v. *fugel*, 2.a.; 12 occ.[e] (*DOEPN*, s.v. *fugles-bēan*).
LATIN EQUIV. in the manuscripts: *pisa agrestis*, *vicia*.
ETYMOLOGY: OE formation; MORPHOLOGY: N_{gen}+N; exocentric, metonymic compound. According to Bierbaumer (1, 2, 3, s.v. *bēan*), the OE lemmas always refer to the seeds of Vicia faba L. and not to the plant itself. The use of *bēan* in the present compound may be seen as metonymic if we accept that with *fuglesbēan* the plant as such is meant and not the seeds only: a part of the plant (the seeds) is used to refer to the whole plant (*pars pro toto*). MOTIVATION: animal & part of plant. Viccia cracca L. is a creeping plant, a common weed growing predominantly in corn fields. Its dark seeds, which grow in pods, are loved by all kinds of birds. Similar names in other Germanic languages are OHG *vogalchrut*, G *Vogelerbse*, Du *vogelwikke*, etc. (Marzell,

s.v. *Viccia cracca* L.). ASSOCIATIVE RELATION: conceptual contiguity (meronymy).
▫ →*bēan*, →*fugel*

fugles-lēac, lit. 'bird's plant'
'? Vicia cracca L. / (tufted) vetch / Vogelwicke'
Not a lemma on its own in the *DOE*, but attested s.v. *fugel*, 2.b.: 2 occ. (*Brussels Glossary*, 300/6; *Durham Plant Name Glossary*, no. 337).
LATIN EQUIV. in the manuscripts: ? *sempervivum*, ? *vicia*.
ETYMOLOGY: OE formation; MORPHOLOGY: N_{gen}+N; endocentric, rectional. MOTIVATION: animal & botanical information. OE *lēac* had a wide range of meanings: from the very general as 'herb, plant' (hence *lēactūn* 'kitchen-garden, garden of herbs' and *lēacweard* 'gardener') to the various kinds of Alliums that it could denote (see →*lēac*). Following Bierbaumer (3, s.v. *fugleslēac*), who equates *fugleslēac* with OHG *vogalchrut*, the Old English plant name denotes Vicia cracca L. and the second element has the general meaning 'herb, plant'; the first element indicates that birds like this kind of plant. ASSOCIATIVE RELATIONS: conceptual contiguity & taxonomic subordination.
▫ →*fugel*, →*lēac*

fugles-pise (-? *wise*), lit. 'bird's pea'
'Delphinium ajacis L. / larkspur / Garten-Rittersporn'
Not a lemma on its own in the *DOE*, but attested s.v. *fugel*, 2.b. (*Brussels Glossary*, 298/24; *Durham Plant Name Glossary*, no. 144; Voss. 1996: 185/20; note 20).
LATIN EQUIV. in the manuscripts: *delphinion*.
ETYMOLOGY: OE formation; MORPHOLOGY: N_{gen}+N; exocentric, metonymic compound. MOTIVATION: shape & part of plant. The motivation of the first element is probably provided by the shape of the flowers of Delphinium: they resemble bird's feet (Bierbaumer 3, s.v. *fugleswīse*). As to the second element, the use of both *wīse* and of *pise* is justified: the former means 'shoot, sprout, stalk', the latter reminds one of the seeds of a related kind of Delphinium described by Dioscorides which resemble peas (cf. Bierbaumer 3, s.v. *fugleswīse*). More probably, however, is that *fugleswīse* is erroneously spelled for *fuglespise*, as the wynn and the grapheme <p> were so similar that they could easily be confused (Voss 1996: 189, note 20; *DOE*, s.v. *fugel*, 2c.) ASSOCIATIVE RELATIONS: conceptual contiguity & conceptual contiguity (meronymy).
▫ →*fugel*, →*pise*, →*wīse*

fūl, adjective
'FOUL / faul, faulig'
DOE (s.v. *fūl* adj.): c. 450 occ.
▫ →*fūla bēam*, →*fūla wermōd*, →*fūle trēow*

fūla bēam, lit. 'foul tree'
'Frangula alnus Miller / alder buckthorn, butcher's prickwood / Faulbaum, Pulverholz'
Not a lemma on its own in the *DOE*, but attested s.v. *fūl*, 1.d.ii.; hapax legomenon (*Læcebōc*, 24/29).
ETYMOLOGY: Independent formation in WGmc; cf. OHG *fuelboum*, OLG *fulbom* (Björkman 1902: 305); MORPHOLOGY: Adj+N; endocentric, copulative:

attributive. MOTIVATION: smell & botanical information. The fresh bark of Frangula alnus emanates a foul, mouldy scent (Marzell, s.v. *Frangula alnus* L.); cf. also *fūletrēow*. ASSOCIATIVE RELATIONS: conceptual contiguity & taxonomic sub-ordination.
▫ →*bēam*, →*fūl*

fūla wermōd, lit. 'foul wormwood'
unidentified plant; perhaps 'Artemisia campestris L. / wormwood / Wermut'
Not a lemma on its own in the *DOE*, but attested s.v. *fūl*, 1.c.; hapax legomenon (*Lǣcebōc*, 96/10).
According to Cockayne (*Leechdoms*, 3: 348) *fūla wermōd* denotes 'Artemisia campestris L. / field wormwood / Feld-Beifuß', but Bierbaumer (1, s.v. *wermōd*) rejects this identification. Cockayne also suggests 'Anthemis cotula L. / stinking camomile / Hundskamille' (*Leechdoms*, 2: 410). Following the *DOE*, the adjective in the noun phrase *se fula wermod* is descriptive, i.e. not part of the plant name: "? offensive to the taste, foul-tasting, bitter, here referring to wormwood" (*DOE*, s.v. *fūl*, 1c.).
ETYMOLOGY: OE formation. MORPHOLOGY: Adj+N; endocentric, copulative: attributive. MOTIVATION: taste & botanical information. In case it denotes Artemisia campestris, it probably encodes the bitter taste the plant is famous for in the first element. ASSOCIATIVE RELATIONS: conceptual contiguity & taxonomic sub-ordination.
▫ →*fūl*, →*wermōd*

fūle-trēow, lit. 'foul tree'
'Frangula alnus Miller / alder buckthorn, berry-bearing elder / Faulbaum'
Not a lemma on its own in the *DOE*, but attested s.v. *fūl*, 1.d.ii.; 4 occ.[e]
LATIN EQUIV. in the manuscripts: *alneus*.
ETYMOLOGY: OE formation; MORPHOLOGY: Adj+N; endocentric, copulative: attributive. MOTIVATION: smell & botanical information. The fresh bark of Frangula alnus emanates a foul, mouldy scent (Marzell, s.v. *Frangula alnus* L.); cf. also →*fūla bēam*. ASSOCIATIVE RELATIONS: conceptual contiguity & taxonomic sub-ordination'.
▫ →*fūl*, →*trēow*

furh, noun
'FURROW / Furche'
DOE (s.v. *furh*): c. 130 occ.
▫ →*furhwudu*

furh-wudu, lit. 'furrow-wood'
'Pinus L. / pine / Kiefer, Föhre'
DOE (s.v. *furh-wudu*): 2 occ.[e] (*Corpus Glossary*, no. 1590; Meritt (1945: 16, no. 8/4).
LATIN EQUIV. in the manuscripts: *pinus*.
ETYMOLOGY: OE formation; MORPHOLOGY: N+N; exocentric, metonymic compound, partial metaphor. MOTIVATION: texture & texture. In both glossaries, *furhwudu* glosses L *pīnus* 'pine-tree'. The bark of the pine is heavily furrowed, hence the first element of the tree's name, *furh* 'furrow'. The second element, *wudu* 'wood' refers to the texture, too – it indicates that this plant name denotes a tree and not a shrub or a flower (which do not have woody stems). In Old English, *wudu* denoted, above all, 'wood, forest, grove, timber' (*ClH*, s.v. *wudu*); cf. compounds such as →*wuduæppel* 'crab-apple', G *Holzapfel*

lit. ‘wood-apple’, or *wudubāt* ‘wooden boat’, where the sense ‘wooden’ is encoded, and *wudublǣd* ‘forest blossom’, *wudubucca* ‘wild buck’, where the sense ‘forest’ is encoded. ASSOCIATIVE RELATIONS: metaphorical similarity & conceptual contiguity.

▫ *→furh, →wudu*

fyrgen-, noun
‘mountain / Berg’
DOE (s.v. *fyrgen-*) calls it a ‘nominal affix’, a nominal first element in compound nouns, which is not attested independently.

▫ *→fyrgenbēam*

fyrgen-bēam, lit. ‘mountain-tree’
‘mountain-tree / Baum in den Bergen’
DOE (s.v. *fyrgen-bēam*): 1 occ. (*Beowulf*, ed. Jack, l. 1412); not in Bierbaumer (1, 2, 3).
ETYMOLOGY: OE compound. MORPHOLOGY: N+N; endocentric, rectional; *fyrgen-* is not attested independently, but in other complex formations such as *fyrgenbeorg* ‘mountain cliff’, *fyrgenbucca* ‘mountain goat’, etc.; see *DOE* (s.vv. *fyrgenbeorg, fyrgenbucca*). MOTIVATION: habitat & botanical information; ‘a tree growing in the mountains’. ASSOCIATIVE RELATIONS: conceptual contiguity & taxonomic subordination.

▫ *→bēam, →fyrgen*

G

gār, noun
(i) ‘spear / Speer’; (ii) name of a runic letter; (iii) element of personal names; (iv) element of place names
DOE (s.v. *gār*): c. 55 occ.

▫ *→gārclīfe, →gārlēac*

gār-clīfe, ***gār-clife***, lit. ‘spear-cleavers’
‘Agrimonia eupatoria L. / agrimony / Odermenning’
DOE (s.v. *gār-clife*): 24 occ.
ILLUSTRATION in manuscript London, British Library, Cotton Vitellius C. iii, fol. 31v (Ohlgren 168(24)).
LATIN EQUIV. in the manuscripts: *agrimonia*.
ETYMOLOGY: OE formation; MORPHOLOGY: N+N; endocentric, rectional; partial metaphor. MOTIVATION: shape & botanical information. The plant has a high and pointed stem that reminds one of a spear (OE *gār*), and the seeds cling to the garments of passers-by (Marzell, s.v. *Agrimonia eupatoria* L.); hence OE *clīfe* ‘burdock’ < *clīfan, clifian* ‘(to) cleave’. ASSOCIATIVE RELATIONS: metaphorical similarity & taxonomic sub-ordination.

▫ *→clīfe, →gār*

gār-lēac, lit. ‘spear-leek’
‘Allium sativum L. / GARLIC / Knoblauch’
DOE (s.v. *gār-lēac*): c. 35 occ.[e]
LATIN EQUIV. in the manuscripts: *alium*.

ETYMOLOGY: OE formation; according to Carr (1939: 32), OE *gārlēac* was borrowed into Old Norse as *geirlaukr*; MORPHOLOGY: N+N; endocentric, rectional; partial metaphor. MOTIVATION: shape & botanical information. This kind of *lēac* 'leek' has leaves which are spear-shaped (Marzell, s.v. *Allium sativum* L.). ASSOCIATIVE RELATIONS: metaphorical similarity & taxonomic sub-ordination.

▫ →*gār*, →*lēac*

gas-, ?
unidentified
Not an entry in the *DOE*.

▫ →*gasdie*

gas-die (= ? *glæs-die*), lit. 'glass-?'
'? Parietaria officinalis L. / upright pellitory / Glaskraut'
Not in the *DOE*; hapax legomenon (Varnhagen 1902: 13). Bierbaumer (3, s.v. *gasdie*) suggests emendation of *gas-* to *glæs-* 'glass'.
LATIN EQUIV. in the manuscripts: *vitreolus*
ETYMOLOGY: OE formation; MORPHOLOGY: Because of the uncertainty as regards the identification of the elements (see above), it is unanalysable and should, perhaps, be regarded as simplex. MOTIVATION: ? colour & ?. When we assume that *gas-* is a miswritten form for *glæs* 'glass' (see above), the fact that Parietaria officinalis has leaves which are shiny like glass (Marzell, s.v. *Parietaria officinalis* L.), would explain the use of this element. ASSOCIATIVE RELATIONS: metaphorical similarity & ?.

▫ →*-die*, →*gas-*

gāt, noun
(i) 'GOAT / Geiß'; (ii) element in place names
DOE (s.v. *gāt*): c. 100 occ.

▫ →*gātetrēow*

gāte-trēow, lit. 'goat's tree'
'Cornus sanguinea L. / common dogwood, dogberry, gatter / Roter Hartriegel'
DOE (s.v. *gāte-trēow*): 1 occ. (*Lǣcebōc*, 27/11).
ETYMOLOGY: OE formation; MORPHOLOGY: N_{gen}+N; endocentric, rectional. MOTIVATION: food & botanical information. The leaves of the plant are eaten by goats (Marzell, s.v. *Cornus sanguinea* L.). ASSOCIATIVE RELATIONS: conceptual contiguity & taxonomic sub-ordination.

▫ →*gāt*, →*trēow*

gærs, noun
'GRASS / Grass'
DOE (s.v. *gærs, græs*): c. 100 occ.
ETYMOLOGY: < Gmc **grasa-* (*Kluge*, s.v. *Gras*).

▫ →*gærswyrt*

▫ →*swīnesgærs*

gærs-wyrt, lit. 'grass-plant'
'grass, herb, hay / Gras, Kraut, Heu'
DOE (s.v. *gærs-wyrt*): 1 occ., attested in *Arundel-Psalter* (36, 2) as "wyrta gærswyrta". Bierbaumer (3, s.v. *gærs-wyrt*) questions the existence of this compound, suggesting a double glossing, *gærs* and *wyrta*. The *DOE* gives it as a headword, i.e. regards it as a compound.
ETYMOLOGY: OE formation; MORPHOLOGY: N+N; endocentric, copulative: subsumptive (pleonastic). MOTIVATION:

plant & botanical information. If it is a compound, the plant's name expresses that it is a kind of plant which is, more specifically, grass. This, however seems rather unlikely, as *gærs* is of Germanic origin and was in all likelyhood a well-known plant (c. 100 occurences according to the *DOE*, s.v. *gærs*), and would not need an explanatory second element ASSOCIATIVE RELATIONS: conceptual identity & taxonomic sub-ordination.

▫ →*gærs*, →*wyrt*

gēac, noun

(i) 'cuckoo / Kuckuck'; (ii) element in place names

DOE (s.v. *gēac*): c. 35 occ.

▫ →*gēaces sūre*

gēaces sūre, lit. 'cuckoo-sorrel'

'Oxalis acetosella L. / wood-sorrel, shamrock / Hasenklee, Kuckucksklee'

Not an entry on its own in the *DOE*, but s.v. *gēac*, 3; 20 occ.[e] according to the *DOEPN* (s.v. *gēaces sūre*).

LATIN EQUIV. and OE synonyms in the manuscripts: *acidula*, *calciculium*, *trifolium* (*silvaticum*); *þrilēfe*.

ETYMOLOGY: OE formation; MORPHOLOGY: N_{gen}+N; endocentric, rectional; partial metaphor. MOTIVATION: time & botanical information. "Die Bezeichnung 'Kuckucks(sauer)ampfer bezieht sich in ihrem ersten Teil auf die frühe Blütezeit, da der Kuckuck als Frühlingskünder schlechthin gilt [...], während der zweite Teil (s.auch [sic!] sūre, s.v.) auf den sauren Geschmack der Blätter Bezug nimmt" ['Cuckoo-sorrel relates with its first element to its early blooming period, as the cuckoo is considered to be THE spring herald, while the second part [...] refers to the sour taste of the leaves'] (Bierbaumer 1, s.v. *gēaces sūre*). Cf. Marzell (s.v. *Oxalis acetosella* L). ASSOCIATIVE RELATIONS: metaphorical similarity & taxonomic sub-ordination.

▫ →*gēac*, →*sūre*

gealla, noun

'GALL / Galle'

DOE (s.v. *gealla*): c. 130 occ.

▫ →*eorþgealla*, →*meargealla*, →*ramgealla*

geard, noun

'YARD, enclosure / Garten, Umzäunung'

DOE (s.v. *geard*): c. 50 occ.

▫ →*wīngeard*

gearwe, noun

'Achillea millefolium L. / YARROW / Schafgarbe'

DOE (s.v. *gearwe* noun[2]): 55 occ.; homonym with *gearwe* 'clothing, apparel' attested 6 times (cf. *DOE*, s.v. *gearwe* 1).

ILLUSTRATION in manuscript London, British Library, Cotton Vitellius C. iii, fol. 46r (Ohlgren 168(53)).

LATIN EQUIV. in the manuscripts: *achillea*, *centifolium*, *millefolium*, *myriophyllon*.

ETYMOLOGY: Of uncertain etymology, but of Germanic origin (*OED*, s.v. *yarrow*; *Kluge*, s.v. *Garbe*).

▫ →*rēade gearwe*

gecyrnlod, adjective

'granulated / gekörnt, granuliert'

ETYMOLOGY: derived from *corn* 'CORN'.

DOE (s.v. *ge·cyrnled*, *ge·cyrnlod*): 7 occ.

LATIN EQUIV. in the manuscripts: *granatus*.
▫ →*gecyrnlod æppel*

gecyrnlod æppel, lit. 'apple (or fruit) full of seeds'
'fruit of Punica granatum L. / pomegranate / Granatapfel'
DOEC: 6 occ.
LATIN EQUIV. in the manuscripts: *malum granatum*.
ETYMOLOGY: Independent formation in WGmc; OHG *gicurnit ephele* (Björkman 1901: 210); loan translation of *mālum grānātum*. MORPHOLOGY: Adj+N; endocentric, copulative: attributive. MOTIVATION: part of plant & botanical information. The first element refers to the juicy, dark red seeds of the pomegranate, which is appreciated not because of its flesh, but because of its seeds; the second element indicates that it is a kind of apple, or a kind of fruit. ASSOCIATIVE RELATIONS: conceptual contiguity (meronymy) & taxonomic sub-ordination.
▫ →*æppel*, →*gecyrnlod*

gelod(u), noun
'spine / Rückgrat'
ETYMOLOGY: ? derived from OE *leodan* 'to spring up, grow' (*Holthausen*, s.v. *gelodwyrt*, *gelodu*).
▫ →*gelodwyrt*

gelod-wyrt, lit. '? spinewort'
'Potentilla anserina L. / silverweed / Gänserich, Silberkraut'
Not in the *DOE*; 15 occ.[e] (*DOEPN*, s.v. *gelod-wyrt*).
LATIN EQUIV. in the manuscripts: *heptaphyllon*, *plantago*, *vii folia*.
ETYMOLOGY: WGmc compound; MHG *lodwurz*, MLG *lotwort* (Carr 1939: 101); MORPHOLOGY: N+N; endocentric, rectional; partial metaphor. MOTIVATION: shape & botanical information. Following *Holthausen* (s.v. *gelodwyrt*), Bierbaumer (1, s.v. *gelodwyrt*) suggests that the first element of the plant name is *(ge)lodr* 'backbone, spine', possibly derived from the verb *lēodan* 'to grow'. This assumption is based on the eye-catching back of the leaves: the leaf veins are arranged in the form of a ladder (Marzell, s.v. *Potentilla anserina* L., 4). ASSOCIATIVE RELATIONS: metaphorical similarity & taxonomic sub-ordination.
▫ →*gelodu*, →*wyrt*

georman-, ?
unidentified
Not attested independently.
▫ →*geormanlēaf*

georman-lēaf (*eorme-*, *gearwan-*, *-letic*), lit. '? leaf'
'Malva silvestris L. / common mallow, high mallow / Wilde Malve'
DOE (s.v. *georman-lēaf*): 16 occ.[e]
LATIN EQUIV. in the manuscripts: *malva* (2x with *erratica*).
ETYMOLOGY: OE formation; MORPHOLOGY: unanalysable formation; ? + noun; exocentric, metonymic compound. MOTIVATION: ? & part of plant. The first element is unclear and several, rather unsatisfactory, explanations have been put forward. One proposal is to connect the plant name to Gmc **germana* 'spike', another one to connect the first element to the root of G *gären* 'to make slimy, to ferment' (Pheifer 1974: 103, no. 656). In any

case, the leaf – a part of the plant – stands for the whole. ASSOCIATIVE RELATIONS: ? & conceptual contiguity (meronymy).

▫ →*georman-*, →*lēaf*

gescād-, ? noun
'? information / Bescheid'
Not attested independently.
ETYMOLOGY: ? derived from *sceadan* '(to) divide, separate, etc.'

▫ →*gescādwyrt*

gescād-wyrt, lit. '? informing plant'
'Verbascum thapsus L. / great mullein / Kleinblumige Königskerze'
DOEPN (s.v. *gescādwyrt*): 7 occ.[e]
LATIN EQUIV. in the manuscripts: ? *talumbus*, *verbascum*.
ETYMOLOGY: OE formation; MORPHOLOGY: N+N; endocentric, rectional; *gescād-* is not attested independently. MOTIVATION: tool & botanical information. As Verbascum thapsus played a major role in Anglo-Saxon oracles, Bierbaumer (1, s.v. *gescādwyrt*) equates the first element, *gescad-*, semantically with G *Bescheid* 'advice, information': 'a plant which informs'; ASSOCIATIVE RELATIONS: conceptual contiguity & taxonomic sub-ordination.

▫ →*gescād-*, →*wyrt*

giþ-, noun
'Agrostemma githago L. / corn-cockle / Gewöhnliche Kornrade'
Not attested independently.
ETYMOLOGY: < L *gith, git, gicti, githago* (Bierbaumer 1, s.v. *giþcorn*).

▫ →*giþcorn*, →*giþrīfe*

giþ-corn (*gīþ-*), lit. '*gith*-corn'
(i) 'Euphorbia lathyris L. / caper-spurge / Spring-Wolfsmilch', or (ii) 'Daphne laureola L. / spurge laurel / Lorbeer-Seidelbast'
DOE (s.v. *giþ-corn*): 15 occ.
ILLUSTRATIONS in manuscripts London, British Library, Cotton Vitellius C. iii, fol. 53r (Ohlgren 168(66)), and Oxford, Bodleian Library, Ashmole 1431 (S.C. 7523), fol. 27v (Ohlgren 221(102)).
LATIN EQUIV. in the manuscripts: *amygdaloides*, *citocatia*, *herba munda* (? for *mundatoria*), *lacterida*, *lacteris*.
ETYMOLOGY: OE formation; hybrid compound (OE *giþ-* < L *gith, git, gicti, githago*; see →*giþ-*). MORPHOLOGY: N+N; exocentric, metonymic compound. MOTIVATION: plant & part of plant. *Giþ-* denotes the corn-cockle (Agrostemma githago L.), but the plant obviously referred to in the Old English texts must be either the caper-spurge or the spurge laurel (*DOE*, s.v. *giþ-corn*, Bierbaumer 1, 2, 3, s.v. *giþcorn*); the motivation of the first element is, therefore, unclear. All three plants differ clearly in their appearance. The second element refers to the small, corn-like fruits both caper-spurge and spurge-laurel have (see also →*lybcorn*; →*springwyrt*). ASSOCIATIVE RELATIONS: ? & conceptual contiguity (meronymy).

▫ →*corn* 1, →*giþ-*

giþ-rīfe (*-hrōfe*), lit. '*gith*-noxious (plant)'
'? Agrostemma githago L. / corn-cockle / Kornrade'
DOE (s.v. *giþ-rife*): 35 occ.
LATIN EQUIV. in the manuscripts: *laser*.
ETYMOLOGY: OE formation; hybrid compound (OE *giþ-* < L *gith, git, gicti,*

githago; cf. also →*giþ*- and *giþcorn*). According to Bierbaumer (2, s.v. *giþrife*) *giþ-hrōfe* (attested, e.g., in *Lācnunga*, 210) is a form of *giþrīfe*, changed by folk etymology owing to the phonological similarity of *-hrōfe* (e.g. in *wuduhrōfe*) and *-rīfe*. MORPHOLOGY: $N+N_{deadj}$; exocentric, metonymic compound. *giþ-* is not attested independently; MOTIVATION: plant & pathogen. The second element, *-rīfe*, alludes, according to *Holthausen* (s.v. *giþrife*), to the lacerating barbs of the plant. Since the corn-cockle has no barbs I rather believe that the plant name refers to the high toxicity of the seeds of the corn-cockle: the Old English adjective *rīf* not only means 'wild, violent', but also 'noxious' (*ClH*, s.v. *rīf*). ASSOCIATIVE RELATIONS: conceptual identity & conceptual contiguity.

▫ →*giþ-*, →*rīfe*

glōf, *glōfa*, *glōfe*, noun
'GLOVE / Handschuh'
DOE (s.v. *glōf*, *glōfe*, *glōfa*): c. 30 occ.

▫ →*glōfwyrt*
▫ →*foxesglōfa*

glōf-wyrt, lit. 'glove-wort'
'Convallaria majalis L. / lily of the valley / Maiglöckchen'; or (ii) 'Digitalis purpurea L. / foxglove / Roter Fingerhut'
DOE (s.v. *glōf-wyrt*): 10 occ.
ILLUSTRATIONS in manuscripts London, British Library, Cotton Vitellius C. iii, fol. 29r (Ohlgren 168(19)), and Oxford, Bodleian Library, Ashmole 1431 (S.C. 7523), fol. 11r (Ohlgren 221(22)).
LATIN EQUIV. in the manuscripts: *apollinaris*.
ETYMOLOGY: OE formation; MORPHOLOGY: N+N; endocentric, rectional; partial metaphor. MOTIVATION: shape & botanical information. Literally 'glove-wort', referring to the glove-shaped flowers both the 'lily of the valley' and the 'foxglove' have; perhaps influenced by L *manicon* – the synonym of *apollinaris* (one of the Latin lemmas in the Old English glossaries) in the Latin *Herbarium Apuleii*; see Bierbaumer 3, s.v. *glōfwyrt*) – which was confused with *manica* 'glove'. ASSOCIATIVE RELATIONS: metaphorical similarity & taxonomic sub-ordination.

▫ →*glōf*, →*wyrt*

gōd, adjective
'GOOD / gut'
DOE (s.v. *gōd* adj.): c. 2500 occ.

▫ →*gōdæppel*

gōd-æppel, lit. 'good fruit' or 'good apple'
'fruit of Cydonia oblonga Mill. / quince / Quitte'
DOE (s.v. *gōd-æppel*): 2 occ.[e] (*Corpus Glossary*, no. 477; *Cleopatra Glossary* (Stryker), 88, no. 92).
LATIN EQUIV. in the manuscripts: *cydonium*.
ETYMOLOGY: OE formation; folk etymology; see MOTIVATION. MORPHOLOGY: Adj+N; endocentric, copulative: attributive. MOTIVATION: secondary, phonological and semantic motivation for the first element: sound pattern & taste; second element: botanical information. This plant name is the result of the folk etymological alteration of *coddæppel* lit. 'husk-fruit' in analogy to OE *gōd* 'good' (Bierbaumer 3, s.v. *gōd-æppel*), based on the

phonological similarity of OE *codd* 'husk' and OE *gōd* 'good', possibly also supported by the semantic motivation of quinces tasting *gōd* 'good'. ASSOCIATIVE RELATIONS: phonological similarity & taxonomic sub-ordination.
▫ →*æppel*, →*gōd*

gold, noun
'GOLD / Gold'
DOE (s.v. *gold*): c. 725 occ.
▫ →*goldblōma*, →*goldwyrt*

gold-blōma, lit. 'gold- ? blossom'
(i) 'gold-bloom, mass of gold'; ? (ii) 'Calendula officinalis L. / marigold / Ringelblume'
DOE (s.v. *gold-blōma*): 2 occ., however, not as a plant name, but in a homily, where it denotes 'gold-bloom', as an epithet for Christ; or 'mass of gold' (see also below).
ETYMOLOGY: Independent formation in Gmc; OHG *goldblômo*, MDu *goutbloeme*, Sw *guldblomma*, Dan *guldblomme* (Carr 1939: 72); MORPHOLOGY: N+N; exocentric, metonymic compound; partial metaphor. MOTIVATION: colour & part of plant. OE *goldblōma* is not attested with the sense 'marigold', but cf. OE *golde* glossing *solsequia* in *Brussels Glossary* (301/5) and *goldwyrt*, glossing *solsequium* in Napier (1900: 194/36), which both denote 'marigold' (*DOE*, s.vv. *golde*, *gold-wyrt*; see also *DOE*, s.v. *blōma*, and *OED*, s.v. *marigold*). Assuming the sense 'marigold' for *goldblōma*, the plant name refers to the golden or yellow flowers; cf. also ModE *marigold*, G *Goldblume*, etc. (see ETYMOLOGY above; Marzell, s.v. *Calendula officinalis* L.). ASSOCIATIVE RELATIONS: metaphorical similarity & conceptual contiguity (meronymy).
▫ →*blōma*, →*gold*

gold-wyrt, lit. 'gold-wort'
'Calendula officinalis L. / marigold / Ringelblume'
DOE (s.v. *gold-wyrt*): 1 occ. (Napier 1900: 194/36).
LATIN EQUIV. in the manuscripts: *solsequium*.
ETYMOLOGY: WGmc compound; OHG *goldwurz*, MLG *goltwort*, MDu *gouldwortel* (Carr 1939: 99); MORPHOLOGY: N+N; endocentric, rectional; partial metaphor. MOTIVATION: colour & botanical information. The first element refers to the golden colour of the plant's blossoms (Marzell, s.v. *Calendula officinalis* L.; cf. also ModE *marigold*). ASSOCIATIVE RELATIONS: metaphorical similarity & taxonomic sub-ordination.
▫ →*gold*, →*wyrt*

gorst, noun
(i) 'Juniperus communis L. / juniper / Gewöhnlicher Wacholder'; (ii) 'Ulex europaeus L. / GORSE / Gewöhnlicher Stechginster'
DOE (s.v. *gorst*): 19 occ.
ILLUSTRATION in manuscript London, British Library, Cotton Vitellius C. iii, fol. 61v (Ohlgren 168(83)), and Oxford, Bodleian Library, Ashmole 1431 (S.C. 7523), fol. 35v (Ohlgren 221(148)).
LATIN EQUIV. in the manuscripts: *ac(c)idinetum*, *herba iras*, *juniperus*, *rhamnus*, *rubus*, *sentis*, *tribulosus*, *tribulus*, *voluma*.
ETYMOLOGY: < IE *g^h*rz-do* (*OED*, s.v. *gorse*).
▫ →*gorstbēam*

gorst-bēam, lit. 'gorse-tree'
'Ulex europaeus L. / gorse, furze-bush / Stechginster'
DOE (s.v. *gorst-bēam*): 1 occ. (*Gospels*: *Mark* 12, 26); not in Bierbaumer (1, 2, 3).
ETYMOLOGY: OE formation; MORPHOLOGY: N+N; endocentric, copulative, subsumptive (pleonastic). MOTIVATION: plant & botanical information. The second element serves as a clarifying element: the gorse is a large shrub; it can grow c. 6.6 ft. (2 m) high, and resembles, then, a tree (cf. ModE *gorse-tree*, *OED*, s.v. *gorse*). ASSOCIATIVE RELATIONS: conceptual identity & taxonomic sub-ordination.
▫ *→bēam*, *→gorst*

gōt-, ?
unidentified
Not attested independently.
▫ *→gōtwōþe*

gōt-wōþe, lit. '? shedding of songs'
unidentified plant; "plant-name of uncertain meaning and etymology; by folk etymology 'goatweed' and 'goutweed' have been suggested" (*DOE*, s.v. *gōt-woþe*).
DOE (s.v. *gōt-woþe*): 9 occ.
ETYMOLOGY: OE formation; MORPHOLOGY: N+N; structure completely unclear as the meaning is uncertain (see above). MOTIVATION and ASSOCIATIVE RELATIONS: unclear.
▫ *→gōt*, *→wōþe*

grēat, adjective
'GREAT, coarse / groß, grob'
DOE (s.v. *grēat*): c. 150 occ.
▫ *→grēate bānwyrt*, *→grēate net(e)l(e)*, *→grēate wyrt*

grēate bānwyrt, lit. 'tall bone-wort'
'Symphytum officinale L. / common comfrey, consound / Große Wallwurz, Beinwell'
Not an entry on its own in the *DOE*, but subsumed s.v. *grēat*, 2.e.; hapax legomenon (*Lǣcebōc*, 96/10).
ETYMOLOGY: OE formation; MORPHOLOGY: Adj+N; endocentric, copulative: attributive. MOTIVATION: size & botanical information. The first element is used presumably to draw attention to a particular size or thickness of the plant: this particular species is bigger than the *bānwyrt*; cf. Bierbaumer (1, s.v. *bānwyrt*). ASSOCIATIVE RELATIONS: conceptual contiguity & taxonomic sub-ordination.
▫ *→bānwyrt*, *→grēat*

grēate net(e)l(e), lit. 'tall nettle'
'Urtica dioica L. / stinging nettle, GREAT NETTLE / Große Brennnessel'
Not an entry on its own in the *DOE*, but subsumed s.v. *grēat*, 2.e.; hapax legomenon (*Lǣcebōc*, 27/15).
ETYMOLOGY: OE formation; MORPHOLOGY: Adj+N; endocentric, copulative: attributive. MOTIVATION: size & botanical information. The premodifying adjective in the noun phrase *grēate net(e)l(e)* marks the distinctive feature in comparison to the smaller species of the genus Urtica L., namely, *smāle net(e)l(e)* 'Urtica urens L., small nettle'. ASSOCIATIVE RELATIONS: conceptual contiguity & taxonomic sub-ordination.
▫ *→grēat*, *→net(e)l(e)*

grēate wyrt, lit. 'tall wort'
(i) 'Colchium autumnale L. / meadow saffron / Herbstzeitlose'; (ii) 'Primula

veris L. / cowslip, paigle / Frühlingsschlüsselblume, Arzneiprimel'
Not an entry on its own in the *DOE*, but subsumed s.v. *grēat*, 2.e.i; c. 7 occ.
ILLUSTRATION in manuscript London, British Library, Cotton Vitellius C. iii, fol. 29r (Ohlgren 168(19)).
LATIN EQUIV. in the manuscripts: *hierobulbum*.
ETYMOLOGY: OE formation; MORPHOLOGY: Adj+N; endocentric, copulative: attributive. MOTIVATION: size & botanical information. According to Bierbaumer (1, s.v. *grēate wyrt*), the name does not refer to the size of the plant but rather to its bulbous roots, as the plant name renders *hierobulbum* lit. 'sacred bulb' in the *Durham Plant Name Glossary* (no. 198). ASSOCIATIVE RELATIONS: conceptual contiguity & taxonomic sub-ordination.
▫ →*grēat*, →*wyrt*

grēne, adjective
(i) 'GREEN / grün', (ii) flourishing / blühend, florierend'; (iii) 'living, fresh / lebend, lebendig'; (iv) 'not prepared for use / unfertig, ungar'
DOE (s.v. *grēne*): c. 275 occ.
▫ →*grēne helde*
▫ →*singrēne*

grēne helde, lit. 'green tansy'
'? Potentilla L. / cinquefoil / Fingerkraut, Gänsekraut'
Not an entry in the *DOE*; 2 occ. (*Peri Didaxeon*, 39/1, 39/7).
ETYMOLOGY: OE formation; MORPHOLOGY: Adj+N; endocentric, copulative: attributive. MOTIVATION: colour & botanical information. *Helde* is the Old English name of Tanacetum vulgare L., the tansy. Tanacetum vulgare and Potentilla anserina share the same medieval Latin names: *tanacetum*, *tanacetum anserina*, etc. (Marzell, s.v. *Tanacetum vulgare* L., s.v. *Potentilla anserina* L.), so that, probably, *grēne helde* 'Potentilla' was seen as a kind of *helde* 'Tanacetum'. A German name for Potentilla anserina is *Wilder Rainfarn*; it corresponds to ModE *wild tansy* and refers, according to Marzell (s.v. *Potentilla anserina* L.), to the shape and form of the leaves of the two plants, which closely resemble one another. ASSOCIATIVE RELATIONS: conceptual contiguity & taxonomic sub-ordination.
▫ →*grēne*, →*helde*

grund, Boden
'GROUND, bottom / Grund'
DOE (s.v. *grund*): c. 300 occ.
▫ →*grundeswelige*

grunde-swelige (*-swylige*), lit. 'ground-devouring (plant)'
'Senecio vulgaris L. / (common) GROUNDSEL / Gemeines Kreuzkraut'
DOE (s.v. *gunde-swelge*, *grunde-swelge*, *grund-swelge*): c. 40 occ.[e]
ILLUSTRATION in manuscript London, British Library, Cotton Vitellius C. iii, fol. 42v (Ohlgren 168(46)).
LATIN EQUIV. in the manuscripts: *hydrogeron*, *senecio*.
ETYMOLOGY: OE formation; folk etymological processes (see below). MORPHOLOGY: synthetic, N+$N_{\text{deverbal}(\textit{-e})}$; endocentric, rectional. MOTIVATION: (a) illness & action; The original name was probably *gundeswelige* (Sauer 2000: 30): since the medical plant is used against pus (OE *gunde*), the original motivation was probably 'pus-

devourer' (attested in the *Épinal-Erfurt Glossary*, no. 976); (b) secondary, phonological and semantic motivation of the first element: sound pattern & habitat; second element: action. Then, the name was remotivated to 'ground-devourer' because of the phonological similarity of OE *gunde* 'pus' and OE *grund* 'ground', an interpretation that is supported by the fact that Senecio vulgaris L. is one of the most widespread weeds (Bierbaumer 1, s.v. *grundeswelge*). ASSOCIATIVE RELATIONS: phonological similarity, supported by conceptual contiguity.
▫ →*grund*, →*sweligean*

gyr, noun
'Abies L. / fir / Tanne'
▫ →*gyrtrēow*

gyr-trēow, lit. 'fir-tree'
'Abies L. / fir / Tanne'
DOE (s.v. *gyr-trēow*): 1 occ. (*Antwerp-London Glossaries*, 122/17).
LATIN EQUIV. and OE synonym in the manuscript: *abies*; *sæppe*.
ETYMOLOGY: OE formation; MORPHOLOGY: N+N; endocentric, copulative, subsumptive (pleonastic). MOTIVATION: plant & botanical information. ASSOCIATIVE RELATIONS: conceptual identity & taxonomic sub-ordination.
▫ →*gyr*, →*trēow*

H

hafoc, noun
'HAWK / Falke'
▫ →*hafocwyrt*

hafoc-wyrt, lit. 'hawk's-wort'
'? Hieracium L. / hawkweed / Habichtskraut'; (ii) '? Lactuca serriola L. / prickly lettuce / Stachellattich'; (iii) '?? Matricaria chamomilla L. / wild camomile / Echte Kamille'; (iv) 'Anthemis nobilis L. / roman camomile / Römische Kamille'
DOEPN (s.v. *hafoc-wyrt*): 2 occ. (*Lǣcebōc*, 18/14; and Meritt 1945: 73, Dc 187/4).
LATIN EQUIV. in the manuscripts: ? *camimula*.
ETYMOLOGY: OE formation; probably loan formation modelled on Gk *ιεράκιον* (see below); MORPHOLOGY: N+N; endocentric, rectional; MOTIVATION: habitat & botanical information; According to Marzell (s.v. *Hieracium* L.), L *hierācium* goes back to Gk *ἱεακία*, *ἱεράκιον*, which is a derivative from Gk *ἱέραξ* 'hawk' and thus corresponds to OE *hafoc* (cf. also *Genaust*, s.v. *Hieracium*). The use of 'hawk' can be explained in two ways: it either refers to the preferred habitat of the plant, which grows on rocks so high that only hawks can access them – i.e. a figurative (metonymic) use of *hafoc* – or to the myth that hawks sharpen their eyes with the milky sap of the plant (Marzell, s.v. *Hieracium L.*). For the identification of *hafocwyrt* as Lactuca serriola or as Matricaria chamomilla, no plausible explanation can be found. ASSOCIATIVE RELATIONS: conceptual contiguity & taxonomic sub-ordination.
▫ →*hafoc*, →*wyrt*

haga, *hagu*, *hege*, noun
'HEDGE / Hecke'
▫ →*haguþorn*, →*hegeclīfe*, →*hegehymele*, →*hegerīfe*

hagu-þorn (*hæg-*), lit. 'hedge-thorn-bush'
'Crataegus oxyacantha L. / common HAWTHORN / Stumpfgelappter Weißdorn'
DOEPN (s.v. *hagu-þorn*): 16 occ.[e]
LATIN EQUIV. in the manuscripts: *spina*, *spina alba*, *tribulus*.
ETYMOLOGY: Gmc compound; ON *hagþorn*, OHG *hagedorn*, OLG *haginthorn*, MDu *hagedorn*, Fris *hagedoarn* (Carr 1939: 48); < Gmc **haga-þurna* (*Kluge*, s.v. *Hagedorn*). MORPHOLOGY: N+N; endocentric, rectional. MOTIVATION: tool & botanical information; The first element *haga*, *hæg* 'hedge, raw' indicates that the hawthorn, a thorny bush (hence the second element *-þorn* 'thorn, thorn-bush'), was often used to enclose land or property (cf. also *Kluge*, s.v. *Hagedorn*: "Dornstrauch für Hecken" ['thorn-bush for hedges']). ASSOCIATIVE RELATIONS: conceptual contiguity & taxonomic sub-ordination.
▫ →*hagu*, →*þorn*

hālig, adjective
'HOLY / heilig'
▫ →*holihocce*

hals, noun
'throat / Hals'
▫ →*halswyrt*

hals-wyrt, lit. 'throat-wort'
(i) '? Campanula trachelium L. / throatwort, bats-in-the-balfry / Nessel-Glockenblume'; (ii) '? Narcissus poeticus L. / pheasant's eye narcissus / Weiße Narzisse'; (iii) '? Symphytum officinale L. / comfrey / Beinwell'
DOEPN (s.v. *hals-wyrt*): 22 occ.
ILLUSTRATIONS in manuscripts London, British Library, Cotton Vitellius C. iii, fol. 37v (Ohlgren 168(36)), fol. 56r (Ohlgren 168(72)), and Oxford, Bodleian Library, Ashmole 1431 (S.C. 7523), fol. 16r (Ohlgren 221(46)).
LATIN EQUIV. in the manuscripts: *aurisfolia*, *auris leporis*, *bulbus*, *caepa canina*, *epicurium*, ? *gahell*, *narcissus*, *scillodes* (? = *scilla*), *symphytum album*.
ETYMOLOGY: OE formation; MORPHOLOGY: N+N; endocentric, rectional. MOTIVATION: illness & botanical information. Bierbaumer (2 and 3, s.vv. *halswyrt*, *healswyrt*) suggests that the first element of the plant name, *hals* 'throat' is motivated by the long and round flowers of Campanula trachelis, which resemble a throat. This interpretation is supported by Modern English and German vernacular plant names, which verbalise the concept THROAT in G *Halskraut*, *Halswurtzel*, ModE *throatwort*, Du *halskruid*; see Marzell (s.v. *Campanula trachelium* L.); these plant names also seem to relate to the healing virtues of the plant: "Der aus der Pflanze bereitete Absud diente als Gurgelmittel bei Halskrankheiten" ['The abstract made from the plant served as a gargle for sore throats']. The plant name, probably, shows loan influence, as the first element of L *trachelium maius* – the vernacular name of Campanula trachelium (cf. Marzell, s.v. *Campanula trachelium* L.) – derives from L *trachia* 'throat' (ultimately a loan from Greek). The motivation for the identification of *halswyrt* as Narcissus poeticus L. and Symphytum officinale L. is not clear to me. Narcissus poeticus L. seems to have no healing virtues at all and

Symphytum officinale L. is used only for broken bones and as a haemostatic remedy (see Marzell, s.v. *Symphytum officinale* L.) and there is no outward similarity in appearance between the plants, either. ASSOCIATIVE RELATIONS: conceptual contiguity & taxonomic sub-ordination.

▫ →*hals*, →*wyrt*

hām, noun
'HOME, Heim'

▫ →*hāmwyrt*

hamor, noun
'HAMMER / Hammer'

▫ →*hamorsecg*, →*hamorwyrt*

▫ →*dȳþhamor*

hamor-secg, lit. '? hammer-sedge, ? rock-sedge'

(i) '? Carex hirta L. / goose-grass / Haar-Segge'; or (ii) '? Typha L. / reed-mace, locks / Rohrkolben'

DOEPN (s.v. *hamorsecg*): 2 occ. (*Lǣcebōc*, 23/34, 39/6).

ETYMOLOGY: OE formation; MORPHOLOGY: N+N; endocentric, rectional; partial metaphor. MOTIVATION: shape & botanical information. The relation between the plant and its name is rather unclear (Bierbaumer 1, s.v. *hamorsecg*). According to Bierbaumer (1, s.v. *hamorwyrt*), OE *hamor* denotes 'hammer' and 'rock' likewise, and *secg* can denote either 'sedge' or 'reed', so that the literal translation of the plant name is either 'hammer-sedge' / 'rock-sedge', or 'hammer-reed' / 'rock-reed'.[308] Since the association of the flowers of the reedmace with hammers is verbalised in the plant name OE *dȳþhamor* 'reedmace', too, I favour the explanation that *hamorsecg* is a kind of sedge with flowers that flare into a bulb-shaped form, i.e. they resemble hammers. ASSOCIATIVE RELATIONS: metaphorical similarity & taxonomic sub-ordination.

▫ →*hamor*, →*secg*

hamor-wyrt, lit. '? hammer-wort, ? Christ-hellebore-wort'

(i) 'Hellborus niger L. / Christ-hellebore / Nieswurz'; (ii) '? Veratrum album L. / white hellebore / Weißer Germer'; (iii) 'Parietaria officinalis L. / pellitory of the wall / Aufrechtes Glaskraut'

DOEPN (s.v. hamor-wyrt): 6 occ.

LATIN EQUIV. in the manuscripts: *perdicalis*.

ETYMOLOGY: OE formation; OE *hamor-* in *hamorwyrt* is probably a form of Gmc **hamiro* (< IE **kemero*) 'Christ-hellebore', which was changed to *hamor* in analogy to OE *hamor* 'hammer' (Bierbaumer 1, s.v. *hamor wyrt*), based on the phonological similarity between the two lexemes, yet without any secondary semantic motivation. MORPHOLOGY: N+N; endocentric, rectional. MOTIVATION: secondary, phonological motivation: sound pattern & botanical information. There is no semantic motivation to the first element *hamor-* (see ETYMOLOGY above). ASSOCIATIVE RELATIONS: phonological similarity & taxonomic sub-ordination.

▫ →*hamor*, →*wyrt*

308 The denotation of *hamor* given in *ClH* (s.v. *hamor*) is 'hammer'.

hām-wyrt, lit. 'home-wort'
'Sempervivum tectorum L. / house-leek, barren privet / Echte Hauswurz'
DOEPN (s.v. *hām-wyrt*): 8 occ.
ETYMOLOGY: Independent formation in WGmc (Carr 1939: 138); cf. OHG *heimwuurz*, etc. (Björkman 1902: 290); MORPHOLOGY: N+N; endocentric, rectional. MOTIVATION: habitat & botanical information. The plant name alludes to the belief that these plants, when planted on the roofs of houses, can protect them from storm damage, esp. from being struck by lightening; cf. the German names *Hauswurz*, *Dachwurzel*, *Donnerkraut* (Marzell, s.v. *Sempervivum tectorum* L.) and OE *þunorwyrt*. ASSOCIATIVE RELATIONS: conceptual contiguity & taxonomic sub-ordination.
▫ *→hām*, *→wyrt*

hār, adjective
'HOAR / weißlich, gräulich'
▫ *→hāra(n)sp(r)ecel*, *→hāranwyrt*, *→hār(e)-hūne*, *→hāreminte*
▫ *→wermōd se hāra*

hara, noun
'HARE / Hase'
▫ *→haranhige*

haran-hige, lit. 'hare's foot'
'Trifolium arvense L. / hare's foot clover / Hasenklee'
DOEPN (s.v. *haran-hige*): 6 occ.
ILLUSTRATIONS in manuscripts London, British Library, Cotton Vitellius C. iii, fol. 39r (Ohlgren 168(39)), and Oxford, Bodleian Library, Ashmole 1431 (S.C. 7523), fol. 17r (Ohlgren 221(51)).
LATIN EQUIV. in the manuscripts: *pes leporis*.
ETYMOLOGY: OE formation; probably a loan translation of L *pēs leporinum* or *leporis pēs* (see below). MORPHOLOGY: N_{gen}+N; exocentric, metaphorical compound; the second element, *-hige*, is not attested independently; MOTIVATION: shape or texture & shape. The OE plant name is probably a loan translation of L *leporis pēs* or *pēs leporinum*, which is motivated by the softness and the shape of the flowers of Trifolium arvense; cf. ON *hêrafôtr*, OHG *hasafuoz*, MLG *hasenvôt* (Carr 1939: 73); for similar names in other languages, see Marzell (s.v. *Trifolium arvense* L.). If we regard the OE plant name as a loan translation, the second element *-hige* must be analysed as a verb-derived noun referring to feet (< OE *higian* '(to) hurry'), which is not attested elsewhere in Anglo-Saxon texts, but cf. G *Lauf* 'foot of game', derived from the verb *laufen* '(to) run'. Lindheim (*Durham Plant Name Glossary*, no. 218) suggests a connection to OE *hī(e)g* 'hay', which is, however, rejected by Bierbaumer (2, s.v. *haran hige*). ASSOCIATIVE RELATION: metaphorical similarity.
▫ *→hara*, *→-hige*

hāra(n)-sp(r)ecel, lit. 'hoary / whitish speck'
'Echium vulgare L. / viper's grass, viper's bugloss / Blaue Ochsenzunge, Gemeiner Natternkopf'
DOEPN (s.v. *hāran-sprecel*): 14 occ.
LATIN EQUIV. in the manuscripts: *echios*.
ETYMOLOGY: OE formation; MORPHOLOGY: Adj+N; exocentric, metaphorical compound; The optional *-n-* was, orginally, an inflectional ending: "[ein] erstarrter Akkusativ einer *sē hāra

sprecel anzusetzenden Form" ['a frozen accusative of a form that must be assessed as *sē hāra sprecel*'] (Bierbaumer 1, s.v. *hāransprecel*). MOTIVATION: colour & part of plant. Bierbaumer (1, s.v. *hāran-sprecel*) connects the first element, *hāra(n)-*, to the whitish, hoary stems and leaves of Echium vulgare (Marzell, s.v. *Echium vulgare* L.); the second element, *sp(r)ecel* 'speck, spot', is perhaps in allusion to the numerous tiny flowers of Echium vulgare. ASSOCIATIVE RELATIONS: conceptual contiguity & metaphorical similarity.

▫ →*hār*, →*sp(r)ecel*

hār(e)-hūne, lit. 'hoary horehound'
'Marrubium vulgare L. / HOARHOUND, HOREHOUND / Weißer Andorn, Gemeiner Andorn'
DOEPN (s.v. *hare-hune*): 19 occ.
ILLUSTRATIONS in manuscript London, British Library, Cotton Vitellius C. iii, fol. 35v (Ohlgren 168(32)), and Oxford, Bodleian Library, Ashmole 1431 (S.C. 7523), fol. 13v (Ohlgren 221(35)).
LATIN EQUIV. in the manuscripts: *marrubium*, *prassium*.
ETYMOLOGY: OE formation; MORPHOLOGY: Adj+N; endocentric, copulative: attributive. MOTIVATION: colour & botanical information. The first element refers to the hoary (grey, old) colour of the pubescent stem; the second element denotes the plant OE *hūne* 'horehound', which resembles Marrubium vulgare L.; i.e. *hār(e)hūne* is a kind of *hūne* (Bierbaumer 1, s.v. *hāre hūne*). ASSOCIATIVE RELATIONS: conceptual contiguity & taxonomic sub-ordination.

▫ →*hār*, →*hūne*

hāre-minte (-*winta*), lit. 'hoary mint'
'? Mentha longifolia Huds. / wild mint / Roßminze'
DOEPN (s.v. *hāre-minte*): 4 occ.
LATIN EQUIV. in the manuscripts: *colocasia*.
ETYMOLOGY: OE formation; hybrid formation (OE *minte* < L *ment(h)a*; see →*minte*). MORPHOLOGY: Adj+N; endocentric, copulative: attributive. MOTIVATION: colour & botanical information. According to Marzell (s.v. *Mentha longifolia* Huds.), the first element indicates the colour of the plant's leaves: "Die Blätter [...] sind weiß- oder graufilzig behaart. Die Blüten stehen in langen, dichten, grau- bis weißfilzigen Scheinähren" ['The leaves are grey or white and hairy. The flowers grow from long, thick, felted whitish-grey spikes']; *hāreminte* is a kind of *minte* 'mint'. ASSOCIATIVE RELATIONS: conceptual contiguity & taxonomic sub-ordination.

▫ →*hār*, →*minte*

hāre-wyrt, hāran-wyrt, lit. 'hoary wort'
'Filago germanica L. / cudweed / Deutsches Fadenkraut, Schimmelkraut'
DOEPN (s.v. *hāre-wyrt*): 3 occ. (*Lǣcebōc*, 105/12, 47/5, 90/13).
ETYMOLOGY: OE formation; according to Bierbaumer (1, s.v. *hārewyrt*), not cognate with OHG *hasanwurz* lit. 'hare-wort', etc. (Björkman 1902: 289); MORPHOLOGY: Adj+N; endocentric, copulative: attributive. MOTIVATION: colour & botanical information. Bierbaumer rejects the literal translation 'hare's wort' on the basis of occurences such as "sio hare wyrt" (*Lǣcebōc*, 90/13) and "harewyrt" (*Lǣcebōc*, 105/12), where, he argues, the first element can only be the adj. *hār* 'hoary', i.e. 'grey-haired, greyish-

white' (Bierbaumer 1, s.v. *hārewyrt*). According to Carr (1939: 293), *hare-wyrt* is a form of *haranwyrt* lit. 'hare's wort' with loss of inflectional *-n-*. Both the compound (*hāre-wyrt*) and the syntactic group (*hāran-wyrt*) existed side by side. *ClH* (s.v. *hāranwyrt* = *hārewyrt*), Bierbaumer (1, s.v. *hāre-wyrt*) and the *TOE* (1: 114) take the plant name to be a compound. *BT* (s.v. ~) translates *haran-*, *hare-* as 'hare'. The first element alludes to the whitish and hairy stems and leaves of the plant Filago germanica (Bierbaumer 1, s.v. *hāranwyrt*). ASSOCIATIVE RELATIONS: conceptual contiguity & taxonomic sub-ordination.

▫ →*hār*, →*wyrt*

hārewinta →*hāreminte*

hǣlan, verb
'HEAL / heilen'

▫ →*selfhǣle*

-hǣleþe →*heolaþ*

hæsel, noun
'Corylus avellana L. / HAZEL / Gewöhnliche Hasel(nuss)'
DOEPN (s.v. *hæsel*): 20 occ.
LATIN EQUIV. in the manuscripts: *abellana*, *corylus*.
ETYMOLOGY: Gmc **hasalo-z* < preGmc **kósolos* (*OED*, s.v. *hazel*).

▫ →*hæselwyrt*
▫ →*hwīt hæsel*

hæsel-wyrt, lit. 'hazel-wort'
'Asarum europaeum L. / asarabacca, HAZELWORT / Haselwurz'
Hapax legomenon (Oxford, Bodleian Library, Ashmole 1431 (S.C. 7523), fol. 42v; Gough 1974: 279), glossing *purpurae violae folia*, "with some species of which hazelnut shares a purple flower and a similarly shaped leaf" (*DOE*, s.v. *hæselwyrt* [to appear][309]).
LATIN EQUIV. in the manuscripts: *viola*.
ETYMOLOGY: WGmc compound; cf. OHG *haselwurz*, etc. (Björkman 1902: 289); according to Bierbaumer (3, s.v. *hæselwyrt*), a folk etymological change of L *asarum* (< Gk ἄσαρον 'an undefined species') in analogy to OE *hæsel* 'hazel', based on the phonological similarity of the Latin and the Old English noun, and motivated semantically by the fact that hazelworts grow preferably under hazel bushes (see MOTIVATION).[310] MORPHOLOGY: N+N; endocentric, rectional; MOTIVATION: secondary, phonological and semantic motivation: sound pattern, habitat & botanical information; According to Marzell (s.v. *Asarum europaeum L.*), the hazelwort grows preferably under hazel bushes, i.e. 'this

---

309 This information was kindly made available by Prof. Dr. Helmut Gneuss.

310 *Genaust* (s.v. *Asarum*) argues that G *Haselwurz* lit. 'hazel-wort' is a loan rendition of L *beta leporina*, lit. 'hare's beetroot' – a hapax legomenon that refers to the beet-like roots of Asarum europaeum L. – and that the German term *Haselwurz* may further be motivated by the similarities of G *Hase* 'hare' and *Hasel* 'hazel'. As OE *hara* 'hare' shows the change of Gmc /z/ > WGmc /r/ (/z/ owing to the operation of Verner's Law; cf. Campbell 1959: §§ 398 (2, 3), 404), while G *Hase* does not (in accordance with Verner's Law; cf. Campbell 1959: § 398 (2)), the explanation is not an option for the Old English plant name.

is a kind of *wyrt* 'plant' that grows under hazel(bushes)'.

When 'hazelwort' is used to refer to the 'violet' (see above) because both have some species with purple flowers and similarly shaped leaves, it is a case of co-taxonomic similarity, which is based on the outward appearances of the two plants. ASSOCIATIVE RELATIONS: conceptual contiguity & taxonomic sub-ordination.

▫ →*hæsel*, →*wyrt*

hǣþ, noun

'Erica L. / HEATH / Heide, Heidekraut'

DOEPN (s.v. *hǣþ*): 17 occ.

LATIN EQUIV. in the manuscripts: ? *alucurus*, *brogus*, *calamaucus*, *edulium*, *myrica*, *thymum*.

ETYMOLOGY: WGmc **haiþjo* (*Kluge*, s.v. *Heide*).

▫ →*hǣþber(i)(g)e*

hǣþ-ber(i)(g)e, lit. 'heath-berry'

'Vaccinium myrtillus L. / whortleberry, blue berry / Blaubeere, Heidelbeere'

DOEPN (s.v. *hǣþ-berie*): 1 occ. (*Lǣcebōc*, 105/13).

ETYMOLOGY: WGmc compound; OHG *heidbere*, Du *heidebes*, Fris *heideber* (Carr 1939: 99); MORPHOLOGY: N+N; exocentric, metonymic compound; MOTIVATION: habitat & part of plant; The first element derives from the fact that the plant grows in woods and heathland; cf. OHG *heitperi*, MHG *heitber* (Marzell, s.v. *Vaccinium myrtillus* L., 1); the second refers to the black, aromatic berries (cf. also *Kluge*, s.v. *Heidelbeere*); ASSOCIATIVE RELATIONS: conceptual contiguity & conceptual contiguity (meronymy).

▫ →*ber(i)(g)e*, →*hǣþ*

hǣwen, adjective

'grey, blue, green, azure / grau, blau, grün, azurblau'[311]

▫ →*hǣwenh(n)ydele*

hǣwen-h(n)ydele, lit. 'grey, green, blue ?'

(i) 'Cochlearia officinalis L. / scurvy-grass / Echtes Löffelkraut'; (ii) 'Rumex hydrolapathum L. / great water-dock / Grind-Ampfer'

DOEPN (s.v. *hǣwen*): 8 occ.

LATIN EQUIV. and OE synonyms in the manuscripts: *britannica*; *witmǣres wyrt*.

ETYMOLOGY: OE formation; MORPHOLOGY: unanalysable formation; adjective + ?. MOTIVATION: colour & ?. The first element denotes a colour; it refers, perhaps, to the white flowers of Cochlearia officinalis, or to the greyish (sometimes reddish or greenish) flowers of Rumex hydrolapathum; the second element is not analysable as it is not attested elsewhere and not transparent in Old English either; it may be connected to Alem *nüttel* 'penis' or to OE *hnossian* '(to) beat' (*Holthausen*, s.v. *h(n)ydela*). ASSOCIATIVE RELATIONS: ? conceptual contiguity & ?.

▫ →*hǣwen*, →*-h(n)ydele*

hēafdeht, adjective

'having a head / mit Kopf'

▫ →*hēafdeht por(r)*

311 For a detailed investigation into Old English colour adjectives, among which is OE *hǣwen*, see Biggam (1997) and (1998).

hēafdeht por(r), lit. 'leek with umbel'
'Allium porrum L. / common leek / Porree'
1 occ. (*Lǣcebōc*, 96/20).
ETYMOLOGY: OE formation; hybrid formation (OE *por(r)* < L *porrum*; →*por(r)*). MORPHOLOGY: Adj+N; endocentric, copulative: attributive; *hēafdeht* is derived from OE *hēafod* 'head'. MOTIVATION: shape & botanical information. The common leek has an eye-catching umbel which is mounted head-like on a high stem, consists of many small (usually purple) flowers and is perfectly round (cf. →*croplēac*). ASSOCIATIVE RELATIONS: conceptual contiguity & taxonomic sub-ordination.
▫ →*hēafdeht*, →*por(r)*

hēafod, noun
'HEAD / Haupt'
▫ →*hundeshēafod*

hēah, adjective
'HIGH / hoch'
▫ →*hēahheoloþe*, →*hēahhymele*

hēah-heoloþe (= ? *eh-heoloþe*), lit. 'high elecampane'
'Inula helenium L. / elecampane / Echter Alant'
DOEPN (s.v. *hēah-heoloþe*): 6 occ.
ETYMOLOGY: OE formation; MORPHOLOGY: Adj+N; endocentric, copulative: attributive. MOTIVATION: size & botanical information. The *hēahheoloþe* is a tall species (Marzell, s.v. *Inula helenium* L.). ASSOCIATIVE RELATIONS: conceptual contiguity & taxonomic sub-ordination.
▫ →*hēah*, →*heoloþe*

hēah-hymele (= ? *hege-hymele*), lit. 'high hop'
'Bryonia L. / bryony / Zaunrübe'
Hapax legomenon (Meritt 1945: 59, no. 69/13).
LATIN EQUIV. in the manuscripts: *bryonia*.
ETYMOLOGY: OE formation; hybrid formation (OE *hymele* < MedL **humilus* < L *humulus*; see →*hymele*). MORPHOLOGY: Adj+N; endocentric, copulative: attributive. MOTIVATION: habitat & botanical information. OE *hēahhymele* looks like an Adj+N compound, the motivation being the height of the plant, expressed in the modifying adjective *hēah-* 'high'. Bryony can get as 'high' as c. 9.8 ft. (3 m) creeping upwards, but, as it is a creeping plant (Marzell, s.v. *Bryonia* L.), the length of the plant may not be its main characteristics; rather its habitat: the bryony grows mainly on hedges and fences. It is, therefore, also plausible to interpret *hēahhymele* not as 'high' *hymele* but as a form of *hege-hymele* 'hedge'-*hymele*, that is, with *hēah* changed from *hege* based on the phonological similarity of the two lexemes and supported by the semantic motivation that Bryony grows on hedges (cf. Bierbaumer 1, 2, 3, s.v. *hēahhymele*, *hegehymele*); cf. →*hege-hymele*. ASSOCIATIVE RELATIONS: conceptual contiguity & taxonomic sub-ordination.
▫ →*hēah*, →*hymele*

healf, numeral
'HALF / halb'
▫ →*healfwudu*

healf-wudu, lit. 'half-wood'
'? Satureja calamintha (L.) Schule / calamint, field-balm / Echte Bergminze, Bergmelisse'
Hapax legomenon (*Lǣcebōc*, 36/15).
ETYMOLOGY: OE formation; MORPHOLOGY: Num+N; exocentric, metonymic compound. MOTIVATION: number & texture. The Old English plant name accounts for the fact that Satureja calamintha L. is a shrub with a woody stalk (Bierbaumer 1, s.v. *healfwudu*), that is, the first element *healf* 'half' indicates that it is not 'real' wood the shrub's stalks are made of, they are only 'woody'. ASSOCIATIVE RELATION: conceptual contiguity.
▫ →*healf*, →*wudu*

heard, adjective
'HARD / hart'
▫ →*īsenhearde*

hege →*haga*

hege-clīfe, lit. 'hedge-clivers'
'Galium aparine L. / cleavers, goose-grass / Klebkraut'
DOE (s.v. *hege-clīfe* [to appear]): 5 occ.
ETYMOLOGY: OE formation; MORPHOLOGY: N+N; endocentric, rectional. MOTIVATION: habitat & botanical information. This kind of *clīfe* is a climbing plant which prefers growing along hedges, fences and bushes (Bierbaumer 1, s.v. *hege-clīfe*). ASSOCIATIVE RELATIONS: conceptual contiguity & taxonomic sub-ordination.
▫ →*clīfe*, →*hege*

hege-hymele, lit. 'hedge-hop'
(i) 'Humulus lupulus L. / hops / Hopfen'; (ii) '? Bryonia L. / bryony / Zaunrübe'
DOEPN (s.v. *hege-hymele*): 2 occ. (*Lācnunga*, 51, and *Brussels Glossary*, 302/5).
LATIN EQUIV. in the manuscripts: *humulus*.
ETYMOLOGY: OE formation; hybrid compound (OE *hymele* < VL **humilus* ~ L *humulus*; see →*hymele*). MORPHOLOGY: N+N; endocentric, rectional. MOTIVATION: habitat & botanical information. Both hops and bryony are climbing plants which can grow rather high: while bryony can get as long as c. 9.8 ft. (3 m), hops can reach even a height or length of c. 19.7 ft. (6 m) and more. They also resemble each other with regard to the shape of their leaves and they both prefer climbing on hedges, fences, bushes (Marzell, s.vv. *Humulus lupulus* L., *Bryonia* L.); the motivation as '*hymele* that grows on hedges and fences, etc.' is, therefore, most plausible. ASSOCIATIVE RELATIONS: conceptual contiguity & taxonomic sub-ordination.
▫ →*hege*, →*hymele*

hege-rīfe, lit. 'hedge-lacerating (plant)'
'Galium aparine L. / HAIRIF, HAYRIF, cleavers, goose-grass / Kleb-kraut, Labkraut'
c. 12 occ. (Bierbaumer 1 and 3, s.v. *hegerife*).
ETYMOLOGY: OE formation; MORPHOLOGY: N+N_{deadj}; exocentric, metonymic compound. MOTIVATION: habitat & 'effect'. OE *-rīfe*, derived from the adjective *rīf* 'wild, rapid, lacerating', alludes to the barbs of cleavers (*Holthausen*, s.v. *giþrife*); The plant Galium

aparine grows in fields, but also in hedges (OE *hege*), hence the use of the first element; it also grows along the borders of woods. ASSOCIATIVE RELATIONS: conceptual contiguity & conceptual contiguity.

▫ *→hege, →-rīfe*

helde, noun
'Tanacetum vulgare L. / tansy / Rainfarn, Wurmkraut'
DOEPN (s.v. *helde*): 15 occ.
LATIN EQUIV. in the manuscripts: *artemisia, artemisia ? tagante, ? tanacetum*.
ETYMOLOGY: ? IE **kel-* (Bierbaumer 1, s.v. *helde*).

▫ *→grēne helde*

henn, noun
'HEN /Henne'

▫ *→hennebelle, →hennewōl*

henne-belle, lit. 'hen's bell'
'Hyoscyamus niger L. / henbane / Schwarzes Bilsenkraut'
DOEPN (s.vv. *hænnebelle, hennebelle*): 11 occ.
ILLUSTRATION in manuscript London, British Library, Cotton Vitellius C. iii, fol. 23v (Ohlgren 168(8)).
LATIN EQUIV. in the manuscripts: ? *gingeralis, jusquiamus, symphoniacus*.
ETYMOLOGY: OE formation; MORPHOLOGY: N_{gen}+N; exocentric, metaphorical compound. MOTIVATION: animal & shape. According to Marzell (s.v. *Hyoscyamus niger* L.), the plant is highly noxious, even for (or especially for?) hens; corresponding German names are, e.g., *Hühnertod* lit. 'hen's death', *Hühnergift* lit. 'hen's poison', etc. Hence also its Old English name *hennewōl*, lit. 'hen's disease', which occurs twice in the extant Old English texts (see *→hennewōl*). The element, *-belle*, refers to the shape of the plant's calyx, which resembles a bell. ASSOCIATIVE RELATIONS: conceptual contiguity & meta-phorical similarity.

▫ *→belle, →henn*

henne-wōl, lit. 'hen's disease'
'Hyoscyamus niger L. / henbane / Schwarzes Bilsenkraut'
2 occ. (Gough 1974: 275; *Durham Plant Name Glossary*, no. 205).
ILLUSTRATION in Oxford, Bodleian Library, Ashmole 1431 (S.C. 7523), fol. 6v (Ohlgren 221(5)).
LATIN EQUIV. in the manuscripts: *jusquiamus, symphoniacus*.
ETYMOLOGY: OE formation; MORPHOLOGY: N_{gen}+N; exocentric, metonymic compound. MOTIVATION: animal & illness. Bierbaumer (3, s.v. *hennebelle*) regards *hennevole* and *hænnewol* as variants of *hennebelle*. I propose to regard both *hennebelle* and *hennewōl* as separate plant names for the same plant Hyoscyamus niger L. – one based on the shape of the plant, the other one on its noxious effects on hens – especially since Bierbaumer (1977: 116-7) stresses explicitly that one gloss (*Durham Plant Name Glossary*, no 205) must be read as "*hennevole*". This reading is supported by the fact that *wōl* 'pestilence, disease' is a common Old English lexeme and that German plant names exhibit the same motives (*Hühnertod*, etc., cf. Marzell, s.v. *Hyoscyamus niger* L.). See also Carr (1939: 318), who lists *hennewol* as an independent compound. ASSOCIATIVE

RELATION: conceptual contiguity.
▫ →*henn*, →*wōl*

heolaþ, *heoloþe*, *hioloþe*, noun
'Inula helenium L. / elecampane / Echter Alant'
DOEPN (s.v. *heoloþe*): 2 occ.
ETYMOLOGY: of unclear origin (Bierbaumer 1, s.v. *hioloþe*).
▫ →*hindheolaþ*, →*hēahheoloþe*

hēop, noun
'fruit of Rosa canina L. / HIP, HEP / Hagebutte'
DOEPN (s.v. *hēope*): 7 occ.
LATIN EQUIV. in the manuscripts: ? *boto* (*botonus*), *rubus*, *sycomorus* (? for *morus*).
ETYMOLOGY: Gmc **heupon-* (*OED*, s.v. *hip*, *hep*).
▫ →*hēopbrēmel*

hēop-brēmel, lit. 'hip-bramble' or 'hip thorn-bush'
'Rosa canina L. / dog-rose / Hundsrose'
DOEPN (s.v. *hēop-brēmel*): 3 occ.
LATIN EQUIV. in the manuscripts: *rubus*, *rumex*.
ETYMOLOGY: OE formation; MORPHOLOGY: N+N; endocentric, rectional. MOTIVATION: part of plant & botanical information. OE *brēmel* is polysemous; it can refer to (a) 'bramble, briar'; (b) 'blackberry bush', (c) 'dogrose'; (d) 'thorny bush'; the motive is therefore 'a kind of thorny bush which carries hips (OE *hēop* 'hip'). ASSOCIATIVE RELATIONS: conceptual contiguity (meronymy) & taxonomic subordination.
▫ →*brēmel*, →*hēop*

heorot, noun
'HART / Hirsch'
▫ →*heorotber(i)(g)e*, →*heorotbrēmel*, →*heorotbrēr*, →*heorotclǣfer*, →*heorotcrop*

heorot-ber(i)(g)e, lit. 'hart's berry'
(i) 'Rubus fruticosus L. / bramble / Brombeere', or (ii) 'fruit of ?? Rhamnus cathartica L. / berries of the hart's thorn / Beeren des Echten Kreuzdorns'
DOE (s.v. *heorot-berie*, *heort-berige* [to appear]): 5 occ.[e]
LATIN EQUIV. in the manuscripts: *celsa agreste* (? for *agrestis*), *fragus* (= *fraga*), *mora*.
The identification as Rubus fruticosus is based on the Latin lemmas, which all denote the bramble, while the identification as Rhamnus cathartica L. (Bierbaumer 3, s.v. *heorotberge*) is based on the Modern English plant name *hart's thorn* 'Rhamnus cathartica L.'. According to the *OED* (s.v. *hart's berry*), ModE *hart's berry* is a dialectal name for the bilberry 'Vaccinium L.', and goes back to OE *heorot brembel leaf* (*Lǣcebōc* 101/23), while Bierbaumer (1, s.v. *heorotbrēmel*) identified the hapax legomenon *heorotbrēmel* as 'Rhamnus cathartica L.', again on the basis of ModE *hart's thorn* and because an old name of Rhamnus cathartica was *spīna cervalis* lit. 'hart's thorn(-bush)'. The *DOE* (s.v. *heorot-berie, heort-berige* [to appear]) believes ModE *hart-berry* 'bilberry' to be an independent formation.

Because of the morphological and semantic parallels of →*heorotbrēmel* 'hart-bramble', →*heorotbrēr* 'hart-bramble' and *heorotber(i)(g)e* 'hart-berry', which describe bushes whose thorn-armed boughs resemble the antlers of roe-bucks, and because of the Latin lemmas they gloss, I would suggest that these plant names denote most probably brambles (Rubus fruticosus), while →*heorotcrop* 'hart-cluster (of berries)' is not entirely clear, but possibly denotes the bramble (Rubus fruticosus), the raspberry (Rubus idaeus) or the hartwort (Laserpitium latifolium) – all thorny bush which strongly resemble each other.

ETYMOLOGY: OE formation; possibly a loan rendition of L *spīna cervalis* in analogy to →*heorotbrēmel*. MORPHOLOGY: N+N; exocentric, metonymic compound; partial metaphor. MOTIVATION: part of plant & part of plant; a bush whose thorn-armed boughs resemble the antlers of roe-bucks (in contrast to →*hindber(i)(g)e* lit. 'hind-berry', where the bush is not armed with thorns); the second element refers to the berries of the plant. ASSOCIATIVE RELATIONS: metaphorical similarity & conceptual contiguity (meronymy).

▫ →*ber(i)(g)e*, →*heorot*

heorot-brēmel, lit. 'hart's bramble' or 'hart's thorn-bush'

(i) 'Rubus fruticosus L. / bramble / Brombeere', or (ii) '? Rhamnus cathartica L. / hart's thorn / Kreuzdorn'

DOE (s.v. *heorot-brembel-lēaf*): 1 occ. (*Lǣcebōc*, 101/23).

ETYMOLOGY: OE formation; possibly a loan rendition of L *spīna cervalis*, an old name of Rhamnus cathartica L. (Bierbaumer 1, s.v. *heorotbrembel*); attested only as *heorotbrēmellēaf* lit. 'leaf of the hart's bramble'; see also →*heorotber(i)(g)e*. MORPHOLOGY: N+N; endocentric, rectional, partial metaphor. MOTIVATION: part of plant & botanical information; for the first element, see →*heorotber(i)(g)e*; the second element denotes the kind. ASSOCIATIVE RELATIONS: metaphorical similarity & taxonomic sub-ordination.

▫ →*brēmel*, →*heorot*

heorot-brēr, lit. 'hart's briar' or 'hart thorn-bush'

'Rubus fruticosus L. / bramble / Brombeere'

2 occ. (*Gospels*, *Luke* (Li), 17, 6 and *Luke* (Ru), 17, 6); for the identification see →*heorotber(i)(g)e*.

LATIN EQUIV. in the manuscripts: *morus*.

ETYMOLOGY: OE formation; possibly a loan rendition of L *spīna cervalis* in analogy to →*heorotbrēmel*. MORPHOLOGY: N+N; endocentric, rectional; partial metaphor. MOTIVATION: part of plant & botanical information. See →*heorotber(i)(g)e*. ASSOCIATIVE RELATIONS: metaphorical similarity & taxonomic sub-ordination.

▫ →*brēr*, →*heorot*

heorot-clǣfer, lit. 'hart's clover'

(i) '? Melilotus officinalis Lam. / common melliot / Echter Steinklee' or (ii) 'Teucrium chamaedrys L. / ribbed melilot / Echter Gamander' or (iii) Medicago maculata L. / spotted medick, heart clover / Dornenkrone'

DOEPN (s.v. *heorot-clǣfre*): 9 occ.

ILLUSTRATION in manuscript London, British Library, Cotton Vitellius C. iii, fol. 29v (Ohlgren 168(20)).

LATIN EQUIV. in the manuscripts: *chamaedrys*, *chamaeleon*, *cynocephalion*.

ETYMOLOGY: OE formation; MORPHOLOGY: N+N; endocentric, rectional. MOTIVATION: animal & botanical information. According to R. Turner, *Botanalogia* (1644), "[...] Kings Claver, and Harts Claver, because Deer delight to feed upon it" (quoted in the *OED*, s.v. *hart-clover, hart's clover*). The leaves of Medicago species and those of Melilotus officinalis strongly resemble those of the clover (Marzell, s.vv. *Medicago* L., *Melilotus officinalis* Lam.). ASSOCIATIVE RELATIONS: conceptual contiguity & taxonomic subordination.

▫ *→clǣfer, →heorot*

heorot-crop, lit. 'hart's cluster (of berries)'

(i) ? 'Vaccinium L. / huckleberry, blueberry, cranberry / Heidelbeere, Preiselbeere' (*OED*, s.v. *hart's crop*), or (ii) 'Rubus fruticosus L. / bramble / Brombeere', or (iii) 'Rubus idaeus L. / raspberry / Himbeere', or (iv) 'Laserpitium latifolium L. / hartwort / Breitblättriges Laserkraut' (*DOE*, s.v. *heorot-crop* [to appear])

DOEP (s.v. *heorot-crop* [to appear]): 1 occ.; (*Lǣcebōc*, 16/24); see also *→heorotber(i)(g)e*.

ETYMOLOGY: OE formation; MORPHOLOGY: N+N; exocentric, metonymic compound; partial metaphor. MOTIVATION: part of plant & part of plant. The identification of this hapax legomenon is not entirely clear. Marzell (s.v. *Vaccinium uliginosum* L.), notes in connection with the plant names *Bullgräwen, Bullgrafen*, etc.: "Von Bulle 'taurus' wie lat. vaccinia von vacca, so hier der Bulle von der Hirschkuh. *-graven* ist das ags. *cropp* 'uvae', ags. *heort-crop* 'vaccinia'." ['From bull 'taurus' like L vaccinia from vacca, here the bull from the hind. *-graven* is OE *cropp* 'uvae'; Anglo-Saxon *heort-crop* 'vaccinia']. The problem is that OE *heorotcrop* occurs only once and does not gloss L *vaccinia*. On the other hand, the OE plant names *→heorotber(i)(g)e*, *→heorotbrēmel* and *→heorotbrēr* all refer or can refer to Rubus fruticosus L., the bramble, G *Brombeere*, which is a thorny bush with black fruits (alluded to metaphorically by the element *heorot*), similar to raspberry, huckleberry or cranberry bushes.

Laserpitium latifolium, in contrast, carries impressive inflorescences, which may be referred to by the element *crop*; parallel names in German – *Hirschwurz*, etc. – are explained as possibly describing the leaves, which are petiolated, with a prominent central rib (Marzell, s.v. *Laserpitium latifolium* L.). ASSOCIATIVE RELATIONS: metaphorical similarity & conceptual contiguity (meronymy).

▫ *→crop, →heorot*

hierde, noun

'shepherd / Hirte'

▫ *→hierdewyrt*

hierde-wyrt, lit. 'shepherd wort'

'Centaurium umbellatum Gilibert / common centaury / Tausendgüldenkraut'

DOEPN (s.v. *hirde-wyrt*): 4 occ.

LATIN EQUIV. and OE synonyms in the manuscripts: *centaurea*; *eorþgealla*, *felterre*.

ETYMOLOGY: OE formation. MORPHOLOGY: N+N; endocentric, rectional. MOTIVATION: ? habitat & botanical information. The motivation of the plant is a puzzle. Does the plant name perhaps refer to the fact that the common centaury grows in meadows where shepherds graze their sheep herds? ASSOCIATIVE RELATIONS: ? conceptual contiguity & taxonomic sub-ordination.

▫ →*hierde*, →*wyrt*

▫ →*hierdewyrt sēo læsse*

hierdewyrt sēo læsse, lit. 'lesser centaury'
'Centaurium umbellatum Gilibert / common centaury / Echtes Tausendgüldenkraut'
DOEPN (s.v. *hirdewyrt, læsse*): 1 occ. (*Lǣcebōc*, 75/20).
ETYMOLOGY: OE formation; possibly a loan rendition of L *centauria minor* in analogy to the rendition of *centauria maior* as *curmealle sēo māre* and of *centauria minor* as *curmealle sēo læsse* as the structure of the Old English plant name corresponds to that of the Latin model. MORPHOLOGY: postmodified noun phrase: noun + demonstrative + adjective. MOTIVATION: botanical information & size. A small kind of *hierdewyrt*. ASSOCIATIVE RELATIONS: taxonomic sub-ordination & conceptual contiguity.

▫ →*hierdewyrt*, →*lȳtel*

-hige, noun
'foot of game / Lauf'
ETYMOLOGY: *-hige*, derived from the verb OE *higian* '(to) hurry' is not attested independently in Old English, but cf. G *Lauf* 'foot of game' and the Latin lemma *leporis pēs* lit. 'hare's foot', translated by OE *haranhige*.

▫ →*haranhige*

hind, noun
'HIND / Hindin'

▫ →*hindber(i)(g)e*, →*hindbrēr*, →*hindheolaþ*

hind-ber(i)(g)e, lit. 'hind's berry'
(i) 'Rubus idaeus L. / raspberry plant / Himbeere'; (ii) 'fruit of Rubus idaeus L. / raspberry / Himbeere'
DOEPN (s.v. *hind-berie*): 19 occ.[e]
LATIN EQUIV. in the manuscripts: *acinus* (*acinum*), *eremio*, *fraga*.
ETYMOLOGY: WGmc formation (Carr 1939: 100); cf. OHG *hintbere*, etc. (Björkman 1902: 306). MORPHOLOGY: N+N; exocentric, metonymic compound, MOTIVATION: habitat & part of plant. The element *hind* 'hind, female deer' possibly indicates that the *hindberge* is not a thorny bush, which is in contrast to *heorotberge*, where *heorot* 'hart, stag' expresses the analogy between the antlers of stags and the thorny twigs (cf. Bierbaumer 1, s.v. *hindberge*). *Kluge* (s.v. *Himbeere*) does not see any convincing reason for the use of the concept HIND in OHG *hintberi*, MHG *hintber* and OE *hindber(i)(g)e*, and, indeed, raspberry bushes have thorny twigs, too (Marzell, s.v. *Rubus idaeus* L.; see also →*hindbrēr*). The motivation for the first element is, therefore, possibly, the typical habitat of this plant: 'a kind of berry that grows where hinds live'; cf. →*hindheolaþ*. ASSOCIATIVE RELATIONS: as regards the raspberry plant: conceptual contiguity & conceptual contiguity (meronymy); as regards the fruit: conceptual contiguity & taxonomic sub-ordination.

▫ →*ber(i)(g)e*, →*hind*

hind-brēr, lit. 'hind's briar'
'Rubus idaeus L. / raspberry plant / Himbeere'
DOEPN (s.v. *hind-brēr*): 3 occ.[e] (*Épinal-Erfurt Glossary*, no. 352; *Harley Glossary*, 156, no. 327; *Lācnunga*, 236).
LATIN EQUIV. in the manuscripts: *eremius*.
ETYMOLOGY: OE formation; MORPHOLOGY: N+N; endocentric, rectional; MOTIVATION: habitat & botanical information. The *hindbrēr* is a *brēr* 'thorn-bush'. The interpretation of the element *hind* 'hind' as indicating metaphorically that the bush is 'not thorny' (see →*hindberge*), in contrast to the element *heorot* 'hart' in →*heorotbrēr*, has to be rejected as the plant name would be a contradiction in terms: 'a thorn-bush that is not thorny'. The motivation is possibly 'a kind of thorn-bush hat grows where hinds live'; cf. →*hindber(i)(g)e* and →*hindheolaþ*. ASSOCIATIVE RELATIONS: conceptual contiguity & taxonomic sub-ordination.
▫ →*brēr*, →*hind*

hind-heolaþ (*-hioloþe, -hǣleþe*), lit. 'hind-elecampane'
(i) 'Eupatorium cannabium L. / water-hemp / Gemeiner Wasserdost', or (ii) 'Teucrium scolodonia L. / wood germander, wood sage / Salbei-Germander'
DOEPN (s.v. *hind-heoloþe*): 36 occ.
LATIN EQUIV. in the manuscripts: *ambrosia* (with *agrestis*).
ETYMOLOGY: OE formation; MORPHOLOGY: N+N; endocentric, rectional; the variant *hǣleþe* (attested, e.g. in *Lācnunga*, 236) is probably changed in analogy to OE *hǣlan* '(to) heal' from *heolaþ*, etc. (Bierbaumer 1, s.v. *hioloþe*; Sauer 1992b: 396). MOTIVATION: habitat & botanical information. Apparently, both Eupatorium cannabium and Teucrium scolodonia grow in the vicinity of deer, so that for both plants the habitat motivated the plant name. Accordingly, *hindheolaþ* is a 'kind of elecampane that grows where hinds live' (Bierbaumer 1, s.v. *hindheolaþ*). ASSOCIATIVE RELATIONS: conceptual contiguity & taxonomic sub-ordination.
▫ →*heolaþ*, →*hind*

-hioloþe →*heolaþ*

hlǣdder, noun
'LADDER / Leiter'
▫ →*hlǣdderwyrt*

hlǣdder-wyrt, lit. 'ladder-wort'
(i) '? Polemonium caeruleum L. / Jacob's ladder / Jakobsleiter'; (ii) '? Polygonum multiflorum All. / Solomon's seal / Salomonsiegel'
Hapax legomenon (*Lācnunga*, 80).
ETYMOLOGY: OE formation; MORPHOLOGY: N+N; endocentric, rectional; partial metaphor. MOTIVATION: shape & botanical information. Since the identification of the plant is not clear (Bierbaumer 2, s.v. *hlǣdderwyrt*), the motivation and associative relations cannot be ascertained. However, the first element of the plant name, *hlǣdder* 'ladder', might point to a plant with veins on the leaves which are arranged in the form of a ladder; see also →*gelodwyrt*. ASSOCIATIVE RELATIONS: metaphorical similarity & taxonomic sub-ordination.
▫ →*hlǣdder*, →*wyrt*

hnut-bēam, lit. 'nut-tree'
'Juglans regia L. / walnut / Welsche Nuß, Edler Walnußbaum'
DOEPN (s.v. *hnut-bēam*): 16 occ.[e]
LATIN EQUIV. and OE synonyms in the manuscripts: *nucarius*, *nux*; *wealhnutu*.
ETYMOLOGY: Independent formation in WGmc; OHG *nuzboum*, MLG *notebôm*, Du *notenboom*, Fris *nutebeam* (Carr 1939: 132); MORPHOLOGY: N+N; endocentric, rectional. MOTIVATION: part of plant & botanical information. This kind of *bēam* 'tree' bears nuts (Marzell, s.v. *Juglans regia* L.). ASSOCIATIVE RELATIONS: conceptual contiguity (meronymy) & taxonomic sub-ordination.
▫ →*bēam*, →*hnutu*
▫ →*easterne hnutbēam*

hnutu, noun
(i) 'Corylus avellana L. / HAZEL / Gewöhnliche Haselnuss'; (ii) 'NUT / Nuss'
DOEPN (s.v. *hnutu*): 16 occ.
LATIN EQUIV. in the manuscripts: *abellana*, *culliolum*, *ficus*, *juglans*, *nux*.
ETYMOLOGY: Gmc **hnut-*, prob. from IE **kneu-* (*Kluge*, s.v. *Nuß*).
▫ →*hnutbēam*
▫ →*eorþhnutu*, →*frencisc hnutu*, →*wealhhnutu*

-h(n)ydele, ?
unidentified
▫ →*hǣwenh(n)ydele*

hoc, noun
'Malva sylvestris L. / common mallow / Wilde Malve'
DOEPN (s.v. *hoc*): 8 occ.
LATIN EQUIV. in the manuscripts: *malva*.
ETYMOLOGY: of unknown origin (*OED*, s.v. *hock*).
▫ →*hoclēaf*
▫ →*holihocce*

hoc-lēaf, lit. 'mallow leaf'
(i) 'Malva silvestris L. / common mallow / Wilde Malve, Roßpappel'; (ii) 'Malva neglecta Wallr. / dwarf mallow / Kleine Malve'
DOEPN (s.v. *hoc-lēaf*): 8 occ.
ILLUSTRATIONS in manuscripts London, British Library, Cotton Vitellius C. iii, fol. 34r (Ohlgren 168(29)) and Oxford, Bodleian Library, Bodley 130, fol. 4v (Ohlgren 222(7)).
LATIN EQUIV. in the manuscripts: *malva*, *malva erratica*.
ETYMOLOGY: OE formation; MORPHOLOGY: N+N; exocentric, metonymic compound. MOTIVATION: plant & part of plant. *Hoc* is of unknown origin, and the two species of Malva are seldom distinguished (Bierbaumer 1, s.v. *hoc*; Marzell, s.v. *Malva silvestris* L.). *Hoc-lēaf* may have been named after the prominent form and size of its leaves in comparison to the much smaller blossoms, but also because of the healing virtues of the leaves: they contain vitamin C and can be used in salads or can be brewed to anti-inflammatory and mucolytic teas (Rausch and Lotz 2003: 166ff.). ASSOCIATIVE RELATIONS: conceptual identity & conceptual contiguity (meronymy).
▫ →*hoc*, →*lēaf*

hōfe, noun
'Glechoma hederacea L. / HOVE, ivy / Gewöhnlicher Gundermann'
DOEPN (s.v. *hōfe*): 19 occ.

LATIN EQUIV. and OE synonym in the manuscripts: *viola*; *bānwyrt*.
ETYMOLOGY: possibly derived from OE *hebban* '(to) heave, lift'.
▫ →*merschōfe*, →*rēade hōfe*, →*tūnhōfe*

hol, adjective
'HOLLOW / hohl'
▫ →*hole cærse*, →*hollēac*

hole cærse, lit. 'hollow cress' or 'hollow plant'
'? Gentiana campestris L. / baldmony, bitterwort / Feld-Enzian'
DOEPN (s.v. *hole cærse*): 3 occ. (*Lǣcebōc*, 11/26, 25/2, 42/10-1).
ETYMOLOGY: OE formation; MORPHOLOGY: Adj+N; endocentric, copulative: attributive. MOTIVATION: shape & botanical information. The first element, *hole* 'hollow' probably refers to the shape of the flower of the baldmony (*DOEPN*, s.v. *hole cærse*). As baldmony shows no conceptual relation to the cress (Bierbaumer 1, s.v. *hole cerse*), we can perhaps assume that the meaning of *cærse* was generalised to 'plant'; see also →*lēac*. ASSOCIATIVE RELATIONS: conceptual contiguity & taxonomic sub-ordination.
▫ →*cærse*, →*hol*

holen, noun
'Ilex aquifolium L. / HOLLY / Gewöhnliche Stechpalme'
DOEPN (s.v. *holen*): 14 occ.
LATIN EQUIV. in the manuscripts: *acrifolium*, *ruscus*, ? *ulcia* (? for *hussus* 'holly').
ETYMOLOGY: Gmc **ak*- 'acute, peaked' (*Kluge*, s.v. *Hulst*).
▫ →*cnēowholen*

holi →*hālig*

holi-hocce, lit. 'holy mallow'
'Althaea officinalis L. / marshmallow, HOLLY-HOCK / Eibisch'
Hapax legomenon (Gough 1974: 286); late Old English or early Middle English word, where the development of OE *ā* > ME *ō* /ɔ:/ south of the Humber was already under way and can be seen in writing (text dated to the 12th century; see Ker *Catalogue*, no. 218, Gneuss *Handlist*, no. 527).
LATIN EQUIV. in the manuscripts: *althaea*.
ETYMOLOGY: OE formation; possibly a loan translation of L **malva benedicta*, see MOTIVATION; MORPHOLOGY: Adj+N; endocentric, copulative: attributive. MOTIVATION: healing virtues & botanical information. According to the *OED*, the plant name is "[...] evidently of hagiological origin; cf. the Welsh name *hocys bendigaid*, which appears to translate a med.L. **malva benedicta*. [...] The guess that 'the hollyhock was doubtless so called from being brought from the Holy Land' has been offered in ignorance of the history of the word" (*OED*, s.v. *hollyhock*). The first element indicates that it is a blessed, a healing herb: "Medicinally, the hollyhock has been used as a diuretic, an emollient, and to solve chest problems" (Ward 1999: 199). ASSOCIATIVE RELATIONS: conceptual contiguity & taxonomic sub-ordination.
▫ →*hoc*, →*hol*

hol-lēac, lit. 'hollow leek'
'? Allium ascalonium L. / shallot / Schalotte'
DOEPN (s.v. *hol-lēac*): 6 occ.

LATIN EQUIV. in the manuscripts: *duricorium*.
ETYMOLOGY: WGmc compound; (Carr 1939: 118); cf. OHG *hollouch* (Björkman 1901: 208); MORPHOLOGY: Adj+N; endocentric, copulative: attributive. MOTIVATION: shape & botanical information. The Old English plant name refers to some hollow parts of the plant: The leaves of most Alliums (e.g. A. ascalonium, A. cepa) are hollow; most characteristically, however, are the leaves of Allium schoenoprasum L., which are erect and tubular, so that OE *hollēac*, but also OHG *holloich*, *hollouch*, etc. (Björkman 1902: 306), refer rather to Allium schoenoprasum than to A. ascalonium; Marzell lists the Old High German lexemes under the headword Allium ascalonium L., yet concedes that they may refer to other species of the genus Allium, as well (Marzell, s.v. *Allium ascalonium* L.). ASSOCIATIVE RELATIONS: conceptual contiguity & taxonomic sub-ordination.
▫ →*hol*, →*lēac*

hors, noun
'HORSE / Pferd, Ross'
▫ →*horselene*, →*horsminte*

hors-elene, lit. 'horse-elecampane'
'Inula helenium L. / elecampane / Echter Alant'
DOEPN (s.v. *hors-elene*): 11 occ.
LATIN EQUIV. in the manuscripts: *inula* (*hinula*), *hinula campana*, *mula*, *origanum*.
ETYMOLOGY: OE formation; hybrid compound (OE *elene* < L *inula*; see →*elene*). MORPHOLOGY: N+N; endocentric, rectional; partial metaphor. MOTIVATION: inferiority & botanical information. According to Bierbaumer (3, s.v. *horselene*), the addition of the element *hors*- to *elene* is probably based on a mistake on the part of the translator, who interpreted *hinula* (possibly a folk etymological rendition of *inula* in analogy to *hinulus* 'mule') as 'horse'. The *OED* (s.v. *horseheal*) suggests that *hors*- means probably not the animal, but was used metaphorically to mean 'wild' or 'coarse', which would stress the difference between *elene* and the inferior, coarser *horselene*, cf. the parallel formations →*horsminte* and →*tūnminte*. ASSOCIATIVE RELATIONS: metaphorical similarity & taxonomic sub-ordination.
▫ →*elene*, →*hors*

hors-minte, lit. 'horse-mint'
'Mentha longifolia Hudson / HORSE-MINT / Roßminze'
DOEPN (s.v. *hors-minte*): 2 occ. (*Herbarium*, no. 92; *Lācnunga*, 983).
ILLUSTRATION in manuscript London, British Library, Cotton Vitellius C. iii, fol. 47r (Ohlgren 168(54)).
ETYMOLOGY: OE formation; hybrid compound (OE *minte* < L *ment(h)a*; see →*minte*). MORPHOLOGY: N+N; endocentric, rectional; partial metaphor. MOTIVATION: inferiority & botanical information. *Hors*- is used metaphorically to indicate that it is a coarser kind of mint than the *minte* or the *tūnminte*, i.e. a wild species. Cf. also →*horselene*. ASSOCIATIVE RELATIONS: metaphorical similarity & taxonomic sub-ordination.
▫ →*hors*, →*minte*

hræfn, noun
'RAVEN / Rabe'
▫ →*hræfnes fōt*, →*hræfnes lēac*

hræfnes fōt, lit. 'raven's foot'
(i) 'Ranunculus L. / crowfoot, buttercups / Hahnenfuß'; (ii) 'Potentilla reptans L. / fiveleaf / Kriechendes Fünffingerkraut'
DOEPN (s.v. *hræfnes fōt*): 25 occ.[e] (see also *DOE*, s.v. *fōt*, 5.a.).
ILLUSTRATION in manuscript London, British Library, Cotton Vitellius C. iii, fol. 30v (Ohlgren 168(22)).
LATIN EQUIV. in the manuscripts: *chamaedaphne*, *pentaphyllon*, *polypodium*, *quinquefolium*.
ETYMOLOGY: OE formation; MORPHOLOGY: N_{gen}+N; exocentric, metaphorical compound. MOTIVATION: animal & shape. Originally, the plant name referred to those species of Ranunculus L. that had divided leaves, i.e. resembled a crow's (or a raven's) foot. Later on, the name was extended as a book-name to denote the whole genus (*OED*, s.v. *crowfoot*). Potentilla reptans L. has divided leaves, too, and resembles Ranunculus L. (see →*fīf-lēafe*, and Bierbaumer 2, s.v. *hræfnes fōt*). ASSOCIATIVE RELATIONS: metaphorical similarity.
▫ →*fōt*, →*hræfnes*

hræfnes lēac, lit. 'raven's leek'
(i) 'Orchis L. / orchis, wild orchid / Knabenkraut'; or (ii) 'Allium ursinum L. / crowleek / Bärlauch'
DOEPN (s.v. *hræfnes lēac*): 5 occ.
ILLUSTRATION in manuscript London, British Library, Cotton Vitellius C. iii, fol. 27r (Ohlgren 168(15)).
LATIN EQUIV. in the manuscripts: *satyrion*.
ETYMOLOGY: OE formation; MORPHOLOGY: N_{gen}+N; endocentric, rectional; partial metaphor. MOTIVATION: inferiority & botanical information. Bierbaumer (1, s.v. *hræfnes lēac*) is quite sure about the identification of *hræfnes lēac* as the orchid, but he does not find any reasonable motive for the plant's name in Old English. I would suggest that, since in Middle English *saturion* could be glossed by *crouelek* 'crowleek' (*MED*, s.v. *croue*) and since MedL *saturion minor* often denoted the wild hyacinth or bluebell (Endymion non-scriptus L.), a plant that grows preferably in woods, OE *hræfnes lēac* did not refer to Orchis L., but to Allium ursinum L., crowleek, a plant very similar in appearance to the bluebell. In addition, Allium ursinum was also named →OE *crāw(e)(an)lēac*, a synonym of OE *hræfnes lēac* as regards the significations of the plant names: they can be rendered both literally as 'crow's leek' and 'raven's leek' (cf. ModE *crow* and *raven*). ASSOCIATIVE RELATIONS: metaphorical similarity & taxonomic sub-ordination.
▫ →*hræfnes*, →*lēac*

hrætel, noun
'RATTLE / Rassel'
▫ →*hrætelwyrt*

hrætel-wyrt, lit. 'rattle-wort'
'? Alectorolophus Böhmer / rattle / Ratteltopf'
Hapax legomenon (*Brussels Glossary*, 301/3).
LATIN EQUIV. in the manuscripts: *hierobotanum*.
ETYMOLOGY: OE formation; MORPHOLOGY: N+N; endocentric, rectional; partial metaphor. MOTIVATION: sound & botanical information. The seeds of Alectorolophus rattle in the capsule when ripe (Marzell, s.v. *Alectoropholus*

Böhmer); cf. also G *Ratteltopf* lit. 'rattle-pot'. ASSOCIATIVE RELATIONS: metaphorical similarity & taxonomic sub-ordination.

▫ →*hrætel*, →*wyrt*

hrōf, adjective
'strong, invigorating / stark, belebend'

▫ →*wuduhrōfe*

-hūl, noun
'? reed / Schilf'
ETYMOLOGY: *-hūl* is not attested independently and its origins are not transparent; possible explanations are: (i) *-hūl* is a corrupted form of *þūl-* (cf. OFris *dūla* 'reeds'), or (ii) *-hūl* is derived from the adjective *hūl* 'hollow, hohl' and refers to reeds, or (iii) *-hūlas* is a corrupted form of *-þȳflas* (plural of *þȳfel* 'bush') (Bierbaumer 3, s.v. *scaldhulas*).

▫ →*scaldhūlas*

hund, noun
'dog, HOUND / Hund'

▫ →*hundesber(i)(g)e*, →*hundescwelcan*, →*hundeshēafod*, →*hundesmicge*, →*hundestunge*

hundes-ber(i)(g)e, lit. 'dog's berry'
'Solanum nigrum L. / morel, garden nightshade / Schwarzer Nachtschatten'
Hapax legomenon (Meritt 1945: 62, no. 73 Dc 185/8).
LATIN EQUIV. in the manuscript: *uva canina*.
ETYMOLOGY: OE formation; loan translation of *uva canina*. MORPHOLOGY: N_{gen}+N; exocentric, metonymic compound; partial metaphor. MOTIVATION: threat or pathogen & part of plant. A number of vernacular plant names denoting Solanum nigrum contain an animal name (wolf, fox, dog) in order to indicate that the plants or parts of it are threatening or even poisonous in comparison to other species of the genus (Marzell, s.v. *Solanum nigrum* L., 3). The second element refers to the poisonous black berries; *ber(i)(g)e* is used metonymically: it stands for the whole plant. ASSOCIATIVE RELATIONS: metaphorical similarity & conceptual contiguity (meronymy).

▫ →*ber(i)(g)e*, →*hund*

hundes-cwelcan (pl.), lit. 'dog's ?'
(i) '? berries of Viburnum opulus L. / guelder rose / Beeren des Gemeinen Schneeballs'; (ii) '? berries of Viburnum lantana L. / wayfaring tree / Beeren des Wolligen Schneeballs'
Hapax legomenon (*Cleopatra Glossary* (Stryker), 89, no. 106).
LATIN EQUIV. in the manuscript: *colocynthida*.
ETYMOLOGY: OE formation; possibly a loan word from Old Norse (see Hoops 1889: 76, and MOTIVATION below), but *Heizmann* does not record plant names that could be connected to this Old English lexeme. MORPHOLOGY: N_{gen}+N; endocentric, rectional; partial metaphor; *-cwelcan* (pl.) is not attested independently in Old English. MOTIVATION: pathogen or inferiority & botanical information. The second element *-cwelcan* (pl.) is not attested independently in Old English. It is possibly a cognate to OHG *svelcbom*, *suelebom* (Björkman 1901: 219), Dan *kvalkved*, *hvalkentræ*, Sw *qualkebär* which may be connected to MHG *swelc* 'withered, brittle' (Marzell, s.v. *Viburnum opulus* L.; Björkman 1902:

305). The first element, *hundes-*, is probably motivated by the fact that the berries of Solanum are poisonous, and toxicity is often expressed by names of (threatening) animals such as wolves or foxes (see also →*hundesber(i)(g)e*). ASSOCIATIVE RELATIONS: metaphorical similarity & taxonomic sub-ordination.
▫ → *-cwelcan* (pl.), →*hund*

hundes-hēafod, lit. 'dog's head'
'Antirrhinum orontium L. / small snapdragon, calf-snout / Acker-Löwenmaul'
Bierbaumer (2, s.v. *hundes hēafod*): 3 occ. (*Herbarium*, no. 88).
ILLUSTRATION in manuscript London, British Library, Cotton Vitellius C. iii, fol. 45v (Ohlgren 168(52)).
LATIN EQUIV. in the manuscripts: *canis caput.*
ETYMOLOGY: Independent formation in WGmc; a loan translation of L *canis caput* < Gk *κυνοκέφαλος*; cf. OHG *hunthoubit(o)*, *hundhoubet* (Carr 1939: 155, 167). MORPHOLOGY: N_{gen}+N; exocentric, metaphorical compound. "The formation may be considered as a Bahuvrîhi or as a metaphorical transference of a determinative like *hundes-tunge*" (Carr 1939: 155). MOTIVATION: animal & overall appearance. As to the motivation, Marzell (s.v. *Antirrhinum orontium* L.) comments: "Die reife Fruchtkapsel springt mit drei Löchern auf, so daß sie eine gewisse Ähnlichkeit mit einem Tier- oder Menschenschädel hat" ['When the ripe capsule breaks open, it forms three holes, resembling the skull of a human or an animal']. ASSOCIATIVE RELATION: metaphorical similarity.
▫ →*hēafod*, →*hund*

hundes-micge, lit. 'dog's urine'
(i) 'Cynoglossum officinale L. / dog's tongue / Gemeine Hundszunge', or (ii) 'Hyoscyamus niger L. / henbane / Schwarzes Bilsenkraut'
Hapax legomenon (*Lācnunga*, 799-800).
ETYMOLOGY: OE formation; MORPHOLOGY: N_{gen}+N; exocentric, metaphorical compound. MOTIVATION: animal & smell. Both Cynoglossum officinale and Hyoscyamus niger emanate a rather unpleasant smell. According to Marzell (s.v. *Hyoscyamus niger* L., 7), Hyoscyamus niger reeks of dog's urine and grows on hedges and fences, i.e. where dogs urinate. The compound can, therefore, be seen either as a metonymic formation (based on conceptual contiguity, i.e. 'where dog's urinate') or, rather, as a metaphorical formation, based on the metaphorical similarity between the plant's reek and dog's urine. ASSOCIATIVE RELATION: metaphorical similarity.
▫ →*hund*, →*micge*

hundes tunge, lit. 'dog's tongue'
(i) 'Cynoglossum officinale L. / dog's tongue / Gemeine Hundszunge'; (ii) Lycopsis arvensis L. (Gunther) / bugloss / Acker-Ochsenzunge'
DOEPN (s.v. *hundes tunge*): 13 occ.; identifications follow de Vriend (1984: 298, no. 42).
ILLUSTRATION in manuscript London, British Library, Cotton Vitellius C. iii, fol. 34v (Ohlgren 168(30)).
LATIN EQUIV. in the manuscripts: *buglossa*, *lingua canis*, *lanceolatus*, ? *venebula.*
ETYMOLOGY: Independent formation in WGmc (Carr 1939: 132); cf. OHG *hunteszunga*, etc. (Björkman 1902:

292); loan translation of L *lingua canis* < Gk *κυνόγλωσσον* (Bierbaumer 1, s.v. *hundestunge*). MORPHOLOGY: N_{gen}+N; exocentric, metaphorical compound. "The compound is a determinative used metaphorically, hardly a Bahuvrîhi formation" (Carr 1939: 132). MOTIVATION: animal & shape. OE *hundestunge* is a loan translation of L *lingua canis*, based on the fact that the plant has coarse, tongue-shaped leaves, like the tongue of a dog (Marzell, s.v. *Cynoglossum officinale* L., 1). ASSOCIATIVE RELATIONS: metaphorical similarity.

▫ →*hund*, →*tunge*

hūne, noun
'Ballota nigra L. / horehound / Langzähnige Schwarznessel'
DOEPN (s.v. *hūne*): 25 occ.
LATIN EQUIV. and OE synonym in the manuscripts: *marrubium*, *prassium*; *bēowyrt*.
ETYMOLOGY: Of uncertain origin (*OED*, s.v. *horehound*; see also *Holthausen*, s.v. *hār(e)hūne*).

▫ →*hār(e)hūne*

hunig, noun
'HONEY / Honig'

▫ →*hunigsūge*

hunig-sūge, lit. 'honey-suckle'
'Lonicera periclymenum L. / HONEYSUCKLE / Geißblatt'
DOEPN (s.v. *hunig-sūge*): 7 occ.[e]
LATIN EQUIV. in the manuscripts: *ligustrum* (but not denoting Ligustrum vulgare L.; cf. Marzell, s.v. *Lonicera periclymenum* L., 5.).
ETYMOLOGY: OE formation; MORPHOLOGY: synthetic, $N+N_{deverbal(-e)}$; endocentric, rectional. MOTIVATION: product & action. Lonicera periclymenum is a plant with beautiful, flavoursome flowers, frequented by insects with long tongues which suck the pollen: "Honigreiche Blüten, die oft von Kindern ausgesogen werden. [...] Sachlich ist zu bemerken, daß das Geißblatt von Schwärmern (Schmetterlingen) bestäubt wird [...]" ['Flowers rich in honey, which are often sucked by children. In actual fact, it has to be mentioned that honeysuckle is pollinated by swarmers (butterflies).'] (Marzell, s.v. *Lonicera periclymenum* L., 5); cf. also OHG *binisuga*. ASSOCIATIVE RELATIONS: conceptual contiguity & taxonomic sub-ordination.

▫ →*hunig*, →*sūgan*

hweorfan, verb
'(to) turn / drehen'

▫ →*sigelhweorfa*

hwer, noun
'pot, bowl / Topf'

▫ →*hwerhwette*

hwer-hwette, lit. 'pot-whetting (plant)'
'Cucumis sativus L. / cucumber / Gurke'
DOEPN (s.v. *hwer-hwette*): 16 occ.
ILLUSTRATION in manuscript London, British Library, Cotton Vitellius C. iii, fol. 53v (Ohlgren 168(67)).
LATIN EQUIV. in the manuscripts: *cucumis*, *cucumis silvatica*.
ETYMOLOGY: OE formation; according to Bierbaumer (3, s.v. *hwerhwette*) and Sauer (1992b: 396), the Old English

plant name is a folk etymological change of the loan word →*cyrfette* (< L *curcubita*), based on the phonological similarity between the loan word *cyrfette* and the Old English terms OE *hwer* 'pot, bowl' and *hwettan* 'to whet, to sharpen'. MORPHOLOGY: synthetic, $N+N_{deverbal(-e)}$. As explained above, a change based on the phonological similarity between a part of the Latin loan and OE *hwet* 'pot', respectively, *hwettan* '(to) whet'. This was, possibly, supported by the fact that people made vessels and water-bottles out of the fruits of the gourd, which, when hollowed out and dried, develop firm 'shells'; cf. the meaning 'bottle, cup (of any material)' which is attested from the 14th century onwards for *gourd* (*OED*, s.v. *gourd*, 5 (a)). The cucumber is part of the family of the Cucurbitaceae, i.e. related taxonomically to the gourd. MOTIVATION: sound pattern & sound pattern. ASSOCIATIVE RELATIONS: phonological similarity & phonological similarity.

▫ →*hwer*, →*hwettan*

hwettan, verb
'(to) WHET / wetzen'
ETYMOLOGY: Folk etymology: change of the submorphemic unit *-fette* of the loan *cyrfette* (< L *curcubita*; see →*cyrfet*), in analogy to OE *hwettan* '(to) whet, (to) sharpen' (see →*hwer-hwette*).

▫ →*hwerhwette*

hwistle, noun
'WHISTLE / Pfeife, Flöte'

▫ →*wōde(h)wistle*

hwīt, adjective
'WHITE / weiß'

▫ →*hwītcwidu*, →*hwīteclǣfre*, →*hwīt hæsel*, →*hwītlēac*, →*hwīt mæg(e)þ(e)*, →*hwīt mæringc*, →*hwīt popig*, →*hwīt wīngeard*

hwīt-cwidu (*-cudu*), lit. 'white mastic'
'resin, mastic of Pistacia lentiscus L. / mastic / Harz des Mastixstrauches'
DOEPN (s.v. *cwidu, hwīt*): 21 occ.[e]
LATIN EQUIV. in the manuscripts: *mastix*.
ETYMOLOGY: OE formation; MORPHOLOGY: Adj+N; endocentric, copulative: attributive. MOTIVATION: colour & botanical information. The first element refers to the ivory-coloured resin of Pistacia lentiscus; the second element, *cwidu* 'mastic', refers to the fact that the resin of Pistacia lentiscus was chewed in order to protect the teeth from decaying (Marzell, s.v. *Pistacia lentiscus* L.). ASSOCIATIVE RELATIONS: conceptual contiguity & taxonomic sub-ordination.

▫ →*hwīt*, →*cwidu*

hwīte blōstman (pl.), lit. 'white blossoms'
'blossoms of Ligustrum vulgare L. / privet / Rainweide'
Hapax legomenon (*Herbarium*, no. 94). The identification is based on an Aldhelm gloss (cf. *Cleopatra Glossary* (Stryker), 282, no. 252, the *DOE* s.v. *blōstm*, *blōstma*, 2 and the *DOEPN*, s.v. *blōstma*).
LATIN EQUIV. in the manuscript: *ligustrum*.
ETYMOLOGY: OE formation; MORPHOLOGY: Adj+N; endocentric, copulative: attributive. MOTIVATION: colour & botanical information. A transparent formation, where the first element

refers to the colour of the blossoms. ASSOCIATIVE RELATIONS: conceptual contiguity & taxonomic sub-ordination.
▫ →*blōstm*, →*hwīt*

hwīte-clǣfre, lit. 'white clover'
'Trifolium repens L. / WHITE CLOVER / Weißklee, Kriechklee'
DOEPN (s.v. *clǣfre, hwīte*): 8 occ.[e]
LATIN EQUIV. in the manuscripts: *calcesta*.
ETYMOLOGY: OE formation; MORPHOLOGY: Adj+N; endocentric, copulative: attributive. MOTIVATION: colour & botanical information. Trifolium repens is a clover with white flowers (Marzell, s.v. *Trifolium repens* L.). ASSOCIATIVE RELATIONS: conceptual contiguity & taxonomic sub-ordination.
▫ →*clǣfer*, →*hwīt*

hwīting-trēow, lit. '? whiting-tree'
(i) '? Viburnum opulus L. / guelder rose / Gemeiner Schneeball', or (ii) '? Viburnum lantana L. / wayfaring tree / Wolliger Schneeball'
Hapax legomenon (*Antwerp-London Glossaries*, 128/18). The identification is based on ModE *whitten-tree* (*OED*, s.v. *whitten-tree*).
LATIN EQUIV. in the manuscript: ? *variculus*.
ETYMOLOGY: OE formation; MORPHOLOGY: N+N; endocentric, rectional; partial metaphor. MOTIVATION: colour & botanical information. Most species of Viburnum have white or bright pink flowers (Marzell, s.vv. *Viburnum lantana* L., *Viburnum opulus* L., 2). Some of these species start blossoming at the end of January and in February, i.e. when there is still a lot of snow north of the Alps. OE *hwīting* n. 'whiting, chalk' (< *hwīt*(*an*) '(to) whiten' + *-ing*) probably refers metaphorically to the white blossoms, comparing them to the snow; cf. also G *Schneeball* lit. 'snowball' for Viburnum. ASSOCIATIVE RELATIONS: metaphorical similarity & taxonomic sub-ordination.
▫ →*hwīting*, →*trēow*

hwīt hæsel, lit. 'white hazelnut'
(i) '? Ulmus montana With. / witch elm, witch hazel / Bergulme' or (ii) '? Ulmus campestris L. / field elm / Feldulme'
Hapax legomenon (*Antwerp-London Glossaries*, 121/10).
LATIN EQUIV. in the manuscript: *saginus*.
ETYMOLOGY: OE formation; MORPHOLOGY: Adj+N; endocentric, copulative: attributive. MOTIVATION: colour & botanical information. The blossoms of Ulmus campestris L. have white borders while the blossoms of Ulmus montana L. have reddish borders (Marzell, s.v. *Ulmus* L.). It seems, therefore, plausible that, if the identification of the plant as a species of Ulmus is right, the Old English plant name denotes the Ulmus campestris L. rather than the Ulmus montana, because of the element *hwīt* 'white'. The association of the elm with the hazel is perhaps due to the leaves, which are quite similar in shape. ASSOCIATIVE RELATIONS: conceptual contiguity & taxonomic sub-ordination.
▫ →*hæsel*, →*hwīt*

hwīting, noun
'WHITING, chalk'
▫ →*hwītingtrēow*

hwīt-lēac, lit. 'white leek'
'Allium L. / leek / Lauch'
4 occ. (*Cleopatra Glossary* (Stryker), 63, no. 494; *Cleopatra Glossary* (Quinn), no. 39/12; *Antwerp-London Glossaries*, 135/214; cf. Voss 1989; *Laud Herbal Glossary*, no. 49).
According to Bierbaumer (3, s.v. *hwītlēac*; 2003: 157), the original Latin text did not read *alba cepa* 'white leek' but *alba cera* 'white wax', which was confused by the compiler with *alba cepa* 'white leek', and translated accordingly. Recently, however, Bierbaumer (personal communication) reconsidered the matter and concluded that, in view of ON *hvítr laukr* (*Heizmann*, 27), OE *hwītlēac* may have been a real plant name.
LATIN EQUIV. in the manuscripts: *caepa alba*, ? *poleus* (? for *propolis*).
ETYMOLOGY: Gmc formation (cf. *hvítr laukr* in Old West Norse, cf. *Heizmann*, 27); loan translation of L *alba cēpa*. MORPHOLOGY: Adj+N; endocentric, copulative: attributive. MOTIVATION: colour & botanical information. The lower parts of all species of Alliums are white (and not green like the upper parts). ASSOCIATIVE RELATIONS: conceptual contiguity & taxonomic subordination.
▫ *→hwīt, →lēac*

hwīt mæg(e)þ(e), lit. 'white camomile'
'Chrysanthemum leucanthemum L. / ox-eye daisy / Wucherblume'
DOEPN (s.v. *mageþe, hwīt*): 2 occ.
LATIN EQUIV. in the manuscripts: *buphthalmon*.
ETYMOLOGY: OE formation; MORPHOLOGY: Adj+N; endocentric, copulative: attributive. MOTIVATION: colour & botanical information. Chrysanthemum leucanthemum has white flowers, hence the use of the first element. As the plant resembles the camomile, especially with regard to its white flowers cf. also G *Wilde Kamille* 'wild camomile', Du *groote kamille* lit. 'tall camomile', Lith *baltíjei ramùmiai* lit. 'white camomile' (compare Marzell, s.v. *Chrysanthemum leucanthemum* L., 11), the two plants were regarded as taxonomically related. ASSOCIATIVE RELATIONS: conceptual contiguity & taxonomic sub-ordination.
▫ *→hwīt, →mæg(e)þ(e)*

hwīt mæringc, lit. 'white ?'
(i) ? 'Ocimum basilicum L. / basil / Basilikum', or (ii) ? 'Primula veris L. / cowslip / Echte Schlüsselblume' (cf. Biggam 1997: 172)
1 occ. (*Lācnunga*, 16).
ETYMOLOGY: OE formation; *-mæringc* is of unclear origin and not attested independently in Old English. MORPHOLOGY: Adj+N; endocentric, copulative: attributive. MOTIVATION: colour & botanical information. As the identification of the plant is unclear, the motivation for this plant name remains unclear, as well. The flowers of basil can be whitish, which would explain the use of the first element (Marzell, s.v. *Ocimum basilicum* L.), but the second element is not clear. ASSOCIATIVE RELATIONS: conceptual contiguity & ? taxonomic sub-ordination.
▫ *→hwīt, →-mæringc*

hwīt popig, lit. 'white poppy'
'Papaver somniferum L. / opium poppy / Schlafmohn'
DOEPN (s.v. *popig, hwīt*): 4 occ.

LATIN EQUIV. in the manuscripts: *meconium*.
ETYMOLOGY: OE formation; loan translation of L *papāver album* (see MOTIVATION); hybrid formation (OE *popig* < L *papāver*; see →*popig*). MORPHOLOGY: Adj+N; endocentric, copulative: attributive. MOTIVATION: colour & botanical information. The plant name refers to the bright seeds of white poppy (MedL *papaver album*), in contrast to the dark ones of the black poppy (MedL *papaver nigrum*; cf. Marzell, s.v. *Papaver somniferum* L.). ASSOCIATIVE RELATIONS: conceptual contiguity & taxonomic sub-ordination.
▫ →*hwīt*, →*popig*

hwīt wīngeard, lit. 'white vine'
(i) 'vine / Weinstock'; (ii) 'Bryonia alba L. / bryony / Zaunrübe'
3 occ. (*Antwerp-London Glossaries*, 119/12, 143/14, 146/1).
LATIN EQUIV. and OE synonyms in the manuscripts: *ampelos leuce*, *bryonia*, *aminaea vitis*; *wilde cyrfet*.
ETYMOLOGY: OE formation. MORPHOLOGY: Adj+N; endocentric, copulative: attributive. MOTIVATION: colour & botanical information. The flowers of Bryonia alba are white. With its tendrils, characteristic leaves and dark fruits (Marzell, s.v. *Bryonia alba* L.), it closely resembles the grape and in bounds, →*wīngeard* 'vine' is used to denote the bryony; cf. →*blæc wīngeard* 'black bryony' lit. 'black vine' and →*wilde wīngeard* 'black bryony' lit. 'wild vine' ASSOCIATIVE RELATIONS: conceptual contiguity & taxonomic sub-ordination.
▫ →*hwīt*, →*wīngeard*

hyl, noun
'HILL / Hügel'
▫ →*hylwyrt*

hyl-wyrt, lit. 'hill-wort'
'Mentha pulegium L. / pennyroyal / Poleiminze'
DOEPN (s.v. *hyl-wyrt*): 11 occ.
LATIN EQUIV. and OE synonyms in the manuscripts: *puleium*, ? *samum*; *brōþorwyrt*, *dweorged(w)os(t)(l)e*.
ETYMOLOGY: Old English compound; MORPHOLOGY: N+N; endocentric, rectional. MOTIVATION: habitat & botanical information. Although the plant Mentha pulegium prefers moist sites as a habitat, its Old English vernacular name encodes the concept HILL. Bierbaumer (2, s.v. *hylwyrt*) therefore suggests that *hylwyrt* relates rather to Calamintha officinalis Moench, G *Bergminze*, a plant that grows in the mountains, than to Mentha pulegium. Another possible explanation is that the name is due to a misinterpretation of L *origanum* (the old Latin name of Mentha pulegium L.) as a cognate to Gk *ὄρος* 'hill, mountain' (Bierbaumer 3, s.v. *hylwyrt*). ASSOCIATIVE RELATIONS: conceptual contiguity & taxonomic sub-ordination.
▫ →*hyl*, →*wyrt*

hymele, noun
'Humulus lupulus L. / hop / Hopfen'
DOEPN (s.v. *hymele*): 11 occ.
ILLUSTRATIONS in manuscripts London, British Library, Cotton Vitellius C. iii, fol. 37r (Ohlgren 168(35)), fol. 40v (Ohlgren 168(42)), and Oxford, Bodleian Library, Ashmole 1431 (S.C. 7523), fol. 15v (Ohlgren 221(45)).

LATIN EQUIV. in the manuscripts: *bryonia, oenanthe, volvola* (= *convolvulus*).
ETYMOLOGY: Early Latin loan, < VL **humilus* ~ L *humulus* (Wollmann 1990: 164; see also *Holthausen*, s.v. *hymele*; *Genaust*, s.v. *humulus*); substitution of native suffix *–el* for Latin -**(i)lus* (Campbell 1959: § 518); possibly from Slavonic (Serjeantson 1935: 279).
▫ →*ēowuhymele*, →*hēahhymele*, →*hegehymele*

I

īfig, noun
'Hedera helix L. / IVY / Efeu'
DOEPN (s.v. *īfig*): 22 occ.
ILLUSTRATIONS in manuscripts London, British Library, Cotton Vitellius C. iii, fol. 55r (Ohlgren 168(70)), and Oxford, Bodleian Library, Bodley 130, fol. 55r (Ohlgren 222(111)).
LATIN EQUIV. in the manuscripts: *hedera, hedera nigra*.
ETYMOLOGY: Gmc **if-heȝ* < Gmc **ibahaujo* (*OED*, s.vv. *ivy*, *hay*).
▫ →*blæc īfig*, →*eorþīfig*

ilugsecg →*eolhsecg*

īsen, *īren*, noun
'IRON / Eisen'
▫ →*īsenhearde*

īsen-hearde, lit. 'iron-hard (plant)'
'? Centaurea jacea L. / knapweed, vervain / Wiesenflockenblume' or another species of Centaurea.
2 occ. in *Lācnunga* (40, 236); identification based on ModE *iron-weed* (Bierbaumer 2, s.v. *īsenhearde*).
ETYMOLOGY: WGmc compound (Carr 1939: 120); cf. OHG *ysenhard*, *isinhart*, etc. (Björkman 1902: 292); MORPHOLOGY: $N+N_{deadj}$; exocentric, metonymic compound; partial metaphor. MOTIVATION: texture & texture. The plant name probably refers to the tough, stringy stems of Centaurea (Bierbaumer 2, s.v. *īsenhearde*), which are compared to iron: 'a plant with stems hard as iron'; cf. also OHG *ysenhard*, *isinhart*, etc. ASSOCIATIVE RELATIONS: metaphorical similarity & conceptual contiguity.
▫ →*heard*, →*īsen*

L

lactuce, noun
'Lactuca L. / lettuce / Salat'
ILLUSTRATION in manuscript London, British Library, Cotton Vitellius C. iii, fol. 53r (Ohlgren 168(66)).
LATIN EQUIV. in the manuscripts: (*herba*) *lactuca leporina*, *sarrabum* (? = *scariola*).
ETYMOLOGY: Later Latin loan; < L *lactūca* (Campbell 1959: § 545; Bierbaumer 1, s.v. *lactūca*).
▫ →*wilde lactuce*

lamb, noun
'LAMB / Lamm'
▫ →*lambescærse*

lambes-cærse, lit. 'lamb's cress'
(i) '? Cardamine hirsuta L. / touch-me-not / Haariges Schaumkraut'; (ii) 'Capsella bursa-pastoris Med. / shepherd's purse / Hirtentäschel'; (iii) 'Thlaspi arvense L. / field-pennycress / Acker-Hellerkraut'
DOEPN (s.v. *lambes-cærse*): 4 occ.
LATIN EQUIV. in the manuscripts: *thlaspi*.
ETYMOLOGY: OE formation; MORPHOLOGY: N_{gen}+N; endocentric, rectional. MOTIVATION: habitat & botanical information. Cardamine hirsuta, Capsella bursa-pastoris and Thlapsi arvense strongly resemble each other: they all have tiny white flowers and small, purse-like capsules (Marzell, s.vv. *Capsella Bursa-pastoris* Med., *Thlaspi arvense* L.). Cardamine hirsuta strongly resembles cress and can also be eaten like cress (*Botanica*, s.v. *Cardamine*), which explains the use of the second element of the Old English plant name. They all grow in fields and pastures and are therefore easily associated with the sheep found there. Either 'a kind of *cærse* that grows where sheep live' (cf. the many names that relate to shepherds: G *Hirtentäschel*, ModE *shepherd's bag*, Du *herderstaschje*, Fr *bourse à pasteur*, Rum *traista ciobanului*, etc.; Marzell, s.v. *Capsella Bursa-pastoris* Med., 1) or 'a kind of *cærse* that is eaten by sheep' (cf. G *Schofskreidl*; Marzell, s.v. *Capsella Bursa-pastoris* Med., 3). ASSOCIATIVE RELATIONS: conceptual contiguity & taxonomic sub-ordination.
▫ →*cærse*, →*lamb*

lang, adjective
'LONG / lang'
▫ →*lange clite*, →*lang pipor*

lange clite (clīte), lit. 'long cleavers'
'Galium aparine L. / cleavers, goosegrass / Klebkraut, Labkraut'
Hapax legomenon (*Lācnunga*, 223).
ETYMOLOGY: OE formation; MORPHOLOGY: Adj+N; endocentric, copulative: attributive. MOTIVATION: size & botanical information. The plant is named so because of its long stems: "'Lange Klette' paßt gut zu den langen Stengeln von Galium aparine" ['Long cleavers' suits the long stems of Galium aparine well'] (Bierbaumer 2, s.v. *clite*). ASSOCIATIVE RELATIONS: conceptual contiguity & taxonomic sub-ordination.
▫ →*clite*, →*lang*

lang pipor, lit. 'long pepper'
'Piper longum L. / LONG PEPPER / Langer Pfeffer'
Hapax legomenon (*Lǣcebōc*, 56/16).
ETYMOLOGY: OE formation; hybrid formation (OE *pipor* < L *piper*; see →*pipor*). MORPHOLOGY: Adj+N; endocentric, copulative: attributive. MOTIVATION: size & botanical information. The fruits of the long pepper can reach the size of a match, i.e. about 0.16 ft. (5 cm) of length, while, for instance, the fruits of Piper nigrum are grain-sized (Marzell, s.vv. *Piper longum* L., *Piper nigrum* L.). ASSOCIATIVE RELATIONS: conceptual contiguity & taxonomic sub-ordination.
▫ →*lang*, →*pipor*

lāþ(e), noun
'pain, harm, annoyance / Schmerz, Schaden, Ärger'
▫ →*āt(t)orlāþe*

lāwer, *laur*, noun
'Laurus nobilis L. / laurel / Loorbeerbaum'
DOEPN (s.v. *laur*): 12 occ.
ETYMOLOGY: Early Latin loan; *lāwer* with *w*-insertion, while *laur* retained L *au* from L *laurus* (cf. Campbell 1959: § 509; Serjeantson 1935: 284; *OED*, s.v. *laure*); ModE *laure* and *laurel* go back to OFr *laurier* and *lorier* (*OED*, s.v. *laure*, *laurel*).
▫ →*lāwerbēam*, →*lāwertrēow*

lāwer-bēam, lit. 'laurel-tree'
'Laurus nobilis L. / laurel / Lorbeerbaum'
DOEPN (s.v. *laur-bēam*): 6 occ.
LATIN EQUIV. in the manuscripts: *daphnis*, *laurus*.
ETYMOLOGY: Independent formation in WGmc; cf. OHG *lôrboum*, MLG *lôrberenbôm*, MDu *lauwerboom* (Carr 1939: 133); hybrid compound (OE *lāwer* < L *laurus*; →*lāwer*). MORPHOLOGY: N+N; endocentric, copulative: subsumptive (pleonastic). MOTIVATION: plant & botanical information. The laurel is a kind of tree. ASSOCIATIVE RELATIONS: conceptual identity & taxonomic sub-ordination.
▫ →*bēam*, →*lāwer*

lāwer-trēow, lit. 'laurel-tree'
'Laurus nobilis L. / laurel / Lorbeerbaum'
DOEPN (s.v. *laur-trēow*): 8 occ.
ETYMOLOGY: Independent formation in WGmc (Carr 1939: 133); hybrid compound (OE *lāwer* < L *laurus*; →*lāwer*). MORPHOLOGY: N+N; endocentric, copulative: subsumptive (pleonastic). MOTIVATION: plant & botanical information. The laurel is a kind of tree. ASSOCIATIVE RELATIONS: conceptual identity & taxonomic sub-ordination.
▫ →*lāwer*, →*trēow*

lǣce, noun
'LEECH, physician / Arzt'
▫ →*lǣcewyrt*

lǣce-wyrt, lit. 'leech-wort'
(i) 'Plantago lanceolata L. / rib-wort / Spitzwegerich'; (ii) 'Agrostemma coronaria L. / rose campion / Stechnelke, Vexiernelke'
DOEPN (s.v. *lǣce-wyrt*): 13 occ.[e]
ILLUSTRATION in manuscript London, British Library, Cotton Vitellius C. iii, fol. 58r (Ohlgren 168(76)).
LATIN EQUIV. in the manuscripts: *lychnis* (*Stephanice*), *quinquenervia*.
ETYMOLOGY: OE compound; MORPHOLOGY: N+N; endocentric, rectional. MOTIVATION: healing virtue & botanical information. Plantago lanceolata is an ancient remedy, already known to and used by the Germanic tribes. Since Agrostemma coronaria was described by Dioscorides as a plant frequently used by doctors, too, both plants can be associated with physicians, i.e. can have been called 'leech-wort' (Marzell s.v. *Plantago lanceolata* L.; Bierbaumer 1, s.v. *lǣcewyrt*). ASSOCIATIVE RELATIONS: conceptual contiguity & taxonomic sub-ordination.
▫ →*lǣce*, →*wyrt*

læsse →*lȳtel*

(sēo) læsse bisceopwyrt, lit. 'lesser, smaller marshmallow'
'Stachys officinalis Trevisan / betony / Betonie'
DOEPN (s.v. *bisceop-wyrt*, *læsse*): 3 occ.
LATIN EQUIV. in the manuscripts: *vettonica*.
ETYMOLOGY: OE formation; MORPHOLOGY: Adj+N; endocentric, copulative: attributive. MOTIVATION: size & botanical information. Both Althaea officinalis (OE *bisceopwyrt*) and Stachys officinalis (*læsse bisceopwyrt*) have heart-shaped leaves, but those of Stachys officinalis Trevisan are smaller than those of the *bisceopwyrt* (Althaea officinalis L.). ASSOCIATIVE RELATIONS: conceptual contiguity & taxonomic sub-ordination.
▫ →*bisceopwyrt*, →*læsse*

lǣstan, verb
'LAST, follow, pursue endure / dauern'
▫ →*ēfenlāste*

le- = ? *lēah*, ? noun
'? LEA, meadow / Wiese' (cf. *Épinal Erfurt Glossary*, 99, no. 606).
▫ →*leloþre*

lēac, noun
(i) 'Allium L. / LEEK / Lauch'; (ii) 'garden-herb, plant / (Garten-)Pflanze'
Most compounds with *lēac* denote Allium species; compounds such as *lēactūn* 'kitchen garden' and *lēacweard* 'gardener', however, "suggest that *lēac* varieties were the prototypical garden produce [...]. This indicates the importance of alliums to the Anglo-Saxons, presumably because of their strong flavour" (Biggam 2011: 32). The denotation of *lēac* was extended to a generic use to denote all kinds of garden herbs and plants.
DOEPN (s.v. *lēac*): 34 occ.
LATIN EQUIV. in the manuscripts: *alium*, *caepe acerbum*, *ambila* (? for *alium*), *cartilago*.
ETYMOLOGY: Gmc **lauka-* (*Kluge*, s.v. *Lauch*).
▫ →*lēaccærse*
▫ →*brādelēac*, →*cīpelēac*, →*crāwanlēac*, →*crāw(e)lēac*, →*croplēac*, →*fugleslēac*, →*gārlēac*, →*hollēac*, →*hræfnes lēac*, →*hwītlēac*, →*por(r)lēac*, →*risclēac*, →*secglēac*, →*toscanlēac*, →*wudulēac*, →*ynnelēac*

lēac-cærse, lit. 'leek-cress', or 'herb-cress'
(i) 'Alliaria officinalis Andrz. / garlic mustard, English treacle / Knoblauchsrauke, Lauchkraut'; (ii) 'Lepidium sativum L. / garden cress / Gartenkresse'
DOEPN (s.v. *lēac-cærse*): 6 occ.[e]
LATIN EQUIV. in the manuscripts: *nasturtium*.
ETYMOLOGY: OE formation; MORPHOLOGY: N+N; endocentric, rectional; partial metaphor. MOTIVATION: ad (i) smell & botanical information. ad (ii) botanical information & botanical information. ad (i) A kind of cress with an onion-like smell (Bierbaumer 1, s.v. *lēaccærse*; Sauer 2000: 22); compare also G *Knoblauchskraut*, *Knoblauchsrauke*, etc. (Marzell, s.v. *Alliaria officinalis* Andrz.), which indicate that it is a kind of plant which smells like an onion without being a real onion. ad (ii) lit. 'herb-cress'. OE *lēac* can denote the

leek, but can also refer in a very general way to 'garden-herbs'. ASSOCIATIVE RELATIONS: ad (i): metaphorical similarity & taxonomic sub-ordination. ad (ii): taxonomic sub-ordination & taxonomic sub-ordination.

▫ →*cærse*, →*lēac*

lēaf, noun

(i) 'LEAF, shoot / Blatt'; (ii) 'sheet of paper / Papierblatt'; (iii) 'leave, permission / Erlaubnis'

DOEPN (s.v. *lēaf*): 117 occ.

LATIN EQUIV. in the manuscripts: *antes*, *folium*, *folium virens*, *holus herbarum*, *stipula*.

ETYMOLOGY: Gmc **lauba*- < IE *laubo*- (*Kluge*, s.v. *Laub*; *OED*, s.v. *leaf*).

▫ →*æppellēaf*, →*fīflēafe*, →*geormanlēaf*, →*hoclēaf*, →*līþelēaf*, →*seofonlēafe*, →*twīlēafa*, →*þrilēfe*

leahtric (*leahtroc*, *leactrog*, ? *lectere*), noun

'Lactuca sativa L. / lettuce / Lattich'

DOEPN (s.v. *leac-trog*): 7 occ.

LATIN EQUIV. in the manuscripts: *lactuca*; *þūþistel*.

ETYMOLOGY: Early Latin loan, < L *lactūca* (Wollmann 1990: 177); the forms *-lectric* (in *wudulectric*; see →*wuduleahtric*) and *lectere* are possibly changed in analogy to *lēac* 'leek, plant' (Campbell 1959: § 544), while *leahtroc* and *leactrog* seem to be, according to Campbell (1959: 241, fn. 2) different words.

▫ →*fenlectere*, →*wuduleahtric*

lectere →*leahtric*

le-loþre (? = *lēah*-), lit. 'meadow-burr'

'Rumex L. / dock / Ampfer'

DOEPN (s.v. *leloþre*): 8 occ.[e]

LATIN EQUIV. in the manuscripts: *lapathium*, *rhododaphne*.

ETYMOLOGY: Old English formation; MORPHOLOGY: Form and structure are not analysable; MOTIVATION: Unclear. According to Pheifer (*Épinal Erfurt Glossary*, 99, no. 606), *le*- is a corrupt form of *lēah* 'meadow' and *loþre* denotes the plant burr. No entry '*loþre*' in *ClH* and *Holthausen*; *BT* (s.v. *leloþre*, following Cockayne, *Leechdoms* 2: 78) supposes that it is an erroneous form of *gelodwyrt*. ASSOCIATIVE RELATIONS: unclear.

▫ →*le*-, →*-loþre*

lēo, noun

'lion / Löwe'

ETYMOLOGY: Later Latin loan (Serjeantson 1935: 285); < L *leo*, acc. sg. L *leōnem* 'lion' < Gk *λέων* (*OED*, s.v. *lion*); ModE *lion* < OFr *liun* < L *leo*, acc. sg. L *leōnem* (*OED*, s.v. *lion*).

▫ →*lēonfōt*

lēon-fōt, lit. 'lion's foot'

'Alchemilla vulgaris L. / lady's mantle / Frauenmantel'

DOEPN (s.v. *lēon-fōt*): 9 occ.

ILLUSTRATION in manuscript London, British Library, Cotton Vitellius C. iii, fol. 24v (Ohlgren 168(10)).

LATIN EQUIV. in the manuscripts: *leontopodion*, *pes leonis*.

ETYMOLOGY: Independent formation in WGmc (Carr 1939: 138); cf. OHG *lewenfuoz*, *lewinfuoz* (Björkman 1901: 209; Carr 1939: 138); hybrid formation (OE *lēo* < L *leo*, cc. . *leōnem*), loan translation of L *pēs leōnis* or

leontopodium (see MOTIVATION). MORPHOLOGY: N_{gen}+N; exocentric, metaphorical compound. MOTIVATION: animal & shape. The Old English plant name is a loan translation and refers to the plant's leaves, which are formed like a lion's foot and perhaps to the yellow flowers that resemble lions in colour (Marzell, s.v. *Alchemilla vulgaris* L). ASSOCIATIVE RELATIONS: conceptual contiguity & metaphorical similarity.
▫ →*fōt*, →*lēo*

leoþo, noun
'soap / Seife'
▫ →*leoþowyrt*

leoþo-wyrt (? = **leaþorwyrt*), lit. 'soap-wort', or →*liþwyrt.*
'? Saponaria officinalis L. / soapwort, latherwort / Gemeines Seifenkraut'
1 occ. (*Lǣcebōc*, 14/14). The identification is unclear and based on the emendation of *leoþowyrt* to *leaþorwyrt* lit. 'soap-wort'; for the emendation to *liþwyrt* lit. 'joint/limb-wort', see →*liþwyrt.*
ETYMOLOGY: OE formation; MORPHOLOGY: N+N; endocentric, rectional. MOTIVATION: product & botanical information. The plant's roots secrete an agent which creates foam when it comes into contact with water (Marzell, s.v. *Saponaria officinalis*); *leoþowyrt* is, thus, 'a plant that produces soap'. ASSOCIATIVE RELATIONS: conceptual contiguity & taxonomic sub-ordination.
▫ →*leoþo*, →*wyrt*

lida, noun
'sailor / Seemann'
▫ →*sumerlida*

lifer, noun
'LIVER / Leber'
▫ →*liferwyrt*
▫ →*ēalifer*

lifer-wyrt, lit. 'liver-wort'
'Anemone hepatica L. / LIVERWORT / Leberblümchen'
Not in Bierbaumer (1, 2, 3); hapax legomenon (Napier 1890: 325-6).
ETYMOLOGY: Gmc compound; OHG *lebarwurzkrût*, etc., MLG *leverwort*, Dan *leverurt*, Sw *lefverört*, Icel *lifrarurt* (Carr 1939: 50); MORPHOLOGY: N+N; endocentric, rectional; partial metaphor. MOTIVATION: shape & botanical information. The plant's leaves are liver-shaped; cf. OHG *leberwurtzkraut*, *liberblume*, *leb(er)blum*, etc. (Marzell, s.v. *Anemone hepatica* L.). Carr gives an alternative explanation: "The plant was so called because it was used against liver disease" (Carr 1939: 50). ASSOCIATIVE RELATIONS: metaphorical similarity & taxonomic sub-ordination.
▫ →*lifer*, →*wyrt*

līn, noun
'flax / Flachs'
DOEPN (s.v. *līn-*): 15 occ.
ETYMOLOGY: Either from Gmc **leina-* (*Kluge*, s.v. *Lein*) or **līnam-* (*ODEE*, s.v. *line*[1], or a loan from Latin (L *līnum*; cf. *ODEE*, s.v. *line*[1]).
▫ →*līnwyrt*

līn-wyrt, lit. 'flax-wort'
'Linum usitatissimum L. / line, flax / Lein, Flachs'
DOEPN (s.v. *līn-wyrt*): 15 occ.
ETYMOLOGY: OE formation; it is not entirely clear whether *līn* is a loan from Latin (L *līnum*) or whether both go

back to an Indo-European etymon (see →*līn*, cf. also *Kluge*, s.v. *Lein*; *ODEE*, s.v. *line*[1], *OED*, s.v. *linen*). MORPHOLOGY: N+N; endocentric, copulative: subsumptive (pleonastic). MOTIVATION: plant & botanical information. The *līn* 'flax' is a kind of *wyrt* 'plant'. ASSOCIATIVE RELATIONS: conceptual identity & taxonomic sub-ordination.
▫ →*līn*, →*wyrt*

liþ, noun
'limb, joint / Gliedmaße, Gelenk'
▫ →*liþwyrt*

līþe, adjective
'mild, gentle, soft, calm / mild, weich, ruhig'
▫ →*līþelēaf*

līþe-lēaf, lit. 'mild leaf'
unidentified plant
Hapax legomenon (Meritt 1945: 61, no. 73, Dc 185; Varnhagen 1902: 14).
LATIN EQUIV. in the manuscript: ? *aviane*.
ETYMOLOGY: OE formation; MORPHOLOGY: Adj+N; exocentric, metonymic compound. MOTIVATION: healing virtues & part of plant. Neither the lemma *aviane*, nor the Old English gloss have been identified yet (no entry in the *DMLBS*; Bierbaumer 2, s.v. *līþelēaf*). Gardner (1968: 221) equates *līþelēaf* with 'camomile'. This identification would fit the assumption that the first element, *līþe-*, is not the noun denoting the 'limb, joint, etc.', but the adjective *līþe* 'gentle, soft, calm, etc.' (*ClH*, s.v. *līþe*): the camomile (Matricaria chamomila L.) is a soothing, antiseptic plant. ASSOCIATIVE RELATIONS: conceptual contiguity & conceptual contiguity (meronymy).
▫ →*lēaf*, →*līþe*

liþ-wyrt, lit. 'joint/limb-wort'
(i) 'Sambuculus ebulus L. / dwarf-elder / Zwergholunder', (ii) '? Galium L. / bedstraw / Labkraut'; (iii) '? Rubia tinctorum L. / madder / Färberröte'; (iv) '? Viburnum lantana L. / cotton-tree, wayfaring-tree / Wolliger Schneeball'; (v) '? Ruta graveolens L. / common rue / Raute'
DOEPN (s.v. *liþwyrt*): 19 occ.; →*leoþowyrt* is possibly a variant of *liþwyrt*.
ILLUSTRATIONS in manuscript London, British Library, Cotton Vitellius C. iii, fol. 30v (Ohlgren 168(22)), fol. 56r (Ohlgren 168(72)).
LATIN EQUIV. in the manuscripts: *eriphion*, *ostriago*.
ETYMOLOGY: In view of OHG *lidewrtz*, *lidwrtz* (Björkman 1902: 294), possibly a formation in West Germanic; MORPHOLOGY: N+N; endocentric, rectional. MOTIVATION: shape or illness & botanical information. Since Sambucus ebulus and Viburnum lantana have flexible stems, the element *liþ-* may encode the shape of the stems (Bierbaumer 3, s.v. *liþwyrt*; Bierbaumer 1, s.v. *leoþuwyrt*). On the other hand, *Lǣcebōc* (40/14) recommends the use of *liþwyrt* for rheumatic pains: "With liþwærce cnua liþwyrt [...]" ['Against pain in the limbs pound *liþwyrt*']. This fits with the plants Galium and Rubia tinctorum since these plants not only have heavily structured stems (made up of several limbs / joints; cf. Bierbaumer 1, 2 and 3, s.v. *liþwyrt*), but are also recommended for rheumatic pains according to the *signatura rerum*.

ASSOCIATIVE RELATIONS: conceptual contiguity & taxonomic sub-ordination.
▫ →*liþ*, →*wyrt*

-loþre, noun
'? Rumex L. / burr / Ampfer' (cf. *Épinal Erfurt Glossary*, 99, no. 606)
Not attested independently.
▫ →*leloþre*

lufe-stice, lit. 'love-stitch'
'Levisticum officinale Koch 1824 / lovage / Liebstöckel'
DOEPN (s.v. *lufestice*): 22 occ.
LATIN EQUIV. in the manuscripts: *levisticus*, *panacea* (*panaces*).
ETYMOLOGY: OE formation; Early Latin loan changed by folk etymology; < MedL *levisticum*, *libestica*, *livisticum*, *lufesticum*, etc. < CL *ligusticum* (*OED*, s.v. *lovage*; cf. also Wollmann 1990: 164; *Kluge*, s.v. *Liebesstöckel*); Birkhan (2012: 136) notes that the Latin plant name is, ultimately, a loan from Gk *λιγύστικον*, *λιγύστικος* 'the one originating from Liguria' (see also *OED*, s.v. *lovage*; and *Lewis and Short*, s.v. *Ligures*, *ligusticum*). MORPHOLOGY: N+N; ? exocentric; MOTIVATION: secondary, phonological motivation in both elements: sound pattern & sound pattern. MedL *levisticum* was borrowed into OE and then changed formally on the basis of phonological similarity between the element *levi-* and OE *lufu* 'love' and between *-sticum* and OE *stice* 'stitch', but there is no plausible extra-linguistic, botanical motivation for the plant name; similar development of the Latin loan in Old High German, where the plant name is attested as *liubistekal*, *lubestechel*, *lobstech*, *lubestekil*, etc.; G *Liebstöckel* (Björkman 1904: 188; see also Birkhan 2012: 137). ASSOCIATIVE RELATIONS: phonological similarity & phonological similarity.
▫ →*lufu*, →*stice*

lufian, verb
'(to) LOVE / lieben'
▫ →*menlufigende*

lufu, noun
'LOVE / Liebe'
▫ →*lufestice*

lungen, noun
'LUNG / Lunge'
▫ →*lungenwyrt*

lungen-wyrt, lit. 'lung-wort'
(i) 'Hieracium murorum L. / golden LUNGWORT, wall hawkweed / Mauer-Habichtskraut'; (ii) 'Helleborus niger L. / black hellebore / Schwarzer Nieswurz'
DOEPN (s.v. *lungen-wyrt*): 2 occ.
LATIN EQUIV. in the manuscripts: *helleborus niger*.
ETYMOLOGY: Independent formation in WGmc (Carr 1939: 133), cf. OHG *lungvurtz*, etc. (Björkman 1902: 294); loan rendition of L *pulmonaria* (Björkman 1902: 294). MORPHOLOGY: N+N; endocentric, rectional. MOTIVATION: illness & botanical information. Both Hieracium murorum and Helleborus niger were used for diseases of the lungs; cf. also Fr *herbe aux poumons*, Dan *lungeurt* (17th c.), ModE *golden lungwort*, Du *gulden longekruid*, etc. (Marzell, s.vv. *Hieracium murorum* L., *Helleborus niger* L.). ASSOCIATIVE

RELATIONS: conceptual contiguity & taxonomic sub-ordination.
▫ →*lungen*, →*wyrt*

lūs, noun
'LOUSE / Laus'
▫ →*lūsþorn*

lūs-þorn, lit. 'louse thorn-bush'
'Evonymus europaeus L. / spindle-tree / Spindelbaum, Spindelstrauch'
2 occ. (*Charters* (Earle), 445); not in Bierbaumer (1, 2, 3); identification according to *ClH* (s.v. *lūsþorn*) and Hooke (2003: 28).
ETYMOLOGY: OE formation; MORPHOLOGY: N+N; endocentric, rectional. MOTIVATION: illness & botanical information. The spindle-tree is a bush whose oil, which is extracted from the fruits, can be used against lice and fleas; cf. Du *luizenboom*, G *Lausbaum, Luusbeer*, etc. (Marzell, s.v. *Euonymus europaeus* L.). ASSOCIATIVE RELATIONS: conceptual contiguity & taxonomic sub-ordination.
▫ →*lūs*, →*þorn*

lust, noun
'LUST / Lust'
▫ →*lustmoce*

lust-moce, lit. 'lust-slime'
'Cardamine pratensis L. / lady's mock, milkmaid / Wiesen-Schaumkraut'
4 occ., all in *Lǣcebōc* (Bierbaumer 1, s.v. *lustmoce*).
ETYMOLOGY: OE formation; MORPHOLOGY: N+N; exocentric, metonymic compound; The second element, *-moce* is not attested in Old English outside this compound, but has cognates in other Germanic languages, e.g. Goth *mūks*, OIcel *mjūkr*, OFris *mūka*, etc. (*Holthausen*, s.v. *moce*; not in *ClH*). MOTIVATION: sound pattern & part of plant. When cicada larvae suck the stems of Cardamine pratensis, a foamy, mucuous substance is secreted (Marzell, s.v. *Cardamine pratensis* L., 5). Because of the first element, OE *lust* 'lust', Bierbaumer (1, s.v. *lustmoce*) suggests the allusion to sperm. However, since both lice and larvae are tiny creatures and both of them cling to stems and leaves of plants, sucking them, I would suggest that the first element, *lust*, is only a folk etymological change of originally *lūs* 'louse' to *lust* 'louse' based on the phonological similarity of *lust* and *lūs*. The motivation and the associative relation for the first element are based on this assumption. ASSOCIATIVE RELATIONS: phonological similarity & conceptual contiguity (meronymy).
▫ →*lust*, →*-moce*

lyb(b), noun
'poison, charm / Gift, Zauber'
ETYMOLOGY: Gmc word, but ultimate origin unclear (cf. *OED*, s.v. *lib*, *cheeselip*, *-lep*).
▫ →*lybcorn*
▫ →*oxnalybb*

lyb-corn, lit. 'poison-corn'
'Euphorbia lathyris L. / caper-spurge / Spring-Wolfsmilch'
DOEPN (s.v. *lyb-corn*): 33 occ.[e]
LATIN EQUIV. in the manuscripts: *cartamus*, *citocatia*, *lathyris*, *tithymalus*, *tithymalus calatides* (? for *galactites*).

ETYMOLOGY: OE formation; MORPHOLOGY: N+N; exocentric, metonymic compound. MOTIVATION: pathogen & part of plant. Euphorbia lathyris is a poisonous plant with peculiar, caper-like fruits (Marzell, s.v. *Euphorbia lathyris* L.), which are expelled explosively from the ripe capsule (see →*springwyrt*); both elements of the plant name refer to the fruits: they are poisonous and round. ASSOCIATIVE RELATIONS: conceptual contiguity & conceptual contiguity (meronymy).

▫ →*corn* 1, →*lyb*

lȳtel, comparative *læsse*, adjective
'small, LITTLE / klein'

▫ →*læsse bisceopwyrt*, →*lȳtle culmille*
▫ →*curmealle sēo læsse*, →*hierdewyrt sēo læsse*

lȳtle culmille, lit. 'small centaury'
'Centaurium umbellatum Gilibert / common centaury / Echtes Tausendgüldenkraut'
Hapax legomenon (*Lǣcebōc*, 19/1-2).
ETYMOLOGY: OE formation; possibly a loan translation of L *centauria minor*; →*curmealle sēo læsse*. MORPHOLOGY: Adj+N; endocentric, copulative: attributive. MOTIVATION: size & botanical information. *Lȳtle culmille* denotes a smaller species of *curmealle*. ASSOCIATIVE RELATIONS: conceptual contiguity & taxonomic sub-ordination.

▫ →*lȳtle*, →*culmille*

M

magdala-, noun
'fruit of Prunus amygdalus Batsch 1801 / almond / Mandel'
Not attested independently.
ETYMOLOGY: Later Latin loan < OE *magdala-* < OE *amigdal* (Serjeantson 1935: 284); < L *amygdala, amygdalum* < Gk *ἀμυγδάλη, ἀμύγδαλον, ἀμύγδαλος* (cf. *Genaust*, s.v. *amygdálinus*; a loan word from an unknown language into Greek; Feulner 2000: 72-3).

▫ →*magdalatrēow*

magdala-trēow, lit. 'almond-tree'
'Prunus amygdalus Batsch 1801 / almond-tree / Mandelbaum'
Hapax legomenon (*Antwerp-London Glossaries*, 129/11).
LATIN EQUIV. in the manuscript: *amygdala, nucicula*.
ETYMOLOGY: OE compound; hybrid compound (OE *magdala-* < L *amygdala, amygdalum*; →*magdala-*). MORPHOLOGY: N+N; endocentric, rectional. MOTIVATION: part of plant & botanical information. Two interpretations are possible: (1) The *magdala-* 'almond tree' is a *trēow* 'tree', or (2) This kind of tree bears *magdala-* 'almonds'. Since the Latin loan *amigdal* used to refer to the almonds only, I favour the second motivation, assuming that *magdala-* was borrowed to denote 'almonds', too. ASSOCIATIVE RELATIONS: conceptual contiguity

(meronymy) & taxonomic sub-ordination.
▫ →*magdala-*, →*trēow*

man, *mon*, noun
'MAN, human being / Mann, Mensch'
▫ →*menlufigende*, →*monnes sūre*

mapul-, noun
'Acer campestre L. / MAPLE / Ahorn'
Not attested independently.
ETYMOLOGY: Origin unclear; not attested independently; probably shortened (back-formation) from *mapuldor* in analogy to *apul* 'apple' and *apuldor* 'apple-tree' (*OED*, s.v. *maple tree*; *Kluge*, s.v. *Maßholder*); OE *mapul-* > ModE *maple*.
▫ →*mapultrēow*

mapul-trēow, lit. 'maple-tree'
'Acer campestre L. / MAPLE (TREE) / Feldahorn'
3 occ. (*Charters* (Birch), 1: no. 216; Hearne 1723: 245-6; *Bodley 730*, no. 184).
LATIN EQUIV. in the manuscripts: *acerabulus*.
ETYMOLOGY: OE formation; MORPHOLOGY: N+N; endocentric, copulative: sub-sumptive (pleonastic). MOTIVATION: plant & botanical information. The *mapul-* is a kind of tree. ASSOCIATIVE RELATIONS: conceptual identity & taxonomic sub-ordination.
▫ →*mapul-*, →*trēow*

mædere, noun
'Rubia tinctorum L. / madder / Echte Färberröte'
DOEPN (s.v. *mædere*): 19 occ.
ILLUSTRATIONS in manuscripts London, British Library, Cotton Vitellius C. iii, fol. 36v (Ohlgren 168(34)), and Oxford, Bodleian Library, Ashmole 1431 (S.C. 7523), fol. 14v (Ohlgren 221(40)).
LATIN EQUIV. in the manuscripts: *anchorum*, *grias*, *rubia*, (*herba*) *sandix*, *hyacinthinus*, *veneria*, *vermiculus*.
ETYMOLOGY: Of Gmc, perhaps even IE origin, but the ultimate etymology is unclear (*OED*, s.v. *madder*).
▫ →*feldmædere*

mæg(e)þ(e), noun
'Chamaemelum nobile (L.) All. / camomile, / Magerwiesen-Margerite'; (ii) 'maiden, virgin / Jungfrau, junge Frau'
DOEPN (s.v. *mageþe*): 31 occ.
ILLUSTRATIONS in manuscripts London, British Library, Cotton Vitellius C. iii, fol. 29v (Ohlgren 168(20)), and Oxford, Bodleian Library, Bodley 130, fol. 44r (Ohlgren 222(86)).
LATIN EQUIV. in the manuscripts: *beneolens*, *caluna*, *chamaemelon*, *ophthalmus*, *putridus*.
ETYMOLOGY: Gmc **magaþi-* 'maiden, virgin' < IE **mak-* or **maki-* (*Kluge*, s.v. *Magd*).
▫ →*hwīt mæg(e)þ(e)*, →*rēade mæg(e)þ(e)*, →*wilde mæg(e)þ(e)*

-mæringc, noun
'? Ocimum basilicum L. / sweet basil / Basilikum'
Attested only in the bounds *mæringcwudu* '? mastic of sweet basil' (*ClH*, s.v. *mæringcwudu*) and →*hwīt mæringc*. For the identification see *BT* (s.v. *mæring*); cf. also *DOEPN* (s.v. *hwīt mæringc*).
▫ →*hwīt mæringc*

mæsten, noun
'MAST / Mast'
▫ →*mæstentrēow*

mæsten-trēow, lit. 'mast tree'
'Quercus suber L. / cork oak / Korkeiche'
Hapax legomenon (*Antwerp-London Glossaries*, 121/15).
LATIN EQUIV. in the manuscripts: *suber* (*suberies*).
ETYMOLOGY: OE formation; MORPHOLOGY: N+N; endocentric, rectional. MOTIVATION: food & botanical information. Isidore's remarks concerning the cork oak may serve as an explanation for the Old English plant name: "Suberies arbor ex qua ualidissimus cortex natatorius extrahitur. Et ideo appellata suberies eo quod fructus eius sues edunt. Porcorum enim sunt alimenta, non hominum; et dicta suberies quasi subedies" ['The cork oak from which we can have a very strong bark which helps floating [on water]. And it is called suberies because the pigs eat its fruits. Namely, they are the food of pigs, not of humans, and *suberies* is said as if *subedies*'] (*Etymologiae*, 17: 7, 27). Here, an analogy is drawn between *suberies* and *subedies* which is based on the phonological similarity of the two Latin terms and which is commented upon ironically by the editor of the Isidore's *Etymologies*, J. André, as a means of creating "de «brillantes» etymologies" [''brilliant' etymologies'].[312] ASSOCIATIVE RELATIONS: conceptual contiguity & taxonomic sub-ordination.
▫ →*mæsten*, →*trēow*

mealwe, noun
'Malva sylvestris L. / MALLOW / Malve'
DOEPN (s.v. *mealwe*): 7 occ.
LATIN EQUIV. in the manuscripts: *malva*; Greek equivalent *μαλάχη*.
ETYMOLOGY: Early Latin loan (Wollmann 1990: 177); < L *malva* (*OED*, s.v. *mallow*); according to Wollmann (1990: 177), L *malva* < Gk *μαλάχη* (see also *Lewis and Short*, s.vv. *malachē*, *molochē*, *malva*). According to *Genaust* (s.v. *Málva*), however, L *malva* is not from the Greek loan *μαλάχη* but was borrowed with the Greek cognate from an unknown Mediterranean language.
▫ →*merscmealwe*, →*wilde mealwe*

mear →*mearh*

mear-gealla (*mer-gelle*), lit. '? horse-gall'
'? Gentiana pneumonanthe L. / lung-flower / Lungen-Enzian'
As a simplex, attested in *Lācnunga* (238). For attestations in bounds, see →*merscmeargealla*.
ETYMOLOGY: OE formation; MORPHOLOGY: N+N; exocentric, metaphorical compound. MOTIVATION: animal & taste. Neither the identification of the plant nor the linguistic analysis of the Old English plant name is entirely clear. The first element may be read as

312 "Sur la correspondance phonétique d/r admise par Isidore pour laudea/laurea, et auricula/audicula, qui permet de ‹‹brillantes›› étymologies, cf. *etym.* 17, 7, 2 et note 172" ['For the phonetic rapport between d and r, assumed by Isidore for *laudea/laurea* and *auricula/audicula*, which allows to create 'brilliant etymolgies, see *etym.* 17, 7, 2 and footnote 172'] (*Etymologiae* 17: 104, fn. 237).

OE *mearh*, ModE *mare* (cf. G *Mähre*), the second element, *gealla* 'gall', can refer metaphorically to the bitter taste of the plant, but can also be interpreted as 'tumour'. I favour the interpretation as 'bitter taste' in analogy to OE *eorþgealla* and *ramgealla*, where *gealla* refers to the bitterness of the plants. ASSOCIATIVE RELATION: metaphorical similarity.
▫ →*mearh*, →*gealla*
▫ →*merscmeargealla*

mearh, *mear-*, noun
'MARE / Mähre, Pferd'
▫ →*meargealla*

medu, noun
'MEAD / Met'
▫ →*meduwyrt*

medu-wyrt (*mede-*, *medo-*), lit. 'mead-wort'
(i) 'Filipendula ulmaria Maxim. / meadow-sweet / Mädesüß, Wiesenkönigin'; (ii) 'Melissa officinalis L. / balm / Zitronenmelisse'; (iii) '? Melilotus officinalis Lam. / ribbed melilot / Echter Steinklee'
DOEPN (s.v. *medo-wyrt*): 20 occ.
LATIN EQUIV. in the manuscripts: ? *axica*, *melitina*, *papamo*, *rubia*.
ETYMOLOGY: Gmc compound; OHG *medewrz*, *medewrt*, *metewurz*, etc. (Björkman 1902: 306); MLG *medewurt*, Dan *mjørdurt*, Sw *mjödört*, Icel *mjaðurt* (Carr 1939: 50). MORPHOLOGY: N+N; endocentric, rectional; partial metaphor. MOTIVATION: taste & botanical information. *Medu* 'mead' indicates that the flowers of the plant were used as sweetening agents in beverages: the stems of Filipendula ulmaria are sweet in taste (cf. also G *Mädesüß*, ModE *meadow-sweet*) and Melissa officinalis is still used in soft drinks. ASSOCIATIVE RELATIONS: metaphorical similarity & taxonomic subordination.
▫ →*medu*, →*wyrt*

melde, noun
'Atriplex hortensis L. / orache / Spanischer Spinat, Gartenmelde'
Bierbaumer (2, s.v. *melde*): 2 occ. (*Lācnunga*, 48, 791).
ETYMOLOGY: ? cognate of Gmc **melwa-* 'flour' (*Kluge*, s.v. *Mehl*).
▫ →*tūnmelde*

mele, noun
'flour / Mehl'
▫ →*meledēaw*

mele-dēaw (*mil-*), lit. 'flour-dew'
(i) 'MILDEW / Mehltau'; (ii) honey-dew / Honigtau'; (iii) 'nectar / Nektar'
2 occ. (*Cleopatra Glossary* ((Stryker), 326, no. 143); according to *DOEPN* (s.v. *mildēaw*) also attested in manuscript Oxford, Bodley 730, but this is not confirmed by the edition (see *Bodley 730*).
LATIN EQUIV. in the manuscripts: *nectar*.
ETYMOLOGY: WGmc compound (Carr 1939: 101); OHG *militou*, OLG *milidou*. MORPHOLOGY: N+N; exocentric, metaphorical compound. MOTIVATION: colour & overall appearance. Mildew is white as flour and covers the leaves of plants like dew. ASSOCIATIVE RELATIONS: metaphorical similarity & metaphorical similarity.
▫ →*dēaw*, →*mele*

men-lufigende, lit. 'men-loving (plant)'
'Galium aparine L. / cleavers, goose-grass / Klebkraut, Labkraut'
Hapax legomenon (*Herbarium*, no. 174).
ILLUSTRATION in manuscript London, British Library, Cotton Vitellius C. iii, fol. 68r (Ohlgren 168(96)).
EQUIV. in the manuscript: *fhylantropos* (for Gk *φιλάνθρωπος*).
ETYMOLOGY: OE formation; loan translation of Gk *φιλάνθρωπος* without Latin mediation (Sauer 1992b: 390): "Đās wyrte man fhylantropos nemneþ þæt ys on ure geþeode menlufigende [...]" ['The plant named *philanthropos*, is, in our language, *menlufigende*'] (*Herbarium*, no. 174). MORPHOLOGY: synthetic formation, $N{+}N_{\text{deverbal}(\textit{-end})}$; endocentric, rectional. MOTIVATION: person & action. Cleavers cleave to the passers-by, i.e. humorously expressed, they seem to 'love' men: "[...] for þȳ heo wyle hrædlice to þ m men geclyfian" ['because it will cleave promptly to men'] (*Herbarium*, no. 174). ASSOCIATIVE RELATIONS: conceptual contiguity & taxonomic subordination.
▫ *→lufian, →men*

merce, noun
'Apium graveolens L. / wild celery / Sellerie'
DOEPN (s.v. *merce*): 72 occ.
ILLUSTRATIONS in manuscripts London, British Library, Cotton Vitellius C. iii, fol. 54v (Ohlgren 168(69)), and Oxford, Bodleian Library, Ashmole 1431 (S.C. 7523), fol. 28v (Ohlgren 221(109)).
LATIN EQUIV. in the manuscripts: *apium, menta, mercurialis*.
ETYMOLOGY: Possibly derived from *mere* 'sea' (Bierbaumer 1, s.v. *merce*); < ? Gmc **mariska-* 'belonging to the sea' (cf. *Kluge*, s.v. *Marsch*[2]).
▫ *→stānmerce, →swīnesmerce, →wudu-merce*

mersc, noun
'MARSH / Moor, Sumpf'
▫ *→merschōfe, →merscmealwe, →mersc-meargealla*

mersc-hōfe, lit. 'marsh-hove'
'Glechoma hederaceum L. / ground-ivy / Gundelrebe'
Hapax legomenon (*Lǣcebōc*, 29/27).
ETYMOLOGY: OE formation; MORPHOLOGY: N+N; endocentric, rectional. MOTIVATION: habitat & botanical information. The plant *merschōfe* is a kind of *hōfe* 'hove, ivy' that prefers marshy grounds. ASSOCIATIVE RELATIONS: conceptual contiguity & taxonomic sub-ordination.
▫ *→hōfe, →mersc*

mersc-mealwe, lit. 'marshmallow'
'Althaea officinalis L. / MARSH-MALLOW / Eibisch'
DOEPN (s.v. *mersc-mealwe*): 14 occ.
ILLUSTRATION in manuscript London, British Library, Cotton Vitellius C. iii, fol. 33v (Ohlgren 168(28)).
LATIN EQUIV. and OE synonyms in the manuscripts: *althaea, fraga, hibiscum*; *eorþber(i)(g)e, strēawber(i)(g)e*.
ETYMOLOGY: OE formation; hybrid compound (OE *mealve* < L *malva*; see *→mealwe*). MORPHOLOGY: N+N; endocentric, rectional. MOTIVATION: habitat & botanical information. This kind of *mealwe* 'mallow' prefers moist ground, such as can be found in the vicinity of marshes (*DOEPN*, s.v. *mersc-mealwe*).

ASSOCIATIVE RELATIONS: conceptual contiguity & taxonomic sub-ordination.
▫ *→mealwe*, *→mersc*

mersc-meargealla (*-mergylle*), lit. 'marsh-gentian'
'? Gentiana pneumonanthe L. / lung-flower / Lungenenzian'
DOEPN (s.vv. *mersc-mear-gealla, mersc-mear-gylle*): 4 occ.
ETYMOLOGY: OE formation; MORPHOLOGY: N+N; endocentric, rectional. MOTIVATION: habitat & botanical information. This kind of *meargealla* is a plant which grows in moist ground, i.e. in the vicinity of marshes (Marzell, s.v. *Gentiana pneumonante* L.). ASSOCIATIVE RELATIONS: conceptual contiguity & taxonomic sub-ordination.
▫ *→meargealla*, *→mersc*

mete, noun
'food / Essen, Futter'
▫ *→meteswamm*, *→meteþistel*
▫ *→cicena mete*

mete-swamm, lit. 'food-mushroom'
'edible mushroom / eßbarer Pilz'
Hapax legomenon (*Antwerp-London Glossaries*, 112/9).
LATIN EQUIV. in the manuscript: *fungus*.
ETYMOLOGY: OE formation; MORPHOLOGY: N+N; endocentric, rectional. MOTIVATION: food & botanical information. As the element mete 'food' indicates, this kind of mushroom is edible. ASSOCIATIVE RELATIONS: conceptual contiguity & taxonomic sub-ordination.
▫ *→mete*, *→swamm*

mete-þistel, lit. 'food-thistle'
'? Carlina acaulis L. / carline thistle / Stengellose Eberwurz'
2 occ. (*Durham Plant Name Glossary*, no. 86; no. 139).
LATIN EQUIV. in the manuscript: ? *camerion.*
ETYMOLOGY: OE formation; MORPHOLOGY: N+N; endocentric, rectional. MOTIVATION: food & botanical information. The lower part of the fruit of Carlina acaulis, which is a kind of *þistel* 'thistle', is edible, hence the first element *mete* 'food' (Marzell, s.v. *Carlina acaulis* L.). ASSOCIATIVE RELATIONS: conceptual contiguity & taxonomic sub-ordination.
▫ *→mete*, *→þistel*

micel, comparative *māra*, *māre*, adjective
'big / groß'
▫ *→micel fearn*, *→micel þung*, *→micel worþignet(e)l(e)*
▫ *→curmealle sēo māre*

micel fearn, lit. 'big fern'
'? Pteris aquilina L. / bracken / Adlerfarn'
Hapax legomenon (*Lǣcebōc*, 39/2).
ETYMOLOGY: OE formation; MORPHOLOGY: Adj+N; endocentric, copulative: attributive. MOTIVATION: size & botanical information. Pteris aquilina can grow as high as c. 6.6 ft. (2 m), i.e. is a 'big kind of fearn' (see Bierbaumer 1, s.v. *fearn*). ASSOCIATIVE RELATIONS: conceptual contiguity & taxonomic sub-ordination.
▫ *→fearn*, *→micel*

micel þung, lit. large poisonous plant'
'? Aconitus napellus L. / wolf's bane / Eisenhut, Sturmhut'
Hapax legomenon (*Lǣcebōc*, 46/26).
ETYMOLOGY: OE formation; MORPHOLOGY: Adj+N; endocentric, copulative: attributive. MOTIVATION: size & botanical information. According to Hoops (1889: 83-4), *þung* denoted in the first place all kinds of poisonous plants; *se micla þung* implies that the Anglo-Saxons knew two different species of *þung*, one which was smaller, the other one bigger, among other kinds of →*þung*. ASSOCIATIVE RELATIONS: conceptual contiguity & taxonomic sub-ordination.
▫ →*micel*, →*þung*

micel worþignet(e)l(e), lit. 'big farm nettle'
'Urtica dioica L. / great nettle, stinging nettle / Große Brennnessel'
Hapax legomenon (*Lǣcebōc*, 35/34).
ETYMOLOGY: OE formation; MORPHOLOGY: Adj+N; endocentric, copulative: attributive. MOTIVATION: size & botanical information. Urtica dioica can reach the height of c. 4.9 ft. (1.5 m) and grows predominantely along the walls of houses and along fences, that is in cultivated areas (Marzell, s.v. *Urtica dioica* L.); cf. also →*grēate net(e)l(e)*. ASSOCIATIVE RELATIONS: conceptual contiguity & taxonomic sub-ordination.
▫ →*micel*, →*worþignet(e)l(e)*

micge, noun
'urine / Urin'
▫ →*hundesmicge*

milsc, adjective
'sweet / süß'
▫ →*milscapuldor*, →*sūrmilscapuldor*

milsc-apuldor (*mirc-*), (*-dre*), lit. 'sweet apple-tree'
'Malus communis L. / apple-tree / Apfelbaum'
DOEPN (s.v. *apuldor, milsc*): 5 occ. [e]
LATIN EQUIV. in the manuscripts: *melarium*.
ETYMOLOGY: OE formation; MORPHOLOGY: Adj+N; endocentric, copulative: attributive. MOTIVATION: taste & botanical information. The native apple-tree of England was not Malus communis with sweet fruit, but Malus silvestris, which produces very sour fruit. "In theory, cultivated varieties of apple trees, which generally yield sweeter fruit, could have been introduced by the Romans, and could have formed hybrid varieties with crab-apple trees, but there is no conclusive evidence" (Biggam 2011: 35). A species of *apuldor* (Malus) with *milsc* 'sweet' fruits. ASSOCIATIVE RELATIONS: conceptual contiguity & taxonomic sub-ordination.
▫ →*apuldor*, →*milsc*
▫ →*sūrmilscapuldor*

mircapuldor →*milscapuldor*

minte, noun
'Mentha L. / MINT / Minze'
DOEPN (s.v. *minte*): 37 occ.
ILLUSTRATION in manuscript London, British Library, Cotton Vitellius C. iii, fol. 55r (Ohlgren 168(70)).
LATIN EQUIV. in the manuscripts: *menta, mentastrum, pulegium*.

ETYMOLOGY: Early Latin loan (Wollmann 1990: 153); < L *ment(h)a* (*OED*, s.v. *mint*); L *ment(h)a* and Gk *μίνθη* are both loans from an unknown, probably a Mediterranean language (*Genaust*, s.v. *méntha*).

▫ →*brōcminte*, →*feldminte*, →*fenminte*, →*hāreminte*, →*horsminte*, →*sǣminte*, →*swearte minte*, →*tūnminte*, →*wilde minte*

mistel, noun

(i) 'Viscum album L. / MISTLE / Mistel', (ii) ? 'birdlime / Vogelleim'

DOEPN (s.v. *mistel*): 12 occ.

ILLUSTRATION in manuscript London, British Library, Cotton Vitellius C. iii, fol. 54v (Ohlgren 168(69)).

LATIN EQUIV. in the manuscripts: *ocimum*, *viscum*.

ETYMOLOGY: Gmc **mistilo*, either from **mihs-tlo-* (*Kluge*, s.v. *Mistel*) or derived from Gmc **mihstu-* 'MIX(EN), dung, filth (cf. G *Mist*)' + *-el* (*OED*, s.v. *mistle*); cf. also Sauer (2012: 218, 223).

▫ →*misteltān*

▫ →*ācmistel*, →*eorþmistel*

mistel-tān, lit. 'mistle-twig'

'Viscum album L. / MISTLETOE / Weiße Mistel'

DOEPN (s.v. *mistel-tān*): 4 occ.

LATIN EQUIV. in the manuscripts: *chamaeleon*, *viscerago*, *viscum*.

ETYMOLOGY: Gmc compound (Carr 1939: 55); cf. OWN *mistilteinn* (Björkman 1901: 223); MORPHOLOGY: N+N; exocentric, metonymic compound; MOTIVATION: plant & part of plant; Originally, the mistletoe was denoted by *mistel*, and *misteltān*, which developed into ModE *mistletoe* (due to reanalysis of the *-n* as plural marker), denoted 'the twig of the mistletoe'. ASSOCIATIVE RELATIONS: conceptual identity & conceptual contiguity (meronymy).

▫ →*mistel*, →*tān*

mixen, noun

'MIX(EN), dung / Mist'

▫ →*mixenplante*

mixen-plante, lit. 'dung-plant'

'Solanum nigrum L. / morel, garden nightshade / Schwarzer Nachtschatten'

DOEPN (s.v. *mixen-plante*): 2 occ. (*Laud Herbal Glossary*, no. 1324; *Lǣcebōc*, 39/28).

LATIN EQUIV. and OE synonym in the manuscripts: *solatrum* (? for *solanum*); *āt(t)orlāþe*.

ETYMOLOGY: OE formation; hybrid compound (OE *plante* < L *planta*; →*plante*). MORPHOLOGY: N+N; endocentric, rectional. MOTIVATION habitat & botanical information. Morels are often found on mixens (Bierbaumer 1, s.v. *myxenplante*). ASSOCIATIVE RELATIONS: conceptual contiguity & taxonomic sub-ordination

▫ →*mixen*, →*plante*

-moce, noun

'slime / Schleim'

▫ →*lustmoce*

mōd, noun

'MOOD / Gemütslage, Laune'

▫ →*wermōd*

mōder, noun

'MOTHER / Mutter'

▫ →*mōderwyrt*

mōder-wyrt, lit. 'mother-wort'
(i) '? Mentha aquatica L. / water-mint / Bachminze'; or (ii) '? Chrysanthemum parthenium Bernh. / feverfew / Fieber-kraut, Mutterkraut'; or (iii) '? Artemisia vulgaris L. / wormwood, mugwort / Beifuß'
Hapax legomenon (*Laud Herbal Glossary*, no. 235).
LATIN EQUIV. in the manuscript: *balsamita*, *palma domini* (cf. *DMLBS*, s.v. *palma*, 7: *palma Christi* 'mother-wort'), *sisymbrium*.
ETYMOLOGY: OE formation; MORPHOLOGY: N+N; endocentric, rectional; partial metaphor. MOTIVATION: illness & botanical information. Latin names of Chrysanthemum parthenium and Artemisia vulgaris are *matricaria* and *mater herbarum* (Marzell, s.vv. *Chrysanthemum parthenium* Bernh., *Artemisia vulgaris* L.); they refer to the healing virtues of these plants, which were used against pains in the uterus. The different kinds of Mentha were used against women's diseases, too (Marzell, s.v. *Mentha crispa* L.), hence the element 'mother'. ASSOCIATIVE RELATIONS: metaphorical similarity & taxonomic sub-ordination.
▫ →*mōder*, →*wyrt*

mold, noun
'sand / Sand'
▫ →*moldcorn*

mold-corn, lit. 'sand-corn'
'? Saxifraga granulata L. / saxifrage / Körniger Steinbrech'
2 occ. (*Brussels Glossary*, 301/18 and *Lācnunga*, 176).
LATIN EQUIV. in the manuscripts: ? *vulnetrum*.
ETYMOLOGY: OE formation; MORPHOLOGY: N+N; exocentric, metonymic compound. MOTIVATION: illness & part of plant. Saxifraga granulata L. was recommended as a remedy against bladder stones, hence the Latin plant name *saxifraga*, lit. 'cracking rocks' (Marzell, s.v. *Saxifraga granulata* L.; see also →*sundcorn*). In addition, the plant's tubers may be compared to 'corns' or 'stones', used to break up bladder stones, which agrees with the principle of curing diseases with similar remedies (*similia similibus*). ASSOCIATIVE RELATIONS: conceptual contiguity & conceptual contiguity (meronymy).
▫ →*corn* 1, →*mold*

monnes sūre, lit. 'man's sorrel'
'Rumex acetosa L. / sorrel / Sauer-ampfer'
Hapax legomenon (*Lǣcebōc*, 38/25).
ETYMOLOGY: OE formation; MORPHOLOGY: N_{gen}+N; endocentric, rectional; partial metaphor. MOTIVATION: height & botanical information. The first element of *monnes sūre* probably refers to the fact that Rumex acetosa with its c. 3.2 ft. (1 m) height is rather tall (almost as tall as a man), and thus much taller than Oxalis acetosella L. (OE *sūre*); Oxalis acetosella is named *Zwerg-Sauerampfer* lit. 'dwarf-sorrel' in German, also in allusion to its low height (cf. Marzell, s.v. *Rumex acetosa* L.). ASSOCIATIVE RELATIONS: metaphorical similarity & taxonomic sub-ordination.
▫ →*man*, →*sūre*

mōr 1, noun
'MOOR / Moor'
▫ →*mōrsecg*, →*mōrwyrt*

mōr- 2, noun
'Morus nigra L. / mulberry (tree) / Maulbeere, Maulbeerbaum'
Not attested independently.
ETYMOLOGY: Later Latin loan according to Wollmann (1990: 172); early Latin loan according to Serjeantson (1935: 280); < L *mōrum* 'mulberry', *mōrus* 'mulberry tree' (*OED*, s.v. *mulberry*).
▫ →*mōrbēam*

mōr-bēam, lit. 'mulberry (tree) tree'
'Morus nigra L. / mulberry tree / Maulbeerbaum'
DOEPN (s.v. *mōr-bēam*): 7 occ.
LATIN EQUIV. in the manuscripts: *morus*, *rubus*.
ETYMOLOGY: Indepent formation in WGmc (Carr 1939: 133); hybrid compound (see MOTIVATION). MORPHOLOGY: N+N; endocentric, copulative: subsumptive (pleonastic). MOTIVATION: plant & botanical information. The first element, which is not attested independently in Old English, is derived from L *mōrum* 'mulberry tree' (perhaps a Greek loan word, cf. Feulner 2000: 393); the second element explains that the *mōr-* is a kind of tree. ASSOCIATIVE RELATIONS: conceptual identity & taxonomic sub-ordination.
▫ →*bēam*, →*mōr-* 2

more, *moru*, noun
(i) 'Daucus carota L. / carrot / Möhre'; (ii) 'root / Wurzel'
more (i): 8 occ. (*DOEPN*, s.v. *more* (1)); *more* (ii) 38 occ. (*DOEPN*, s.v. *more* (2)); according to Banham (1990: 199-200; quoted in *DOEPN*, s.v. *more* (2)), "where the word *more* occurs without qualification, it seems to mean 'carrot/parsnip' rather than just 'root'. But it is impossible to tell which individual plant is meant, or if they were being distinguished at all."
LATIN EQUIV. in the manuscripts: *carota*, *pastinaca*.
ETYMOLOGY: < WGmc **murhon* (*Kluge*, s.v. *Möhre*).
▫ →*englisc more*, →*feldmore*, →*wealhmore*, →*wīelisc moru*

mōr-secg, lit. 'moor-sedge'
'Carex L. / sedge / Schilfgras'
Hapax legomenon (*Peri Didaxeon*, 51/15).
ETYMOLOGY: OE formation; MORPHOLOGY: N+N; endocentric, rectional. MOTIVATION: habitat & botanical information. Sedge grows in the immediate vicinity of moors. ASSOCIATIVE RELATIONS: conceptual contiguity & taxonomic sub-ordination.
▫ → *mōr* 1, →*secg*

mōr-wyrt, lit. 'moor-wort'
'? Drosera rotundifolia L. / sun-dew, moorgrass / Rundblättriger Sonnentau'
Hapax legomenon (*Lǣcebōc*, 39/15).
ETYMOLOGY: OE formation; MORPHOLOGY: N+N; endocentric, rectional. MOTIVATION: habitat & botanical information. A plant growing near moors. ASSOCIATIVE RELATIONS: conceptual contiguity & taxonomic sub-ordination.
▫ → *mōr* 1, → *wyrt*

mucg-wyrt, lit. 'midge-wort'
'Artemisia vulgaris L. / wormwood, MUGWORT / Beifuß'
DOEPN (s.v. *mucg-wyrt*): 40 occ.
ILLUSTRATIONS in manuscripts London, British Library, Cotton Vitellius C. iii, fol. 25v (Ohlgren 168(12)), fol. 26r (Ohlgren 168(13)), and Oxford, Bodleian Library, Ashmole 1431 (S.C. 7523), fol. 8r (Ohlgren 221(10)).
LATIN EQUIV. in the manuscripts: *artemisia*, *artemisia* ? *tagante*, *mater herbarum*, *leptophyllon*, ? *lioparis*, (*artemisia*) *monoclonos*.
ETYMOLOGY: OE formation; MORPHOLOGY: N+N; endocentric, rectional. MOTIVATION: animal & botanical information. This plant attracts flies and midges: they like to sit on it (Bierbaumer 1, s.v. *mucgwyrt*). ASSOCIATIVE RELATIONS: conceptual contiguity & taxonomic sub-ordination.
▫ →*mycg*, →*wyrt*

mūs, noun
'MOUSE / Maus'
▫ →*mūsepise*

mūse-pise, lit. 'mouse-pea'
(i) 'Vicia cracca L. / (tufted) vetch / Vogelwicke'; (ii) 'Lathyrus montanus Bernh. / bitter vetch, heath-pea / Berg-Platterbse'
Hapax legomenon (*Antwerp-London Glossaries*, 143/6).
ETYMOLOGY: OE formation; hybrid compound (OE *pise* < L *pīsum*; →*pise*). MORPHOLOGY: N_{gen}+N; exocentric, metonymic compound; partial metaphor. MOTIVATION: size & part of plant. The seeds of Vicia cracca and those of Lathyrus are so small that they are referred to metaphorically as mice peas – 'peas as small as to fit mice' (cf. also G *Muustort* lit. 'mouse-tart', *Myzeârt* lit. 'mouse-pea', etc.; Marzell, s.v. *Vicia cracca* L., 6); alternatively, one could argue that the element *mūs* is used to indicate the inferiority of bitter vetch in comparison to 'real' peas (Pisum sativum L.) ASSOCIATIVE RELATIONS: metaphorical similarity & conceptual contiguity.
▫ →*mūs*, →*pise*

mycg, noun
'MIDGE / Mücke'
▫ →*mucgwyrt*

N

nafela, noun
'NAVEL /Nabel'
▫ →*eorþnafela*

nǣdder, *nǣdre*, noun
'ADDER / Natter'
▫ →*nǣdrewinde*, →*nǣdrewyrt*

nǣdre-winde, lit. 'adder-winder'
'Polygonum bistorta L. / bistort, snakeroot / Schlangenknöterich'
Hapax legomenon (*Brussels Glossary*, 287/16).
LATIN EQUIV. in the manuscript: *viperina*.
ETYMOLOGY: Independent formation in WGmc; cf. OLG *nâdarwinda* (Carr 1939: 142-3); MORPHOLOGY: synthetic, N+$N_{deverbal(-e)}$; endocentric, rectional, partial metaphor. MOTIVATION: shape &

action. Both *nǣdre* and *-winde* (< OE *windan* '(to) wind') relate to the winding roots of the plant; see also →*nǣdrewyrt*. ASSOCIATIVE RELATIONS: metaphorical similarity & taxonomic sub-ordination.

▫ →*nǣdder*, →*windan*

nǣdre-wyrt, lit. 'adder-wort'
(i) 'Polygonum bistorta L. / bistort, snake-root / Schlangenknöterich'; (ii) '? Ocimum basilicum L. / sweet basil / Basilie'
DOEPN (s.v. *nædderwyrt*): 19 occ.
ILLUSTRATIONS in manuscripts London, British Library, Cotton Vitellius C. iii, fol. 24r (Ohlgren 168(9)), fol. 57r (Ohlgren 168(74)), and Oxford, Bodleian Library, Ashmole 1431 (S.C. 7523), fol. 7r (Ohlgren 221(6)), fol. 30r (Ohlgren 221(120)).
LATIN EQUIV. in the manuscripts: *basilicon*, *viperina*.
ETYMOLOGY: Independent formation in WGmc (Carr 1939: 134); cf. OHG *naderawrz*, *nateruurz*, etc. (Björkman 1902: 306); cf. also below; loan translation of L *herba vīperīna*. MORPHOLOGY: N+N; endocentric, rectional; partial metaphor. MOTIVATION: shape & botanical information. In contrast to Sauer (1992b: 402), who connects the first element with the habitat of the plant (i.e. 'plant which grows where a certain snake, the basilisk, lives'), I believe the plant name to have been motivated by the plant's roots, which are snaky, serpentine; cf. also MedL *viperina, serpentaria*, etc. and OHG *naderawrz*, *nateruurz*, *nâterwurz*, *natrwrze*, etc. (Marzell, s.v. *Polygonum bistorta* L.). ASSOCIATIVE RELATIONS: metaphorical similarity & taxonomic sub-ordination.

▫ →*nǣdder*, →*wyrt*

nægl, noun
'NAIL / Nagel'

▫ →*coltnægl*

nægles →*cuneglæsse*

nǣp, noun
'Brassica napus L. var. arvensis (Lam.) Thellung / (wild) rape / Wilder Raps'
DOEPN (s.v. *nǣp*): 12 occ.
LATIN EQUIV. in the manuscripts: *napus*, *rapa*.
ETYMOLOGY: Early Latin loan; < L *nāpus* (Campbell 1959: § 504) < Gk *σίναπι* 'mustard' (Serjeantson 1935: 246); according to *Genaust* not deriving from but cognate with Gk *νᾶπυ* 'mustard' (an old form for *σίναπι* 'mustard', acc. to *Lewis and Short*, s.v. *nāpy*), and with Arm *niv* 'wild turnip' (*Genaust*, s.v. *nápus*).

▫ →*englisc nǣp*, →*tūnnǣp*, →*wilde nǣp*

net(e)l(e), noun
'Urtica L. / NETTLE / (Brenn-)Nessel'
DOEPN (s.v. *netele*): 29 occ.
ILLUSTRATION in manuscript London, British Library, Cotton Vitellius C. iii, fol. 72r (Ohlgren 168(103)).
LATIN EQUIV. in the manuscripts: *urtica*, ? *verticeta*.
ETYMOLOGY: WGmc **natilon* < Gmc **naton* + *-el* (*Kluge*, s.v. *Nessel*; *OED*, s.v. *nettle*); cf. also Sauer (2012: 218, 223).

▫ →*blindenet(e)l(e)*, →*grēate net(e)l(e)*, →*rēade net(e)l(e)*, →*smale net(e)l(e)*, →*worþignet(e)l(e)*

niht, noun
'NIGHT / Nacht'
▫ →*nihtscada*

niht-scada, lit. 'night-shade'
'Solanum nigrum L. / morel, garden NIGHTSHADE / Schwarzer Nachtschatten'
2 occ. (Förster 1917: 140; *Laud Herbal Glossary*, no. 1335)
LATIN EQUIV. in the manuscripts: *strumus*, *uva lupina*.
ETYMOLOGY: WGmc compound (Carr 1939: 102); cf. OHG *nachtschato*, *nahtscade*, etc. (Björkman 1902: 306); MORPHOLOGY: N+N; exocentric, metonymic compound. MOTIVATION: time & pathogen. The motivation is not entirely clear. According to *Kluge* (s.v. *Nachtschatten*), G *Nachtschatten* lit. 'nightshade' originally referred to smaller animals, which were, then, associated with the dark berries and flowers of Solanum nigrum to an extent that the name was taken over for the plant. Marzell (s.v. *Solanum nigrum* L.), in contrast, cites a number of texts where the second element is associated with G *Schaden* 'harm, damage' and is explained as the damage (in the form of headache, nightmares, sleepwalking) caused by the flower's sweet scent, which they distribute generously during the night. This interpretation, however, is probably based on secondary, phonological and semantic motivation, as G *Schaden* and *Schatten* (cognate with OE *sceadu*) have different etymologies (cf. *Kluge*, s.vv. *Schaden*, *Schatten*). ASSOCIATIVE RELATIONS: conceptual contiguity & conceptual contiguity.
▫ →*niht*, →*scada*

O

ofersǣwisc, adjective
'from overseas, foreign / Übersee-, fremd'
▫ →*ofersǣwisc rind*

ofersǣwisc rind, lit. 'overseas-bark'
'bark of Cinnamomum cassia Blume / cinnamon / Zimt'
Hapax legomenon (*Lǣcebōc*, 16/39-17/1).
ETYMOLOGY: OE formation; MORPHOLOGY: Adj+N; endocentric, copulative: attributive. MOTIVATION: origin & botanical information. This plant is described as a 'bark from overseas', that is, a foreign one: Cinnamomum cassia is an evergreen tree native to Asia, especially to Southern China (Marzell, s.v. *Cinnamomum cassia* Blume); see also →*sūþerne rind*. ASSOCIATIVE RELATIONS: conceptual contiguity & taxonomic sub-ordination.
▫ →*ofersǣwisc*, →*rind*

open, adjective
'OPEN / offen'
▫ →*openærs*

open-ærs, lit. 'open-arse'
'Mespillus germanica L. / medlar / Mispel'
2 occ. (*Antwerp-London Glossaries*, 122/6; *Bodley 730*, no. 176).
LATIN EQUIV. in the manuscripts: *aesculum*, *mespila*.

ETYMOLOGY: WGmc compound; MLG *apenärseken* (Carr 1939: 118); MORPHOLOGY: Adj+N; exocentric, metaphorical compound: "The compound is hardly a Bahuvrîhi formation, but a metaphorical transference of a determinative" (Carr 1939: 118). MOTIVATION: shape. The fruits of the medlar seem to be open at the bottom; hence they are compared to an (open) bottom; cf. also LG *Apenärschen*, *Apenoors*, etc., Fris *Iepenearske* (Marzell, s.v. *Mespillus germanica* L.). ASSOCIATIVE RELATION: metaphorical similarity.

▫ *→ærs, →open*

oxa, noun
'OX / Ochs'

▫ *→oxanslyppe, →oxantunge, →ox(e)nalybb*

oxan-slyppe, lit. 'ox-pat/-dung'
'Primula elatior (L.) Schreber / OXLIP, true cowslip / Hohe Schlüsselblume'
DOEPN (s.v. *oxan-slyppe*): 2 occ. (*Lācnunga*, 523; *Lǣcebōc*, 11/17).
ETYMOLOGY: OE formation; MORPHOLOGY: N_{gen}+N; exocentric, metaphorical compound. MOTIVATION: animal & shape. The withered, brownish leaves of the plant resemble the dung of cows and oxen. Compare also *→cūslyppe*. ASSOCIATIVE RELATION: metaphorical similarity.

▫ *→oxa, →slyppe*

oxan-tunge, lit. 'ox-tongue'
'Anchusa officinalis L. / alkanet / Echte Ochszunge'
Hapax legomenon (*Durham Plant Name Glossary*, no. 221).
LATIN EQUIV. in the manuscripts: *lingua bubali*.
ETYMOLOGY: Independent formation in WGmc, cf. OHG *ohsenzunga* (Björkmann 1901: 209); the compound is not recorded in Carr (1939); as both OE *oxantunge* and *ohsenzunga* have coalesced into compounds from syntactic phrases as loan translations of L *lingua bovis* or *lingua būbula* (*OED*, s.v. *ox-tongue*; Björkman 1901: 209), they probably do not go back to an original West Germanic formation, but independent formations in the WGmc languages. MORPHOLOGY: N_{gen}+N; exocentric, metaphorical compound. MOTIVATION: animal & shape. The Old English plant name, a translation of L *lingua būbula* or *lingua bovis*, is based on the shape of the plant's leaves, which are coarse, hairy and tongue-shaped. Cf. OHG *ochsenzunga*, etc. (Björkman 1902: 296). ASSOCIATIVE RELATION: meta-phorical similarity.

▫ *→oxa, →tunge*

ox(e)na-lybb, lit. 'ox's drug/herb'
'? Helleborus foetidus L. / setterwort, stinking hellebore / Stinkende Nieswurz'
DOEPN (s.v. *oxna-lybb*): 3 occ.
LATIN EQUIV. and OE synonym in the manuscripts: *heliotropium*; *coþuwyrt*.
ETYMOLOGY: OE formation; MORPHOLOGY: N_{gen}+N; endocentric, rectional. MOTIVATION: animal & botanical information. All parts of Helleborus foetidus are highly noxious, so that Cockayne's (*Leechdoms* 3: 341) interpretation as 'healing wort for oxen, Ochsenheilkraut' is not really convincing. On the other hand, ModE *oxheal* seems to point to the same virtues and the *DOEPN* (s.v. *oxna-lyb*) quote "Britten / Holland (1886, 364): 'In English Oxeheele or Setterwoort, which names

are taken from his vertues, in curing oxen and such like cattell.' Ger. 825."
ASSOCIATIVE RELATIONS: conceptual contiguity & taxonomic sub-ordination.
▫ →*lybb*, →*oxa*

P

palm, noun
'Arecaceae L. / PALM, palm tree / Palme'
DOEPN (s.v. *palm*): 20 occ.
LATIN EQUIV. in the manuscripts: *palma*, ? *palmites*.
ETYMOLOGY: Later Latin loan (Serjeantson 1935: 284); < L *palma* (*Kluge*, s.v. *Palme*).
▫ →*palmtrēow*

palm-trēow, lit. 'palm-tree'
'Arecaceae L. / PALM TREE / Palme'
DOEPN (s.v. *palm-trēow*): 2 occ.
LATIN EQUIV. in the manuscripts: *palma*.
ETYMOLOGY: OE formation (Carr 1939: 146); hybrid compound (OE *palm* < L *palma*; →*palm*). MORPHOLOGY: N+N; endocentric, copulative: subsumptive (pleonastic). MOTIVATION: plant & botanical information. The *palm* 'palm-tree' is a kind of *trēow* 'tree'. ASSOCIATIVE RELATIONS: conceptual identity & taxonomic sub-ordination.
▫ →*palm*, →*trēow*

persoc, noun
'fruit of Prunus persica Sieb. et Zucc. / peach / Pfirsich'
DOEPN (s.v. *persoc*): 4 occ.
LATIN EQUIV. in the manuscripts: *persicus*.
ETYMOLOGY: Early Latin loan (Wollmann 1990: 158), < L *mālus Persica* 'peach tree', *(mālum) Persicum* 'peach' < Gk *περσαία* 'peach tree' < Gk *περσικός* 'Persian' (*Genaust*, s.v. *pérsica*; *Kluge*, s.v. *Pfirsich*); according to Campbell (1959: § 518) with suffix substitution (native *-oc* for L *-ic*). On the Germanic diminutive suffix OE *-oc* see *ODEE* (s.v. *-ock*), *OED* (s.v. *-ock*).
▫ →*persoctrēow*

persoc-trēow, lit. 'peach-tree'
'Prunus persica Sieb. et Zucc. / peach-tree / Pfirsichbaum'
Hapax legomenon (*Antwerp-London Glossaries*, 122/8).
LATIN EQUIV. in the manuscripts: *persicaria*.
ETYMOLOGY: OE formation; hybrid compound (OE *persoc* 'peach' < L *(mālum) persicum*; Wollmann 1990: 158). MORPHOLOGY: N+N; endocentric, rectional. MOTIVATION: part of plant & botanical information. This kind of tree bears peaches. ASSOCIATIVE RELATIONS: conceptual contiguity (meronymy) & taxonomic sub-ordination.
▫ →*persoc*, →*trēow*

pīn, noun
'Pinus L. / PINE / Fichte, Föhre'
DOEPN (s.v. *pīn*): 1 occ. (Ker *Catalogue*, no. 371; cf. Gneuss *Handlist*, no. 902).
LATIN EQUIV. in the manuscripts: *pinus*.
ETYMOLOGY: Early Latin loan; < L *pīnus* (Wollmann 1990: 156) < Gk *πίτυς* (*Lewis and Short*, s.v. *pīnus*).
▫ →*pīnbēam*, →*pīntrēow*

pīn-bēam, lit. 'pine tree'
'Pinus L. / PINE tree / Kiefer, Föhre'
Not yet in the *DOE*, but a link is provided s.v. *bēam*; 2 occ. (Ælfric's *Homilies* II.34.162; Ælfric's *Lives of Saints* (Martin), 408); not in Bierbaumer (1, 2, 3).
ETYMOLOGY: Independent formation in WGmc (Carr 1939: 134); hybrid compound (< L *pīnus*; →*pīn*). MORPHOLOGY: N+N; endocentric, copulative: subsumptive (pleonastic). MOTIVATION: plant & botanical information. The *pīn* 'pine' is a kind of tree. ASSOCIATIVE RELATIONS: conceptual identity & taxonomic sub-ordination.
▫ →*bēam*, →*pīn*

pīn-trēow, lit. 'pine tree'
'Pinus L. / PINE TREE / Föhre'
DOEPN (s.v. *pīn-trēow*): 8 occ.
LATIN EQUIV. in the manuscripts: *pinus*.
ETYMOLOGY: OE formation; hybrid compound (< L *pīnus*; →*pīn*). MORPHOLOGY: N+N; endocentric, copulative: subsumptive (pleonastic). MOTIVATION: plant & botanical information. The *pīn* 'pine' is a kind of tree. ASSOCIATIVE RELATIONS: conceptual identity & taxonomic sub-ordination.
▫ →*pīn*, →*trēow*

pipor, noun
'Piper nigrum L. / Brown's PEPPER / Echter Pfeffer'
DOEPN (s.v. *pipor*): 79 occ.
ETYMOLOGY: Early Latin loan; < L *piper* < Gk *πέπερι* < OInd (Sanscrit) *pippali* (*Genaust*, s.v. *píper*; Wollmann 1990: 153; *Lewis and Short*, s.v. *piper*); according to Campbell (1959: § 518) with suffix substitution (native *-or* for L *-er*).
▫ →*blæc pipor*, →*lang pipor*

pise, noun
'Pisum sativum L. / PEA / Erbse'
DOEPN (s.v. *pise*): 27 occ.
LATIN EQUIV. in the manuscripts: *cicer*, *lens*, *lenticula*, *pisum*, *siliqua*.
ETYMOLOGY: Early Latin loan (Wollmann 1990: 154); < L *pīsum* < Gk *πισόν*, a loan word of unclear origin (*Genaust*, s.v. *písum*).
▫ →*eallhwīte pise*, →*mūsepise*, →*wilde pise*

plante, noun
'PLANT / Pflanze'
DOEPN (s.v. *plante*): 8 occ.
LATIN EQUIV. in the manuscripts: *planta*.
ETYMOLOGY: Early Latin loan, < L *planta* (Wollmann 1990: 175).
▫ →*mixenplante*

plūm(e), noun
'fruit of Prunus domestica L. / PLUM / Pflaume'
DOEPN (s.v. *plūm*): 8 occ.
LATIN EQUIV. in the manuscripts: *plumus*, *prunum*, *prunus lignum*, *prunum fructus*.
ETYMOLOGY: Early Latin loan (Wollmann 1990: 166, 167); < MedL *prunum* 'plum', *prunus* 'plum tree' < Gk *προύνη*, poss. of Syrian origin (*Genaust*, s.v. *prúnus*; Feulner 2000: 408-10); CL *prūnus* 'plum-tree' was borrowed into OE as *plȳme* (Feulner 2000: 408-11).
▫ →*plūmslā*, →*plūmtrēow*

plūm-slā, lit. 'plum-sloe'
'Prunus insitia Juslen / wild plum / Kriechenpflaume'
Hapax legomenon (*Antwerp-London Glossaries*, 129/5).
LATIN EQUIV. in the manuscript: *pruniculus*.
ETYMOLOGY: OE formation; hybrid compound (OE *plūm(e)* < L *prūnum*; →*plūm*). MORPHOLOGY: N+N; exocentric, metonymic compound. MOTIVATION: part of plant & part of plant. Both OE *plūm(e)* and OE *slā(h)* denote the fruits of the sloethorn – 'sloe' aquired the meaning of 'Prunus' in the 18th century, only (*OED*, s.v. *sloe* 2.a.); the compound *plūmslā* is, thus, a *pars-pro-toto* formation denoting the sloethorn (Bierbaumer 3, s.v. *plūmslā*). ASSOCIATIVE RELATIONS: conceptual contiguity (meronymy) & conceptual contiguity (meronymy).
▫ *→plūm, →slā(h)*

plūm-trēow, lit. 'plum-tree'
'Prunus domestica L. / PLUM-TREE / Pflaumenbaum'
DOEPN (s.v. *plūm-trēow*): 7 occ.[e]
LATIN EQUIV. in the manuscripts: *coccymella*, *nixa* (= *myxa*), *plumus*, *prunum*.
ETYMOLOGY: OE formation; hybrid compound (OE *plūm(e)* < L *prūnum*; →*plūm*). MORPHOLOGY: N+N; endocentric, rectional. MOTIVATION: part of plant & botanical information. *Plūm-trēow* is a tree that bears plums. ASSOCIATIVE RELATIONS: conceptual contiguity (meronymy) & taxonomic sub-ordination.
▫ *→plūm(e), →trēow*

popig, noun
'Papaver L. / POPPY / Mohn'
DOEPN (s.v. *popig*): 27 occ.
ILLUSTRATION in manuscript London, British Library, Cotton Vitellius C. iii, fol. 37v (Ohlgren 168(36)).
LATIN EQUIV. in the manuscripts: *codia*, *cucumis*, *meconium*, *papaver*.
ETYMOLOGY: Early Latin loan, < L *papāver*, with suffix substitution: native *-ig* for L *-āver* (Campbell 1959: § 518; see also *OED*, s.v. *poppy*).
▫ *→basu popig, →hwīt popig, →sūþerne popig, →wilde popig*

por(r), noun
'Allium porrum L. / leek / (Winter-) Lauch'
DOEPN (s.v. *por*): 8 occ.
LATIN EQUIV. in the manuscripts: *porrum*.
ETYMOLOGY: Early Latin loan (Wollmann 1990: 169); < L *porrum* (*OED*, s.v. *porre*).
▫ *→por(r)lēac*
▫ *→hēafdeht por(r)*

por(r)-lēac, lit. 'leek-plant' or 'leek-leek'
'Allium porrum L. / common leek / Porree'
DOEPN (s.v. *por-lēac*): 2 occ. (Förster 1917: 136/216; *Lācnunga*, 83).
LATIN EQUIV. in the manuscripts: *porrum*.
ETYMOLOGY: WGmc formation (Carr 1939: 112); hybrid compound (OE *por(r)* < L *porrum*; →*por(r)*). MORPHOLOGY: N+N; endocentric, copulative: subsumptive; when interpreted as lit. 'leek-leek' it must be classified as tautologic; however, this interpretation is less likely, as *lēac* was polysemous and could denote, very generally,

'garden-herb, plant' (see →*lēac*). MOTIVATION: plant & botanical information. The *por(r)* 'leek' is a kind of *lēac* 'garden-herb, plant'. ASSOCIATIVE RELATIONS: conceptual identity & conceptual identity.

▫ →*lēac*, →*por(r)*

prician, verb

'(to) PRICK, (to) pierce / piksen, stechen'

▫ →*pricþorn*

pric-þorn, lit. 'pricking thorn'

'Ilex aquifolium L. / holly / Stechpalme'

Not in Bierbaumer (1, 2, 3); hapax legomenon (*Charters* (Birch), 3: 123, no. 945). The identification is based on Hooke (2003: 28).

ETYMOLOGY: OE formation; MORPHOLOGY: V_{stem}+N; endocentric, rectional. MOTIVATION: action & botanical information. Ilex aquifolium is a thorn-bush with thorny, prickly leaves (Marzell, s.v. *Ilex aquifolium* L.). ASSOCIATIVE RELATIONS: conceptual contiguity & taxonomic sub-ordination.

▫ →*prician*, →*þorn*

pūl, noun

'POOL / Becken, Bassin'

▫ →*pūlsper*

pūl-sper, lit. 'pool-spear'

'reed / Schilfrohr'

Hapax legomenon (*Gospels*: *Matthew* (Li), 11, 7); not in Bierbaumer (1, 2, 3).

ETYMOLOGY: OE formation; MORPHOLOGY: N+N; exocentric, metaphorical compound. MOTIVATION: habitat & shape. Reed looks like spears growing around pools; semantic disambiguation of *sper* by the first element. ASSOCIATIVE RELATIONS: conceptual contiguity & metaphorical similarity.

▫ →*pūl*, →*sper*

R

ram, noun

'RAM / Schafbock'

▫ →*ramgealla*

ram-gealla, lit. 'ram-gall'

'? Menyanthes trifoliata L. / buck-bean, bog-bean / Sumpfbitterklee'

DOEPN (s.v. *ram-gealla*): 2 occ.

ETYMOLOGY: OE formation; MORPHOLOGY: N+N; exocentric, metaphorical compound. MOTIVATION: animal & taste. The second element, *gealla* 'gall', refers metaphorically to the bitter taste of the plant. Why the first element – - *ram* 'ram' – is used, is unclear; Marzell (s.v. *Menyanthes trifoliata* L.) offers different explanations, but is not convinced of any of them; there are semantically similar plant names: G *Bocksbohnen*, *Scharbocksklee*, *Geißblatt*, etc. and ModE *buck-bean* (Marzell, s.v. *Menyanthes trifoliata* L.). Do rams like the plant? ASSOCIATIVE RELATIONS: ? & metaphorical similarity.

▫ →*gealla*, →*ram*

rǣdic, noun

'Raphanus sativus L. / garden RADISH / Radieschen'

DOEPN (s.v. *rǣdic*): 52 occ.
LATIN EQUIV. in the manuscripts: *hierobotanum*, *radix*, *raphanus*, *verbenaca*.
ETYMOLOGY: Early Latin loan (Campbell 1959: § 504); < L *rādīx* (*OED*, s.v. *radish*).
▫ *→sūþerne rǣdic*, *→wilde rǣdic*

rēad, adjective
'RED / rot'
▫ *→rēad æppel*, *→rēad clǣfre*, *→rēad docce*, *→rēade fille*, *→rēade gearwe*, *→rēade hōfe*, *→rēade mæg(e)þ(e)*, *→rēade net(e)l(e)*, *→rēad sealh*, *→rēad secg*, *→rēade wudufille*

rēad æppel, lit. 'red apple'
'fruit of Punica granatum L. / pomegranate / Granatapfel'
DOEPN (s.v. *æppel, rēad*): 2 occ.
LATIN EQUIV. in the manuscripts: *malum punicum*.
ETYMOLOGY: Independent formation in WGmc; OHG *rotes apholes* (Björkman 1901: 210). MORPHOLOGY: Adj+N; endocentric, copulative: attributive. MOTIVATION: colour & botanical information. The pomegranate is a kind of apple with dark red seeds that contain a red juice; cf. OHG *rotes apholes*, etc. for *mala punica* (Björkman 1901: 210). ASSOCIATIVE RELATIONS: conceptual contiguity & taxonomic sub-ordination.
▫ *→rēad*, *→æppel*

rēad clǣfre, lit. 'red clover'
'Trifolium pratense L. / RED CLOVER, purple clover / Rotklee, Wiesenklee'
DOEPN (s.v. *clǣfre, rēade*): 9 occ.[e]
LATIN EQUIV. in the manuscripts: *calt(h)a*, *trifolium rubrum*.
ETYMOLOGY: OE formation; loan translation of *trifolium rubrum*; MORPHOLOGY: Adj+N; endocentric, copulative: attributive. MOTIVATION: colour & botanical information. A kind of clover with red flowers (Marzell, s.v. *Trifolium pratense* L.). ASSOCIATIVE RELATIONS: conceptual contiguity & taxonomic sub-ordination.
▫ *→clǣfre*, *→rēad*

rēad docce, lit. 'red dock'
'Rumex sanguineus L. / bloody dock, redveined dock / Hainampfer'
DOEPN (s.v. *docce, rēade*): 2 occ.
ETYMOLOGY: OE formation; MORPHOLOGY: Adj+N; endocentric, copulative: attributive. MOTIVATION: colour & botanical information. "Der Name bezieht sich auf den rötlich angelaufenen Stengel der Pflanze" ['The name alludes to the reddish stems of the plant'] (Bierbaumer 1, s.v. *docce*). ASSOCIATIVE RELATIONS: conceptual contiguity & taxonomic sub-ordination.
▫ *→docce*, *→rēad*

rēade fille, lit. 'red thyme'
'? Thymus serpyllum L / wild thyme / Feldthymian, Quendel'
Hapax legomenon (*Lācnunga*, 634-5).
ETYMOLOGY: OE formation; despite the fact that OE *fille* 'thyme' ultimately goes back to L *chaer(e)phylla* 'thyme', the plant name *rēade fille* 'red thyme' is not hybrid, since *fille* is a back-formation of the loan OE *cerfille* 'thyme' (< L *chaerephylla*, Campbell 1959: § 544) and, thus, an Old English lexeme. MORPHOLOGY: Adj+N; endocentric, copulative: attributive. MOTIVATION: colour & botanical information. A species of thyme with red or pink

flowers. ASSOCIATIVE RELATIONS: conceptual contiguity & taxonomic sub-ordination.
▫ →*fille*, →*rēad*

rēade gearwe, lit. 'red yarrow, rote Schafgarbe'
'perhaps a sub-species of Achillea millefolium L. / milfoil, common yarrow / Gemeine Schafgarbe'
DOEPN (s.v. *gearwe, rēade*): 2 occ. (*Lācnunga*, 238; *Læcebōc* 108/10).
ETYMOLOGY: OE formation; MORPHOLOGY: Adj+N; endocentric, copulative: attributive. MOTIVATION: colour & botanical information. Achillea millefolium can have a variety of colours (Marzell, s.v. *Achillea millefolium* L.), and *rēade gearwe* probably refers to a variant with red flowers: "Wahrscheinlich eine Unterart mit roten Zungenblüten" ['Most probably a sub-species with red tongue blades'] (Bierbaumer 2, s.v. *gearwe*). ASSOCIATIVE RELATIONS: conceptual contiguity & taxonomic sub-ordination.
▫ →*gearwe*, →*rēad*

rēade hōfe, lit. 'red ground-ivy'
'Glechoma hederaceum L. / ground-ivy / Gundelrebe'
DOEPN (s.v. *hōfe, rēade*): 7 occ.
ETYMOLOGY: OE formation; MORPHOLOGY: Adj+N; endocentric, copulative: attributive. MOTIVATION: colour & botanical information. The flowers of ground-ivy can be of different colours (Marzell, s.v. *Glechoma hederaceum* L.); a 'kind of ground-ivy with red flowers'. ASSOCIATIVE RELATIONS: conceptual contiguity & taxonomic sub-ordination.
▫ →*hōfe*, →*rēad*

rēade mæg(e)þ(e), lit. 'red camomile'
'Anthemis rosa L. / red camomile / Rote Kamille'
DOEPN (s.v. *mageþe, rēade*): 2 occ.
ETYMOLOGY: OE formation; MORPHOLOGY: Adj+N; endocentric, copulative: attributive. MOTIVATION: colour & botanical information. "The red camomile is native to Crete and described in Dioscorides III,154 (and in De herbis femininis XIX; Kästner 1896, 602) and is one of three species of camomile generally referred to in old herbals" (*DOEPN*, s.v. *mageþe, rēade*): a 'kind of *mæg(e)þ(e)* with red flowers'. ASSOCIATIVE RELATIONS: conceptual contiguity & taxonomic sub-ordination.
▫ →*mæg(e)þ(e)*, →*rēad*

rēade net(e)l(e), lit. 'red nettle'
'Lamium purpureum L. / red dead-nettle / Stinkende Taubnessel, Ackerstaubnessel'
DOEPN (s.v. *netele, rēade*): 23 occ.
ETYMOLOGY: OE formation; MORPHOLOGY: Adj+N; endocentric, copulative: attributive. MOTIVATION: colour & botanical information. A kind of nettle with red flowers. ASSOCIATIVE RELATIONS: conceptual contiguity & taxonomic sub-ordination.
▫ →*net(e)l(e)*, →*rēad*

rēade wudufille, lit. 'red wood-chevril'
'Asparagus officinalis L. / asparagus / Spargel'
Hapax legomenon (*Lācnunga*, 737).
ETYMOLOGY: OE formation; OE *fille* in *wudufille* is a shortened form of OE *cerfille* < L *chaer(e)phylla* (Campbell 1959: § 544), but the complex plant name *rēade wudufille* is not a hybrid formation, as *wudufille* is Old English.

MORPHOLOGY: Adj+N; endocentric, copulative: attributive; MOTIVATION: colour & botanical information. The colour 'red' can relate to the red berries produced by the plant ("[...] könnte sich auf die roten kugeligen Beeren von A. officinalis beziehen (vgl. dt. Koralkraut, Marzell I, 465)"; cf. Bierbaumer 2, s.v. *wudufille*). ASSOCIATIVE RELATIONS: conceptual contiguity & taxonomic sub-ordination.

▫ *→rēad, →wudufille*

rēad sealh, lit. 'red sallow'

'? Lythrum salicaria L. / purple loosestrife / Blutweiderich'

Hapax legomenon (*Lācnunga*, 841).

ETYMOLOGY: OE formation; MORPHOLOGY: Adj+N; endocentric, copulative: attributive. MOTIVATION: colour & botanical information. Lythrum salicaria is not related to the Salix, the willow, but, as both grow in moist habitats, along lakes and ponds, and as both can grow high (Lythrum salicaria can get as high as c. 6.56 ft. (2 m)) and have lancet-shaped leaves (Marzell, s.v. *Lythrum salicaria* L.), probably, the two plants were not easily distinguished and the red sallow was regarded as a kind of willow; cf. also ModE *red sally*. ASSOCIATIVE RELATIONS: conceptual contiguity & taxonomic sub-ordination.

▫ *→rēad, →sealh*

rēad secg, lit. 'red sedge'

unidentified plant

Hapax legomenon (*Lǣcebōc*, 31/40).

ETYMOLOGY: OE formation; MORPHOLOGY: Adj+N; endocentric, copulative: attributive. MOTIVATION: colour & botanical information. A kind of sedge with red parts (Bierbaumer 1, s.v. *secg*). ASSOCIATIVE RELATIONS: conceptual contiguity & taxonomic sub-ordination.

▫ *→rēad, →secg*

rīf, adjective

'fierce' (*BT*, s.v. *rif*); 'wild, lacerating, noxious' (*ClH*, s.v. *rīf*)

▫ *→giþrīfe, →hegerīfe*

rima, noun

'RIM, belt / Gürtel'

▫ *→eorþrima*

rind, noun

'RIND, bark / Rinde'

DOEPN (s.v. *rind*): 42 occ.

LATIN EQUIV. in the manuscripts: *codex* (? for *cortex*), *cortex*, *liber*.

ETYMOLOGY: WGmc **rendon* 'rind, bark' (*Kluge*, s.v. *Rinde*).

▫ *→ofersǣwisc rind, →sūþerne rind*

risc, noun

'Juncus L. / RUSH / Binse'

DOEPN (s.v. *risc*): 16 occ.

LATIN EQUIV. in the manuscripts: *juncus*, *papyrus*, *scirpus*.

ETYMOLOGY: Gmc word of unclear origin (*OED*, s.v. *rush*).

▫ *→risclēac*

▫ *→ēarisc*

risc-lēac, lit. 'rush-leek'

'Allium schoenoprasum L. / rush garlic / Schnittlauch'

Hapax legomenon (*Cleopatra Glossary* (Stryker), 72, no. 627).

LATIN EQUIV. in the manuscripts: *alans* (? for *alium*).

ETYMOLOGY: OE formation; MORPHOLOGY: N+N; endocentric, rectional; partial metaphor. MOTIVATION: shape & botanical information. OE *risclēac* is probably named so because its halms resemble the halms of rushes, though rushes are much bigger (see also →*secglēac*); rush garlic is a kind of leek, with its specific taste and smell (Marzell, s.v. *Allium schoenoprasum* L.). ASSOCIATIVE RELATIONS: metaphorical similarity & taxonomic sub-ordination.

▫ →*lēac*, →*risc*

rūde, noun

(i) 'Ruta montana L. / mountain RUE / Kleine Wiesenraute'; (ii) 'Ruta graveolens L. / common RUE / Wein-Raute'

DOEPN (s.v. *rūde*): 106 occ.

ILLUSTRATIONS in manuscript London, British Library, Cotton Vitellius C. iii, fol. 47r (Ohlgren 168(54)), fol. 54r (Ohlgren 168(68)).

LATIN EQUIV. in the manuscripts: ? *fulle ruta*, *ruta*, ? *sarta* (? for *ruta*).

ETYMOLOGY: Early Latin loan, probably 5th century (Wollmann 1990: 167, 591-601); < L *rūta* < Gk *ρυτή*, originally a Peloponnesian word (*Genaust*, s.v. *rúta*; *OED*, s.v. *rue*).

▫ →*feldrūde*, →*wilde rūde*

rūne, noun

'mystery, secret, whisper, runic character, letter / Geheimnis, Raunen, Zeichen (Rune)'

▫ →*dolgrūne*

rūw, adjective

'not even, not smooth, hairy / rau'

▫ →*rūwe wegbrǣde*

rūwe wegbrǣde, lit. 'hairy waybread'

'Plantago media L. / hoary plantain / Mittlerer Wegerich'

DOEPN (s.v. *wegbrāde, rūwe*): 3 occ.

ETYMOLOGY: OE formation; MORPHOLOGY: Adj+N; endocentric, copulative: attributive. MOTIVATION: texture & botanical information. This plant name alludes to the fact that plantains do not have smooth surfaces, or to the flowers, whose anthers look like fine little hairs when ripe; cf. also its ModE name *hoary plantain* (Marzell, s.v. *Plantago lanceolata* L.; Plantago lanceolata was often confused with Plantago major and Plantago media). ASSOCIATIVE RELATIONS: conceptual contiguity & taxonomic sub-ordination.

▫ →*rūw*, →*wegbrǣde*

S

sǣ, noun

'SEA / Meer'

▫ →*sǣminte*, →*sǣwār*

sǣ-minte, lit. 'sea-mint'

(i) 'Nepeta cataria L. / catmint / Katzenminze'; (ii) 'Althaea officinalis L. / marshmallow / Eibisch'

DOEPN (s.v. *sǣ-minte*): 5 occ.

LATIN EQUIV. in the manuscripts: *althaea*, *hibiscum*, *nepeta*.

ETYMOLOGY: OE formation; hybrid compound (OE *minte* < L *ment(h)a*; →*minte*); very probably a loan word from Latin, changed by folk etymology from L *siminta*, *siminza* 'nepeta' to *sǣminte* in analogy to OE *sǣ* 'sea' + *minte* 'mint' (Bierbaumer 1, s.v. *sǣ-*

minte). MORPHOLOGY: N+N; endocentric, rectional. MOTIVATION: secondary, phonological motivation: sound pattern & botanical information. Nepeta cataria does not live in the vicinity of brooks and fens, but Althaea officinalis does, which would explain the assocation of the plant with *sǣ* 'sea'. ASSOCIATIVE RELATIONS: phonological similarity & phonological similarity.
▫ *→sǣ, →minte*

sǣ-wār, lit. 'sea-seaweed'
'seaweed / Tang, Alge'
DOEPN (s.v. *sǣwar*): 2 occ. (*Antwerp-London Glossaries*, 111/10; *Laud Herbal Glossary*, no. 173).
LATIN EQUIV. in the manuscripts: *alga*.
ETYMOLOGY: OE formation; MORPHOLOGY: N+N; endocentric, rectional. MOTIVATION: habitat & botanical information. The plant name indicates that algae, perceived as a kind of weed, grow in the sea. ASSOCIATIVE RELATIONS: conceptual contiguity & conceptual identity.
▫ *→sǣ, →wār*

scada, noun
'SHADOW / Schatten'
▫ *→nihtscada*

scald, adjective
'shallow / seicht'
▫ *→scaldhūlas, →scaldþȳfel*

scald-hūlas (pl.), lit. 'shallow ? reeds'
'? reed, sedge / Schilf'
1 occ.[e] (*Corpus Glossary*, no. 1491).
LATIN EQUIV. in the manuscripts: *papyrus*.
ETYMOLOGY: OE formation; MORPHOLOGY: Adj+N; endocentric, rectional. -*hūl* is otherwise not attested. MOTIVATION: habitat & botanical information. 'Shallow' in the sense of '[plant growing in] shallow water'. The second element probably denotes 'reeds' (see *→hūl*). ASSOCIATIVE RELATIONS: conceptual contiguity & taxonomic subordination.
▫ *→-hūl, →scald*

scald-þȳfel, lit. 'shallow bush'
'reed, sedge / Schilf'
DOEPN (s.v. *scald-þȳfel*): 3 occ.[e] (*Corpus Glossary*, no. 125; 2x in *Épinal-Erfurt Glossary*, no. 58).
LATIN EQUIV. in the anuscripts: *alga*.
ETYMOLOGY: OE formation; MORPHOLOGY: Adj+N; endocentric, rectional. MOTIVATION: habitat & botanical information. Sedges grow in shallow waters. ASSOCIATIVE RELATIONS: conceptual contiguity & taxonomic sub-ordination.
▫ *→scald, →þȳfel*

scearp, adjective
'SHARP / scharf'
▫ *→scearpe docce, →scearpe þistel*

scearpe docce, lit. 'sharp dock'
(i) 'Rumex obtusifolius L. / broad-leaved dock / Stumpfblättriger Ampfer'; (ii) 'Rumex crispus L. / curled dock / Krauser Ampfer'
Attested in *Cleopatra Glossary* ((Stryker), 349, no. 221), *Durham Plant Name Glossary* (no. 264), and, according to Bierbaumer (3, s.v. *docce*), also in manuscript London, British Library, Cotton Otho E.i (resp., Junius 77); this

is not confirmed, however, by Voss (1996).
LATIN EQUIV. in the manuscripts: *oxylapathum*.
ETYMOLOGY: OE formation; loan translation of Gk *ὀξυλαπάθον* or its Latin equivalent *lapathum acutum* (Bierbaumer 3, s.v. *scearpe docce*). Rumex obtusifolius and Rumex crispus are seldom distinguished in folk taxonomies (Marzell, s.v. *Rumex obtusifolius* L.). MORPHOLOGY: Adj+N; endocentric, copulative: attributive. MOTIVATION: texture & botanical information. The first element of *scearpe docce* probably refers to the leaves of the *docce* 'dock', which develop spines during the bloom; the second indicates that this plant is a kind of *docce*. ASSOCIATIVE RELATIONS: conceptual contiguity & taxonomic sub-ordination.
▫ →*docce*, →*scearp*

scearpe þistel, lit. 'sharp thistle'
'? Cirisium arvense (L.) Scop. / creeping thistle / Ackerdistel'
Hapax legomenon (*Lǣcebōc*, 96/27).
ETYMOLOGY: OE formation; MORPHOLOGY: Adj+N; endocentric, copulative: attributive. MOTIVATION: texture & botanical information. Cirisium arvense is a kind of thistle with long, sharp thorns (Marzell, s.v. *Cirisium arvense* L.). ASSOCIATIVE RELATIONS: conceptual contiguity & taxonomic sub-ordination.
▫ →*scearp*, →*þistel*

sealh, noun
'Salix L. / WILLOW / Weide'
DOEPN (s.v. *sealh*): 17 occ.
LATIN EQUIV. in the manuscripts: *amerinus*, *saliuncula*, *salix*.
ETYMOLOGY: IE **salik-* 'willow' (*Kluge*, s.v. *Salweide*; *OED*, s.v. *sallow*).
▫ →*rēad sealh*

sēcan, verb
'(to) SEEK / suchen'
▫ →*sōlsēce*

secg, noun
'Carex L. / SEDGE / Segge'
DOEPN (s.v. *secg*): 28 occ.
LATIN EQUIV. in the manuscripts: *carex*, *gladiolus*, *lisca* (cf. OHG *lisca*, G *Liesch*; *Kluge*, s.v. *Liesch*), ? *sabium*.
ETYMOLOGY: Gmc **sagjo-z*, belongs to IE root **sek-* '(to) cut' (*Kluge*, s.v. *Segge*).
▫ →*secglēac*
▫ →*eolhsecg*, →*hamorsecg*, →*mōrsecg*, →*rēad secg*

secg-lēac, lit. 'sedge-leek'
'Allium schoenoprasum L. / rush garlic / Schnittlauch'
DOEPN (s.v. *secg-lēac*): 2 occ. (*Lācnunga*, 505; *Lǣcebōc*, 39/17).
ETYMOLOGY: OE formation; MORPHOLOGY: N+N; endocentric, rectional; partial metaphor. MOTIVATION: shape & botanical information. OE *secglēac* is probably named so because its halms resemble the halms of rushes, though these are much bigger (see also →*risc-lēac*); rush garlic is a kind of *lēac* 'leek', with its specific smell and taste (Marzell, s.v. *Allium schoenoprasum* L.). ASSOCIATIVE RELATIONS: metaphorical similarity & taxonomic sub-ordination.
▫ →*lēac*, →*secg*

self, pronoun
'SELF / selbst'
▫ →*selfǣte*, →*selfhǣle*

self-ǣte, lit. 'self-eat(ing plant)'
'Senecio vulgaris L. / common groundsel / Gemeines Kreuzkraut'
4 occ. (Bierbaumer 1, s.v. *selfǣte*).
ETYMOLOGY: WGmc compound; OHG *selbeza* (Björkman 1902: 307); as a parallel formation of OHG *selbeza*, *selfǣte* is possibly a loan translation of L *manducantes se ipsos* (Bierbaumer 1, s.v. *selfǣte*). MORPHOLOGY: synthetic; Pron+$N_{deverbal(-e)}$; endocentric, rectional. MOTIVATION: person & action. OHG *selbeza* glosses L *senecio*, which, in turn, is glossed by →*grundeswel(i)ge*. Similar to *grundeswelige*, *selfǣte* seems to refer to the fact that the plant 'swallows' the ground, but why exactly the pronoun *self* is used, is not evident to me. ASSOCIATIVE RELATIONS: ? & taxonomic sub-ordination.
▫ →*etan*, →*self*

self-hǣle, lit. 'self-heal(ing plant)'
'Sanicula europaea L. / wood march / Sanikel'
Hapax legomenon (*Durham Plant Name Glossary*, no. 297).
LATIN EQUIV. in the manuscripts: *sanicula*.
ETYMOLOGY: WGmc compound; cf. also OHG *selbheila*, etc. (Björkman 1902: 298). Lindheim (*Durham Plant Name Glossary*, no. 297) suggests that the Old English plant name is a literal translation of MedL *sanicula* (derived from L *sānāre* '(to) heal' + diminutive suffix *-icula*; cf. *Genaust*, s.v. *Sanícula*); as the Old English plant name contains the element *self*, which is not present in the Latin plant name, it must be classified as a loan rendition of the Latin model. MORPHOLOGY: synthetic, Pron+$N_{deverbal(-e)}$; endocentric, rectional. MOTIVATION: person & action. The plant was renowned for its healing virtues; cf. also MHG *sayleßhail*, Du *heelkruid*, etc. (Marzell, s.v. *Sanicula europaea* L.). Why the pronoun *self* is used as a first element is not evident to me. ASSOCIATIVE RELATIONS: ? & taxonomic sub-ordination.
▫ →*hǣlan*, →*self*

seofon, numeral
'SEVEN / sieben'
▫ →*seofonlēafe*

seofon-lēafe, lit. 'seven-leaf'
'Potentilla erecta (L.) Räuschel / common tormentil / Blutwurz'
DOEPN (s.v. *seofon-lēafe*): 4 occ.
ILLUSTRATION in manuscript London, British Library, Cotton Vitellius C. iii, fol. 54r (Ohlgren 168(68)).
LATIN EQUIV. in the manuscripts: *heptaphyllon*.
ETYMOLOGY: OE formation; loan translation of *heptaphyllon* < Gk *ἑπτάφυλλον* lit. 'seven leaves'; cf. OHG *sibenblat*, etc. (Björkman 1902: 298). MORPHOLOGY: Num+N; exocentric, metonymic compound. MOTIVATION: number & part of plant. The name refers to the plant's leaves, which seem to be divided into seven parts: in fact, they consist of only three to five parts; the religiously loaded number 'seven' is achieved when the stipules are counted as well (Marzell, s.v. *Potentilla erecta* Räuschel, 3). ASSOCIATIVE RELATION: conceptual contiguity (meronymy).
▫ →*lēaf*, →*seofon*

sigel, noun
'sun / Sonne'
▫ →*sigelhweorfa*

sigel-hweorfa, ***-hweorfe***, lit. 'sun-turning (plant)'
(i) 'Calendula officinalis L. / common marigold / Ringelblume'; (ii) 'Cichorium intybus L. / wild chicory / Zichorie, Gemeine Wegwarte'; (iii) 'Taraxacum officinale Weber / dandelion / Gemeiner Löwenzahn'; (iv) 'Hypochoeris glabra L. / smooth catsear / Sand-Ferkelkraut'; (v) '? Heliotropium europaeum L. / heliotrope / Sonnenwende'
DOEPN (s.v. *sigel-hweorfa*): 24 occ.
ILLUSTRATIONS in manuscripts London, British Library, Cotton Vitellius C. iii, fol. 36v (Ohlgren 168(34)), fol. 59v (Ohlgren 168(79)), and Oxford, Bodleian Library, Ashmole 1431 (S.C. 7523), fol. 14v (Ohlgren 221(39)).
LATIN EQUIV. in the manuscripts: *heliotropium*, *nymphaea*, *solsequium*.
ETYMOLOGY: OE formation; probably loan rendition of L *sōlsequium*, the Latin term being a translation of Gk *ἡλιοτρόπιον* (Bierbaumer 3, s.v. *sigel-hweorfa*, Sauer 1992b: 405; Wollmann 1990: 171). MORPHOLOGY: synthetic, $N+N_{deverbal(-e)}$; endocentric, rectional. MOTIVATION: habitat & action. The plants denoted by OE *sigelhweorfa* all grow preferably in sunny places and turn their flowers according to the movement of the sun. Many of them, e.g., common marigold and dandelion, have yellow flowers and thus resemble the sun. ASSOCIATIVE RELATIONS: conceptual contiguity & taxonomic sub-ordination.
▫ →*hweorfan*, →*sigel*

Simǣring, proper noun
▫ →*simǣringwyrt*

simǣring-wyrt, lit. 'Simǣring's wort'
'Malva L. / mallow / Malve'
DOEPN (s.v. *simering-wyrt*): 11 occ.
LATIN EQUIV. in the manuscripts: (*malva*) *crispa*, *victoriola*, *viola*.
ETYMOLOGY: OE formation; MORPHOLOGY: $PropN_{ing}$+N; endocentric, rectional; partial metaphor. MOTIVATION: healing virtues & botanical information. Bierbaumer (3, s.v. *simeringwyrt*) sees a connection to G *Siegmarswurz* (Malva alcea L.), which is a metaphorical allusion to both Simeon of the bible, who was happy that his eyes had seen the Saviour, and to the healing virtue of the plant, which is said to heal diseases of the eyes. ASSOCIATIVE RELATIONS: metaphorical similarity & taxonomic sub-ordination.
▫ →*Simǣring*, →*wyrt*

singrēne, noun
(i) '? Vinca minor L. / SENGREEN, evergreen / Immergrün'; (ii) 'Sempervivum tectorum L. / houseleek / Echte Hauswurz'
DOEPN (s.v. *singrēne*): 20 occ.
ILLUSTRATION in manuscript London, British Library, Cotton Vitellius C. iii, fol. 36v (Ohlgren 168(34)).
LATIN EQUIV. in the manuscripts: *calatides* (? for *galactites*), *temolus* (= *moly*), *tithymalus*.
ETYMOLOGY: Gmc **sengronjo* < **sen-*, *sem-* 'always' (suffix) + **gronjo* 'green'; cf. OHG *sin gruen* (Birkhan 2012: 129).
▫ →*smale singrēne*

slā(h), noun
'fruit of Prunus spinosa L. / SLOE / Schlehe'
DOEPN (s.v. *slāh*): 6 occ.
LATIN EQUIV. in the manuscripts: *bellicum, morus, prunellum.*
ETYMOLOGY: Gmc **slaiwa-* (*Kluge*, s.v. *Schlehe*).
▫ →*slāhþorn*
▫ →*plūmslā*

slāh-þorn, lit. 'sloe thorn-bush'
'Prunus spinosa L. / blackthorn / Schlehdorn'
DOEPN (s.v. *slāh-þorn*): 10 occ.[e]
LATIN EQUIV. in the manuscripts: *spina nigra.*
ETYMOLOGY: WGmc formation (Carr 1939: 113); cf. OHG *slêhthorn* (Björkman 1901: 212); MORPHOLOGY: N+N; endocentric, rectional. MOTIVATION: part of plant & botanical information. This kind of thorn-bush bears sloes. ASSOCIATIVE RELATIONS: conceptual contiguity (meronymy) & taxonomic sub-ordination.
▫ →*slā(h)*, →*þorn*

slyppe, noun
'dung, slime / Dung, Mist, Schleim'
▫ →*cūslyppe*, →*oxanslyppe*

smale, smæl, adjective
'SMALL, narrow / schmal'
▫ →*smale āt(t)orlāþe*, →*smale clāte*, →*smale clīfe*, →*smale cliþwyrt*, →*smale felterre*, →*smale net(e)l(e)*, →*smale singrēne*, →*smale wenwyrt*, →*smælþistel*

smale āt(t)orlāþe, lit. 'small poison-harm'
'? Setaria viridis P.B. / green bristle-grass / Grüner Fennich'
2 occ. (*Lǣcebōc*, 33/40, 35/17).
ETYMOLOGY: OE formation; MORPHOLOGY: Adj+N; endocentric, copulative: attributive. MOTIVATION: size & botanical information. Setaria viridis is the modern name for Panicum viride (Marzell, s.v. *Setaria viridis* P.B.), which is related to Panicum crus galli, the Old English *āt(t)orlāþe*; OE *smale āttorlāþe* is, thus, a smaller kind of *āttorlāþe*. ASSOCIATIVE RELATIONS: conceptual contiguity & taxonomic sub-ordination.
▫ →*āt(t)orlāþe*, →*smale*

smale clāte, lit. 'small cleavers'
'Galium aparine L. / cleavers, goose-grass / Klebkraut, Labkraut'
Hapax legomenon (*Lǣcebōc*, 31/25).
ETYMOLOGY: OE formation; MORPHOLOGY: Adj+N; endocentric, copulative: attributive. MOTIVATION: size & botanical information. "G. aparine wurde im Mittelalter ebenfalls zu den Lappa-Arten gezählt […]. Die Teilfrüchtchen sind um vieles kleiner als die mit Widerhaken versehenen Knöpfchen von A. lappa, daher die 'kleine Klette'" ['In the Middle Ages, G. aparine was percieved as a kind of Lappa. The fruits are much tinier than those of A. lappa, which are also equipped with barbs, hence the 'small cleavers''] (Bierbaumer 1, s.v. *clāte*). ASSOCIATIVE RELATIONS: conceptual contiguity & taxonomic sub-ordination.
▫ →*clāte*, →*smale*

smale clīfe, lit. 'small cleavers'
'Galium aparine L. / cleavers, goosegrass / Klebkraut, Labkraut'
2 occ. (*Lǣcebōc*, 38/11, 8/10).
ETYMOLOGY: OE formation; MORPHOLOGY: Adj+N; endocentric, copulative: attributive. MOTIVATION: size & botanical information. See →*smale clāte*. ASSOCIATIVE RELATIONS: conceptual contiguity & taxonomic sub-ordination.
▫ →*clīfe*, →*smale*

smale cliþwyrt, lit. 'small cleavers'
'Galium aparine L. / cleavers, goosegrass / Klebkraut, Labkraut'
Hapax legomenon (*Lācnunga*, 742).
ETYMOLOGY: OE formation; MORPHOLOGY: Adj+N; endocentric, copulative: attributive. MOTIVATION: size & botanical information. See →*smale clāte*. ASSOCIATIVE RELATIONS: conceptual contiguity & taxonomic sub-ordination.
▫ →*cliþwyrt*, →*smale*

smale felterre, lit. 'small centaury'
'Centaurium umbellatum Gilibert / centaury / Tausendgüldenkraut'
Hapax legomenon (*Lǣcebōc*, 99/20).
ETYMOLOGY: OE formation; hybrid formation (OE *felterre* < L *fel terrae*; →*felterre*); possibly a loan translation of *centauria minor*, in analogy to its synonym →*hierdewyrt sēo læsse*. MORPHOLOGY: Adj+N; endocentric, copulative: attributive. MOTIVATION: size & botanical information. A small kind of *felterre* 'centaury'. ASSOCIATIVE RELATIONS: conceptual contiguity & taxonomic sub-ordination.
▫ →*felterre*, →*smale*

smale net(e)l(e), lit. 'small nettle'
'Urtica urens L. / SMALL NETTLE / Kleine Brennnessel'
Hapax legomenon (*Lǣcebōc*, 21/25).
ETYMOLOGY: OE formation; MORPHOLOGY: Adj+N; endocentric, copulative: attributive. MOTIVATION: size & botanical information. A small kind of nettle, in contrast to →*grēate net(e)l(e)*. ASSOCIATIVE RELATIONS: conceptual contiguity & taxonomic sub-ordination.
▫ →*net(e)l(e)*, →*smale*

smale singrēne, lit. 'small evergreen'
'? Vinca minor L. / common periwinkle / Gemeines Immergrün'
Hapax legomenon (*Lǣcebōc*, 17/24).
ETYMOLOGY: OE formation; MORPHOLOGY: Adj+N; endocentric, copulative: attributive. MOTIVATION: size & botanical information. As its scientific names, Vinca minor and Vinca maior, indicate, as well, Vinca minor is a kind of *singrēne* 'Vinca' which is smaller than Vinca maior. ASSOCIATIVE RELATIONS: conceptual contiguity & taxonomic sub-ordination.
▫ →*singrēne*, →*smale*

smale wenwyrt, lit. 'small wenwort'
unidentified plant; perhaps 'Ranunculus ficaria L. / lesser celandine, figwort, pilewort / Scharbockskraut'
Hapax legomenon (*Lācnunga*, 515-6).
ETYMOLOGY: OE formation; MORPHOLOGY: Adj+N; endocentric, copulative: attributive. MOTIVATION: size & botanical information. Ranunculus ficaria is a low-growing plant. ASSOCIATIVE RELATIONS: conceptual contiguity & taxonomic sub-ordination.
▫ →*smale*, →*wenwyrt*

smæl-þistel, lit. 'narrow thistle'
'? Cnicus ferox L. / prickly cnicus / Hohe Bergdistel'
DOEPN (s.v. *smælþistel*): 4 occ.
LATIN EQUIV. in the manuscripts: *acanthus leuce*, *carduus*.
ETYMOLOGY: OE formation; MORPHOLOGY: Adj+N; endocentric, copulative: attributive. MOTIVATION: size & botanical information; 'This kind of thistle has narrow leaves' – in reference to Dioscorides' description of *acantha* as having narrow leaves (Bierbaumer 3, s.v. *smælþistel*); see also *→unbrāda þistel*. ASSOCIATIVE RELATIONS: conceptual contiguity & taxonomic subordination.
▫ *→smale*, *→þistel*

smeoru, noun
'grease / Schmiere, Fett'
▫ *→smeoruwyrt*

smeoru-wyrt, lit. 'grease-wort'
(i) 'Aristolochia rotunda L. / SMEARWORT / Rundblättrige Osterluzei'; (ii) 'Brassica rapa L., var. silvestris / turnip / Rübenkohl'
DOEPN (s.v. *smeru-wyrt*): 20 occ.[e]
ILLUSTRATIONS in manuscripts London, British Library, Cotton Vitellius C. iii, fol. 28r (Ohlgren 168(17)), and Oxford, Bodleian Library, Ashmole 1431 (S.C. 7523), fol. 10v (Ohlgren 221(19)).
LATIN EQUIV. in the manuscripts: *aristolochia*, *febrifuga*, *nap silvatica*, ? *uncteolam*, *veneria*.
ETYMOLOGY: WGmc compound (Carr 1939: 103); MORPHOLOGY: N+N; endocentric, rectional. MOTIVATION: product & botanical information. Brassica rapa yields oil; cf. also G *Speckrübe*, *Schmalzrübe* (Bierbaumer 1, s.v. *smeoruwyrt*). Further, Aristolochia rotunda has been used in ointments since the antiquity: "plant the sap of which is smeared on wounds" (Carr 1939: 103). ASSOCIATIVE RELATIONS: conceptual contiguity & taxonomic sub-ordination.
▫ *→smeoru*, *→wyrt*

smēþe, adjective
'smooth / glatt, weich'
▫ *→smēþe wegbrǣde*

smēþe wegbrǣde, lit. 'smooth waybread'
'Plantago major L. / greater plantain / Großer Wegerich, Wegbreite'
Hapax legomenon (*Lǣcebōc*, 107/5).
ETYMOLOGY: OE formation; MORPHOLOGY: Adj+N; endocentric, copulative: attributive. MOTIVATION: texture & botanical information. Obviously, the adjective 'smooth' is used to distinguish this kind of *wegbrǣde* from the *rūwe wegbrǣde* 'hoary plantain', which is, as the name says, *rūw* 'hairy' (see *→rūwe wegbrǣde*). ASSOCIATIVE RELATIONS: conceptual contiguity & taxonomic sub-ordination.
▫ *→smēþe*, *→wegbrǣde*

snīþan, verb
'(to) cut / schneiden'
▫ *→snīþstrēo*

snīþ-strēo, lit. 'cut-straw'
(i) '? Carlina acaulis L. / carline thistle, stemless carline / Stengellose Eberwurz, Silberdistel'; (ii) '? some kind of sedge / eine Schilfart'
DOEPN (s.v. *snīþstrēo*): 4 occ.[e]

LATIN EQUIV. in the manuscripts: ? *gacila*, ? *sisca*, ? *sista*.
ETYMOLOGY: OE formation; MORPHOLOGY: V_{stem}+N; exocentric, metaphorical compound. MOTIVATION: action & overall appearance. Carlina acaulis, the carline thistle is marked by dry, straw-like petals which seem to be very sharp (OE *snīðan* '(to) cut, (to) lance'); cf. Marzell (s.v. *Carlina acaulis* L.). ASSOCIATIVE RELATIONS: conceptual contiguity & metaphorical similarity.
▫ →*snīþan*, →*strēaw*

sōl, noun
'sun / Sonne'
▫ →*sōlsēce*

sōl-sēce, lit. 'sun-seeking (plant)'
(i) 'Calendula officinalis L. / common marigold / Ringelblume'; (ii) 'Cichorium intybus L. / wild chicory / Zichorie, Gemeine Wegwarte'; (iii) 'Taraxacum officinale Weber / dandelion / Gemeiner Löwenzahn'; (iv) 'Hypochoeris glabra L. / smooth catsear / Sand-Ferkelkraut'; (v) 'Heliotropium europaeum L. / heliotrope / Sonnenwende'; (vi) 'Atropa belladonna L. / deadly nightshade / Tollkirsche'; (vii) '? Solanum nigrum L. / black nightshade / Schwarzer Nachtschatten'
DOEPN (s.v. *sōl-sēce*): 10 occ.
ILLUSTRATION in manuscript London, British Library, Cotton Vitellius C. iii, fol. 42v (Ohlgren 168(46)).
LATIN EQUIV. in the manuscripts: *heliotropium*, *solata* (? for *solago* or for *solanum*), *solsequium*.
ETYMOLOGY: OE formation; in *sōlsēce*, several etymological processes combine. According to Wollmann (1990: 171), it is not an early borrowing of L *sōlsequium*, as was suggested by Serjeantson (1935: 280) and Campbell (1959: 511), but a later, learned loan word, as the native rendering of L *sōlsequium*, he argues, was OE →*sigelhweorfe*. The second element of the loan word was changed on the basis of the phonological similarity into an agentive derivative of the OE verb *sēcan* 'to seek', which, at the same time, translates the Latin verb (cf. also →*fēferfuge*); loan translation (see MOTIVATION). MORPHOLOGY: synthetic, N+$N_{deverbal(-e)}$; endocentric, rectional. MOTIVATION: habitat & action. Literally, a 'sun-seeking plant': the flowers of the plants in question turn their heads according to the movement of the sun, i.e. they literally 'seek the sun'. ASSOCIATIVE RELATION: conceptual contiguity.
▫ →*sēcan*, →*sōl*

sper(e), noun
'SPEAR / Speer'
▫ →*sperewyrt*
▫ →*pūlsper*

spere-wyrt, lit. 'spear-wort'
(i) 'Inula helenium L. / SPEARWORT, elecampane / Echter Alant'; (ii) 'Iris pseudacorus L. / gladdon, flag / Wasser-Schwertlilie'
DOEPN (s.v. *spere-wyrt*): 10 occ.[e]
ILLUSTRATIONS in manuscripts London, British Library, Cotton Vitellius C. iii, fol. 49r (Ohlgren 168(58)), and Oxford, Bodleian Library, Ashmole 1431 (S.C. 7523), fol. 25r (Ohlgren 221(86)).
LATIN EQUIV. and OE synonym in the manuscripts: *inula campana*, *nap silvatica*, *veneria*; *wilde nǣp*.
ETYMOLOGY: OE formation; MORPHOLOGY: N+N; endocentric, rectional;

partial metaphor. MOTIVATION: shape & botanical information. Bierbaumer (2, s.v. *sperewyrt*) connects the element *spere* with the roots of Inula helenium, yet in view of the height of the plant – it can reach a height of c. 3.3 ft. (c. 1 m) (*Botanica*, s.v. *Iris pseudacorus*) – it seems to be more likely that the association was made between the high stems of Inula and spears. Iris pseudacorus has lancet-shaped leaves; cf. also the German name *Schwertlilie* (Marzell, s.v. *Iris pseudacorus* L.). ASSOCIATIVE RELATIONS: metaphorical similarity & taxonomic sub-ordination.
▫ →*spere*, →*wyrt*

sp(r)ecel, noun
'speck / Fleck'
▫ →*hara(n)sp(r)ecel*

springan, verb
'(to) SPRING, (to) jump / springen'
▫ →*springwyrt*

spring-wyrt, lit. 'jump(ing) plant'
'Euphorbia lathyris L. / caper-spurge / Spring-Wolfsmilch'
DOEPN (s.v. *spring-wyrt*): 20 occ.
LATIN EQUIV. and OE synonym in the manuscripts: ? *atulenta*, *pis lingua* (? for *bislingua*; cf. Bierbaumer 3, s.v. *springwyrt*); *wildewise* (? = *wildepise*).
ETYMOLOGY: WGmc compound (Carr 1939: 103), cf. OHG *springwrz*, etc. (Björkman 1902: 299-300); MORPHOLOGY: V_{stem}+N; endocentric, rectional. According to Carr (1939: 103) "the first part of this compound is not a verbal stem: see OHG. *springa*, wild caper." As **springa* 'wild caper' is not attested in Old English, and the fruits of the plant explosively expel the seeds when ripe, I take the first element of the plant name for a verbal stem (see MOTIVATION). MOTIVATION: action & botanical information. The seeds of the caper-plant are shot out of the ripe capsules (see also Birkhan 2012: 174 with reference to OHG *springwurtz*). ASSOCIATIVE RELATIONS: conceptual contiguity & taxonomic sub-ordination.
▫ →*springan*, →*wyrt*

stān, noun
'STONE / Stein'
▫ →*stāncrop*, →*stāncysten*, →*stānmerce*

stān-crop, lit. 'stone-umbel'
'Sedum acre L. / bitter STONECROP, ginger / Scharfer Mauerpfeffer'
DOEPN (s.v. *stān-crop*): 3 occ.
ETYMOLOGY: OE formation; MORPHOLOGY: N+N; exocentric, metonymic compound. MOTIVATION: habitat & part of plant. The plant likes stony habitats (walls, stony places, cf. G *Mauerpfeffer*) and produces big umbels (Marzell, s.v. *Sedum acre* L.). ASSOCIATIVE RELATIONS: conceptual contiguity & conceptual contiguity (meronymy).
▫ →*crop*, →*stān*

stān-cysten (*-ceastel*), lit. 'stone-chestnut'
'Castanea sativa Mill. / chestnut tree / Kastanie'
Not in Bierbaumer (1, 2, 3); hapax legomenon (*Codex* (Kemble), 4: 8, no. 730); identification based on *BT* (s.v. *stan-cist, cisten*); see also *TOE* (1: 101).
ETYMOLOGY: OE formation; hybrid compound (OE *cysten* < L *castanea*; →*cisten*; Wollmann 1990: 178).

MORPHOLOGY: N+N; endocentric, rectional; partial metaphor. MOTIVATION: part of plant & botanical information. The chestnuts are compared to stones; cf also G *Stänäbbel* (Marzell, s.v. *Castanea sativa* L.). ASSOCIATIVE RELATIONS: metaphorical similarity & taxonomic sub-ordination.

▫ *→cysten, →stān*

stān-merce, lit. 'stone-celery'

'Petroselinum hortense Hoffm. / parsley / Petersilie'

DOEPN (s.v. *stān-merce*): 2 occ. (Förster 1917: 137/224, and *Brussels Glossary*, 299/27).

LATIN EQUIV. in the manuscripts: *petroselinum*; *sigsonte* '? a plant name'. ETYMOLOGY: OE formation; ? loan translation of L *petroselīnum* (< Gk *πετροσέλινον*, *Genaust*, s.v. *Petrorhagia*). MORPHOLOGY: N+N; endocentric, rectional. MOTIVATION: habitat & botanical information. *Stān* 'stone' translates L *petr(o)-* 'stone' and *merce* L *-selīnum* 'celery'. According to Pliny (*Naturalis historia*, 20, 12): "*petroselinum* quidam appellant in saxis natum", i.e. parsley is named so because it grows out of stony ground. According to *Genaust* (s.v. *Petrorhagia*), however, the name is based on its healing virtues for bladder stones: "[er] deutet [...] auf die Verwendung [...] als Heiplpfl. gegen Nieren- und Blasensteine (so schon bei Diosk. und noch bei Alb. Magn [...])" ['it points to its being used as a medical plant for kidney and bladder stones (already used in this way by Dioscorides and Albertus Magnus)']. ASSOCIATIVE RELATIONS: conceptual contiguity & taxonomic sub-ordination.

▫ *→merce, →stān*

stæl, noun

'place / Ort, Platz'

▫ *→stælwyrt*

stæl-wyrt, lit. 'place-wort'

(i) '? Asplenium trichomans L. / common maidenhair / Brauner Milzfarn', or (ii) '? Adiantum capillus veneris L. / maidenhair / Frauenhaar'

Hapax legomenon (*Brussels Glossary*, 299/5).

LATIN EQUIV. in the manuscript: *callitrichum*.

ETYMOLOGY: OE formation; MORPHOLOGY: N+N; endocentric, rectional. MOTIVATION: habitat & botanical information. Literally, the plant name means 'place-wort', but there is no obvious reason to describe the Asplenium trichomans or the Adiantum capillus as 'place-wort' (Bierbaumer, 3, s.v. *stælwyrt*). ASSOCIATIVE RELATIONS: ? & taxonomic sub-ordination.

▫ *→stæl, →wyrt*

stǣþ, noun

'shore, bank / Strand, Ufer, Gestade'

▫ *→stǣþwyrt*

stǣþ-wyrt, lit. 'shore-wort'

'? Limonium vulgare Miller / sea-lavender / Strandnelke'

Hapax legomenon (*Lǣcebōc*, 24/21).

ETYMOLOGY: OE formation; folk etymology (see MOTIVATION). MORPHOLOGY: N+N; endocentric, rectional. MOTIVATION: habitat & botanical information. The motivation of the plant's name and the identification of the plant are not entirely clear. When the plant *stǣþwyrt* is Limonium vulgare, as was proposed by Cockayne (*Leechdoms*, 2:

406), then OE *stǣþ* 'shore, bank / Gestade' refers to the plant's preference for land bordering on water. An alternative or additional explanation can be that the plant's old Latin name, *statice*, was remotivated phonologically and semantically in Old English after OE *stǣþ* 'shore, bank'. ASSOCIATIVE RELATIONS: conceptual contiguity & taxonomic sub-ordination.

▫ *→stǣþ, →wyrt*

stice, noun
'STITCH / Stich'
Derived from *→stician.*

▫ *→lufestice*

stician, verb
'(to) STICK, (to) prick / stechen'

▫ *→sticwyrt*

stic-wyrt, lit. 'stitchwort'
'Stellaria holostea L. / greater stitchwort / Sternmiere'
DOEPN (s.v. *stic-wyrt*): 4 occ.
LATIN EQUIV. in the manuscripts: *agrimonia*, *ostriago*, *sagittula*.
ETYMOLOGY: OE formation; MORPHOLOGY: V_{stem}+N; endocentric, rectional. MOTIVATION: illness & botanical information. The *OED* (s.v. *stitchwort*) quotes from a text from 1657: "COLES *Adam in Eden* cxlvi: It is called in English Stitch-wort, for its property in helping Stitches and pains in the sides"; this explains the use of the plant and its name: 'a plant that helps with stiches'. ASSOCIATIVE RELATIONS: conceptual contiguity & taxonomic sub-ordination.

▫ *→stician, →wyrt*

strǣl, noun
'arrow / Pfeil'

▫ *→strǣlwyrt*

strǣl-wyrt, lit. 'arrow-wort'
'Lycopodium clavatum L. / club-moss / Schlangenmoos, Keulenförmiger Bärlapp'
DOEPN (s.v. *strœl-wyrt*): 2 occ. (*Lǣcebōc*, 16/34, 30/10).
ETYMOLOGY: OE formation; MORPHOLOGY: N+N; endocentric, rectional; partial metaphor. MOTIVATION: shape & botanical information. The plant is probably compared to an arrow because of the arrangement of its spikes (Marzell, s.v. *Lycopodium clavatum* L., 2). ASSOCIATIVE RELATIONS: metaphorical similarity & taxonomic subordination.

▫ *→strǣl, →wyrt*

strēaw, *strēo*, noun
'STRAW / Stroh'
DOEPN (s.v. *strēaw*): 13 occ.
LATIN EQUIV. in the manuscripts: *faenum*, *festuca*.
ETYMOLOGY: Gmc **strawa*- 'straw' (*Kluge*, s.v. *Stroh*).

▫ *→strēawber(i)(g)e*

▫ *→snīþstrēo, →windelstrēaw*

strēaw-ber(i)(g)e, lit. 'strawberry'
'Fragaria vesca L. / STRAWBERRY / Erdbeere'
DOEPN (s.v. *strēaw-berie*): 32 occ.
ILLUSTRATION in manuscript London, British Library, Cotton Vitellius C. iii, fol. 33v (Ohlgren 168(28)).
LATIN EQUIV. and OE synonyms in the manuscripts: ? *fascinium* (? = for *vaccinia*; cf. Bierbaumer 3, s.v. *strēaw*-

berige), *fraga*; *eorþber(i)(g)e*, *mersc-mealwe*.
ETYMOLOGY: OE formation; MORPHOLOGY: N+N; exocentric, metonymic compound. MOTIVATION: habitat & part of plant. In general, Fragaria vesca L. is called *strēowber(i)(g)e* in Old English. There are a number of instances where strawberries and other berries are genitive attributes to the noun *wīse* 'sprout, stalk' (see Bierbaumer 1, s.v. *wīse*), indicating that the author of the texts did not intend to write about the fruits or the flowers of the respective plant, but rather about the stem and the leaves, or about the plant as a whole.

The *OED* notes that the motivation of *strawberry* is unclear (*OED*, s.v. *strawberry*). According to Bender (1934), berries were not cultivated in northern Europe until after the Middle Ages (the cultivation of strawberry beginning about the end of the 16th century); he suggests that the name derives from the fact that "the wild strawberry grows chiefly in grassy places and hay fields. It ripens at the time of the hay harvest and the red berries are very frequently found in the stubble under the mown hay" (Bender 1934: 73). Sauer (1992b: 401), however, argues that the Anglo-Saxon gardeners, indeed, knew the plant (cf. also →*eorþber(i)(g)e*) and might already have known the procedure of protecting the fruits from decay by putting straw around the plant – as gardeners nowadays do (see also Durkin 2009: 57-8). ASSOCIATIVE RELATIONS: conceptual contiguity & conceptual contiguity (meronymy).
▫ →*ber(i)(g)e*, →*strēaw*

sūgan, verb
'(to) suck / saugen'
▫ →*hunigsūge*

sumer, noun
'SUMMER / Sommer'
▫ →*sumerlida*

sumer-lida, lit. 'summer-sailor'
'shoot / Schössling'
Hapax legomenon (*Cleopatra Glossary* (Stryker), 314, no. 367).
LATIN EQUIV. in the manuscripts: *malleolus*.
ETYMOLOGY: OE formation; folk etymology (see MOTIVATION). MORPHOLOGY: N+N; endocentric, rectional. MOTIVATION: time & secondary, phonological motivation sound pattern. According to Bierbaumer 3 (s.v. *sumer-lida*), a folk etymological change of **sumer-loda* lit. 'summer-shoot' (cf. OHG *sumarlota* 'shoot') into OE *sumer-lida* 'summer sailor' based on the phonological similarity of the original plant name and *lida* 'sailor'. ASSOCIATIVE RELATIONS: conceptual contiguity & phonological similarity.
▫ →*lida*, →*sumer*

sun, noun
'SUN / Sonne'
▫ →*sundēaw*, →*sunfolgend*, →*sunnancorn*, →*suntrēow*

sund, noun
'sea / Meer'
▫ →*sundampre*, →*sundcorn*

sund-ampre (*-ompre*), lit. 'sea-dock'
'Rumex maritimus L. / golden dock / Strand-Ampfer'
Hapax legomenon (*Lǣcebōc*, 36/4).
ETYMOLOGY: OE formation; MORPHOLOGY: N+N; endocentric, rectional. MOTIVATION: habitat & botanical information. This kind of dock grows preferably at the sea-side (Marzell, s.v. *Rumex maritimus* L.). ASSOCIATIVE RELATIONS: conceptual contiguity & taxonomic sub-ordination.
▫ *→ampre, →sund*

sund-corn, lit. 'sea-corn'
'Saxifraga granulata L. / saxifrage / Körniger Steinbrech'
DOEPN (s.v. *sund-corn*): 18 occ.
ILLUSTRATIONS in manuscripts London, British Library, Cotton Vitellius C. iii, fol. 49v (Ohlgren 168(59)), and Oxford, Bodleian Library, Ashmole 1431 (S.C. 7523), fol. 25v (Ohlgren 221(88)).
LATIN EQUIV. in the manuscripts: *lithospermon*, *saxifragus*.
ETYMOLOGY: OE formation; MORPHOLOGY: N+N; exocentric, metonymic compound. MOTIVATION: illness & part of plant. As it is unclear why Saxifraga granulata L., a plant that generally grows in the meadows, should verbalise *sund* 'sea', Bierbaumer (2, s.v. *sundcorn*) suggests that the Old English plant name is a remotivated form of *sunnancorn* lit. 'sun's corn', for Lithospermum officinale. Bierbaumer argues on the basis of explanations given by Marzell (s.v. *Lithospermium officinale* L.). However, Marzell does not take into consideration that OE *sund*, *sunt* meant 'water', too (see *ClH*, s.v. *sund*). This, in turn, indicates another possible etymology of the Old English plant name: when we consider that both Lithospermium officinale and Saxifraga granulata were recommended as remedies against bladder stones – (hence the Latin *saxifraga*, lit. 'cracking rocks'; see Marzell, s.v. *Saxifraga granulata* L., and *Lǣcebōc* 98/3-5): "Gif man ne mæge gemigan 7 him weaxan stanas on þære blædran wyl sundcorn on ealaþ ond petersilian sele him drincan" ['If one cannot urinate and stones are growing in his bladder, boil in ale sundcorn and parsley and give him to drink'] – it seems possible that the name *sundcorn* relates to the remedy function of the plant. As the doctrine of signatures (*signatura rerum*) recommends healing after the principle *similia similibus*, the literal translation of *sundcorn* as 'water-corn' could refer to the urine containing bladder stones, and *corn* to either (a) the bladder stones the plant breaks up or (b) to the plant's tubers, compared to 'corns' or 'stones' and used to break up the bladder stones. Sauer (1992b: 401) give an alternative explanation, proposing to regard *sundcorn* as a 'plant, which makes people healthy', i.e. OE (*ge*)*sund* 'healthy'. ASSOCIATIVE RELATIONS: conceptual contiguity & conceptual contiguity (meronymy).
▫ *→corn, →sund*

sun-dēaw (? = *sund-*), lit. 'sea-dew'
'Rosmarinus officinalis L. / rosemary / Rosmarin'
DOEPN (s.v. *sun-dēaw*): 2 occ.
LATIN EQUIVALENTS and OE synonyms in the manuscripts: *ros marinus*; *boþen* 'rosemary', *feldmædere*.
ETYMOLOGY: OE formation; loan translation of *ros marinus* lit. 'sea-dew'. MORPHOLOGY: N+N; exocentric compound. MOTIVATION: habitat & ?. The

first element probably did not refer originally to the sun but to the sea (OE *sund*; *DOEPN*, s.v. *sun-dēaw*; *BT*, s.v. *sund-dēaw*) since the Old English plant name is a loan translation of *rōs marīnus* lit. 'sea-dew'. There is no obvious reason why the Romans should call the rosemary literally 'sea-dew': the plant cannot even survive in the immediate vicinity of sea spray, because sea water is salted. ASSOCIATIVE RELATIONS: conceptual contiguity & ?.

▫ →*dēaw*, →*sun*

sun-folgend, lit. 'sun-following (plant)'
'Calendula officinalis L. / common marigold / Ringelblume' or one of the plants identified by *sigelhweorfa*.
1 occ.[e] (*Corpus Glossary*, no. 1880).
LATIN EQUIV. in the manuscripts: *solsequium*.
ETYMOLOGY: OE formation; possibly loan translation of L *sōlsequium*; cf. →*sigelhweorfa* and →*sōlsēce*. MORPHOLOGY: synthetic formation, $N+N_{deverbal(-end)}$; endocentric, rectional. MOTIVATION: habitat & action. Calendula officinalis and the other plants in question need sunny places to grow; the plants turn their heads to the sun, i.e. they 'follow the sun'. ASSOCIATIVE RELATION: conceptual contiguity.

▫ →*folgian*, →*sun*

sunnan-corn, lit. 'sun's corn'
'Lithospermum officinale L. / common gromwell / Echter Steinsame'
DOEPN (s.v. *sunnan-corn*): 3 occ. (Meritt 1945: 58, no. 67; *Herbarium*, no. 180; *Laud Herbal Glossary*, no. 897).
LATIN EQUIV. in the manuscripts: *lithospermon*, *saxifragus*.
ETYMOLOGY: Independent formation in WGmc (Carr 1939: 140); loan translation of L *grānum sōlis*; cf. OHG *sunnenkorn*, etc. (Björkman 1902: 301). MORPHOLOGY: N_{gen}+N; exocentric, metonymic compound, partial metaphor. MOTIVATION: colour & part of plant. OE *sunnancorn* is a loan translation of L *grānum sōlis* lit. 'corn of the sun' or *milium sōlis* lit. 'millet of the sun' (cf. Bierbaumer 2, s.v. *sunnancorn*, and Carr 1939: 140). According to Bierbaumer (2, s.v. *sunnancorn*), the elements *sunnen-* and *corn* are motivated by the plant's seeds, which are white and shiny. ASSOCIATIVE RELATIONS: metaphorical similarity & conceptual contiguity (meronymy).

▫ →*corn*, →*sun*

sun-trēow, lit. 'sun-tree'
unidentified plant; according to the *DMLBS* (s.v. *oryza*), the Old English gloss to *oryza* 'rice' is probably not a compound but the noun phrase *sum trēow* 'some tree'.
Hapax legomenon (*Cleopatra Glossary* (Quinn), 67/1).
LATIN EQUIV. in the manuscript: *oryza*.
ETYMOLOGY: OE formation; MORPHOLOGY: N+N; endocentric, rectional. MOTIVATION: habitat & botanical information. Whether the gloss *sun treow* is considered to be the compound *sun-trēow* or the syntactic group *sum trēow*, its relation to *oryzo* 'rice' is not clear: Oryza sativa L. is a cereal and not tree. Taken as a compound, the literal translation of the gloss indicates that it relates to a tree which perhaps prefers

sunny places (cf., for instance, L *arbor solis* 'a fabulous oriental tree' lit. 'sun tree'; *DMLBS*, s.v. *arbor*). ASSOCIATIVE RELATIONS: conceptual contiguity & taxonomic sub-ordination.

▫ →*sun*, →*trēow*

sūr, adjective
'SOUR / sauer'

▫ →*sūrapuldor*, →*sūre*, →*sūr æppeltrēow*, →*sūrmilscapuldor*

sūr-apuldor (*-dre*), lit. 'sour apple-tree'
'Malus communis Lam. subsp. silvestris / crab-tree / Wilder Apfelbaum, Holzapfel'
DOEPN (s.v. *apuldor-rind, sūre*): 2 occ. (*Lācnunga*, 129; *Lǣcebōc*, 80/18).
ETYMOLOGY: Independent formation in West Germanic; cf. OHG *surephele* (Björkman 1901: 210); MORPHOLOGY: Adj+N; endocentric, copulative: attributive. MOTIVATION: taste & botanical information. The wild subspecies of Malus communis is sour in taste; cf. OHG *surephele* for *mala acida* (Björkman 1901: 210). ASSOCIATIVE RELATIONS: conceptual contiguity & taxonomic sub-ordination.

▫ →*apuldor*, →*sūr*

▫ →*wudusūræppel*

sūr æppeltrēow, lit. 'sour apple-tree'
'Malus communis Lam. subsp. silvestris / crab-tree / Wilder Apfelbaum, Holzapfel'
Not in Bierbaumer (1, 2, 3); hapax legomenon (Sawyer 1968: 879). The identification is based on Hooke (2003: 30); the plant name also matches perfectly the structurally slightly different *sūrapuldor*.
ETYMOLOGY: Independent formation in West Germanic; cf. OHG *surephele* (Björkman 1901: 210); MORPHOLOGY: Adj+N; endocentric, copulative: attributive. MOTIVATION: taste & botanical information. The wild subspecies of Malus communis is sour in taste; cf. OHG *surephele* for *mala acida* (Björkman 1901: 210). ASSOCIATIVE RELATIONS: conceptual contiguity & taxonomic sub-ordination.

▫ →*æppeltrēow*, →*sūr*

sūre, noun
'Rumex acetosa L. / common sorrel, (wood) sorrel / Großer Sauerampfer'
DOEPN (s.v. *sūre*): 10 occ.
LATIN EQUIV. in the manuscripts: ? *calcir* (? for *calciculium*), *saliunca*, *salsa*.
ETYMOLOGY: derived from the adjective *sūr* 'sour'.

▫ →*gēaces sūre*, →*monnes sūre*

sūrmilsc-apuldor (*-dre*), lit. 'sour-sweet apple-tree'
(i) 'Malus communis L. / apple-tree / Apfelbaum'; or (ii) 'Malus communis Lam. subsp. silvestris / crab-tree / Wilder Apfelbaum'
Hapax legomenon (*Antwerp-London Glossaries*, 122/4).
LATIN EQUIV. in the manuscripts: *malus Matianus*.
ETYMOLOGY: OE formation; MORPHOLOGY: Adj+N; endocentric, copulative: attributive. One possible analysis is to split the plant name into two parts which both are complex; the first is, then, a combination of two adjectives: *sūr* 'sour' + *milsc* 'sweet', and the second a derivative (→*apuldor*). It is, however, also possible to split the plant

name into *sūr* 'sour' + *milscapuldor* 'sweet apple-tree' (→*milscapuldor*). MOTIVATION: taste & botanical information. In either case of the morphological analysis, the first element refers to the taste of the fruits, the second to the tree species. ASSOCIATIVE RELATIONS: conceptual contiguity & taxonomic sub-ordination.

▫ →*apuldor*, →*milsc*, →*milscapuldor*, →*sūr*

sūþern, adjective
'SOUTHERN / südlich'

▫ →*sūþerne bisceopwyrt*, →*sūþerne cymen*, →*sūþerne finu(g)le*, →*sūþerne popig*, →*sūþerne rǣdic*, →*sūþerne rind*, →*sūþerne wermōd*, →*sūþerne wudu*

sūþerne bisceopwyrt, lit. 'southern bishop's wort'
unidentified plant
Hapax legomenon (*Lǣcebōc*, 83/4-5).
ETYMOLOGY: OE formation; MORPHOLOGY: Adj+N; endocentric, copulative: attributive. The second element, *bisceopwyrt*, is complex and, originally, a Latin loan word changed by folk etymological processes (→*bisceopwyrt*). MOTIVATION: origin & botanical information. As the plant has not yet been identified, the morphological and semantic analysis cannot go any further than state that this plant is a kind of *bisceopwyrt* that is of southern origin. ASSOCIATIVE RELATIONS: conceptual contiguity & taxonomic sub-ordination.

▫ →*bisceopwyrt*, →*sūþern*

sūþerne cymen, lit. 'southern cumin'
'Cuminum cyminum L. / cumin / Kreuzkümmel, Römischer Kümmel'
Hapax legomenon (*Lǣcebōc*, 55/38).
ETYMOLOGY: hybrid formation (OE *cymen* < L *cumīnum*; →*cymen*). MORPHOLOGY: Adj+N; endocentric, copulative: attributive. MOTIVATION: origin & botanical information. Originally from Turkmenistan, Cuminum cyminum L. now grows all around the Mediterranean Sea; during the Middle Ages, it was grown throughout Europe (Marzell, s.v. *Cuminum cyminum* L.); 'a kind of cymen of southern origin'. ASSOCIATIVE RELATIONS: conceptual contiguity & taxonomic sub-ordination.

▫ →*cymen*, →*sūþerne*

sūþerne finu(g)le, lit. 'southern fennel'
'? Pimpinella anisum L. / anise / Anis'
Hapax legomenon (*Lǣcebōc*, 43/13).
ETYMOLOGY: OE formation; hybrid formation (OE *finu(g)le* < L *fēniculum*; →*finu(g)le*). MORPHOLOGY: Adj+N; endocentric, copulative: attributive. MOTIVATION: origin & botanical information. As a simplex, OE *finu(g)le* denotes Foeniculum; the complex plant name *sūþerne finu(g)le*, however, denotes Pimpinella anisum: this is due to the fact that Foeniculum vulgare and Pimpinella anisum are taxonomically related and, traditionally, share a number of names (Marzell, s.v. *Foeniculum vulgare* Miller). ASSOCIATIVE RELATIONS: conceptual contiguity & taxonomic sub-ordination.

▫ →*finu(g)le*, →*sūþerne*

sūþerne popig, lit. 'southern poppy'
'Papaver somniferum L. / opium poppy / Schlafmohn'
Hapax legomenon (*Lǣcebōc*, 63/37).
ETYMOLOGY: OE formation; hybrid formation (OE *popig* < L *papāver*; →*popig*). MORPHOLOGY: Adj+N; endo-

centric, copulative: attributive. MOTIVATION: origin & botanical information. According to Marzell (s.v. *Papaver sominiferum* L.), opium poppy, originally, grew in Mediterranean regions, a fact that could explain the element *sūþern*; however, the plant spread to northern Europe too early to be still perceived as foreign or southern by the Anglo-Saxons (namely, in the Stone Age, already); and still, this is the signification of *sūþerne popig* lit. 'Southern poppy'. ASSOCIATIVE RELATIONS: conceptual contiguity & taxonomic sub-ordination.

▫ *→popig*, *→sūþern*

sūþerne rǣdic, lit. 'southern radish'
'Raphanus sativus L. / radish / Gartenrettich'
DOEPN (s.v. *rǣdic, sūþerne*): 2 occ. (*Lācnunga*, 1017; *Lǣcebōc*, 19/13).
ETYMOLOGY: OE formation; hybrid formation (OE *rǣdic* < L *rādīx*; *→rǣdic*). MORPHOLOGY: Adj+N; endocentric, copulative: attributive. MOTIVATION: origin & botanical information. *Rǣdic* and *sūþerne rǣdic* refer both to Raphanus sativus L. and there is no subspecies which could be called 'southern'. Marzell (s.v. *Raphanus sativus* L.) remarks, however, that gardeners distinguish between the "spätreifenden Winterrettich und den früher reifenden, im allgemeinen etwas kleineren Sommerrettich sowie die kleinen, fleischigen, zarten Radieschen" ['black Spanish radish, which ripens late, the radish, which ripens earlier and is, in general, somewhat smaller, and the small, succulent, tender small radish']. OE *sūþerne rǣdic* may perhaps have referred to one of these three species. ASSOCIATIVE RELATIONS: conceptual contiguity & taxonomic sub-ordination.

▫ *→rǣdic*, *→sūþern*

sūþerne rind, lit. 'southern rind'
'bark of Cinnamomum cassia Blume / cinnamon / Zimt'
DOEPN (s.v. *rind, sūþerne*): 3 occ. (*Durham Plant Name Glossary*, no. 115 and no. 288; *Harley Glossary*, 74, no. 896).
LATIN EQUIV. in the manuscripts: *cinnamomum*, *cyminum*, *resina*.
ETYMOLOGY: OE formation; MORPHOLOGY: Adj+N; endocentric, copulative: attributive. MOTIVATION: origin & botanical information. With reference to the spice, the Anglo-Saxons used either *→ofersǣwisc rind* or *sūþerne rind* (the plant name is *→swēte wyrt*). *Ofersǣwisc rind* and *sūþerne rind* both allude to the fact that the spice, which is made of the bark of the plant, is of southern or foreign origin. ASSOCIATIVE RELATIONS: conceptual contiguity & taxonomic sub-ordination.

▫ *→rind*, *→sūþern*

sūþerne wermōd, lit. 'southern wormwood'
'Artemisia abrotanum L. / southernwood / Eberraute'
3 occ. (*Lǣcebōc*, 12/3-4, 54/16, 71/17).
ETYMOLOGY: OE formation; MORPHOLOGY: Adj+N; endocentric, copulative: attributive. MOTIVATION: origin & botanical information. A kind of *wermōd* of southern origin: according to Marzell (s.vv. *Artemisia abrotanum* L., *Artemisia absinthium* L.), both species of Artemisia – A. abrotanum and A. absinthium – come from the Middle East and the Mediterranean region, but

were grown in Europe during the Middle Ages. ASSOCIATIVE RELATIONS: conceptual contiguity & taxonomic sub-ordination.

▫ *→sūþern*, *→wermōd*

sūþerne wudu, lit. 'southern-wood'
'Artemisia abrotanum L. / SOUTHERNWOOD / Eberraute'
DOEPN (s.v. *sūþerne wudu*): 16 occ.
ILLUSTRATION in manuscript London, British Library, Cotton Vitellius C. iii, fol. 58v (Ohlgren 168(77)).
LATIN EQUIV. in the manuscripts: *habrotonum*, *heraclea*.
ETYMOLOGY: OE formation; MORPHOLOGY: Adj+N; exocentric, metonymic formation. MOTIVATION: origin & texture. Artemisia abrotanum L. is a woody plant (a shrub) from the Mediterranean region (see Bierbaumer 3, s.v. *sūþerne wudu*) – the first element, thus, refers to the origin of the plant, the second to its texture. ASSOCIATIVE RELATIONS: conceptual contiguity & conceptual contiguity.

▫ *→sūþern*, *→wudu*

swamm, noun
'mushroom / Pilz'
DOEPN (s.v. *swamm*): 19 occ.
LATIN EQUIV. in the manuscripts: *boletus*, *fungus*, *spongia*, *volvola* (= *convolvulus*), *tuber*, *moisseris* (? for *mossicus*); *feldswamm*.
ETYMOLOGY: Of unclear origin; probably not Indo-European (*Kluge*, s.v. *Schwamm*).

▫ *→feldswamm*, *→meteswamm*

swan, noun
'SWAN / Schwan'
or
swān, noun
'herdsman, swineherd / Hirte, Schweinehirt'

▫ *→swanewyrt*

swane-wyrt or ***swāne-wyrt***, lit. ? 'swan's wort' or 'herdsman's wort'
unidentified plant
Hapax legomenon (*Lǣcebōc*, 23/26).
ETYMOLOGY: OE formation; MORPHOLOGY: N_{gen}+N; endocentric, rectional. MOTIVATION: ? & botanical information. Since the identification of the plant and the identity of the first element are unclear, the motivation of the plant name is unclear, too. *Swane-* is either from *swanes-* 'swan's' or from the *swāne-* 'swineherd'. In both cases, the complex lexeme denotes a kind of *wyrt* 'plant'. ASSOCIATIVE RELATIONS: ? & taxono-mic sub-ordination.

▫ *→swan*, *→wyrt*

sweart, adjective
'black / schwarz'

▫ *→swearte bēanen*, *→swearte minte*

swearte bēanen (pl.), lit. 'black beans'
unidentified plant
Hapax legomenon (*Peri Didaxeon*, 5/26).
ETYMOLOGY: OE formation; loan rendition of L *bleta nigra* (see MOTIVATION). MORPHOLOGY: Adj+N; endocentric, copulative: attributive. MOTIVATION: colour & botanical information. L *bleta nigra* is probably due to the contamination of L *beta* (*nigra*) lit. 'black beteroot' with *blitum* 'amaranth',

but why *bleta*, which was interpreted as a plural noun, was translated as *bēan(en)* 'bean(s)' is not entirely clear (Bierbaumer 2, s.v. *bēan*). There are, however, beans of a very dark red colour, which could be interpreted as (almost) 'black' – cf. →*swearte minte*, where the dark red stems were interpreted as 'black' – and the phonological similarity of L *bleta* and OE *bēan* may have furthered this translation. ASSOCIATIVE RELATIONS: conceptual contiguity & taxonomic sub-ordination.

▫ →*bēan*, →*sweart*

swearte minte, lit. 'black mint'

'? Mentha crispa L. / crisped mint, curled mint / Krause Minze'

Hapax legomenon (*Peri Didaxeon*, 35/11).

ETYMOLOGY: OE formation; hybrid formation (OE *minte* < L *ment(h)a*; →*minte*); possibly a loan translation of MedL *menta nigra*. MORPHOLOGY: Adj+N; endocentric, copulative: attributive. MOTIVATION: colour & botanical information. "'Schwarze' Minze bezieht sich anscheinend auf die Pfl.n. mit dunkelrot gefärbten Stengel, die auch im Engl. als 'black mint' bezeichnet wird" [''black' mint relates to the plants with stems of a dark red colour, which are called in English 'black mint''] (Marzell, s.v. *Mentha crispa* L.); cf also G *swartz mintzen* (15th c.) and Dan *sort krusemynte* (18th c.) and MedL *menta nigra*, from which it may have been translated. ASSOCIATIVE RELATIONS: conceptual contiguity & taxonomic sub-ordination.

▫ →*minte*, →*sweart*

swegel, noun

'heaven / Himmel'

▫ →*swegles æppel*

swegles æppel, lit. 'heaven's apple'

'? fruit of Punica granatum L. / pomegranate / Granatapfel'

DOEPN (s.v. *swegles æppel*): 9 occ.; for Bierbaumer (1 and 2, s.v. *swegles æppel*) the identification of this plant is unclear. Cockayne (*Leechdoms*, 1: 23) suggests 'beetlenut'; Storms (1948: 327) 'swail's apple'. For the identification as pomegranate see MOTIVATION below.

ETYMOLOGY: OE formation; MORPHOLOGY: N_{gen}+N; endocentric, rectional. MOTIVATION: ? healing virtue & botanical information. I think that OE *swegles æppel* refers to the pomegranate, since the pomegranate is often associated with heaven. In the Bible, the pomegranate is always mentioned in connection with paradise, fertility and food; it is believed to have 613 kernels, thus corresponding to the numbers of the laws in the *Old Testament*, and is mentioned several times in the Koran. It plays a role in Greek mythology and has been associated with life and fertility ever since. In addition, it is extremely valuable in curing for instance ailments of the coronary system, in curing arthritis and cancer (*Brockhaus*, s.v. *Granatapfel*). ASSOCIATIVE RELATIONS: metaphorical similarity & taxonomic sub-ordination.

▫ →*æppel*, →*swegel*

sweligean, verb

'SWALLOW / schlucken'

▫ →*grundeswelige*

swēte, adjective
'SWEET / sweet'
▫ →*swēte apuldor*, →*swēte wyrt*

swēte apuldor (*-dre*), lit. 'sweet apple-tree'
'Malus communis Lam. / apple-tree / Apfelbaum'
DOEPN (s.v. *apuldor, swēte*): 1 occ. (*Antwerp-London Glossaries*, 122/5).
LATIN EQUIV. in the manuscripts: *melimela.*
ETYMOLOGY: Independent formation in West Germanic; cf. OHG *suozephel* (Björkman 1901: 210); MORPHOLOGY: Adj+N; endocentric, copulative: attributive; the base of the formation is complex (→*apuldor*). MOTIVATION: taste & botanical information. In contrast to the fruits of the wild crab-tree, Malus communis is sweet in taste (cf. →*sūrapuldor*); cf. OHG *suozephel* lit. 'sweet apple' glossing L *malomelum* (Björkman 1901: 210). ASSOCIATIVE RELATIONS: conceptual contiguity & taxonomic sub-ordination.
▫ →*apuldor*, →*swēte*

swēte wyrt, lit. 'sweet wort'
'Cinnamomum cassia Blume / cinnamon / Zimtbaum'
2 occ. (*Lǣcebōc*, 68/22; *Stowe Psalter*, 44, 9).
LATIN EQUIV. in the manuscripts: *cassia.*
ETYMOLOGY: OE formation; MORPHOLOGY: Adj+N; endocentric, copulative: attributive. MOTIVATION: smell & botanical information. The adjective relates to the scent of the plant; cf. "Cassia flores bene olentes unde unguentum fit" ['The blossoms of cassia, which smell good, whence an ointment is made'] (*Glossae Codicis Vaticani 3321* in Goetz 1888-1923, vol. 4: 30/59). ASSOCIATIVE RELATIONS: conceptual contiguity & taxonomic sub-ordination.
▫ →*swēte*, →*wyrt*

swīn, noun
'SWINE / Schwein'
▫ →*swīnesgærs*, →*swīnesmerce*

swīnes-gærs, lit. 'swine's grass'
'Polygonum aviculare L. / common knotgrass / Vogelknöterich'
Hapax legomenon; Gough (1974: 276) reads *swinewert*, but Bierbaumer (1977: 116) argues for *swines gers.*
ETYMOLOGY: OE formation; MORPHOLOGY: N_{gen}+N; endocentric, rectional. MOTIVATION: animal & botanical information. The plant is eaten by pigs. Cf. Marzell (s.v. *Polygonum aviculare* L.), who lists a number of similar names in other European vernaculars, e.g. ME *swynys grees*, ModE *swine's grass*, *hogweed*, *pig-weed*, Dan *svinegræs*, Du *zwienegras*, G *Schweinegras*, etc. ASSOCIATIVE RELATIONS: conceptual contiguity & taxonomic sub-ordination.
▫ →*gærs*, →*swīn*

swīnes-merce, lit. 'swine's cellery'
'Apium graveolens L. / cellery, march parsley, smallage / Echte Sellerie'
Hapax legomenon (*Cleopatra Glossary* (Stryker), 54, no. 376).
LATIN EQUIV. in the manuscripts: *apiaster.*
ETYMOLOGY: OE formation; MORPHOLOGY: N_{gen}+N; endocentric, rectional. MOTIVATION: sound pattern & botanical

information. OE *swīnesmerce* is a synonym for *wudumerce* and for *merce*. The first element is perhaps due to the phonological similarity of the Latin lemma *apiaster* and L *aper* 'boar' (Bierbaumer 3, s.v. *swinesmerce*), the second element provides botanical information: the plant is a kind of *merce*. ASSOCIATIVE RELATIONS: phonological similarity & taxonomic sub-ordination.
▫ →*merce*, →*swīn*

syrfe, noun
'Sorbus domestica L. / sorb, SERVICE-tree / Speierling, Vogelbeere'
Not in the *DOEPN*; not in Bierbaumer (1, 2, 3); attested in *Charters* ((Birch), 2: 490, no. 762; 3: 52, no. 892).
ETYMOLOGY: Early Latin loan (Wollmann 1990: 170); OE *syrfe* < MedL **sorbea* < L *sorbus* (*ODEE*, s.v. *service*; *OED*, s.v. *sorb*).
▫ →*syrftrēow*

syrf-trēow, lit. 'service-tree'
'Sorbus domestica L. / SERVICE TREE / Speierling, Vogelbeere'
2 occ. (*Charters* (Birch), 1: 290, no. 204); not in Bierbaumer (1, 2, 3); identification follows *ClH* (s.v. *syrftrēow*).
ETYMOLOGY: OE formation; hybrid compound (OE *syrfe* < L *sorbus*; →*syrfe*). MORPHOLOGY: N+N; endocentric, copulative: subsumptive (pleonastic). MOTIVATION: plant & botanical information. The *syrf(e)* 'service-tree' is a kind of tree. ASSOCIATIVE RELATIONS: conceptual identity & taxonomic sub-ordination.
▫ →*syrf*, →*trēow*

T

tān, noun
(i) 'twig / Ast, Zweig' (ii) 'lot / Los'
DOEPN (s.v. *tān*): 22 occ.
Latin equivalents in the manuscripts: *antes*, *dactylus*, *frons*, *nicolaus*, *ramusculus*, *sors*, *vimen*, *virgultus*.
ETYMOLOGY: Gmc **taino-z* (*OED*, s.v. *mistletoe*, *teanel*).
▫ →*misteltān*

-tǣsel, noun
'Dipsacus L. / TEASEL, TEAZLE / Karde'
Derived from OE *tǣsan* '(to) tease'.
▫ →*wulfestǣsel*

teors, noun
'penis / Penis'
▫ →*wulfesteors*

tosca, noun
'toad / Kröte'
▫ →*toscanlēac*

toscan-lēac, lit. 'toad's plant'
'Equisetum L. / horsetail / Schachtelhalm'
Hapax legomenon (*Durham Plant Name Glossary*, no. 208).
LATIN EQUIV. in the manuscript: *equisaetum*, *hippuris*.
ETYMOLOGY: OE formation; MORPHOLOGY: N_{gen}+N; endocentric, rectional. MOTIVATION: habitat & botanical information. The first element, *tosca* 'frog,

toad', seems to point metaphorically to a species of Equisetum which grows in marshy lands (Bierbaumer 3, s.v. *toscanlēac*). The second element, →*lēac*, does not refer to a species of Allium, but has the very general meaning of 'plant, garden-herb'. ASSOCIATIVE RELATIONS: conceptual contiguity & taxonomic sub-ordination.

▫ →*lēac*, →*tosca*

træppe, noun
'TRAP, snare / Falle'

▫ →*coltetræppe*

trēow, noun
(i) 'TREE / Baum'; (ii) 'wood / Holz', (iii) 'beam, log, timber / Balken, Bauholz'; (iv) 'truth / Wahrheit' and related concepts

DOEPN (s.v. *trēow*): 170 occ., but possibly much more often attested, as the corpus of the *DOEPN*, which is, largely, the corpus of texts examined by Bierbaumer (1, 2, 3), does not cover all Old English texts (see chapter 1.3.), and as *trēow* 'tree' belongs to the common stock of vocabulary and is not as specific as most of the botanical terms that hardly appear in non-botanical texts (cf. Neville 2003).

LATIN EQUIV. in the manuscripts: *arbor* (*silvestris*), *arbustum*, *cylindrus*, *lignum* (*viridus*), *crux*, *cucurbita*, *sycomorus*, *patibulum*.

ETYMOLOGY: Gmc **trewo*- < IE * *trĕ̄u*- '(to) thrive' (*Pokorny*, s.v. *trĕ̄u*; *OED*, s.v. *tree*).

▫ →*āctrēow*, →*æppeltrēow*, →*bēctrēow*, →*bōctrēow*, →*boxtrēow*, →*byrctrēow*, →*cedertrēow*, →*ciristrēow*, →*corntrēow*, →*cwictrēow*, →*eletrēow*, →*ellentrēow*, →*fīctrēow*, →*fūle trēow*, →*gātetrēow*, →*gyrtrēow*, →*hwītingtrēow*, →*lāwertrēow*, →*magdalatrēow*, →*mapultrēow*, →*mæstentrēow*, →*palmtrēow*, →*persoctrēow*, →*pīntrēow*, →*plūmtrēow*, →*suntrēow*, →*syrftrēow*, →*ulmtrēow*, →*wīntrēow*, →*wīrtrēow*

tūn, noun
'enclosure, garden, TOWN / eingefriedetes Gelände, Garten, Stadt'

▫ →*tūncærse*, →*tūnhōfe*, →*tūnmelde*, →*tūnminte*, →*tūnnǣp*

tūn-cærse, lit. 'garden-cress'
'Lepidium sativum L. / garden-cress / (Garten)Kresse'

DOEPN (s.v. *tūn-cærse*): 11 occ.[e]

ILLUSTRATION in Oxford, Bodleian Library, Ashmole 1431 (S.C. 7523), fol. 11r (Ohlgren 221(20)).

LATIN EQUIV. in the manuscripts: *nasturtium*.

ETYMOLOGY: OE formation; MORPHOLOGY: N+N; endocentric, rectional. MOTIVATION: habitat & botanical information. This kind of cress is grown in gardens, in enclosed places. OE *tūn* may be interpreted as habitat of the plant, but can also be understood metaphorically as 'cultivated' in contrast to 'growing wild, not cultivated by people, i.e. not in gardens' (cf. →*worþigcærse*). ASSOCIATIVE RELATIONS: conceptual contiguity & taxonomic sub-ordination.

▫ →*cærse*, →*tūn*

tunge, noun
'TONGUE / Zunge'

▫ →*hundestunge*, →*oxantunge*

tūn-hōfe, lit. 'garden-hove'
'Glechoma hederaceum L. / ground-ivy / Gundelrebe'

Hapax legomenon (*Lǣcebōc*, 105/5). ETYMOLOGY: OE formation; MORPHOLOGY: N+N; endocentric, rectional. MOTIVATION: habitat & botanical information. A highly nutritious plant that grows in the vicinity of men since it prefers soil rich in nitrates (Marzell, s.v. *Glechoma hederaceum* L.), which is often the case where humans and animals urinate. ASSOCIATIVE RELATIONS: conceptual contiguity & taxonomic sub-ordination.

▫ *→hōfe*, *→tūn*

tūn-melde, lit. 'garden-orache'
'Atriplex hortensis L. / orache / Gartenmelde'
Hapax legomenon (*Harley Glossary*, 113, no. 2099).
LATIN EQUIV. in the manuscript: *atriplex*, *chrysolachana*, *(h)olus*.
ETYMOLOGY: OE formation; MORPHOLOGY: N+N; endocentric, rectional; MOTIVATION: habitat & botanical information. *Orache* was well-known in antiquity as a highly nutritious plant. It was introduced to Europe by the Romans and lost importance in the 12th century, with the growing popularity of spinach (cf. Marzell, s.v. *Atriplex hortense* L.). Since it was cultivated, the plant name indicates the habitat: the plant was grown in gardens and enclosed places. ASSOCIATIVE RELATIONS: conceptual contiguity & taxonomic sub-ordination.

▫ *→melde*, *→tūn*

tūn-minte, lit. 'garden-mint'
'Mentha spicata L. / spear-mint / Grüne Minze'
DOEPN (s.v. *tūn-minte*): 3 occ.
ETYMOLOGY: OE formation; hybrid compound (OE *minte* < L *ment(h)a*; *→minte*). MORPHOLOGY: N+N; endocentric, rectional; MOTIVATION: habitat & botanical information. A kind of mint that was cultivated, i.e. grown in gardens. ASSOCIATIVE RELATIONS: conceptual contiguity & taxonomic sub-ordination.

▫ *→minte*, *→tūn*

tūn-nǣp, lit. 'garden-rape'
'Brassica napus L. / rape / Rapskohl'
Hapax legomenon (*Lācnunga*, 624).
ETYMOLOGY: OE formation; hybrid compound (OE *nǣp* < L *nāpus*; *→nǣp*). MORPHOLOGY: N+N; endocentric, rectional. MOTIVATION: habitat & botanical information; a kind of rape that was grown and cultivated in enclosed places. ASSOCIATIVE RELATIONS: conceptual contiguity & taxonomic sub-ordination.

▫ *→nǣp*, *→tūn*

Tunsing, proper noun

▫ *→tunsingwyrt*

tunsing-wyrt (*tungilsing-*, *tuning-*), lit. 'Tunsing's wort'
'Veratrum album L. / white hellebore / Weißer Germander'
DOEPN (s.v. *tunsing-wyrt*): 8 occ.
ILLUSTRATION in manuscript London, British Library, Cotton Vitellius C. iii, fol. 60v (Ohlgren 168(81)).
LATIN EQUIV. and OE synonyms in the manuscripts: *helleborus* (*albus*), ? *suffunie*; *hramsa* 'ramsons', *wudulēac*.
ETYMOLOGY: OE formation; MORPHOLOGY: PropN+N; endocentric, rectional. MOTIVATION: person & botanical infor-

mation. The etymology of this plant name is unclear. A connection to the Old English toponym *Tunsing*, which became a personal name, and is recorded, e.g., in the *Charters* ((Birch), 3: 90, no. 920) as "Herist of tunsing þere on þe rihte" ['Herist of Tunsing, there on the right'], seems probable (cf. Bier-baumer 1, s.v. *tungilsinwyrt*, *tuning-wyrt*; 2 and 3, s.v. *tunsingwyrt*). ASSO-CIATIVE RELATIONS: ? & taxonomic sub-ordination.
▫ →*Tunsing*, →*wyrt*

twā, twī, numeral
'TWO / zwei'
▫ →*twīlēafa*

twī-lēafa, lit. 'two-leaf'
'Ruscus hypoglossum L. / large butcher's broom / Zungen-Mäusedorn'
Hapax legomenon (*Durham Plant Name Glossary*, no. 338).
LATIN EQUIV. in the manuscript: *vinca pervinca.*
ETYMOLOGY: OE formation; probably a loan rendition of Gk *δίγλωσσος* (see below); MORPHOLOGY: Num+N; exo-centric, metonymic compound. MOTIV-ATION: number & part of plant. The leaves of the Ruscus hypoglossum are formed interestingly: in the middle of every leaf grows another, smaller, leaf. Hence the name Gk *δίγλωσσος* 'double-tongue, Doppelzunge', a synonym of *vinca pervinca* in the Latin *Herbarium Apuleii.* OE *twīlēafa* is probably a direct loan rendition of Gk *δίγλωσσος*. ASSOCIATIVE RELATION: conceptual contiguity (meronymy).
▫ →*lēafa*, →*twī*

Þ

þēfan-, ?
not identified
▫ →*þēfanþorn*

þēfan-þorn (*þȳfe-*, *þēfe-*), lit. '?-thorn'
'Rhamnus cathartica L. / common buckthorn, THEVETHORN / Purgier-Kreuzdorn'
DOEPN (s.v. *þēfan-þorn*): 20 occ.[e]
LATIN EQUIV. in the manuscripts: *grosellarius*, *rhamnus*, *sentis ursinum*.
ETYMOLOGY: WGmc compound; cognate with OHG *depandorn* (Carr 1939: 107). MORPHOLOGY: unclear formation; ? + noun; endocentric. The etymology of the first element is uncertain (*OED*, s.v. *thevethorn*). Pheifer (1974: 98, no. 601) suggests a connection with *þūfe* 'tufted'; →*þūfeþistel*. MOTIVATION: ? & botanical information. ASSOCIATIVE RE-LATIONS: ? & taxonomic sub-ordi-nation.
▫ →*þēfan*, →*þorn*

þelne-, ?
not identified
▫ →*þelneþung*

þelne-þung, lit. '?-poisonous plant'
'? Ranunculus sceleratus L. / celery-leaved crowfoot / Gift-Hahnenfuß'
Hapax legomenon (Meritt 1945: 61, no. 73 Dc185/2).
LATIN EQUIV. and OE synonym in the manuscript: ? *pamfeso*; *clufþung*.

ETYMOLOGY: OE formation; MORPHOLOGY: unclear formation; ?+N; endocentric, rectional. MOTIVATION: ? & botanical information. The first element is not transparent; the second element gives botanical information: this plant is a poisonous one. ASSOCIATIVE RELATIONS: ? & taxonomic sub-ordination.
▫ *→þelne-, →þung*

þēor, noun
'inflammation / Entzündung'
▫ *→þēorwyrt*

þēor-wyrt (*þyor-*), lit. '? inflammation-wort'
'?? Inula conyza DC. / great fleabane, ploughman's spikenard / Dürrwurz'
2 occ. (*Lǣcebōc*, 24/30, 99/20).
ETYMOLOGY: OE formation; MORPHOLOGY: N+N; endocentric, rectional. MOTIVATION: illness & botanical information. The identification and the motivation of the plant name are not clear. Apparently, it is a plant that helps cure inflammations, as indicated by the first element (Bierbaumer 1, s.v. *þēorwyrt*). ASSOCIATIVE RELATIONS: conceptual contiguity & taxonomic sub-ordination.
▫ *→þēor, →wyrt*

þistel, noun
'THISTLE / Distel'
DOEPN (s.v. *þistel*): 15 occ.
LATIN EQUIV. in the manuscripts: *carduus*, *pappus*.
ETYMOLOGY: Gmc **þistila-* < **(s)teig-* '(to) prick' + *-el* (*Kluge*, s.v. *Distel*; *OED*, s.v. *thistle*; Sauer 2012: 223).
▫ *→brādþistel, →meteþistel, →scearpe þistel, →smælþistel, →þūþistel, →unbrāda þistel, →wuduþistel*

-þone, noun
'tendril / Ranke, Dohne'
Not attested independently.
ETYMOLOGY: WGmc **þunon* (*Kluge*, s.v. *Dohne*).
▫ *→ælfþone*

þorn, noun
(i) 'THORN / Dorn', (ii) 'thorn-bush / Dornenstrauch' (iii) 'thorny tree / Baum mit Dornen'
63 occ. in the medico-botanical literature (cf. *DOEPN*, s.v. *þorn*); the *DOEC* (s.v. *þorn*) produces 470 hits, of which c. 330 are attested in the charters (as simplexes and in bounds). According to Hooke (2009: 80), "the thorn stands out as by far the most frequently recorded tree on estate boundaries."
LATIN EQUIV. in the manuscripts: *dumus*, *sentis*, *spica*, *spina*, *tribulus*.
ETYMOLOGY: Gmc **þurnu-* < IE **trn-* 'thorn, point, tip' (*Kluge*, s.v. *Dorn*); in contrast to *þyrne* 'thorn-bush', which is derived (< **þurn+jon*, with *i*-mutation; *→þyrne*), *þorn* 'thorn-bush' is a metonymic extension of the original meaning 'thorn'.
▫ *→æppelþorn, →brēmelþorn, →haguþorn, →lūsþorn, →pricþorn, →slāhþorn, →þēfanþorn*

þrote, noun
'THROAT / Hals, Kehle'
▫ *→æscþrote, →eoforþrote*

þrie, numeral
'THREE / drei'
▫ *→þrilēfe*

þri-lēfe, lit. 'three-leaf'
'Oxalis acetosella L. / wood sorrel, shamrock / Wald-Sauerklee, Hasenklee, Kuckucksklee'
Hapax legomenon (*Antwerp-London Glossaries*, 107/2).
LATIN EQUIV. and OE synonym in the manuscript: *trifolium*; *gēaces sūre*.
ETYMOLOGY: OE formation; probably loan translation of L *trifolium* (Bierbaumer 3, s.v. *þrilēfe*). MORPHOLOGY: Num+N; exocentric, metonymic compound. MOTIVATION: number & part of plant. Since Oxalis acetosella is a kind of trifoliate clover (Marzell, s.v. *Oxalis acetosella* L.), the motivation of the plant name as lit. 'three-leaf' is not surprising. ASSOCIATIVE RELATION: conceptual contiguity (meronymy).
▫ *→lēaf*, *→þrie*

þū- ?
'? tuft / Büschel, Quaste'
▫ *→þūþistel*

þung, noun
(i) 'poison / Gift'; (ii) 'poisonous plant / Giftpflanze'; (iii) 'Aconitus napellus L. / monkshood / Blauer Eisenhut', or (iv) 'Veratrum album L. / white veratrum / Weißer Germer' (cf. Hoops 1889, 83-4; *DOEPN*, s.v. *þung*)
DOEPN (s.v. *þung*): 26 occ.
LATIN EQUIV. and OE synonym in the manuscripts: *aconitum*, *batrachion*, *helleborus*, *mandragoras*, *sole* (? for *solanum* or *solata*), *toxa* (? for *toxicum*); *clufþung*.
ETYMOLOGY: IE **tenk-* '(to) contract' (*Pokorny*, s.v. *tenk-*; Bierbaumer 1, s.v. *þung*).
▫ *→brōcþung*, *→clufþung*, *→micel þung*, *→þelneþung*

þunor, noun
'THUNDER / Donner'
▫ *→þunorclǣfre*, *→þunorwyrt*

þunor-clǣfre, lit. 'thunder-clover'
(i) 'Ajuga reptans L. / common bugle / Kriechender Günsel'; (ii) '? Chrysanthemum leucanthemum L. / ox-eye daisy, marguerite / Wucherblume'
DOEPN (s.v. *þunor-clǣfre*): 1 occ. (*Lācnunga*, 26).
ETYMOLOGY: OE formation; MORPHOLOGY: N+N; endocentric, rectional. MOTIVATION: threat & botanical information. Both plants were thought to help against thunder, hence the use of the first element (see also *→þunor-wyrt*); why the plants were compared to *clǣfre* 'clover' is not evident, but *þunor-clǣfre* was seen as taxonomically sub-ordinated to *clǣfre*. ASSOCIATIVE RELATIONS: conceptual contiguity & taxonomic sub-ordination.
▫ *→clǣfre*, *→þunor*
▫ *→smale þunorclǣfre*

þunor-wyrt, lit. 'thunder-wort'
'Sempervivum tectorum L. / houseleek, barren privet / Echte Hauswurz'
Hapax legomenon (*Lǣcebōc*, 36/16).
ETYMOLOGY: OE formation; OHG *donerswurze* (Björkman 1902: 284); MORPHOLOGY: N+N; endocentric, rectional. MOTIVATION: threat & botanical information. The plant name alludes to the belief that these plants, when planted on the roofs of houses, can protect them from storms, esp. from being struck by lightening, cf. also the German names *Hauswurz*, *Dachwurzel*, *Donnerkraut* (Marzell, s.v. *Sempervivum tectorum* L.); see, further, *→þunorclǣfre* and *→hāmwyrt*. ASSO-

CIATIVE RELATIONS: conceptual contiguity & taxonomic sub-ordination.
▫ →*þunor*, →*wyrt*

þū-þistel (*þūfe-*), lit. '? tuft(ed) thistle'
'Sonchus oleraceus L. Em. Gouan / sow-thistle, hare's colewort / Kohl-Gänsedistel'
DOEPN (s.v. *þū-þistel*): 6 occ.[e]
LATIN EQUIV. and OE synonym in the manuscripts: *lactuca*; *leahtric*.
ETYMOLOGY: WGmc compound; OHG *sûdistel*, OLG *duthistel* (Carr 1939: 104); MORPHOLOGY: unclear formation; noun or adjective + noun; endocentric, rectional. MOTIVATION: part of plant & botanical information. The first element is rather obscure. *þū-* may be shortened for *þūfe-* 'tuft' (see Pheifer 1974: 98, no. 601). Cognate with OHG *dudistel* of unclear origin (*Holthausen*, s.v. *ðuðistel*; Björkman 1901: 226; 1902: 291); according to Carr (1939: 104), the German and English names have been associated with G *Sau*, ModE *sow*. ASSOCIATIVE RELATIONS: conceptual contiguity (meronymy) & taxonomic sub-ordination.
▫ →*þistel*, →*þū*

þȳfel, noun
'thicket / Dickicht'
DOEPN (s.v. *þȳfel*): 17 occ.
LATIN EQUIV. and OE synonym in the manuscripts: *arbustum*, *belsa*, *defrutus .i. vino*, *defrutus medona vino*, *frutectum*, *ramus*, *ruscus*, *spartum*, *spina*, *sentis*; *box*.
ETYMOLOGY: *þūf* 'thicket, bush' + *-el* (cf. *Holthausen*, s.v. *þūf*, *þȳfel*; *OED*, s.v. *thyvel*, *thuvel*); cf. also Sauer (2012: 220).
▫ →*brēmelþȳfel*, →*scaldþȳfel*

þyrne, noun
'bush, thorn-bush / Busch, Dornenbush'
DOEPN (s.v. *þyrne*): 16 occ.
LATIN EQUIV. in the manuscripts: *dumus*, *frutex*, *ramus*, *rubus*, *spina*, *taxus*.
ETYMOLOGY: Gmc **þurnjon*, a derivative from **þurn-* 'thorn' (*OED*, s.v. *thorn*).
▫ →*brēmelþyrne*, →*brērþyrne*

U

ulm-, noun
'Ulmus campestris L. / elm / Ulme'
Not attested independently.
LATIN EQUIV. in the manuscripts: *ulmus*.
ETYMOLOGY: < L *ulmus* (*Kluge*, s.v. *Ulme*); ModE *elm* < OE *elm* < WGmc **elmoz* (*OED*, s.v. *elm*).
▫ →*ulmtrēow*

ulm-trēow, lit. 'elm-tree'
'Ulmus campestris L. / elm-tree / Feldulme'
Hapax legomenon (*Antwerp-London Glossaries*, 122/18); *DOEPN* (s.v. *ulm-trēow*).
ETYMOLOGY: OE formation; hybrid compound (OE *ulm* < L *ulmus*; →*ulm*). MORPHOLOGY: N+N; endocentric, copulative: subsumptive (pleonastic). MOTIVATION: plant & botanical information. The elm is a kind of tree. ASSOCIATIVE RELATIONS: conceptual identity & taxonomic sub-ordination.
▫ →*trēow*, →*ulm-*

unbrād, adjective
'not broad / nicht breit'
▫ →*unbrāda þistel*

unbrāda þistel, lit. 'not broad thistle'
(i) '? Cnicus ferox L. / prickly cnicus / Hohe Bergdistel', or (ii) ? 'Cynara cardunculus L. / Karde / teasel'
2 occ. (*Durham Plant Name Glossary*, no. 312; *Brussels Glossary*, 301/22).
LATIN EQUIV. in the manuscripts: *scolymos*.
ETYMOLOGY: OE formation; MORPHOLOGY: Adj+N; endocentric, copulative: attributive. MOTIVATION: size & botanical information. The lemma *scolymus* in the glosses is one of the MedL name of Cynara cardunculus L. 'teasel', G *Karde* (Marzell, s.v. *Cynara Cardunculus* L.), but this thistle-like plant has broad, fleshy leaves, which can be eaten like the leaves of the artichoke (Cynara scolymos). As this does not fit with the Old English plant name *unbrāda þistel* lit. 'not broad thistle', Bierbaumer (3, s.vv. *þistel*, *smælþistel*) suggests the alternative identification as 'Cnicus ferox, prickly cnicus': this kind of thistle has narrow leaves; see also →*smælþistel*. ASSOCIATIVE RELATIONS: conceptual contiguity & taxonomic sub-ordination.
▫ →*þistel*, →*unbrād*

W

wan, adjective
'flexible / biegbar'
▫ →*wananbēam*

wanan-bēam *(anan-bēam)*, lit. '? flexible tree'
'Evonymus europaeus L. / prickwood, spindle-tree / Pfaffenkäppchen, Gemeiner Spindelstrauch'
DOEPN (s.v. *wanan-bēam*): 8 occ.[e]
LATIN EQUIV. in the manuscripts: *fagus*, *fusarius*.
ETYMOLOGY: OE formation; MORPHOLOGY: Adj+N; endocentric, copulative: attributive. The element *wanan-* is otherwise not attested in Old English, but can be related to an Old Saxon cognate lexeme (see MOTIVATION). MOTIVATION: texture & botanical information. *Holthausen* (s.v. *wananbēam*) and Gardner (1968: 158) suggest that the first element is either an unattested adjective which refers to the bright colour of this plant (cf. OS *wānam* 'shiny'), or a noun related to OIcel *vanir* 'gods'. Bierbaumer (3, s.v. *wanan-bēam*) rejects both interpretations and proposes to regard the first element as the adjective OE **wana* 'flexible, bendable' (related to the IE root **uei-* '(to) bend, (to) twist'), an explanation, which is followed here, as the plant's twigs are quite flexible. ASSOCIATIVE RELATIONS: conceptual contiguity & taxonomic sub-ordination.
▫ →*bēam*, →*wanan*

wār, noun
'seaweed / Seetang'
▫ →*sǣwār*

wæter, noun
'WATER / Wasser'
▫ →*wæterbōg*, →*wæterwyrt*

wæter-bōg (*-bōh*), lit. 'water-bough'
'sucker / Schössling'
Hapax legomenon (*Antwerp-London Glossaries*, 144/13); cf. Bierbaumer (3, s.v. *wæter-bōh*).
LATIN EQUIV. in the manuscripts: *surculus*.
ETYMOLOGY: OE formation; MORPHOLOGY: N+N; endocentric, rectional. MOTIVATION: part of plant & botanical information. Young suckers are full of sap, a fact encoded in *wæter* 'water'. ASSOCIATIVE RELATIONS: conceptual contiguity (meronymy) & taxonomic sub-ordination.
▫ →*bōg*, →*wæter*

wæter-wyrt, lit. 'water-wort'
(i) 'Eupatorium cannabium L. / water-hemp / Gemeiner Wasserdost'; (ii) 'Callitrice L. / starwort / Wasserstern'; (iii) '? Asplenium trichomans L. / common maidenhair / Brauner Milzfarn'; (iv) 'Ranunculus aquatilis L. / water liverwort / Gemeiner Wasser-Hahnenfuß'
DOEPN (s.v. *wæter-wyrt*): 9 occ.
ILLUSTRATION in manuscript London, British Library, Cotton Vitellius C. iii, fol. 36r (Ohlgren 168(33)).
LATIN EQUIV. and OE synonym in the manuscripts: *callitrichum*; *ēalifer*.
ETYMOLOGY: WGmc compound (Carr 1939: 104), OHG *wazaruuurz* (Björkman 1902: 302); MORPHOLOGY: N+N; endocentric, rectional. MOTIVATION: (i) habitat & botanical information; (ii) illness & botanical information. ad (i) Eupatorium cannabium, Callitrice and Ranunculus aquatilis all prefer moist ground, which would explain the use of the first element. ad (ii) Asplenium trichomans is used against dropsy (Bierbaumer 2, s.v. *wæterwyrt*), an illness where water is accumulated beneath the skin, hence the use of *wæter* 'water'. ASSOCIATIVE RELATIONS: in both cases, the associative relations are conceptual contiguity & taxonomic sub-ordination.
▫ →*wæter*, →*wyrt*

weal, noun
'WALL / Wall'
ETYMOLOGY: Early Latin loan, < L *vallum* (Wollmann 1990: 177).
▫ →*wealwyrt*

wealh, noun
'foreigner / Fremder'
▫ →*wealhhnutu*, →*wealhmore*, →*wealhwyrt*

wealh-hnutu, lit. 'foreigner-nut'
'Juglans regia L. / WALNUT / Walnuß'
DOEPN (s.v. *wealh-hnutu*): 1 occ. (*Cleopatra Glossary* (Stryker), 320, no. 52).
LATIN EQUIV. and OE synonym in the manuscripts: *nux*; *hnūtbēam*.
ETYMOLOGY: WGmc compound (Carr 1939: 119); possibly loan rendition of *nux gallica* (see MOTIVATION); MORPHOLOGY: N+N; exocentric, metonymic compound. The explanations of *wealhhnutu* as 'foreign nut, walnut' and of *wealhmore* as 'foreign root, carrot, parsnip' by *BT* presupposes that the first elements of the combinations are adjectives (see also Bierbaumer 1, s.v. *more* and Sauer 1992b: 398). This interpretation does not seem correct to me, as Old English had an adjective *wȳlisc*, which is derived from the noun *wealh* by the adjective suffix *-isc*. In the botanical lexicon, it is extant in the combinations *wīelisc wudu* 'Artemisia

abrotanum L. / southernwood' and *wȳlisc moru* 'Pastinacia sativa L. / common parsnip'. Therefore, the first elements of the combinations *wealh-hnutu* 'Juglans regia L., walnut' and *wealhmore* (i) 'Pastinacia sativa L. / commmon parsnip', (ii) 'Daucus carota L., wild carrot' are not analysed as adjectives, but as nouns denoting the *wealh* 'foreigner'. MOTIVATION: origin & part of plant. Cf. MLG *wallnut*, MDu *walnote*; ON *valhnot* was borrowed from Middle Low German (Carr 1939: 119); MHG *wälhisch nuß*, Flem *walsche not*, G *Welsche Nuß*, etc. (*ODEE*, s.v. *walnut*); Loan rendition of L *nux gallica*, called so probably because the plant was grown predominantly in Gaul (*Kluge*, s.v. *Walnuß*); see also →*frencisc hnutu*; ASSOCIATIVE RELATIONS: conceptual contiguity & conceptual contiguity (meronymy).

▫ →*hnutu*, →*wealh*

wealh-more, lit. 'foreigner-carrot', or 'foreigner root'

(i) 'Pastinacia sativa L. / common parsnip, wild parsnip / Gemeiner Pastinak, Hirschmöhre'; or (ii) 'Daucus carota L. / wild carrot / Gemeine Mohrrübe, Gelbe Rübe'

DOEPN (s.v. *wealh-more*): 14 occ.[e]

LATIN EQUIV. in the manuscripts: *carota, daucus, pastinaca.*

ETYMOLOGY: OE formation; MORPHOLOGY: N+N; endocentric, rectional. For the interpretation of *wealh* as adjective, see →*wealhhnutu*. MOTIVATION: (i) origin & botanical information. (ii) overall appearance. ad (i) Pastinacia sativa L. was introduced to Anglo-Saxon England by the Romans and it was obviously perceived as foreign, hence the use of the first element *wealh-* 'foreigner' (Bierbaumer 1, s.v. *more*). ad (ii) In medieval times, Daucus carota L. (OE *more*), the carrot, was not orange as it is today – it strongly resembled the small and whitish Pastinacia sativa L., the parsnip (Banham 2003: 25), so that the two were easily confused. ASSOCIATIVE RELATIONS: ad (i) conceptual contiguity & taxonomic sub-ordination; ad (ii) co-taxonomic similarity.

▫ →*more*, →*wealh*

wealh-wyrt (*weal-*), lit. 'foreigner-wort'
'Inula helenium L. / elecampane / Echter Alant'

Following Förster (1917: 133f.), I distinguish between *wealhwyrt* 'elecampane' and →*wealwyrt* 'dwarf elder'; 34 occ.[e] in total; they cannot be easily distinguished; *DOEPN* (s.v. *weal-wyrt*).

LATIN EQUIV. in the manuscripts: *intula.*

ETYMOLOGY: OE formation; MORPHOLOGY: N+N; endocentric, rectional. MOTIVATION: origin & botanical information. Inula helenium originally stems from Central Asia, but it was introduced to Europe (first in the Mediterranean region) quite early (Marzell, s.v. *Inula helenium* L.); the foreign origin is, still, encoded in the first element of the plant name. ASSOCIATIVE RELATIONS: conceptual contiguity & taxonomic sub-ordination.

▫ →*wealh*, →*wyrt*

weal-wyrt (*wæl-*), lit. 'wall-wort'
'? Sambucus ebulus L. / dwarf elder / Zwergholunder, Attich'

Following Förster (1917: 133f.) I distinguish between →*wealhwyrt* 'ele-

campane' and *wealwyrt* 'dwarf elder'; *DOEPN* (s.v. *weal-wyrt*): 34 occ.[e] (see →*wealhwyrt*).
ILLUSTRATION in manuscript London, British Library, Cotton Vitellius C. iii, fol. 47v (Ohlgren 168(55)).
LATIN EQUIV. in the manuscripts: *ebulus*.
ETYMOLOGY: OE formation; hybrid compound (OE *weal(l)* < L *vallum*; →*weal*); it can be easily confused with →*wealhwyrt*; MORPHOLOGY: N+N; endocentric, rectional. MOTIVATION: tool & botanical information. Sambucus ebulus was often used to enclose real estate (Bierbaumer 1, s.v. *wealwyrt*), hence the use of the first element *weal* 'wall'. Not convincing to me is the explanation as lit. 'slaughter-plant', OE *wælwyrt*, given by Hoops (1889: 65), as there is no obvious reason to connect the dwarf elder conceptually with slaughter and blood-shedding. ASSOCIATIVE RELATIONS: conceptual contiguity & taxonomic sub-ordination.
▫ →*weal*, →*wyrt*

weax, noun
'sinew / Sehne'
▫ →*wuduweax*

wēde, *wōde*, noun
'madness, fury / Wahnsinn, Tollheit'
▫ →*wēdeber(i)(g)e*, →*wōdehwistle*

wēde-ber(i)(g)e, lit. 'madness-berry'
'Veratrum album L. / white hellebore / Weißer Germander'
DOEPN (s.v. *wede-berie*): 10 occ.[e]
LATIN EQUIV. in the manuscripts: *helleborus*.
ETYMOLOGY: OE formation; see MOTIVATION. MORPHOLOGY: N+N; exocentric, metonymic compound. MOTIVATION: illness & part of plant. As the hellebore had been associated with madness since antiquity, d'Aronco (1988: 30) suggests that OE *wēdeberge* is a loan translation based on L *elleborum album*. Since it does not correspond formally or semantically to its Latin model, in my view, OE *wēdeberge* is not a loan translation and not even a loan creation based on L *elleborum*, but rather an expression reflecting the knowledge about the noxiousness of the berries of the Veratrum album, a knowledge which was transmitted from classical antiquity to medieval times and led, eventually, to the combination of the elements *wēde* 'madness, fury, rage' and *ber(i)(g)e* 'berry', i.e. 'berries that induce madness'. ASSOCIATIVE RELATIONS: conceptual contiguity & conceptual contiguity (meronymy).
▫ →*ber(i)(g)e*, →*wēde*

weg, noun
'WAY / Weg'
▫ →*wegbrǣde*

weg-brǣde (*-brāde*), lit. 'way-broad (plant)'
'Plantago maior L. / greater plantain, WAYBREAD / Wegbreite'
DOEPN (s.v. *weg-brāde*): 92 occ.[e]
ILLUSTRATIONS in manuscripts London, British Library, Cotton Vitellius C. iii, fol. 21v (Ohlgren 168(5)), and Oxford, Bodleian Library, Ashmole 1431 (S.C. 7523), fol. 5r (Ohlgren 221(2)).

LATIN EQUIV. in the manuscripts: *arnoglossa*, *cynoglossa*, *lapathium*, *plantago*, *septemnerva*.
ETYMOLOGY: WGmc compound (Carr 1939: 104); OHG *wegebreita* (Björkman 1901: 230); MORPHOLOGY: N+N_{deadj}; exocentric, metonymic compound. MOTIVATION: habitat & size. Plantago major grows mainly along roads and its leaves are much broader than those of its co-taxonym Plantago lanceolata L. Cf. also OHG *wegebreita*, G *Wegbreite*. Obviously, it is so significantly often found alongside roads, that modern archeologists rely on the absence or presence of its pollen to determine prehistoric paths and roads (Birkhan 2012: 171). ASSOCIATIVE RELATIONS: conceptual contiguity & conceptual contiguity.
▫ →*brǣde*, →*weg*
▫ →*rūwe wegbrǣde*, →*smēþe wegbrǣde*

wen, noun
'WEN, tumour / Tumor'
▫ →*wenclufu*, →*wenwyrt*

wen-clufu, lit. 'tumour-clove'
'? Ranunculus ficaria L. / pilewort, figwort / Feigwurz, Scharbockskraut'
Hapax legomenon (*Laud Herbal Glossary*, no. 642).
LATIN EQUIV. and OE synonym in the manuscript: ? *emigrana*; *wenwyrt*.
ETYMOLOGY: OE formation; MORPHOLOGY: N+N; exocentric, metonymic compound. MOTIVATION: illness or part of plant & part of plant. The roots of Ranunculus ficaria are bulbous, clove-like. Ranunculus ficaria was applied to hemorrhoids (piles), which look like 'tumours' (Marzell, s.v. *Ranunculus ficaria* L., 4); hence, the use of the first element OE *wen* 'tumour' and the second element *clufu* 'clove'; see also →*fīcwyrt* and →*wenwyrt*. ASSOCIATIVE RELATIONS: conceptual contiguity & conceptual contiguity (meronymy).
▫ →*clufu*, →*wen*

wen-wyrt, lit. 'tumour-wort'
'? Ranunculus ficaria L. / pilewort, figwort / Feigwurz, Scharbockskraut'
DOEPN (s.v. *wen-wyrt*): 12 occ.
LATIN EQUIV. and OE synonym in the manuscripts: ? *emigrana*; *wenclufu*.
ETYMOLOGY: OE formation; MORPHOLOGY: N+N; endocentric, rectional. MOTIVATION: illness or part of plant & botanical information. Ranunculus ficaria has small bulbous, clove-like roots, which explains the use of the second element, *clufu* 'clove'; these can be associated with tumours on the basis of their shape (OE *wen* 'tumour'), but the first element of the plant name may also refer to tumour-like hemorrhoids (piles) for which the plant was used (Marzell, s.v. *Ranunculus ficaria* L., 4); →*fīcwyrt* and →*wenclufu*. ASSOCIATIVE RELATIONS: conceptual contiguity & taxonomic sub-ordination.
▫ →*wen*, →*wyrt*
▫ →*clufihte wenwyrt*, →*cnēowihte wenwyrt*, →*smale wenwyrt*

weoþo-, *wiþo-*, noun
'band, rope / Band, Seil'
▫ →*weoþobend*, →*wiþowinde*

weoþo-bend (*wiþo-*), lit. 'band, rope-binding plant'
'Lonicera periclymenum L. / woodbine, honeysuckle / Geißblatt'
Hapax legomenon (*Lǣcebōc*, 96/5).

ETYMOLOGY: OE formation; MORPHOLOGY: synthetic, $N+N_{deverbal(-e)}$; endocentric, rectional. MOTIVATION: shape & action. According to Bierbaumer (3, s.v. *wiþowinde*), *weoþo-* (*wiþo-*) in *weoþobend* and *wiþo-* in →*wiþowinde* combine two Germanic morphemes: **widu-* 'wood' (OE *wudu*) and **wiþi-* 'rope, cord' (OE *wīþī(g)e* 'willow, band, garland') as these were often confused and assimilated due to their phonological similarity. It cannot be ascertained whether the plant name refers to the plant binding together woody rope-like plants (cf. →*bere-winde*) or whether the first element refers to the fact that the plant itself is made up as if of ropes. ASSOCIATIVE RELATIONS: conceptual contiguity & taxonomic sub-ordination.

▫ →*bindan*, →*weoþo-*

wer, noun
'man / Mann'

▫ →*wermōd*

wer-mōd, lit. '? man- ?mood'
'Artemisia absinthium L. / WORMWOOD, absinth / Wermut, Absinth'
DOEPN (s.v. *wermōd*): 104 occ.[e]
ILLUSTRATION in manuscript London, British Library, Cotton Vitellius C. iii, fol. 50v (Ohlgren 168(61)).
LATIN EQUIV. in the manuscripts: *absinthius*, *aloxis*.
ETYMOLOGY: WGmc lexeme of unclear origin (*Kluge*, s.v. *Wermut*); OHG *werm(u)ota*, *werim(u)ota*, *wermuot*, *wormota*, etc. (Marzell, s.v. *Artemisia absinthium* L.); possibly the result of folk etymology. MORPHOLOGY: N+N; analysis not possible, as the motivation is not transparent (see below). MOTIVATION: The elements of the composite, *wer* 'man' and *mōd* 'mood' are identifiable, but how and why these two concepts have been brought together is not transparent and not explainable at all from the features of the plant; their form is probably due to folk etymology. Originally from WGmc **wermōda-*, it has been suggested that the plant name is cognate with C **swerwo-* and OIr *serb-* 'bitter', as the plant is known, among others, for its bitter taste (*Kluge*, s.v. *Wermut*); ASSOCIATIVE RELATIONS: unclear.

▫ →*mōd*, →*wer*
▫ →*wermōd se hāra*
▫ →*fūla wermōd*, →*sūþerne wermōd*

wermōd se hāra, lit. 'hoary wormwood'
'? Artemisia maritima L. / sea-wormwood / Strand-Beifuß'
Hapax legomenon (*Lācnunga*, 527).
ETYMOLOGY: OE formation; MORPHOLOGY: postmodified noun phrase: noun + demonstrative + adjective. MOTIVATION: botanical information & colour. According to Bierbaumer (2, s.v. *wermōd*), the first element aludes to the hoary leaves of sea-wormwood. ASSOCIATIVE RELATIONS: taxonomic subordination & conceptual contiguity.

▫ →*hāra*, →*wermōd*

wīelisc (*wȳlisc*), adjective
'foreign / fremd'

▫ →*wȳlisc more*, →*wīelisc* [*wudu*]

***wīelisc* [*wudu*]**, lit. 'foreign [wood]'
'? Artemisia abrotanum L. / southernwood / Eberraute'
It occurs in *Cleopatra Glossary* ((Stryker), 349, no. 217), and, accord-

ing to Bierbaumer (3, s.v. *wǣlisc*), also in manuscript London, British Library, Cotton Otho E.i (resp., Junius 77); this is not confirmed, however, by Voss (1996).
LATIN EQUIV. in the manuscripts: *habrotonum*.
ETYMOLOGY: OE formation; MORPHOLOGY: Adj+N; exocentric, metonymic formation. MOTIVATION: origin & texture. Artemisia abrotanum L. is a woody plant (a shrub) from the Mediterranean region (see Bierbaumer 3, s.v. *wǣlisc*; s.v. *sūþerne wudu*). ASSOCIATIVE RELATIONS: conceptual contiguity & conceptual contiguity.
▫ →*wīelisc*, →*wudu*

wȳlisc moru, lit. 'foreign (Welsh) carrot'
'Pastinacia sativa L. / common parsnip, wild parsnip / Gemeiner Pastinak, Hirschmöhre'
Hapax legomenon (*Lǣcebōc*, 96/9).
ETYMOLOGY: OE formation; MORPHOLOGY: Adj+N; endocentric, copulative: attributive. MOTIVATION: origin & botanical information. Pastinacia sativa L. was introduced to Anglo-Saxon England by the Romans (see also →*wealhmore*; Bierbaumer 1, s.v. *more*) and it was obviously perceived as foreign; hence the naming 'foreign carrot'. ASSOCIATIVE RELATIONS: conceptual contiguity & taxonomic subordination.
▫ →*moru*, →*wīelisc*

wild, adjective
'WILD / wild'
▫ →*wilde caul*, →*wilde cyrfet*, →*wilde elebēam*, →*wilde eletrēow*, →*wilde lactuce*, →*wilde mæg(e)þ(e)*, →*wilde mealwe*, →*wilde minte*, →*wilde nǣp*, →*wilde pise*, →*wilde popig*, →*wilde rǣdic*, →*wilde rūde*, →*wilde wīngeard*

wilde caul, lit. 'wild cabbage'
'Brassica rapa L. / wild cabbage / Rübenkohl'
1 occ. (*Laud Herbal Glossary*, no. 210); not a lemma on its own in the *DOE*, but it appears subsumed to the entry *cāwel*, 1.a. "in combinations: *brad cawel*, *read cawel*, *wilde cawel*"; not a lemma in Bierbaumer, but appears as attestation for *cāwel* (Bierbaumer 3, s.v. *cāwel*).
The identification presented here rests on the Latin lemma in *Laud Herbal Glossary* (no. 210): "brassica i. caulis siluatica", which is glossed by *wilde caule*; cf. also G *Wilder Kohl*, Dan *vild kål*, Sw *vildkaal* (Marzell, s.v. *Brassica Rapa* L.).
LATIN EQUIV. in the manuscript: *caulis silvatica*.
ETYMOLOGY: OE formation; probably a loan translation of L *caulis siluatica*; hybrid compound (OE *caul* < L *caulis*; see also →*cāwel*). MORPHOLOGY: Adj+N; endocentric, copulative: attributive. MOTIVATION: wild & botanical information. While the second element, the head of the combination, indicates the taxonomic rank – the plant is a kind of *cāwel* 'cabbage' – the first element refers to the fact that it is a wild species: "Die wild wachsende Pfl., die als Ackerunkraut, auf Schutt usw. vorkommt […]" ['The plant, which grows wildly, and grows like weeds on debris'] (Marzell, s.v. *Brassica Rapa* L.). ASSOCIATIVE RELATIONS: conceptual contiguity & taxonomic sub-ordination.
▫ →*cāwel*, →*wild*

wilde cyrfet, lit. ‘wild gourd’
(i) ‘Cucurbita L. / gourd / Kürbis’; or (ii) ‘Citrullus colocynthis Schrad. / colocynth / Kolokynthe’; or (iii) ‘Bryonia L. / bryony / Zaunrübe’
DOEPN (s.v. *cyrfet, wilde*): 2 occ. (Förster 1917 124/169, and *Antwerp-London Glossaries*, 119/12).
LATIN EQUIVALENTS and OE synonyms in the manuscripts: *bryonia, colocynthida*; *hwīt wīngeard*.
ETYMOLOGY: Independent formation in WGmc; OHG *wildekvirbs*, *wildekurbebez*, etc. (Björkman 1904: 186); hybrid formation (OE *cyrfet* < L *cucurbita*; →*cyrfet*); probably a loan rendition of L *cucurbita agrestis* (Bierbaumer 3, s.v. *cyrfet*). MORPHOLOGY: Adj+N; endocentric, copulative: attributive. MOTIVATION: wild & botanical information. As the plant name is a loan translation of *cucurbita agrestis*, its motivation is ‘a kind of *cyrfet* that grows wildly, in the fields.’ ASSOCIATIVE RELATIONS: conceptual contiguity & taxonomic sub-ordination.
▫ *→cyrfet, →wild*

wilde elebēam, lit. ‘wild olive-tree’
wild form of ‘Olea europaea L. subsp. silvestris Rouy / olive-tree / Olivenbaum’
1 occ. (*Lǣcebōc*, 28/26); not a plant name in Bierbaumer (1, 2, 3), but subsumed s.v. *elebēam*.
LATIN EQUIV. in the manuscript: *oleaster*.
ETYMOLOGY: WGmc formation; OHG *wildiolepoume*, *wildoleboum*, etc. (Björkman 1904: 191); MORPHOLOGY: Adj+N; endocentric, copulative: attributive. MOTIVATION: wild & botanical information. A wild species of the olive-tree, as is also indicated by the addition ‘silvestris’ in the scientific botanical name. ASSOCIATIVE RELATIONS: conceptual contiguity & taxonomic sub-ordination.
▫ *→elebēam, →wild*

wilde eletrēow *(windel-trēow)*, lit. ‘wild olive-tree’
wild form of ‘Olea europaea L. subsp. silvestris Rouy / olive-tree / Olivenbaum’
1 occ. (*Cleopatra Glossary* (Stryker), 339, no. 106; spelling *windeltreow*, probably erroneously for *wilde eletrēow*); not a lemma in Bierbaumer (1, 2, 3), but attested s.v. *eletrēow*.
LATIN EQUIV. in the manuscript: *oleaster*.
ETYMOLOGY: OE formation. MORPHOLOGY: Adj+N; endocentric, copulative: attributive. MOTIVATION: wild & botanical information; A wild species; see *→elebēam*. ASSOCIATIVE RELATIONS: conceptual contiguity & taxonomic sub-ordination.
▫ *→eletrēow, →wild*

wilde lactuce, lit. ‘wild lettuce’
‘Lactuca L. / lettuce / Lattich’
2 occ. (*Antwerp-London Glossaries*, 146/5; *Laud Herbal Glossary*, no. 1338); not a lemma in Bierbaumer (1, 2, 3), but found s.v. *lactūca*.
LATIN EQUIV. in the manuscripts: *sarrabum* (? = *scariola*).
ETYMOLOGY: OE formation; parallel formation OHG *wildilateche*, glossing *lactuca siluatica* (Björkman 1904: 187); hybrid formation (OE *lactuce* < L *lactūca*; *→lactuce*). MORPHOLOGY: Adj+N; endocentric, copulative: attributive. MOTIVATION: wild & botanical information. A kind of lettuce that is

not cultivated. ASSOCIATIVE RELATIONS: conceptual contiguity & taxonomic sub-ordination.
▫ →*lactuce*, →*wild*

wilde mæg(e)þ(e), lit. 'wild camomile'
(i) 'Matricaria chamomilla L. / wild camomile, common camomile / Echte Kamille, Mägdeblume'; (ii) 'Anthemis arvensis L. / field-camomile / Wilde Kamille, Acker-Hundskamille'
DOEPN (s.v. *mageþe, wilde*): 1 occ. (*Lǣcebōc*, 62/12).
ETYMOLOGY: OE formation; MORPHOLOGY: Adj+N; endocentric, copulative: attributive. MOTIVATION: wild & botanical information; A kind of camomile that grows wildly. ASSOCIATIVE RELATIONS: conceptual contiguity & taxonomic sub-ordination.
▫ →*mæg(e)þ(e)*, →*wild*

wilde mealwe, lit. 'wild mallow'
'Malva sylvestris L. / mallow / Rosspappel'
Hapax legomenon (*Lǣcebōc*, 64/26); *DOEPN* (s.v. *mealwe, wilde*).
ETYMOLOGY: OE formation; hybrid formation (OE *mealwe* < L *malva*; →*mealwe*). MORPHOLOGY: Adj+N; endocentric, copulative: attributive. MOTIVATION: wild & botanical information. Malva sylvestris grows wildly (but it can also be cultivated). ASSOCIATIVE RELATIONS: conceptual contiguity & taxonomic sub-ordination.
▫ →*mealwe*, →*wild*

wilde minte, lit. 'wild mint'
'Mentha L. / mint / Minze'
Hapax legomenon (*Lǣcebōc*, 71/20).
ETYMOLOGY: WGmc formation; OHG *wildeminza*, etc. (Björkman 1904: 190); hybrid formation (OE *minte* < L *ment(h)a*; →*minte*). MORPHOLOGY: Adj+N; endocentric, copulative: attributive. MOTIVATION: wild & bot-anical information. Apparently a common name for various species of Mentha (Bierbaumer 1, s.v. *minte*), indicating that it grows in the wild. ASSOCIATIVE RELATIONS: conceptual contiguity & taxonomic sub-ordination.
▫ →*minte*, →*wild*

wilde nǣp, lit. 'wild rape'
'Brassica rapa L. var. silvestris / rape / Rapskohl'
1 occ. (*Antwerp-London Glossaries*, 118/14).
LATIN EQUIV. and OE synonym in the manuscript: *nap silvatica*; *wilde nǣp*.
ETYMOLOGY: OE formation; hybrid formation (OE *nǣp* < L *nāpus*; →*nǣp*). MORPHOLOGY: Adj+N; endocentric, copulative: attributive. MOTIVATION: wild & botanical information. A wild variant of Brassica rapa L. ASSOCIATIVE RELATIONS: conceptual contiguity & taxonomic sub-ordination.
▫ →*nǣp*, →*wild*

wilde [p]ise (*wilde wise*), lit. 'wild pea'
'? Lathyrus L. / WILD PEA / Wald-Platterbse'
Hapax legomenon (Gough 1974: 279); the manuscript has *wildewise*, but Bierbaumer (3, s.v. *wildewise*) considers it to be an erroneous gloss.
LATIN EQUIV. and OE synonym in the manuscript: ? *atulenta*, *pis lingua* (? for *bislingua*; cf. Bierbaumer 3, s.v. *springwyrt*); *springwyrt*.

ETYMOLOGY: OE formation; hybrid formation (OE *pise* < L *pīsum*; →*pise*). MORPHOLOGY: Adj+N; endocentric, copulative: attributive. MOTIVATION: wild & botanical information. The plants of the genus Lathyrus L. have pea-like fruits and were therefore perceived as taxonomically related to the pea 'Pisum sativum L.' (cf. Marzell, s.v. *Lathyrus* L.); see also →*mūsepise*; ASSOCIATIVE RELATIONS: conceptual contiguity & taxonomic sub-ordination.
▫ →*pise*, →*wild*

wilde popig, lit. 'wild poppy'
'Papaver rhoeas L. / common red poppy / Klatschmohn'
Hapax legomenon (*Antwerp-London Glossaries*, 109/10).
LATIN EQUIV. in the manuscripts: *saliunca*.
ETYMOLOGY: OE formation; possibly a loan rendition of L *papāver agreste* (*DOEPN*, s.v. *popig, wilde*) hybrid formation (OE *popig* < L *papāver*; →*popig*). MORPHOLOGY: Adj+N; endocentric, copulative: attributive. MOTIVATION: wild & botanical information. A kind of *popig* that grows wildly. ASSOCIATIVE RELATIONS: conceptual contiguity & taxonomic sub-ordination.
▫ →*popig*, →*wild*

wilde rǣdic, lit. 'wild radish'
'Raphanus raphanistrum L. / WILD RADISH / Hederich'
It occurs in *Durham Plant Name Glossary* (no. 40), and, according to Bierbaumer (3, s.v. *rǣdic*), also in manuscript London, British Library, Cotton Otho E.i (resp., Junius 77); this is not confirmed, however, by Voss (1996).
LATIN EQUIV. in the manuscripts: ? *arboracia*.
ETYMOLOGY: OE formation; hybrid formation (OE *rǣdic* < L *rādīx*; →*rǣdic*). MORPHOLOGY: Adj+N; endocentric, copulative: attributive. MOTIVATION: wild & botanical information; 'a wild kind of radish'. ASSOCIATIVE RELATIONS: conceptual contiguity & taxonomic sub-ordination.
▫ →*rǣdic*, →*wild*

wilde rūde, lit. 'wild rue'
'? Thalictrum flavum L. / fen rue, meadow rue / Gelbe Wiesenraute'
Hapax legomenon (*Lǣcebōc*, 9/12).
ETYMOLOGY: WGmc formation; OHG *wilderute, wilderuch* (Björkman 1904: 194); hybrid formation (OE *rūde* < L *rūta*; →*rūde*); possibly a loan rendition of L *rūta agrestis* (Bierbaumer 1, s.v. *rūde*). MORPHOLOGY: Adj+N; endocentric, copulative: attributive. MOTIVATION: wild & botanical information. Because of the similar leaves and yellow flowers, Thalictrum flavum was seen as a sub-species to Ruta graveolens (Marzell, s.v. *Ruta graveolens* L.), but as a wild, not a cultivated, kind of *rūde* 'rue'. ASSOCIATIVE RELATIONS: conceptual contiguity & taxonomic sub-ordination.
▫ →*rūde*, →*wild*

wilde wīngeard, lit. 'wild vine'
'Tamus communis L. / black bryony / Schmerwurz'
DOEPN, s.v. (*wīn-geard, wilde*): 1 occ. (*Antwerp-London Glossaries*, 106/9).
LATIN EQUIV. in the manuscripts: *labrusca*.
ETYMOLOGY: OE formation; probably a loan rendition of L *vitis agrestis* (cf.

Etymologiae 17: 5, 3). MORPHOLOGY: Adj+N; endocentric, copulative: attributive. MOTIVATION: wild & botanical information. Tamus communis is a creeping plant, just like a vine; it is not cultivated by man, but grows wildly; hence its MedL name *vitis agrestis* (Bierbaumer 3, s.v. *wīngeard*; 2, s.v. *hymele*), the model on which the Old English plant name was probably formed; cf. also *→blæc wīngeard*, *→hwīt wīngeard* ASSOCIATIVE RELATIONS: conceptual contiguity & taxonomic sub-ordination.
▫ *→wild*, *→wīngeard*

wilde wise →wilde pise

wīn, noun
'VINE, vineyard / Weinrebe, Weingarten'
ETYMOLOGY: Early Latin loan (Wollmann 1990: 156) < L *vīnea* (*OED*, s.v. *vine*).
▫ *→wīngeard*, *→wīntrēow*

windan, verb
'(to) WIND / winden'
▫ *→berewinde*, *→nǣdrewinde*, *→wiþowinde*

windel, noun
'basket / Korb'
▫ *→windelstrēaw*

windel-strēaw, lit. 'basket-straw'
'WINDLESTRAW / Grasart mit hohem Stengel'
3 occ. (*Cleopatra Glossary* (Quinn), 45/3; *Cleopatra Glossary* (Stryker), 103, no. 294; *Lǣcebōc*, 14/22); cf. *DOEPN* (s.v. *windel-strēaw*).
LATIN EQUIV. in the manuscripts: *calmum*.
ETYMOLOGY: OE formation; MORPHOLOGY: N+N; endocentric, rectional. MOTIVATION: tool & botanical information. This kind of strawlike grass is used for making baskets. ASSOCIATIVE RELATIONS: conceptual contiguity & taxonomic sub-ordination.
▫ *→strēaw*, *→windel*

windeltrēow →wilde eletrēow

wīn-geard, lit. 'vineyard'
(i) 'Vitis vinifera L. / vine / Weinrebe, Weinstock' (ii) 'vine, vineyard'
DOEPN (s.v. *wīn-geard*): 156 occ.[e] as a plant name.
LATIN EQUIV. in the manuscripts: *antes*, *capreolus*, *palmetum*, *surculus*, *vinea*, *vitis*.
ETYMOLOGY: Independent formation in Gmc or WGmc: "Although the Latin vinum was borrowed into Germanic in the earliest period it is hardly likely that these compounds were formed in Primitive Germanic. It is possible that they existed in West Germanic […]" (Carr 1939: 69-70; 105); semantic loan of L *vīnea* 'vine, vineyard' (see MOTIVATION); hybrid compound (OE *wīn* < L *vīnea*; *→wīn*). MORPHOLOGY: The original compound denoting the 'vineyard' is a N+N, endocentric, rectional compound; when denoting 'vine', it must be regarded as an exocentric, metonymic compound. MOTIVATION: L *vīnea* is polysemous: it refers not only to the vineyard, but also to the vine. Thus, in analogy to L *vīnea*, OE *wīngeard* 'vineyard' also adopted the meaning of 'vine'. It is, therefore, not possible to establish motivations and

ASSOCIATIVE RELATIONS between the elements of the plant name and the plant, as the lexeme as a whole was lexicalised and institutionalised with the newly adopted meaning 'vine'. As a plant name, it served as the basis for further plant names (see references below).

▫ *→geard, →wīn*

▫ *→blæc wīngeard, →hwīt wīngeard, →wilde wīngeard*

wīn-trēow, lit. 'vine-tree'
'Vitis vinifera L. / vine / Weinstock'
c. 9 occ. Bierbaumer (1, 2, 3, s.v. *wīn*).
ETYMOLOGY: Independent formation in Gmc; ON *vîntrê* (Carr 1939: 82); hybrid compound (OE *wīn* < L *vīnea*; *→wīn*). MORPHOLOGY: N+N; endocentric, rectional. MOTIVATION: part of plant or product & botanical information. The first element is a loan word and goes back either to L *vīnum* (> OE *wīn* 'wine, grapes'), thus relating to a product made of Vitis vinifera L., or it goes back to L *vīnea* 'vine, vineyard', but if so, it is not attested independently in Old English. OE *wīn-* may equally refer to the fruits, the grapes. That the vine is characterised as a *trēow* can be attributed perhaps to its woody texture, or to the fact that vines can become quite high, so that it was conceptualised as being similar to other fruit bearing trees. ASSOCIATIVE RELATIONS: conceptual contiguity & taxonomic sub-ordination.

▫ *→trēow, →wīn*

wīr, noun
'Myrica gale L. / bog-myrtle / Moor-Gagelstrauch'
DOEPN (s.v. *wīr*): 8 occ.
LATIN EQUIV. and OE synonym in the manuscripts: *myrtus*; *wīrtrēow*.
ETYMOLOGY: Cognate with MLG *wire*, OIcel *vírr*, Sw *vir*, "probably < an extended form (*r*-extension) of the same Indo-European base as WITHE n." (*OED*, s.v. *wire*).

▫ *→wīrtrēow*

wīr-trēow, lit. 'myrtle-tree'
'Myrica gale L. / bog-myrtle, sweet gale / Gagel, Brabanter Myrte'
2 occ. (*Cleopatra Glossary* (Stryker), 298, no. 132; *Herbarium*, no. 119).
LATIN EQUIV. and OE synonym in the manuscripts: *myrtus*; *wīr*.
ETYMOLOGY: OE formation; MORPHOLOGY: N+N; endocentric, copulative: subsumptive (pleonastic). MOTIVATION: plant & botanical information. The *wīr* 'myrtle' is a kind of tree. ASSOCIATIVE RELATIONS: conceptual identity & taxonomic sub-ordination.

▫ *→trēow, →wīr*

wīse, noun
'sprout, stalk / Spross, Trieb, Halm, Stengel'
DOEPN (s.v. *wīse*): 9 occ.
ETYMOLOGY: Cognate with OIcel *visir*, Lith *veisti*, L *virere* 'to become green' (*Holthausen*, s.v. *wīse*).

▫ *→fugleswīse*

Witmǣr, proper noun

▫ *→witmǣres wyrt*

witmǣres wyrt, lit. '*Witmǣr's* wort'
unidentified plant
2 occ. (*Durham Plant Name Glossary*, no. 68 and *Lǣcebōc*, 11/14).

LATIN EQUIV. and OE synonym in the manuscripts: *britannica*; *hǣwen-h(n)ydele*.
ETYMOLOGY: OE formation; MORPHOLOGY: PropN$_{es}$+N; endocentric, rectional. MOTIVATION: person & botanical information. The plant has not been identified yet (Bierbaumer 3, s.v. *wihtmereswyrt*). According to Cockayne (*Leechdoms*, 2: 411), it is 'Cochlearia anglica L. / spoonwort / Englisches Löffelkraut'; 'a kind of *wyrt* 'plant', which is associated with a person called *Witmǣr*. ASSOCIATIVE RELATIONS: ? & taxonomic sub-ordination.
▫ →*Witmǣr*, →*wyrt*

wiþo-winde, lit. 'band, rope-winder'
(i) 'Convulvulus arvensis L. / small bindweed / Ackerwinde'; (ii) 'Convulvulus sepium L. / great bind-wind / Zaunwinde'
3 occ. (*Lǣcebōc*, 11/33, 17/3, 38/3).
ETYMOLOGY: Gmc compound; ON *viðvindill*, OHG *witewinda*, MLG *wedewinde* (Carr 1939: 53; Björkman 1902: 307). MORPHOLOGY: synthetic, N+N$_{\text{deverbal(-}e\text{)}}$; endocentric, rectional. MOTIVATION: shape & action. For the etymology of the first element see *weoþobend* above and Bierbaumer (3, s.v. *wiþowinde*). The second element, *-winde* (< OE *windan* 'to wind') refers to the fact that the plant is a winder. It is not entirely clear whether the plant winds around woody ropes, e.g. those of the willow-tree (cf. also →*bere-winde*) or whether the first element refers to the fact that the plant itself is made up as if of ropes; for detailed explanations and different interpretations, see Marzell (s.v. *Convulvulus arvensis* L). ASSOCIATIVE RELATIONS: conceptual contiguity & taxonomic sub-ordination.
▫ →*weoþo-*, →*windan*

wōde-(h)wistle, lit. 'madness-whistle'
'Cicuta virosa L. / cowbane / Wasserschierling'
DOEPN (s.v. *wōde-wistle*): 7 occ.[e]
LATIN EQUIV. in the manuscripts: *cicuta*, *helleborus*.
ETYMOLOGY: WGmc compound (Carr 1939: 114; Björkman 1902: 307); MORPHOLOGY: N+N; exocentric, metonymic compound. MOTIVATION: illness & tool. "Der N[ame] bed[eutet] 'Wutflöte,Tollflöte' und bezieht sich im Determinans auf die durch die Pflanze hervorgerufenen Vergiftungserscheinungen (Krämpfe mit Toben, Schreien, Zähneknirschen), im Determinatum aber darauf, daß man aus den hohlen Stengeln Pfeifen verfertigte" ['The plant name means, literally, 'madness-whistle' and refers in its determinant to the symptoms of poisoning (convulsions, with raging, screaming, teeth grinding), in its determinatum, however, to the fact that people used to make whistles out of the hollow stems'] (Bierbaumer 3, s.v. *wōdewistle*). ASSOCIATIVE RELATIONS: conceptual contiguity & conceptual contiguity.
▫ →*hwistle*, →*wēde*

wōl, noun
'pestilence, disease / Krankheit'
▫ →*hennewōl*

-wōp, noun
'? Humulus lupulus L. / hop / Hopfen'
▫ →*feldwōp*

worþig, noun
'farm, enclosed homestead / Bauernhof, eingefriedetes Gehöft'
▫ →*worþigcærse*, →*worþignet(e)l(e)*

worþig-cærse, lit. 'farm-cress'
'Lepidium sativum L. / garden cress / Gartenkresse'
2 occ. (*Cleopatra Glossary* (Stryker), 226, no. 166; *Durham Plant Name Glossary*, no. 190); cf. *DOEPN* (s.v. *worþig-cærse*).
LATIN EQUIV. in the manuscripts: *cresso*, *garinus*.
ETYMOLOGY: OE formation; MORPHOLOGY: N+N; endocentric, rectional; partial metaphor. MOTIVATION: habitat & botanical information; This kind of cress is grown in enclosed habitats, in gardens; see also →*tūncærse*. ASSOCIATIVE RELATIONS: conceptual contiguity & taxonomic sub-ordination.
▫ →*cærse*, →*worþig*

worþig-net(e)l(e), lit. 'farm nettle'
'Urtica L. / nettle / Brennnessel'
Attested only in the phrase *micle worþig netele* (*Lǣcebōc*, 35/34).
ETYMOLOGY: OE formation; MORPHOLOGY: N+N; endocentric, rectional. MOTIVATION: habitat & botanical information. A nettle growing where people and cattle live, since it prefers soils rich in nitrates, i.e. where people and animals urinate (Marzell, s.v. *Urtica dioica* L.). ASSOCIATIVE RELATIONS: conceptual contiguity & taxonomic sub-ordination.
▫ →*net(e)l(e)*, →*worþig*
▫ →*micel worþignet(e)l(e)*

wōþe, noun
'? shedding of songs'
▫ →*gōtwōþe*

wudu, noun
(i) 'WOOD, forest, grove / Wald, Hain'; (ii) 'WOOD / Holz'
DOEPN (s.v. *wudu*): 96 occ.
LATIN EQUIV. in the manuscripts: *brassica silvatica*, *cedrus Libani*, *lignum*, *saltus*, *silva*.
ETYMOLOGY: Gmc **widu-z* (*OED*, s.v. *wood*).
▫ →*wuduæppel*, →*wudubinde*, →*wuducerfille*, →*wuduclāte*, →*wuduclǣfre*, →*wuducunille*, →*wududocce*, →*wudufille*, →*wudu(h)rōfe*, →*wudulēac*, →*wuduleahtric*, →*wudumerce*, →*wudusūræppel*, →*wuduþistel*, →*wuduweax*
▫ →*furhwudu*, →*healfwudu*, →*sūþerne wudu*, →*wīelisc* [*wudu*]

wudu-æppel, lit. 'wood apple'
'fruit of Malus communis Lam. subsp. silvestris / crab apple / Holzapfel'
Hapax legomenon (*Lǣcebōc*, 57/23).
ETYMOLOGY: Independent formation in WGmc (see below); MORPHOLOGY: N+N; endocentric, rectional. MOTIVATION: texture & botanical information. The crab-tree is a wild variant of the apple tree, with smaller, woody fruits; cf. OHG *holzephele* lit. 'wood-apple' glossing *mala agrestia* and OHG *holzappel*, glossing *mala matiana* (Björkman 1901: 210). ASSOCIATIVE RELATIONS: conceptual contiguity & taxonomic sub-ordination.
▫ →*æppel*, →*wudu*

wudu-binde (*-bend*), lit. 'wood-binding plant'
'Lonicera periclymenum L. / WOODBINE, honeysuckle / Geißblatt'
DOEPN (s.v. *wudu-bind*): 41 occ.[e]
ILLUSTRATION in manuscript London, British Library, Cotton Vitellius C. iii, fol. 70r (Ohlgren 168(99)).
LATIN EQUIV. in the manuscripts: *capparis*, *caprifolium*, *hedera* (*nigra*), ? *involuoco* = ? *volvola* (= *convolvulus*), *lignarium*, *viburnum*, *viticella*, *volvola* (= *convolvulus*).
ETYMOLOGY: Gmc compound; ON *viðvindill*, OHG *witewinda*, MLG *wedewinde* (Carr 1939: 53); MORPHOLOGY: synthetic, $N{+}N_{deverbal(-e)}$; endocentric, rectional. MOTIVATION: habitat & action. Honeysuckle is a plant that winds around trees, bushes, etc. (Marzell, s.v. *Lonicera periclymenum* L.). ASSOCIATIVE RELATIONS: conceptual contiguity & taxonomic sub-ordination.
▫ →*bindan*, →*wudu*

wudu-bindel, lit. 'wood-binding plant'
'Lonicera periclymenum L. / woodbine, honeysuckle / Geißblatt'
Hapax legomenon (*Lǣcebōc*, 11/2).
ETYMOLOGY: OE formation; MORPHOLOGY: synthetic, $N{+}N_{deverbal(-e)}$; endocentric, rectional. In *wudubindel*, the second element was enlarged by the suffix *-el*; *-el* is part of the suffix family *-el*, *-il*, etc. (Sauer 2001: 296-300 and Sauer 2012; see also, e.g., →*bisceopwyrtel*). MOTIVATION: habitat & action. Honeysuckle is a plant that winds around, e.g., trees (Marzell, s.v. *Lonicera periclymenum* L.); see →*wudubinde*; ASSOCIATIVE RELATIONS: conceptual contiguity & taxonomic sub-ordination.
▫ →*bindan*, →*wudu*

wudu-cerfille, lit. 'wood-chervil'
'Anthriscus silvestris Hoff. / cow parsley / Wiesenkerbel'
The identification as 'asparagus', which was also proposed, is probably wrong (Bierbaumer 3, s.v. *wuducerfille*).
DOEPN (s.v. *wudu-cerfille*): 13 occ.
ILLUSTRATION in manuscript London, British Library, Cotton Vitellius C. iii, fol. 45r (Ohlgren 168(51)).
LATIN EQUIV. in the manuscripts: *asparagus agrestis*, *brassica* (*silvatica*), *pastinaca*.
ETYMOLOGY: OE formation; hybrid compound (OE *cerfille* < L *chaer(e)phylla*; →*cerfille*). MORPHOLOGY: N+N; endocentric, rectional; MOTIVATION: habitat & botanical information. This kind of *cerfille* is not cultivated, but grows wildly, in the fields and woods. ASSOCIATIVE RELATIONS: conceptual contiguity & taxonomic sub-ordination.
▫ →*cerfille*, →*wudu*

wudu-clāte, lit. 'wood-cleavers'
a kind of 'Aristolochia L., e.g. Aristolochia clematis L. / birthwort / Osterluzei'
Hapax legomenon (Meritt 1945: 62, no. 73 Dc 186/ 9).
LATIN EQUIV. in the manuscripts: *aristolochia*.
ETYMOLOGY: OE formation; MORPHOLOGY: N+N; endocentric, rectional; partial metaphor. MOTIVATION: habitat & botanical information. A kind of *clāte* that grows wildly, a fact expressed metaphorically in the first element →*wudu*. ASSOCIATIVE RELATIONS: con-

ceptual contiguity & taxonomic sub-ordination.

▫ →*clāte*, →*wudu*

wudu-clǣfre, lit. 'wood-clover'
a kind of 'Trifolium L. / clover / Klee'
Hapax legomenon (*Durham Plant Name Glossary*, no. 76).
LATIN EQUIV. in the manuscript: *caltha silvatica*.
ETYMOLOGY: OE formation; possibly a loan rendition of *caltha silvatica*. MORPHOLOGY: N+N; endocentric, rectional. MOTIVATION: habitat & botanical information. A kind of clover that is not cultivated, but grows wildly, in the fields and woods. ASSOCIATIVE RELATIONS: conceptual contiguity & taxonomic sub-ordination.

▫ →*clǣfre*, →*wudu*

wudu-cunille (*-cunelle*), lit. 'wood-thyme'
'Thymus serpyllum L. / wild thyme, breckland thyme / Quendel, Sand-Thymian'
3 occ. (Meritt 1945: 59, no. 69/12; *Lǣcebōc*, 30/21, 98/11-2).
LATIN EQUIV. in the manuscripts: ? *buduca*.
ETYMOLOGY: OE formation; hybrid compound (OE *cunille* < L *cunīla* < Gk *κονίλη*; cf. *Etymologiae* 17: 206, fn. 507; Feulner 2000: 180). MORPHOLOGY: N+N; endocentric, rectional. MOTIVATION: habitat & botanical information. As is expressed in the Modern English plant name 'wild thyme' too, this kind of thyme is not cultivated, but grows wildly, in the fields and woods. ASSOCIATIVE RELATIONS: conceptual contiguity & taxonomic sub-ordination.

▫ →*cunille*, →*wudu*

wudu-docce, lit. 'wood-dock'
(i) 'Rumex acetosa L. / sorrel / Sauerampfer'; (ii) 'Rumex obtusifolius L. / broad-leaved dock / Stumpfblättriger Ampfer'; (iii) 'Rumex crispus L. / curled dock / Krauser Ampfer'
DOEPN (s.v. *docce, wudu*): 3 occ.
ILLUSTRATION in manuscript London, British Library, Cotton Vitellius C. iii, fol. 32r (Ohlgren 168(25)).
LATIN EQUIV. in the manuscripts: *lapathium*.
ETYMOLOGY: OE formation; loan translation of *lapathium silvaticum* lit. 'wood dock'. MORPHOLOGY: N+N; endocentric, rectional; MOTIVATION: habitat & botanical information. As a loan translation of L *lapathium silvāticum*, lit. 'wood dock': a plant that grows wildly, i.e. is not cultivated, and grows in the fields and woods. ASSOCIATIVE RELATIONS: metaphorical similarity & taxonomic sub-ordination.

▫ →*docce*, →*wudu*

wudu-fille, lit. 'wood-thyme'
(i) 'Thymus serpyllum L. / wild thyme / Quendel'; (ii) 'Anthriscus silvestris Hoffm. / cow parsley / Wiesenkerbel'
DOEPN (s.v. *wudu-fille*): 4 occ.
ETYMOLOGY: OE formation; Since *fille* 'thyme' is a shortened form of *cerfille* (< L *chaer(e)phylla*; Campbell 1959: § 544; *OED*, s.v. *fille*), i.e. an Old English formation, *wudufille* is not hybrid. MORPHOLOGY: N+N; endocentric, rectional. MOTIVATION: habitat & botanical information; see →*wuducerfille*. ASSOCIATIVE RELATIONS: conceptual contiguity & taxonomic sub-ordination.

▫ →*fille*, →*wudu*

▫ →*rēade wudufille*

wudu-(h)rōfe, lit. '? wood-strong (plant)'
(i) 'Asperula odorata L. / WOODRUFF-aperule, sweet WOODRUFF / Echter Waldmeister'; (ii) 'Asphodelus L. / asphodel / Affodill'
DOEPN (s.v. *wudu-rōfe*): 30 occ.
ILLUSTRATIONS in manuscript London, British Library, Cotton Vitellius C. iii, fol. 32r (Ohlgren 168(25)), fol. 37r (Ohlgren 168(35)).
LATIN EQUIV. in the manuscripts: *asphodelus*, *astula regia*, *malacinagria* (= *malache*), *ostriolum* (? for *ostriago* or *ostriagum*).
ETYMOLOGY: OE formation; MORPHOLOGY: N+N$_{deadj}$; exocentric, metonymic compound. MOTIVATION: habitat & healing virtue. Asperula odorata is generally found in woods. No explanation was found for the second element. It may perhaps refer to the invigorating, heart-revitalising effect of this plant when it is made into tea, as it contains coumarin; cf. Marzell (s.v. *Asperula odorata* L.); OE *rōf* 'vigorous, strong, etc.' (*ClH*, s.v. *rōf*). ASSOCIATIVE RELATIONS: conceptual contiguity & conceptual contiguity.
▫ →*rōf*, →*wudu*

wudu-lēac, lit. 'wood-leek'
'Allium ursinum L. / ramsons, broad-leaved garlic / Bär(en)lauch'
Hapax legomenon (*Laud Herbal Glossary*, no. 543).
LATIN EQUIV. and OE synonyms in the manuscripts: *helleborus* (*albus*), ? *suffunie*; *hramsa* 'ramsons', *tunsingwyrt*.
ETYMOLOGY: OE formation; MORPHOLOGY: N+N; endocentric, rectional. MOTIVATION: habitat & botanical information. Ramsons, a kind of leek, grows preferably in woods and clearings (Marzell, s.v. *Allium ursinum* L). ASSOCIATIVE RELATIONS: conceptual contiguity & taxonomic sub-ordination.
▫ →*lēac*, →*wudu*

wudu-leahtric (*-lectric*), lit. 'wood-lettuce'
'Lactuca silvatica L. / lettuce / Lattich'
DOEPN (s.v. *wudu-lectric*): 6 occ.
ILLUSTRATION in manuscript London, British Library, Cotton Vitellius C. iii, fol. 31r (Ohlgren 168(23)).
LATIN EQUIV. in the manuscripts: *lactuca silvatica*.
ETYMOLOGY: OE formation; possibly a loan translation of *lactuca silvatica* lit. 'wildly growing lettuce'; hybrid compound (OE *leahtric* < L *lactūca*; →*leahtric*). MORPHOLOGY: N+N; endocentric, rectional; MOTIVATION: habitat & botanical information. A kind of *leahtric* growing wildly, in woods. ASSOCIATIVE RELATIONS: conceptual contiguity & taxonomic sub-ordination.
▫ →*leahtric*, →*wudu*

wudu-merce, lit. 'wood-celery'
(i) 'Apium graveolens L. / celery / Sellerie'; (ii) 'Sanicula europaea L. / wood sanicle, WOOD MARCH / Wald-Sanikel, Waldklette'
DOEPN (s.v. *wudu-merce*): 19 occ.
LATIN EQUIV. in the manuscripts: *ambrosia*, *apiaster*, *apium*, *apium silvaticum*.
ETYMOLOGY: OE formation; MORPHOLOGY: N+N; endocentric, rectional. MOTIVATION: habitat & botanical information. As alluded to also by G *Wald-Sanikel* and E *woodmarch*, this kind of *merce* grows wildly, in woods.

ASSOCIATIVE RELATIONS: conceptual contiguity & taxonomic sub-ordination.
▫ →*merce*, →*wudu*

wudu-sūr-æppel, lit. 'wood crab-apple'
'Malus communis Lam. subsp. silvestris / crab-tree / Holzapfelbaum'
Hapax legomenon (*Lācnunga*, 599).
ETYMOLOGY: OE formation; MORPHOLOGY: N+N; exocentric, metonymic formation; the structural analysis of this plant name is complicated by the fact that the plant name is made up of three morphemes and that neither the combination of the first two – **wudu-sūr* – nor the combination of the last two morphemes – **sūræppel* – are independently attested lexemes of Old English. Thus, the modifier and head of the compound are not entirely clear. I would opt rather for *wudu* as the modifier to the head *sūræppel* in analogy to the tree name →*sūrapuldor* 'crab-tree'. Thus, the whole plant name would be a metonymic extension of lit. 'wood crab-apple', i.e. a crab-tree growing in woods. MOTIVATION: habitat & botanical information; see above; ASSOCIATIVE RELATIONS: conceptual contiguity & conceptual contiguity.
▫ →*sūrapuldor*, →*wudu*

wudu-þistel, lit. 'wood-thistle'
(i) '? Sonchus spec. / sow-thistle / Saudistel'; (ii) '? Atractylis gummifera L. / bird-lime, blue thistle / Mastixdistel'; (iii) 'Cardopatium corymbosum Pers. / black chameleon / Schirmsaflor'
DOEPN (s.v. *wudu-þistel*): 7 occ.
ILLUSTRATION in manuscript London, British Library, Cotton Vitellius C. iii, fol. 52v (Ohlgren 168(65)).
LATIN EQUIV. in the manuscripts: *carduus silvaticus*.
ETYMOLOGY: OE formation; probably a loan translation of *carduus silvāticus* (Bierbaumer 1, 3, s.v. *wuduþistel*). MORPHOLOGY: N+N; endocentric, rectional, MOTIVATION: habitat & botanical information: 'a kind of thistle growing wildly', i.e. in the fields and woods. ASSOCIATIVE RELATIONS: conceptual contiguity & taxonomic subordination.
▫ →*þistel*, →*wudu*

wudu-weax, lit. 'wood-sinew'
'Genista tinctoria L. / dyer's broom, greenwood / Färber-Ginster'
DOEPN (s.v. *wudu-weax*): 8 occ.
ETYMOLOGY: OE formation; MORPHOLOGY: N+N; exocentric, metaphorical compound; partial metaphor. MOTIVATION: texture & shape. The woody stalks of dyers broom (hence the use of *wudu* 'wood') resemble sinews (hence, the use of *weax* 'sinew'; cf. Marzell, s.v. *Genista tinctoria* L., 4). ASSOCIATIVE RELATIONS: conceptual contiguity & metaphorical similarity.
▫ →*weax*, →*wudu*

wulf, noun
'WOLF / Wolf'
▫ →*wulfescamb*, →*wulfestǣsel*, →*wulfesteors*

wulfes-camb, lit. 'wolf's comb'
'Dipsacus silvestris Huds. / wild teasel / Wilde Karde, Kardendistel'
DOEPN (s.v. *wulfes-camb*): 22 occ.[e]
ILLUSTRATIONS in manuscript London, British Library, Cotton Vitellius C. iii, fol. 30r (Ohlgren 168(21)) and Oxford,

Bodleian Library, Bodley 130, fol. 44v (Ohlgren 222(87)).
LATIN EQUIV. in the manuscripts: *chamaedrys*, *chamaeleon*, *cynocephalion*, ? *pentilupi* (? for *lupipecten*).
ETYMOLOGY: OE formation; perhaps a loan translation of L *pecten lupi* lit. 'wolf's comb' (see MOTIVATION). MORPHOLOGY: N_{gen}+N; exocentric, metaphorical compound. MOTIVATION: animal & action. The plant's small hooks seem to be there to 'comb' the wolves's fur when they pass by. Since the Latin lemma in the *Durham Plant Name Glossary* is *pentilupi* (no. 270, and p. 64), which should probably be emended to *pecten lupi* 'wolf's comb', the OE plant name is probably a loan translation. ASSOCIATIVE RELATION: metaphorical similarity.
▫ →*camb*, →*wulf*
▫ →*brāde wulfescamb*

wulfes-tǣsel (*-tǣsl*), lit. 'wolf's teasel'
'Dipsacus silvestris Huds. / WOLF'S TEASEL, wild teasel / Wilde Karde, Kardendistel'
DOEPN (s.v. *wulfes-tǣsel*): 2 occ.
ILLUSTRATION in manuscript London, British Library, Cotton Vitellius C. iii, fol. 65v (Ohlgren 168(91)).
LATIN EQUIV. in the manuscripts: *chamaelea alba*.
ETYMOLOGY: WGmc formation; cf. OHG *wulfeszeisala*, etc. (Björkman 1901: 231). MORPHOLOGY: N_{gen}+N; exocentric, metaphorical compound. MOTIVATION: animal & action. The motivation is the same as for the synonym →*wulfescamb*: The second element, *-tǣs(e)l*, is a noun derived from the verb *tǣsan* 'to tease' (cf. OHG *zeisala*, MHG *zeisel* 'thistle'; Marzell, s.v. *Dipsacus silvester* L.); a plant that seems to 'comb the wolves'. ASSOCIATIVE RELATIONS: metaphorical similarity.
▫ →*-tǣsel*, →*wulf*
▫ →*brāde wulfestǣsel*

wulfes-teors, lit. 'wolf's penis'
'Dipsacus silvestris Huds. / wolf's teasel, wild teasel / Wilde Karde, Kardendistel'
Hapax legomenon (*Laud Herbal Glossary*, no. 322).
LATIN EQUIV. in the manuscripts: *chamaelea alba*.
ETYMOLOGY: OE formation; folk etymology (see MOTIVATION). MORPHOLOGY: N_{gen}+N; exocentric, metaphorical compound. MOTIVATION: animal & secondary, phonological and semantic motivation: sound pattern and shape. *Wulfesteors* is probably a form of *wulfestǣsel*, which was changed formally to match the phonological similarity of OE *tǣsel* and OE *teors* 'penis'; probably supported by the oval shape of the erect flower of the wolf's teasel (cf. Marzell, s.v. *Dipsacus silvester* L.). ASSOCIATIVE RELATIONS: phonological similarity & conceptual contiguity.
▫ →*teors*, →*wulf*

wurma, noun
'Majorana hortensis Moench / marjoram / Majoran'
1 occ. (*Lǣcebōc*, 89/24).
▫ →*feltwurma*

wylle, noun
'WELL / Quelle'
▫ →*wyllecærse*

wylle-cærse, lit. 'well-cress'
(i) 'Nasturtium officinale R.Br. / water-cress / Brunnenkresse'; (ii) 'Trigonella foenum graecum L. / fenugreek / Bockshornklee, Griechisch-Heu'
DOEPN (s.v. *wylle-cærse*): 8 occ. The identification as Trigonella foenum graecum is probably based on a false gloss and is ignored in the semantic analysis of the plant name.
LATIN EQUIV. in the manuscripts: *faenugrecum*, ? *scubricia*, ? *seu britia*.
ETYMOLOGY: OE formation; MORPHOLOGY: N+N; endocentric, rectional. MOTIVATION: habitat & botanical information. The element *wylle* 'well' refers to where this kind of cress – Nasturtium officinale – grows preferably: on moist ground (Marzell, s.v. *Nasturtium officinale* R.Br.); see also →*brōccærse*, →*ēacærse* and →*fencærse*. ASSOCIATIVE RELATIONS: conceptual contiguity & taxonomic sub-ordination.
▫ →*cærse*, →*wylle*

wyrm, noun
'WORM / Wurm'
▫ →*wyrmwyrt*

wyrm-wyrt, lit. 'worm-wort'
'? Sedum album L. / wormgrass, prick madam / Weißer Mauerpfeffer'
4 occ. (*Lǣcebōc* 29/34, 32/10, 39/10 and 95/2).
ETYMOLOGY: WGmc compound (Carr 1939: 105); cf. OHG *wormwrz* (Björkman 1902: 304); MORPHOLOGY: N+N; endocentric, rectional; partial metaphor. MOTIVATION: shape & botanical information. Bierbaumer (1, s.v. *wyrmwyrt*) argues that the plant was probably used to cure intestinal worms. Marzell (s.v. *Sedum album* L.) explains that the tiny leaves of Sedum album were compared with little worms, cf. the MedL plant name *vermicularis* 'vermicular' and G *Wurmerlkrauda*. ASSOCIATIVE RELATIONS: metaphorical similarity & taxonomic sub-ordination.
▫ →*wyrm*, →*wyrt*

wyrt, noun
(i) 'WORT, root / Wurz, Wurzel'; (ii) 'plant / Pflanze'; (iii) 'herb / Kraut'
DOEPN (s.v. *wyrt*): 270 occ.
LATIN EQUIV. in the manuscripts: *faratalis*, *gramen*, *herba*, *herba holera*, *legumen*, *pigmentum*, *sandix*.
ETYMOLOGY: Gmc **wurti-* < IE **wrəd-i-* (*Kluge*, s.v. *Wurz*; *OED*, s.v. *wort*).
▫ →*æscwyrt*, →*æþelferþingwyrt*, →*bānwyrt*, →*bēowyrt*, →*bisceopwyrt*, →*bisceopwyrtel*, →*brōþorwyrt*, →*brūn(e)wyrt*, →*brȳsewyrt*, →*candelwyrt*, →*ceasterwyrt*, →*clǣfrewyrt*, →*clīfwyrt*, →*cliþwyrt*, →*clufwyrt*, →*coþuwyrt*, →*cyningeswyrt*, →*ēagwyrt*, →*ēawyrt*, →*ellenwyrt*, →*feldwyrt*, →*feltwyrt*, →*fīcwyrt*, →*flēaþe wyrt*, →*flēotwyrt*, →*gærswyrt*, →*gelodwyrt*, →*gescādwyrt*, →*glōfwyrt*, →*goldwyrt*, →*grēate wyrt*, →*hafocwyrt*, →*halswyrt*, →*hamorwyrt*, →*hāmwyrt*, →*hārewyrt*, →*hæselwyrt*, →*hierdewyrt*, →*hlǣdderwyrt*, →*hrætelwyrt*, →*hylwyrt*, →*lǣcewyrt*, →*leoþowyrt*, →*liferwyrt*, →*līnwyrt*, →*liþwyrt*, →*lungenwyrt*, →*meduwyrt*, →*mōderwyrt*, →*mōrwyrt*, →*mucgwyrt*, →*nǣdrewyrt*, →*simǣringwyrt*, →*smeoruwyrt*, →*sperewyrt*, →*springwyrt*, →*stælwyrt*, →*stǣþwyrt*, →*sticwyrt*, →*strǣlwyrt*, →*swanewyrt/swānewyrt*, →*swete wyrt*, →*tunsingwyrt*, →*þēorwyrt*, →*þunorwyrt*, →*wæterwyrt*, →*wealhwyrt*, →*wealwyrt*, →*wenwyrt*, →*witmǣres wyrt*, →*wyrmwyrt*

-*wyrtel*, noun
(i) 'plant / Pflanze'; (ii) 'wort / Wurz, Wurzel'
ETYMOLOGY: < OE *wyrt* + *-el* (Sauer 2012: 218).
▫ →*bisceopwyrtel*

Y

ynne-, noun
'Allium cepa L. / onion / Zwiebel'
ETYMOLOGY: Early Latin loan (Wollmann 1990: 166); < L *uniō* (*OED*, s.v. *onion*; Serjeantson 1935: 276); not attested independently.
▫ →*ynnelēac*

ynne-lēac (*enne-*), lit. 'onion-plant' or 'onion-leek'
'Allium cepa L. / onion / Zwiebel'
DOEPN (s.v. *ynne-lēac*): 16 occ.[e]
LATIN EQUIV. in the manuscripts: *aschalonia*, *caepa*, *chaerefolium*, *unnio*.
ETYMOLOGY: WGmc compound (Carr 1939: 114), OHG *unloich*, etc. (Björkman 1904: 197); hybrid compound with the first element, *ynne-*, not attested independently (OE *ynne-* < L *uniō*; →*ynne-*). MORPHOLOGY: N+N; endocentric, copulative: subsumptive; when the second element, *lēac*, is interpreted as 'leek', the plant name must be classified as tautologic; it is more likely, however, that the interpretation of *lēac* as 'garden-herb, plant' (→*lēac*) explains the first element, which, as a loan, was not transparent in Old English and probably not understood by everyone. MOTIVATION: plant & botanical information. *Ynne-* 'onion' is a kind of *lēac* 'garden-herb, plant'. ASSOCIATIVE RELATIONS: conceptual identity & taxonomic sub-ordination.
▫ →*lēac*, →*ynne-*

BIBLIOGRAPHY

PRIMARY SOURCES

André, Jaques, ed. 1981. *Isidorus Hispalensis: Etymologiae XVII*. Paris: Société d'Édition "Les Belles Lettres".

d'Aronco, Maria Amalia and Malcolm L. Cameron, eds. 1998. *The Old English Illustrated Pharmacopoeia: British Library Cotton Vitellius C III*. Early English Manuscripts in Facsimile, 27. Copenhagen: Rosenkilde and Bagger.

Bately, Janet M., ed. 1986. *The Anglo-Saxon Chronicle: A Collaborative Edition. Vol. 3: MS A*. Cambridge: D.S. Brewer.

Beck, Lily Y., transl. 2005. *De Materia Medica by Pedanius Dioscorides*. Altertumswissenschaftliche Texte und Studien, 38. Hildesheim: Olms-Wiedmann.

Birch, Walter de Gray, ed. 1885-1899. *Cartularium Saxonicum: A Collection of Charters Relating to Anglo-Saxon History*. 4 vols. London: Whiting [reprint 1964, New York: Johnson].

Bischoff, Bernhard, Mildred Budny, Geoffrey Harlow et al., eds. 1988. *The Épinal, Erfurt, Werden, and Corpus Glossaries. Epinal Bibliothèque Municipale 72 (2), Erfurt Wissenschaftliche Bibliothek Amplonianus 2° 42, Düsseldorf Universitätsbibliothek Fragm. K 19: Z 9/1, Munich Bayerische Staatsbibliothek Cgm. 187 III (e.4), Cambridge Corpus Christi College 144*. Early English Manuscripts in Facsimile, 22. Copenhagen: Rosenkilde and Bagger.

Bremmer, Rolf H. Jr. and Kees Dekker., eds. 2006. *Manuscripts in the Low Countries*. Anglo-Saxon Manuscripts in Microfiche Facsimile, 13. Tempe: Arizona Center for Medieval and Renaissance Studies.

Chaucer, Geoffrey. "The Parliament of Fowls" in *The Riverside Chaucer*. Ed. Larry D. Benson. 1988. Oxford: Oxford University Press [hardcover edition 1987, Houghton Mifflin Company], 385-94.

Cockayne, Oswald, ed. 1864-1866. *Leechdoms, Wortcunning and Starcraft of Early England*. 3 vols. Rolls Series, 35. London: Longman.

Cornelius Tacitus. *De Origine et Situ Germanorum Liber*. Ed. Henry Furneaux. 1990. Oxford: Clarendon Press [available online at *Perseus Digital Library*. Ed. Gregory R. Crane. 2011-. Tuft University. <http://data.perseus.org/citations/urn:cts:latin Lit:phi1351. phi002.perseus-lat1:1>; last accessed on 27.11.2012].

Crawford, Samuel J., ed. 1922. *The Old English Version of the Heptateuch*, Early English Text Society, Original Series, 160. London: Oxford University Press [reprinted 1969 with additions by N.R. Ker].

Doane, Alger N. and Phillip Pulsiano. 1994-. *Anglo-Saxon Manuscripts in Microfiche Facsimile*. vols 1-4 (Binghamton, NY: State University of New York Press, 1994-1996); vols. 5- (Tempe, Arizona: Arizona Center for Medieval and Renaissance Studies, 1997-).

Earle, John, ed. 1888. *A Hand·Book to the Land·Charters, and other Saxonic Documents*. Oxford: Clarendon Press.

Förster, Max. 1917. "Die Altenglische Glossenhandschrift Plantinus 32 (Antwerpen) und Additional 32246 (London)" in *Anglia* 41: 94-161.

Goetz, Georg. ed. 1888-1923. *Corpus Glossariorum Latinorum a Gustavo Loewe Inchoatum*. Leipzig: Teubner [reprint 1965, Amsterdam].

Godden, Malcolm, ed. 1979. *Ælfric's Catholic Homilies: The Second Series, Text.* London: Early English Text Society.

Gough, John V., ed. 1974. "Some Old English Glosses" in *Anglia* 92: 273-90.

Grattan, J.H.G. and Charles Singer. 1952. *Anglo-Saxon Magic and Medicine: Illustrated Specially from the Semi-Pagan Text 'Lacnunga'.* London: Oxford University Press.

Healey, Antonette diPaolo with John Price Wilkin and Xin Xiang, eds. 2007. *Dictionary of Old English Web Corpus*. <http://tapor.library.utoronto.ca/doe corpus/> [last accessed on 20.11.2012].

Hearne, Thomas, ed. 1723. *Hemingi Chartularium Ecclesiae Wigorniensis*. Oxford.

Hessels, John H. ed. 1890. *An Eight-Century Latin-Anglo-Saxon Glossary*. Cambridge: Cambridge University Press.

Hessels, John Henry. ed. 1906. *A Late Eighth Century Latin-Anglo-Saxon Glossary Preserved in the Library of the Leiden University (MS Voss. Qo. Lat. no. 69)*. Cambridge: Cambridge University Press.

Kemble, John Mitchell, ed. 1839-1845. *Codex Diplomaticus Aevi Anglo-Saxonici*. 3 vols. London: English Historical Society.

Kimmens, Andrew C., ed. 1979. *The Stowe Psalter*. Toronto Old English Series, 3. Toronto: Toronto University Press.

Kindschi, Lowell, ed. 1955. "The Latin-Old English Glossaries in Plantin-Moretus MS 32 and British Museum MS Additional 32, 246". Stanford University, unpublished dissertation.

van Kirk Dobbie, Elliott, ed. 1942. *The Anglo-Saxon Minor Poems*. Anglo-Saxon Poetic Records. A Collective Edition, VI. New York: Columbia University Press [2nd impression 1958].

Kittlick, Wolfgang. 1998. *Die Glossen der Hs. British Library, Cotton Cleopatra A. III. Phonologie, Morphologie, Wortgeographie*. Europäische Hochschulschriften, 14. Angelsächsische Sprache und Literatur, 347. Frankfurt: Peter Lang Verlag.

Koschwitz, Eduard, ed. 1964. *Les plus anciens monuments de la langue française. II. Textes critiques et glossaire*. 6th ed. München: Hueber.

Krapp, George Philip and Elliott van Kirk Dobbie, eds. 1936. *The Exter Book*. Anglo-Saxon Poetic Records. A Collective Edition, III. New York: Columbia University Press.

Lendinara, Patrizia, ed. 1988-1989. "Il glossario del ms. Oxford, Bodleian Library, 163" in *Romanobarbarico* 10: 485-516.

Leonhardi, Günter, ed. 1905. *Kleinere angelsächsische Denkmäler. 1. Das Lǣcebōc. 2. Die Lācnunga mit grammatischer Einleitung. 3. Der Lorica-Hymnus mit der angelsächsischen Glossierung nebst einer Abhandlung über Text u. Sprache des Denkmals. 4. Das Lorica-Gebet und die Lorica-Namen.* Greins Bibliothek der angelsächsischen Prosa, 6. Hamburg: Grand.

Leslie, Roy F., ed. 1961. *Three Old English Elegies: The Wife's Lament, The Husband's Message, The Ruin*. Manchester: Manchester University Press.

Lindheim, Bogislav von, ed. 1941. *Das Durhamer Pflanzenglossar: Lateinisch und Altenglisch*. Bochum-Langendreer: Heinrich Pöppinghaus O.H.-G.

Lindsay, Wallace Martin. 1921. *The Corpus, Épinal, Erfurt and Leyden Glossaries*. Publications of the Philological Society, 8. Oxford: Oxford University Press.

von Linné, Carl. 1753. *Species Plantarum, Exhibentes Plantas Rite Cognitas, Ad Genera Relatas, Cum Differntiis Specificis, Nominibus Trivialibus, Synonymis Selectis, Locis Natalibus, Secundum Systema Sexuale Digestas*. 2 vols. Stockholm: Laurentius Salvius. [available online at < http://www.botanicus.org/title/b12069 590>].

Löweneck, Max, ed. 1896. *Peri Didaxeon. Eine Sammlung medizinischer Texte in englischer Sprache. Nach einer Handschrift des Britischen Museums*. Erlanger Beiträge zur Englischen Philologie, 12. Erlangen: Junge.

Maion, Danielle, ed. 1999. "Eduzione, Traduzione e Commento del Peri Didaxeon". Università degli Studi Roma Tre, unpublished dissertation.

Meritt, Herbert Dean, ed. 1945. *Old English Glosses: A Collection*. New York: Modern Language Association of America [reprint 1971].

Meritt, Herbert Dean, ed. 1961. "Old English Glosses, Mostly Dry Point" in *Journal of English and Germanic Philology* 60: 441-50.

Napier, Arthur S., ed. 1890. "Altenglische Miscellen" in *Archiv für das Studium der neueren Sprachen und Literaturen* 84: 323-7.

Napier, Arthur S., ed. 1900. *Old English Glosses*. Oxford: Clarendon.

Oess, Guido, ed. 1910. *Der altenglische Arundel-Psalter*. Anglistische Forschungen, 30. Heidelberg: Carl Winter.

Olds, Barbara M., ed. 1984. "The Anglo-Saxon Leechbook III: A Critical Edition and Translation". Ann Arbor, Michigan: University Microfilms International.

Oliphant, Robert T., ed. 1966. *The Harley Latin-Old English Glossary*. The Hague: Mouton & Co.

Pettit, Edward, ed. and transl. 2001. *Anglo-Saxon Remedies, Charms, and Prayers from British Library MS Harley 585: The* Lacnunga. 2 vols. Lewiston: The Edwin Mellen Press.

Pheifer, Jason Donovan, ed. 1974. *Old English Glosses in the Épinal-Erfurt Glossary*. Oxford: Clarendon Press.

Pliny the Elder. *Naturalis historia*. Ed. Karl F. Th. Mayhoff. 1906. Leipzig: Teubner.

Porter, David W., ed. 2011a. *The Antwerp-London Glossaries: The Latin and Latin-Old English Vocabularies from Antwerp, Museum Plantin-Moretus 16.2 – London, British Library Add. 32246*. Toronto: Pontifical Institute of Medieval Studies.

Quinn, John Joseph, ed. 1956. "The Minor Latin-Old English Glossaries in MS. Cotton Cleopatra A.III". Ann Arbor: University Microfilms, unpublished dissertation.

de Renzi, Salvatore, ed. 1856. *Collectio Salernitana*. 5 vols. Napoli: Typographie du Filiatre-Sebezio.

Sanborn, Linda, ed. 1983. "An Edition of British Library Ms Harley 6258B: Peri Didaxeon". Ottawa: University of Ottawa, unpublished dissertation.

Sawyer, P.H. 1968. *Anglo-Saxon Charters: An Annotated List and Bibliography*. London: Offices of the Royal Historical Society.

Schauman, Bella and Angus Cameron, eds. 1977. "A Newly-Found Leaf of Old English from Louvain" in *Anglia* 95: 289-312.

Skeat, Walter William, ed. 1871-1887. *The Four Gospels in Anglo-Saxon, Northhumbrian, and Old Mercian Versions*. Cambridge: Cambridge University Press [reprint 1970, Darmstadt].

Skeat, Walter William, ed. 1881-1900. *Ælfric's Lives of Saints*. 4 vols. Early English Text Society, 76, 82, 94, 114. London: Kegan Paul, Trench, Trübner & Co. [reprint in 2 vols. 1966].

Smith, Albert H., ed. 1980. *The Parker Chronicle: 832-900*. 2nd, revised ed. Exeter: University of Exeter [first edition 1935, London: Methuen's Old English Library].

Steinmeyer, Elias and Eduard Sievers, eds. 1879-1922. *Die Althochdeutschen Glossen*. 5 vols. Berlin: Weidmann [reprint 1968-1969].

Steinmeyer, Elias, ed. 1889. "Lateinische und Altenglische Glossen" in *Zeitschrift für deutsches Altertum, Anzeiger* 22: 242-51.

Steinmeyer, Elias, ed. 1916. *Die kleineren althochdeutschen Sprachdenkmäler*. Berlin: Weidmann.

Stracke, J. Richard, ed. 1974. *The Laud Herbal Glossary*. Amsterdam: Rodopi.

Stryker, William G., ed. 1952. "The Latin-Old English Glossary in Ms Cotton Cleopatra A iii". Stanford University, unpublished dissertation.

Sweet, Henry, ed. 1871-1872. *King Alfred's West-Saxon Version of Gregory's Pastoral Care*. Early English Text Society, Original Series, 45, 50. London: N. Trübner & Co. [reprint 1958].

Sweet, Henry, ed. 1978. *A Second Anglo-Saxon Reader, Archaic and Dialectal*. 2nd ed. revised by T.F. Hoad. Oxford: Clarendon Press.

Torkar, Roland. 1976. "Zu den ae. Medizinaltexten in Otho B. XI und Royal 12 D. XVII. Mit einer Edition der Unica (Ker 1990, no. 180 art. 11a-d)" in *Anglia* 94: 319-38.

Varnhagen, Hermann, ed. 1902. *De Glossis Nonnullis Anglicis*. Erlangen: Friedrich Jung.

Voss, Manfred. 1988. "Strykers Edition des alphabetischen Cleopatraglossars: Corrigenda und Addenda" in *Arbeiten aus Anglistik und Amerikanistik* 13: 123-38.

Voss, Manfred. 1989. "Quinns Edition der kleineren Cleopatraglossare; Corrigenda und Addenda" in *Arbeiten aus Anglistik und Amerikanistik* 14: 127-39.

Voss, Manfred. 1996. "Altenglische Glossen aus MS British Library, Cotton Otho E. I", *Arbeiten aus Anglistik und Amerikanistik* 21: 179-203.

de Vriend, Hubert Jan, ed. 1984. *The Old English Herbarium and Medicina de Quadrupedibus*. Early English Text Society, Original Series, 286. London: Oxford University Press.

Walahfrid Strabo. "De cultura hortorum". Ed. Eckehart Stöve [no date], in *Collegium Cartesianum*. <http://www.uni-due.de/collcart/hortus/strabo/HB-hort1 .htm> [last accessed on 22.11.2012].

Whitelock, Dorothy, ed. 1994. *Sweet's Anglo-Saxon Reader in Prose and Verse*. 15th, revised edition. Oxford: Clarendon Press [first edition 1967].

Wimmer, Friedrich, ed. 1842. *Theophrastus Eresii Historia Plantarum*. Bratislava: Ferdinand Hirt [reprint 2009, Whitefish: Kesslinger Publishing].
Wright, Cyril E., ed. 1955. *Bald's Leechbook: British Museum, Royal Manuscript 12 D. xvii.* Early English Manuscripts in Facsimile, 5. With an appendix by Randolph Quirk. Copenhagen: Rosenkilde and Bagger.
Wright, Thomas and Richard Paul Wülcker, eds. 1884. *Anglo-Saxon and Old English Vocabularies*. 2nd ed. by Richard Paul Wülcker. 2 vols. London: Trübner [reprinted 1968, Darmstadt: Wissenschaftliche Buchgesellschaft].
Wynn, John, ed. 1961-1962. "An Edition of the Anglo-Saxon Corpus Glosses". Oxford: Oxford University, unpublished dissertation.
Ziegler, Waltraud. 1981. "Die unveröffentlichten Glossare der Hs Oxford, Bodley 730" in *Arbeiten aus Anglistik und Amerikanistik* 6: 127-54.
Zielkowski, Jan M., ed. and transl. 1994. *The Cambridge Songs*. New York: Garland.
Zupitza, Julius, ed. 1880. *Ælfrics Grammatik und Glossar: Text und Varianten.* Reprint 2001 with an Introduction by Helmut Gneuss. Hildesheim: Weidmann.

SECONDARY SOURCES

DICTIONARIES AND ENCYCLOPAEDIAS

Bierbaumer, Peter and Hans Sauer with Helmut Klug and Ulrike Krischke, eds. 2007-. *The Dictionary of Old English Plant Names*. <http://oldenglish-plantnames.uni-graz.at/> [last accessed on 27.11.2012].

Bierbaumer, Peter. 1975. *Der botanische Wortschatz des Altenglischen, 1. Teil: Das Lǣcebōc*. Bern: Herbert Lang.

Bierbaumer, Peter. 1976. *Der botanische Wortschatz des Altenglischen, 2. Teil: Lācnunga, Herbarium Apuleii, Peri Didaxeon*. Bern: Herbert Lang.

Bierbaumer, Peter. 1979. *Der botanische Wortschatz des Altenglischen, 3. Teil: Der botanische Wortschatz in altenglischen Glossen*. Bern: Herbert Lang.

Bosworth, Joseph. 1882-1989. *An Anglo-Saxon Dictionary*. Based on the Manuscript Collections of the Late Joseph Bosworth, edited and enlarged by Thomas Northcote Toller, Oxford: Oxford University Press [many reprints]. *An Anglo-Saxon Dictionary: Supplement*, by Thomas Northcote Toller. 1908-1921. Oxford: Oxford University Press [reprint 1973]. *An Anglo-Saxon Dictionary: Enlarged Addenda and Corrigenda to the Supplement by Thomas N. Toller*, ed. by Alistair Campbell. 1972. Oxford: Oxford University Press.

Britten, James and Robert Holland. 1886. *Dictionary of English Plant Names*. London: Trübner & Company [reprint 2010, Nabu Press].

Brockhaus – Die Enzyklopädie. 1996. 24 vols. 20th ed. Leipzig: Brockhaus.

Cameron, Angus, Ashley Crandell Amos, Antonette diPaolo Healey et al., eds. 2007-. *Dictionary of Old English: A-G Online*. Toronto: University of Toronto. <http://www.doe.utoronto. ca/>.

Cheers, Gordon, ed. 1999. *Botanica. Das Abc der Pflanzen. 10.000 Arten in Text und Bild*. 3rd ed. Milson Point: Random House Australia.

Clark Hall, J.R. 1960. *A Concise Anglo-Saxon Dictionary*. 4th ed., with a Supplement by Herbert D. Meritt. Toronto: University of Toronto Press [first edition 1894].

Genaust, Helmut. 2005. *Etymologisches Wörterbuch der botanischen Pflanzennamen*. 3rd ed. Hamburg: Nikol.

Heizmann, Wilhelm. 1993. *Wörterbuch der Pflanzennamen im Altwestnordischen*. Berlin: Walter de Gruyter.

Holthausen, Ferdinand. 1934. *Altenglisches etymologisches Wörterbuch*. Heidelberg: Carl Winters Universitätsbuchhandlung.

Jack, George, ed. 1994. *Beowulf. A Student Edition*. Oxford: Oxford University Press.

Kluge, Friedrich. 2002. *Etymologisches Wörterbuch der deutschen Sprache*. Ed. Elmar Seebold. 4th ed. Berlin: de Gruyter.

Kurath, Hans, Sherman Kuhn and Robet E. Lewis, eds. 1952-2001. *Middle English Dictionary*. Ann Arbor: University of Michigan Press. [available online at *Middle English Compendium*: <http://quod.lib.umich.edu/m/mec/>].

Lass, Roger, Rhona Alcorn, Margaret Laing and Keith Williamson, eds. forthc. *A Corpus of Narrative Etymologies from Primitive Old English to Early Middle English and A Corpus of Changes*. Edinburgh: The University of Edinburgh.

Latham, Ronald E. 1965. *Revised Medieval Latin Word-List*. London: Oxford University Press for the British Academy.

Latham, Ronald E., David R. Howlett and Richard K. Ashdowne, eds. 1975-. *Dictionary of Medieval Latin from British Sources: A-Sal*. Oxford: Oxford University Press.

Lewis, Charlton T. and Charles Short. 1879. *A Latin Dictionary*. Oxford: Clarendon Press [reprint 1991].

Maltby, Robert. 1991. *A Lexicon of Ancient Latin Etymologies*. Classical and Medieval Texts, Papers and Monographs, 25. Leeds: Francis Cairns.

Marzell, Heinrich. 1943-1979. *Wörterbuch der deutschen Pflanzennamen*. 5 vols. Stuttgart: S. Hirzel [reprint 2000, Köln: Parkland].

Murray, James A. H., Henry Bradley, Sir William A. Craigie and Charles T. Onions, eds. 1884-1928. *The Oxford English Dictionary*. [formerly *A New English Dictionary on Historical Principles*]. 1933. *Supplement and Bibliography*. Robert W. Burchfield, ed. 1972-1986. *Supplement*. John A. Simpson and Edmund S.C. Weiner, eds. 1989. *The Oxford English Dictionary*. 2nd ed. John A. Simpson, Edmund S.C. Weiner, Michael Proffitt, eds. 1993-1997. *Additional Series*. John A. Simpson, ed. 2000-. *The Oxford English Dictionary*. 3rd ed. [in progress]. <www.oed.com>. Oxford: Oxford University Press.

Onions, Charles T. et al. 1966. *Oxford Dictionary of English Etymology*. Oxford: Clarendon Press.

Pokorny, Julius. 1959. *Indogermanisches Etymologisches Wörterbuch*. 2 vols. Bern: Francke.

Roberts, Jane, Christian Kay and Lynne Grundy. 1995. *A Thesaurus of Old English*. 2 vols. Amsterdam: Rodopi [2nd impr. 2000]. [available online at <http://libra. englang.arts.gla.ac.uk/oe thesaurus/>].

Soukhanov, Anne H. et al., eds. 1992. *The American Heritage Dictionary of the English Language*. 3rd ed. Boston: Houghton Mifflin Company.

De Vaan, Michiel. 2008. *Etymological Dictionary of Latin and the Other Italic Languges*. Leiden Indo-European Etymological Dictionary Series, 7. Leiden: Brill.

The Vocabulary of English Place-Names. Nottingham: Centre for English Place Names. *Fasc. 1: Á–Box*. David N. Parsons and Tania Styles, with Carole Hough. 1997; *Fasc. 2: Brace–Cæster*. David N. Parsons and Tania Styles. 2000; *Fasc. 3: Ceafor–Cock-pit*. David N. Parsons. 2004; Volumes on Co–D, M–O and P–R are in preparation.

GENERAL WORKS

Anderson, John. 2002. "On the Grammar of Names". <http://www.fb10.uni-bremen.de/linguistik/dpng/pdf/Anderson2002OnTheGrammarOfNames.pdf> [last accessed on 30.08.2012].

Anderson, John. 2007. *The Grammar of Names*. Oxford: Oxford University Press. [paperback edition 2008].

d'Aronco, Maria Amalia. 1988. "The Botanical Lexicon of the Old English *Herbarium*" in *Anglo-Saxon England* 17: 15-33.

d'Aronco, Maria Amalia. 2003. "Anglo-Saxon Plant Pharmacy and the Latin Medical Tradition" in Biggam (2003a, ed.), 133-51.

d'Aronco, Maria Amalia. 2007. "The Transmission of Medical Knowledge in Anglo-Saxon England: the Voices of the Manuscripts" in Lendinara, Lazzari and d'Aronco (eds.), 35-58.

d'Aronco, Maria Amalia. 2008. "Gardens on Vellum: Plants and Herbs in Anglo-Saxon Manuscripts" in Dendle and Touwaide (eds.), 101-27.

d'Aronco, Maria Amalia. 2009. "The Edition of the Old English Herbal and *Medicina de Quadrupedibus*: Two Case Studies" in Bierbaumer and Klug (eds.), 35-56.

van Arsdall, Anne. 2002. *Medieval Herbal Remedies. The* Old English Herbarium *and Anglo-Saxon Medicine*. London: Routledge.

van Arsdall, Anne, Helmut W. Klug and Paul Blanz. 2009. "The Mandrake Plant and its Legend: A New Perspective" in Bierbaumer and Klug (eds.), 285-346.

Ascoli, Graziadio. 1864. "Lingue e nazioni" in *Politecnico* 22: 77-100.

Bailey, Charles J. and Karl Maroldt. 1977. "The French lineage of English" in *Langues en Contact – Pidgins – Creoles*. Ed. Jürgen M. Meisel. Tübingen: Narr, 21-53.

Bally, Charles. 1944. *Linguistique générale et linguistique française*. Berne: Francke.

Bandle, Oskar, John Insley and Thorsten Andersson. 2005. "Siedlungsnamen" in *Reallexikon der Germanischen Altertumskunde*. vol. 28. Eds. Heinrich Beck, Dieter Geuenich and Heiko Steuer. Berlin: de Gruyter, 337-64.

Banham, Debby. 1990. "The Knowledge and Uses of Food Plants in Anglo-Saxon England". Cambridge: University of Cambridge, unpublished dissertation.

Banham, Debby. 2003. "Be *hlafum and wyrtum*: Food Plants in Anglo-Saxon Society and Economy" in Biggam (2003a, ed.), 119-31.

Bauer, Laurie. 1983. *English Word-Formation*. Cambridge: Cambridge University Press.

Bauer, Laurie. 2003. *Introducing Linguistic Morphology*. Washington, D.C.: Georgetown University Press.

Bauer, Laurie. 2006. "Compounds and Minor Word-Formation Types" in *The Handbook of English Linguistics*. Eds. Bas Aarts and April McMahon. Malden, MA: Blackwell Publishing, 483-506.

Bauer, Renate and Ulrike Krischke, eds. 2011a. *More Than Words: English Lexicography and Lexicology Past and Present. Essays Presented to Hans Sauer on the Occasion of his 65th Birthday, Part I.* Eds. Renate Bauer and Ulrike Krischke. Frankfurt a.M: Peter Lang.

Bauer, Renate and Ulrike Krischke, eds. 2011b. *Fact and Fiction: From the Middle Ages to Modern Times. Essays Presented to Hans Sauer on the Occasion of his 65th Birthday, Part II.* Frankfurt a.M: Peter Lang.

Baugh, Albert C. and Thomas Cable. 2002: *A History of the English Language.* 5th ed. London: Routledge [many reprints; last reprint 2010].

Bayrhuber, Horst, Ulrich Kull, Ulrich Bäßler and Albert Danzer, eds. 1989. *Linder Biologie.* 20th, revised ed. Stuttgart: Metzler.

Bender, Harold. 1934. "English 'Strawberry'" in *American Journal of Philology* 217: 71-4.

Berlin, Brent, Dennis E. Breedlove, and Peter H. Raven. 1973. "General Principles of Classification and Nomenclature in Folk Biology" in *American Anthropologist* 75 (1): 214-42.

Berlin, Brent, Dennis E. Breedlove, and Peter H. Raven. 1974. *Principles of Tzeltal Plant Classification: An Introduction to the Botanical Ethnography of a Mayan-Speaking People of Highland Chiapas.* New York: Academic Press.

Berschin, Walter. 1988. *Greek Letters and the Latin Middle Ages: From Jerome to Nicholas of Cusa.* Washington, D.C.: Catholic University of America Press.

Betz, Werner. 1949. *Deutsch und Lateinisch: Die Lehnbildungen der althochdeutschen Benediktinerregel.* Bonn: Bouvier.

Bezzo, Luisa. 2007. "Parallel Remedies: Old English «Paralisin Þæt Is Lyftald»" in Lendinara, Lazzari and d'Aronco (eds.), 435-45.

Bierbaumer, Peter. 1974. "Altenglisch *Fornetes Folm* – eine Orchideenart" in *Anglia* 92: 171-6.

Bierbaumer, Peter. 1977. "Zu J. V. Goughs Ausgabe einiger altenglischer Glossen" in *Anglia* 95: 115-21.

Bierbaumer, Peter. 1989. "Small Latin and Less Greek? Zur Kenntnis der Klassischen Sprachen in altenglischer Zeit" in *The History and the Dialects of English. Festschrift for Eduard Kolb.* Ed. Andreas Fischer. Heidelberg: Winter, 79-90.

Bierbaumer, Peter. 2003. "Real and Not So Real Plant-Names in Old English Glosses" in Biggam (2003a, ed.), 153-60.

Bierbaumer, Peter and Helmut W. Klug, eds. 2009. *Old Names - New Growth. Proceedings of the 2nd Symposium of the ASPNS (Anglo-Saxon Plant-Name Survey), 6th - 10th June 2007, University of Graz, Austria, and Related Essays.* Frankfurt a.M.: Peter Lang.

Bierbaumer, Peter, Hans Sauer, Helmut Klug and Ulrike Krischke. 2008. "Old English Plant Names Go Cyber: The Graz - Munich Dictionary Project", in *...Un Tuo Serto di Fiori in Man Recando: Scritti in Onore di Maria Amalia D'Aronco, a Cura di Silvana Serafin e Patrizia Lendinara.* Ed. Patrizia Lendinara. 2 vols., Udine: Forum, vol. 2: 43-62.

Biggam, Carole P. 1994. "*Hæwenhnydele*: An Anglo-Saxon Medicinal Plant" in *Botanical Journal of Scotland* 46.4: 617-22.

Biggam, Carole P. 1997. *Blue in Old English: An Interdisciplinary Semantic Study.* Costerus Essays in English and American Language and Literature N.S. 110. Amsterdam: Rodopi.

Biggam, Carole P. 1998. *Grey in Old English: An Interdisciplinary Semantic Study*. London: Runetree.

Biggam, Carole P., ed. 2003a. *From Earth to Art: The Many Aspects of the Plant-World in Anglo-Saxon England*. Amsterdam: Rodopi.

Biggam, Carole P. 2003b. "The *Æspe* Tree in Anglo-Saxon England" in Biggam (2003a, ed.), 195-230.

Biggam, Carole P. 2011. "The True Staff of Life: The Multiple Roles of Plants" in Clegg Hyer and Owen-Crocker (eds.), 23-48.

Birkhan, Helmut. 2012. *Pflanzen im Mittelalter. Eine Kulturgeschichte*. Wien, Köln, Weimar: Böhlau Verlag.

Björkman, Erik. 1901. "Die Pflanzennamen der althochdeutschen Glossen" in *Zeitschrift für Deutsche Wortforschung* 2: 202-33.

Björkman, Erik. 1902. "Die Pflanzennamen der althochdeutschen Glossen" in *Zeitschrift für Deutsche Wortforschung* 3: 263-307.

Björkman, Erik. 1904. "Die Pflanzennamen der althochdeutschen Glossen" in *Zeitschrift für Deutsche Wortforschung* 4: 174-98.

Blank, Andreas. 1997. *Prinzipien des lexikalischen Bedeutungswandels am Beispiel der romanischen Sprachen*. Tübingen: Max Niemeyer.

Blank, Andreas. 1998. "Kognitive italienische Wortbildungslehre" in *Italienische Studien* 19: 5-27.

Blank, Andreas. 1999. "Co-Presence and Succession: A Cognitive Typology of Metonymy" in *Metonymy in Language and Thought*. Eds. Klaus-Uwe Panther and Günter Radden. Amsterdam: John Benjamins, 168-91.

Blank, Andreas. 2001a. *Einführung in die lexikalische Semantik für Romanisten*. Tübingen: Max Niemeyer.

Blank, Andreas. 2001b. "Words and Concepts in Time: Towards Diachronic Cognitive Onomasiology" in *Metaphorik*. <www.metaphorik.de> [last accessed on 19.10. 2007].

Bloomfield, Leonard. 1933. *Language*. Delhi: Motilal Banarsidass [reprint 2005].

Blumbach, Wolfgang. 1974. "Studien zur Spirantisierung und Entspirantisierung altenglischer Konsonanten: 1. Labiale u. Tektale". Göttingen, unpublished dissertation.

Bonser, Wilfrid. 1963. *The Medical Background of the Anglo-Saxons: A Study in History, Psychology, and Folklore*. London: The Wellcome Historical Medical Library.

Booij, Geert. 2005. *The Grammar of Words: An Introduction to Morphology*. 2nd ed. Oxford: Oxford University Press.

Boretzky, Norbert. 1987. "Lexikalische Natürlichkeit: Benennungsmotive in Pflanzennamen" in *Beiträge zum 3. Essener Kolloquium über Sprachwandel und seine bestimmenden Faktoren vom 30.09-02.10.1987 an der Universität Essen*. Eds. Norbert Boretzky, Werner Enninger and Thomas Stolz. Bochum: Brockmeyer, 53-76.

Brunner, Karl. 1960-1962. *Die englische Sprache*. 2 vols. 2nd ed. Tübingen: Niemeyer.

Cameron, Angus. 1973. "A List of Old English Texts" in *A Plan for the Dictionary of Old English.* Eds. Roberta Frank and Angus Cameron. Toronto: University of Toronto Press, 25-306.

Cameron, Malcolm L. 1983a. "The Sources of Medical Knowledge in Anglo-Saxon England" in *Anglo-Saxon England* 11: 135-55.

Cameron, Malcolm L. 1983b. "Bald's *Leechbook*: Its Sources and their Use in its Compilation" in *Anglo-Saxon England* 12: 153-82.

Cameron, Malcolm L. 1988. "Anglo-Saxon Medicine and Magic" in *Anglo-Saxon England* 17: 191-215.

Cameron, Malcolm L. 1993. *Anglo-Saxon Medicine.* Cambridge Studies in Anglo-Saxon England, 7. Cambridge: Cambridge University Press.

Campbell, Alistair. 1959. *Old English Grammar.* Oxford: Clarendon Press [reprint 2003].

Carr, Charles. 1939. *Nominal Compounds in Germanic.* St. Andrews University Publications, 41. London: Humphrey Milford.

Clark, Cecily. 1992. "Onomastics" in Hogg (1992a, ed.), 452-89.

Clegg Hyer, Maren and Gale Owen-Crocker, eds. 2011. *The Material Culture of Daily Living in the Anglo-Saxon World.* Exeter: University of Exeter Press.

Clyne, Michael. 1991. *Community Languages: The Australian Experience.* Cambridge: Cambridge University Press.

Coseriu, Eugenio. 1981. "Los conceptos de 'dialecto', 'nivel' y 'estilo de lengua' y el sentido propio de la dialectologia" in *Lingüística española actual* 3: 1-32.

Cruse, Alan D. 1986. *Lexical Semantics.* Cambridge: Cambridge University Press.

Cruse, Alan D. 1988. "Word Meaning and Encyclopaedic Knowledge" in Hüllen and Schulze (eds.), 73-84.

Crystal, David. 1995. *Die Cambridge Enzyklopädie der Sprache.* Transl. and revised by Stefan Röhrich. Frankfurt am Main: Campus Verlag.

Danchev, Andrei. 1997. "The Middle English Creolization Hypothesis Revisited" in *Studies in Middle English Linguistics.* Ed. Jacek Fisiak. Berlin: de Gruyter, 79-108.

Dendle, Peter and Alain Touwaide, eds. 2008. *Health and Healing from the Medieval Garden.* Woodbridge: Boydell and Brewer.

Domingue, Nicole Z. (1977). "Middle English: Another Creole?" in *Journal of Creole Studies* 1: 89-100.

Durkin, Philip. 2009. *The Oxford Guide to Etymology.* Oxford: Oxford University Press.

Earle, John. 1880. *Plant Names from the Tenth to the Fifteenth Century.* Oxford: Clarendon Press.

Ebert, Günter, ed. 2005. *Die Schmetterlinge Baden-Württembergs.* Band 10. Stuttgart: Ulmer Verlag.

Ernst, Gerhard, Martin-Dietrich Glessgen, Christian Schmitt and Wolfgang Schweickard, eds. 2003. *Manuel international d'histoire linguistique de la Romania / Ein internationales Handbuch zur Geschichte der romanischen Sprachen.* Vol. 1. Berlin: de Gruyter.

Ernst, Gerhard, Martin-Dietrich Glessgen, Christian Schmitt and Wolfgang Schweickard, eds. 2006. *Manuel international d'histoire linguistique de la Romania / Ein internationales Handbuch zur Geschichte der romanischen Sprachen*. Vol. 2. Berlin: de Gruyter.

Faiss, Klaus. 1978. *Verdunkelte Compounds im Englischen. Ein Beitrag zu Theorie und Praxis der Wortbildung*. Tübinger Beiträge zur Linguistik, 104. Tübingen: TBL Verlag Gunter Narr.

Ferguson, Charles. 1959. "Diglossia", in *Word* 15: 325-40.

Fernie, Eric. 1983. *The Architecture of the Anglo-Saxons*. London: Batsford Ltd.

Feulner, Anna Helene. 2000. *Die griechischen Lehnwörter im Altenglischen*. Münchner Universitäts-Schriften. Texte und Untersuchungen zur englischen Philologie, 21. Frankfurt a. Main: Peter Lang.

Fillmore, Charles C. 1975. "An Alternative to Checklist Theories of Meaning" in *Proceedings of the Berkeley Linguistic Society*. Eds. Cathy Cogen, H. Thompson, G. Thurgood and K. Whistler. Berkely: Berkeley Linguistics Society, 123-31.

Firbas, Jan. 1992. *Functional Sentence Perspective in Written and Spoken Communication*. Cambridge: Cambridge University Press.

Fishman, Joshua. 1967. "Bilingualism with and without Diglossia; Diglossia with and without Bilingualism" in *Journal of Social Issues* 23 (2): 29-38.

Flydal, Leiv. 1951. "Remarques sur certains rapports entre le style et l'état de langue" in *Norsk Tidsskrift for Sprogvidenskap* 16: 240-57.

Gallée, Johan Hendrik. 1993. *Altsächsische Grammatik. Mit Berichtigungen und Literaturnachträgen von Heinrich Tiefenbach*. 3rd ed. Tübingen: Max Niemeyer Verlag.

Gardner, Thomas J. 1968. *Semantic Patterns in Old English Substantival Compounds*. Hamburg, privately printed.

Gärtner, Susanne. 2011. "When Scribe Meets Manuscript: Medieval Scribal Copies as Linguistic Contact Phenomena" in Bauer and Krischke (2011b, eds.), 101-25.

Geeraerts, Dirk. 2002. "The Scope of Diachronic Onomasiology" in *Das Wort. Seine strukturelle und kulturelle Dimension. Festschrift für Oskar Reichmann zum 65. Geburtstag*. Eds. Vilmos Agel, Andreas Gardt, Ulrike Hass-Zumkehr and Thorsten Roelcke. Tübingen: Niemeyer, 29-44 [available online at <http://wwwling.arts. kuleuven.ac.be/qlvl/PDFPublications/02Thescopeofdiachronic.pdf>].

Geeraerts, Dirk and Hubert Cuyckens, eds. 2007. *The Oxford Handbook of Cognitive Linguistics*. Oxford: Oxford University Press.

Glaze, Florence Eliza. 2007. "Master-Student Medical Dialogues: The Evidence of London, British Library, Sloane 2839" in Lendinara, Lazzari and d'Aronco (eds.), 467-94.

Gneuss, Helmut. 1955. *Lehnbildungen und Lehnbedeutungen im Altenglischen*. Berlin: Erich Schmidt Verlag.

Gneuss, Helmut. 1972. "The Origins of Standard Old English and Æthelwold's School at Winchester" in *Anglo-Saxon England* 1: 63-83.

Gneuss, Helmut. 1981. "A Preliminary List of Manuscripts Written or Owned in England up to 1100" in *Anglo-Saxon England* 9: 1-60.

Gneuss, Helmut. 1993. "*Anglicae linguae interpretatio*: Language Contact, Lexical Borrowing and Glossing in Anglo-Saxon England" in *Sir Gollancz Memorial Lecture 1992. Proceedings of the British Academy* 82: 107-48.

Gneuss, Helmut. 2001a. "Einleitung" in *Ælfrics Grammatik und Glossar: Text und Varianten*. Ed. Julius Zupitza. 3rd ed. Hildesheim: Weidmann, iii-xviii.

Gneuss, Helmut. 2001b. *Handlist of Anglo-Saxon Manuscripts. A List of Manuscripts and Manuscript Fragments Written or Owned in England up to 1100*. Medieval and Renaissance Texts and Studies, 241. Tempe: Arizona Center for Medieval and Renaissance Studies.

Gneuss, Helmut. 2003. "Addenda and Corrigenda to the *Handlist of Anglo-Saxon Manuscripts*" in *Anglo-Saxon England* 32: 239-305.

Gneuss, Helmut. 2011. "Second Addenda and Corrigenda to the *Handlist of Anglo-Saxon Manuscripts*" in *Anglo-Saxon England* 40: 293-306.

Görlach, Manfred. 1990. "Middle English – A Creole?" in *Studies in the History of the English Language*. Ed. Manfred Görlach. Heidelberg: Carl Winter, 65-78.

Greenbaum, Sidney and Randolph Quirk. 1990. *A Student's Grammar of the English Language*. Harlow: Longman Group UK Limited [8th impr. 1995].

Grendon, Felix. 1974. *Anglo-Saxon Charms*. Folcroft: Folcroft Library Editions.

Gretsch, Mechthild. 1973. *Die Regula Sancti Benedicti in England und ihre altenglische Übersetzung*. Münchner Universitäts-Schriften. Texte und Untersuchungen zur englischen Philologie, 2. München: W. Fink.

Gretsch, Mechthild. 1999a. "Æthelwold's Translation of the *Regula Sancti Benedicti* and its Latin Exemplar" in *Anglo-Saxon England* 3: 125-51.

Gretsch, Mechthild. 1999b. *The Intellectual Foundation of the English Benedictine Reform*. Cambridge: Cambridge University Press.

Groß, Elvira. 2002. *Pflanzennamen und ihre Bedeutung*. 2nd ed. Köln: DuMont Buchverlag.

Gross, Stephen. 2005. "Natural Kind Terms", in *Encyclopedia of Language and Linguistics*. <http://mysite.verizon.net/vze299db/sitebuildercontent/sitebuilderfiles/naturalkinds.pdf> [last accessed on 27.11.2012].

ten Hacken, Pius. 2000. "Derivation and Compounding" in *An International Handbook on Inflection and Word-Formation / Ein internationales Handbuch zur Flexion und Wortbildung*. Eds. Geert Booij, Christian Lehmann and Joachim Mugdan in collaboration with Wolfgang Kesselheim and Stavros Skopeteas. Berlin: de Gruyter, 349–60.

Hall, Allan R. 2003. "Investigating Anglo-Saxon Plant Life and Plant Use: the Archaeobotanical Angle" in Biggam (2003a, ed.), 101-18.

Hankins, Freda Richards. 1992. "Bald's 'Leechbook' Reconsidered". Ann Arbor: UMI Dissertation Services, unpublished dissertation.

Healey, Antonette diPaolo. 1997. "Words, Story, History: the Mapping of Meaning and Toronto's Dictionary of Old English" in *To Explain the Present: Studies in the Changing English Language in Honour of Matti Rissanen*. Eds. Terttu Nevalainen and Lena Kahlas-Tarkka. Helsinki: Société Néophilologique, 35-49.

Healey, Antonette diPaolo. 2009. "Perplexities About Plant Names in the Dictionary of Old English" in Bierbaumer and Klug (eds.), 99-120.

Herbst, Thomas, Rita Stoll and Rudolf Westermayr. 1991. *Terminologie der Sprachbeschreibung: Ein Lernwörterbuch für das Anglistikstudium*. Ismaning: Max Hueber.

Hofstetter, Walter. 1983. "Zur lateinischen Quelle des altenglischen Pseudo-Dioskurides" in *Anglia* 101: 315-60.

Hofstetter, Walter. 1987. *Winchester und der spätaltenglische Sprachgebrauch. Untersuchungen zur geographischen und zeitlichen Verbreitung altenglischer Synonyme*. Münchner Universitäts-Schriften. Texte und Untersuchungen zur englischen Philologie, 14. München: W. Fink.

Hogg, Richard M. 1992a. ed. *The Cambridge History of the English Language*. Vol. 1: *The Beginnings to 1066*. Ed. Richard M. Hogg. Cambridge: Cambridge University Press.

Hogg, Richard M. 1992b. "Introduction" in Hogg (1992a, ed.), 1-25.

Hogg, Richard M. 1992c. "Phonology and Morphology" in Hogg (1992a, ed.), 67-167.

Hollis, Stephanie and Michael Wright. 1992. *Old English Prose of Secular Learning*. Annotated Bibliographies of Old and Middle English Literature, 4. Cambridge: D.S. Brewer.

Hooke, Della. 2003. "Trees in the Anglo-Saxon Landscape: the Charter Evidence" in Biggam (2003a, ed.), 17-39.

Hooke, Della. 2009. "Trees in Anglo-Saxon Charters: Some Comments and Some Uncertainties" in Bierbaumer and Klug (eds.), 75-97.

Hoops, Johannes. 1889. *Über die altenglischen Pflanzennamen*. Freiburg im Breisgau: Chr. Lehmann.

Hoops, Johannes. 1898. "Meerrettich" in *Beiträge zur Geschichte der Deutschen Sprache und Literatur* (*PBB*) 23: 559-70.

Hoops, Johannes. 1903. "Alte *k*-Stämme unter den germanischen Baumnamen" in *Zeitschrift für Indogermanische Sprach- und Altertumskunde* 14: 478-85.

Hough, Carole. 2003. "Place-Name Evidence for Anglo-Saxon Plant Names" in Biggam (2003a, ed.), 41-78.

Hüllen, Werner. 1999. *English Dictionaries 800-1700: The Topical Tradition*. Oxford: Clarendon Press.

Hüllen, Werner and Rainer Schulze, eds. 1988. *Understanding the Lexicon: Meaning, Sense and World Knowledge in Lexical Semantics*. Linguistische Arbeiten, 210. Tübingen: Max Niemeyer.

Hunt, Tony. 1989. *Plant Names of Medieval England*. Cambridge: D.S. Brewer.

Hutterer, Claus Jürgen. 1975. *Die germanischen Sprachen. Ihre Geschichte in Grundzügen*. Budapest: Akadémiai Kiadó.

International Code of Botanical Nomenclature, or *Vienna Code*. 2005. <http://ibot.sav.sk/icbn/main.htm> [last accessed on 07.11.2009].

Isaac, Graham. 2003. "Diagnosing the Symptoms of Contact: Some Celtic-English Case Histories" in Tristram (2003, ed.), 46-64.

Jackendorff, Ray. 1975. "Morphological and Semantic Regularities in the Lexicon" in *Language* 51: 639-71.

Jacob Grimm. 1878. *Deutsche Mythologie*. 4th ed. by Elard H. Meyer. 3 vols. [first ed. 1844]. Berlin: Ferd. Dümmler [an English version is available online at <http://www.northvegr.org/lore/grimmst/index.php>].

Jespersen, Otto. 1942. *A Modern English Grammar on Historical Principles*. vol. 6: *Morphology*. London: George Allen [reprint 1954].

Johnston, Andrew J. 2011. "Calques and Culture: Revisiting an Issue in Old English Lexical Morphology" in Bauer and Krischke (2011a, eds.), 67-79.

Jolly, Karen Louise. 1996. *Popular Religion in Late Anglo-Saxon England: Elf Charms in Context*. Chapel Hill: University of North Carolina Press.

Jones, Daniel. 1956. "The hyphen as a phonetic sign" in *Daniel Jones. Selected Works*. Ed. Beverley Collins and Inger Mees. London: Routledge, 99-107.

Kastovsky, Dieter. 1968. *Old English Deverbal Substantives Derived by Means of a Zero Morpheme*. Esslingen: B. Langer.

Kastovsky, Dieter. 1982. *Wortbildung und Semantik*. Studienreihe Englisch, 14. Düsseldorf: Pädagogischer Verlag Schwann-Bagel GmbH.

Kastovsky, Dieter. 1988. "Structural Semantics or Prototype Semantics? The Evidence of Word-Formation" in Hüllen and Schulze (eds.), 190-203.

Kastovsky, Dieter. 1992. "Semantics and Vocabulary" in Hogg (1992a, ed.), 290-408.

Kastovsky, Dieter. 2005a. "The 'Invisible Hand', Drifts, and Typological Shifts. Examples from English" in *A Companion to Linguistics. A Festschrift for Anders Ahlqvist on the Occasion of his Sixtieth Birthday*. Eds. Bernardette Smelik, Rijcklof Hofman, Camiel Hamans and David Cram. Nijmegen: Stichting Uitgeverij de Keltische Draak, 286-95.

Kastovsky, Dieter. 2005b. "Hans Marchand and the Marchandeans" in *Handbook of Word-Formation*. Eds. Pavol Stekauer and Rochelle Lieber. Netherlands: Springer, 99-124.

Kastovsky, Dieter. 2006. "Typological Changes in Derivational Morphology" in van Kemenade and Los (eds.), 152-76.

Keller, Wolfgang. 1906. *Angelsächsische Palaeographie*. Berlin: Mayer und Müller.

van Kemenade, Ans and Bettelou Los. 2006. *The Handbook of the History of English*, Malden, MA: Blackwell.

Ker, Neil R. 1957. *Catalogue of Manuscripts Containing Anglo-Saxon*. Oxford: Clarendon Press [repr. 1990].

Klare, Johannes. 1998. *Französische Sprachgeschichte*. Stuttgart: Ernst Klett [reprint 2001].

Knappe, Gabriele. 2004. "*Greyhounds* Are Not *Grey*: On Folk-Etymological Change and Its Role in the History of English" in *Anglistentag 2003 München. Proceedings*. Eds. Christoph Bode, Sebastian Domsch and Hans Sauer. Wissenschaftlicher Verlag Trier: Trier, 491-505.

Koch, Peter. 1999a. "Frame and Contiguity: On the Cognitive Bases of Metonymy and Certain Types of Word-Formation" in *Metonymy in Language and Thought*. Eds. Klaus-Uwe Panther and Günter Radden. Amsterdam: John Benjamins, 139-67.

Koch, Peter. 1999b. "Tree and Fruit: A Cognitive-Onomasiological Approach" in *Studi di Linguistica Teorica e Applicata* 28 (2): 331-47.

Koch, Peter and Wulf Oesterreicher. 1985. "Sprache der Nähe – Sprache der Distanz. Mündlichkeit und Schriftlichkeit im Spannungsfeld von Sprachtheorie und Sprachgeschichte" in *Romanistisches Jahrbuch* 36: 15-43.

Koch, Peter and Wulf Oesterreicher. 1990. *Gesprochene Sprache in der Romania: Französisch, Italienisch, Spanisch*. Romanistische Arbeitshefte, 31. Tübingen: Niemeyer.

Koch, Peter and Wulf Oesterreicher. 2001. "Gesprochene Sprache und geschriebene Sprache / Langage parlé et langage écrit" in *Lexikon der Romanistischen Linguistik*. Eds. Günther Holtus, Michael Metzeltin and Christian Schmitt. 2 vols. Tübingen: Niemeyer, 1: 584-627.

Koch, Peter and Wulf Oesterreicher. 2007. "Schriftlichkeit und kommunikative Distanz" in *Zeitschrift für germanistische Linguistik* 35: 346–75.

Kornexl, Lucia. 2003. "'Unnatural Words'? Loan Formations in Old English Glosses" in *Language Contact in the History of English*. Eds. Dieter Kastovsky and Arthur Mettinger. 2nd ed. Frankfurt: a.M.: Lang, 195-216.

Kortmann, Bernd. 2005. *English Linguistics: Essentials*. Berlin: Cornelsen.

Koskela, Anu. 2005. "On the Distinction Between Metonymy and Vertical Polysemy in Encyclopaedic Semantics", online at *University of Sussex Working Papers in Linguistics and English Language LxWP 19/05*. <http://www.sussex.ac.uk/ linguistics/documents/ak_metonymy.pdf> [last accessed on 16.11.2009].

Koziol, Herbert. 1972. *Handbuch der englischen Wortbildungslehre*. 2nd ed. Heidelberg: Winter.

Krefeld, Thomas. 2003. "Methodische Grundlagen der Strataforschung" in Ernst, Glessgen, Schmitt and Schweickard (2003, eds.), 555-68.

Krischke, Ulrike. 2009. "On the Semantics of Old English Compound Plant Names: Motivations and Associations" in Bierbaumer and Klug (eds.), 211-78.

Krischke, Ulrike. 2011. "A Glimpse at Anglo-Saxon Biosystematics" in Bauer and Krischke (2011a, eds.), 81-101.

Labov, William. 1994. *Principles of Linguistic Change. Vol. 1: Internal Factors*. Language in Society. Oxford: Blackwell.

Laing, Margaret. 2004. "Multidimensionality: Time, Space, and Stratigraphy in Historical Dialectology" in *Methods and Data in English Historical Dialectology*. Eds. Marina Dossena and Roger Lass. Linguistic Insights: Studies in Language and Communication, 16. Bern: Lang, 49-96.

Lapidge, Michael et al., eds. 1999. *The Blackwell Encyclopaedia of Anglo-Saxon England*. Oxford: Blackwell Publishers Inc.

Lee, Christina. 2011. "Earth's Treasures: Food and Drink" in Clegg Hyer and Owen-Crocker (eds.), 142-56.

Leech, Geoffrey N. 1974. *Semantics*. Harmondsworth: Penguin.

Lehmann, Wolfgang. 1907. "Zu ae. *collon-croh*, ahd. *coller-wurz* 'nymphaea'" in *Zeitschrift für deutsche Wortforschung* 9: 23-6.

Lehnert, Martin. 1990. *Altenglisches Elementarbuch*. Sammlung Göschen 2210, 10th ed. Berlin: de Gruyter.

Leisi, Ernst and Christian Mair. 1999. *Das heutige Englisch: Wesenszüge und Probleme*. Sprachwissenschaftliche Studienbücher. 8th ed. Heidelberg: Universitätsverlag C. Winter.

Lendinara, Patrizia. 1999. *Anglo-Saxon Glosses and Glossaries*. Variorum Collected Studies Series. Aldershot: Ashgate.

Lendinara, Patrizia. 2007. "Instructional Manuscripts in England: the Tenth- and Eleventh-Century Codices and the Early Norman Ones" in Lendinara, Lazzari and d'Aronco (eds.), 59-113.

Lendinara, Patrizia, Loredana Lazzari and Maria Amalia D'Aronco, eds. 2007. *Form and Content of Instruction in Anglo-Saxon England in the Light of Contemporary Manuscript Evidence: Papers Presented at the International Conference Udine, 6-8 April 2006*. Turnhout: Brepols.

Lenker, Ursula. 2000. "The Monasteries of the Benedictine Reform and the 'Winchester School': Model Cases of Social Networks in Anglo-Saxon England?" in *Social Network Analysis and the History of English*. Ed. Ingrid Tieken-Boon van Ostade and Terttu Nevalainen. European Journal of English Studies 4.3, 225-38.

Levi, Judith N. 1978. *The Syntax and Semantics of Complex Nominals*. New York: Academic Press.

Lipka, Leonhard. 2002. *English Lexicology*. 3rd ed. Tübingen: Gunter Narr.

Lipp, Frank. 2002. *Kräuterheilkunde: Heilung und Harmonie, Symbolik, Rituale und Folklore, östliche und westliche Traditionen*. Köln: Taschen GmbH.

Lutz, Angelika. 2000. "When Did English Begin?" in *Sounds, Words, Texts and Change*. Eds. Theresa Fanego, Belen Mendez-Naya and Elena Seoane. Amsterdam: Benjamins, 145-71.

Lyons, John. 1977. *Semantics*. 2 vols. Cambridge: Cambridge University Press.

Maion, Danielle. 2007. "The Fortune of the So-Called *Practica Petrocelli Salernitani* in England: New Evidence and Some Consideration" in Lendinara, Lazzari and d'Aroncon (eds.), 495-512.

Marchand, Hans. 1969. *The Categories and Types of Present-Day English Word-Formation: A Synchronic-Diachronic Approach*. 2nd ed. München: C.H. Beck'sche Verlagsbuchhandlung.

Mayer, Erwin. 1962. "Sekundäre Motivation: Untersuchungen zur Volksetymologie und verwandten Erscheinungen im Englischen". Cologne, University of Cologne, unpublished dissertation.

McGurk, Patrick et al., eds. 1983. *An Eleventh Century Anglo-Saxon Illustrated Miscellany: British Library Cotton Tiberius B. V. Part I; Together with Leaves from British Library Cotton Nero D. II*. Copenhagen: Rosenkilde and Bagger.

McIntosh, Angus. 1973. "Word Geography in the Lexicography of Medieval England" in *Annals of the New York Academy of Sciences* 211: 55-66.

McWorther, John. 2002. "What Happened to English?" in *Diachronica* 19 (2): 207-72.

Meaney, Audrey. 1984. "Variant Versions of Old English Medical Remedies and the Compilation of Bald's *Leechbook*" in *Anglo-Saxon England* 13: 235-68.

Mengden, Ferdinand von. 2010. *Cardinal Numerals: Old English From a Cross-Linguistic Perspective*. Berlin u.a.: De Gruyter Mouton.

Mitchell, Bruce. 1985. *Old English Syntax*. 2 vols. Oxford: Clarendon Press.

Mitchell, Bruce and Fred Robinson. 2012. *A Guide to Old English.* 8th ed. Oxford: Blackwell.

Moldenke, Harold N. and Alma L. Moldenke. 1957. *Plants of the Bible.* Waltham: Chronica Botanica Co.

Morgan, Hollie. 2010. "Oxford, St. John's College, 17 + London, British Library, Nero C. vii, fols 80-84" in *The Production and Use of English Manuscripts 1060 to 1220.* Eds. Orietta Da Rold, Takako Kato, Mary Swan and Elaine Treharne. Leicester: Unversity of Leiceister. <http://www.le.ac.uk/english/em1060to1220/mss/CUL.li.1.33.htm> [last accessed 27 March 2013].

Mugglestone, Lynda, ed. 2008. *The Oxford History of English.* Revised paperback edition [first ed. 2006]. Oxford: Oxford University Press.

Neville, Jennifer. 2003. "Leaves of Glass: Plant-Life in Old English Poetry" in Biggam (2003a, ed.), 287-300.

Ogden, Charles Kay and Ivor Armstrong Richards. 1956. *The Meaning of Meaning.* 10th ed. London: Lund Humphries.

Ohlgren, Thomas H., ed. 1986. *Insular and Anglo-Saxon Illuminated Manuscripts. An Iconographic Catalogue, c. A.D. 625 to 1100.* New York & London: Garland Publishing Inc.

Olschansky, Heike. 1996. *Volksetymologie.* Reihe Germanistische Linguistik, 175. Tübingen: Max Niemeyer Verlag.

Pavord, Anna. 2005. *The Naming of Names: The Search for Order in the World of Plants.* London: Bloomsbury Publishing.

Payne, Joseph Frank. 1904. *The Fritz-Patrick Lectures for 1903: English Medicine in the Anglo-Saxon Times.* Oxford: Clarendon Press.

Plag, Ingo. 2003. *Word-Formation in English.* Cambridge: Cambridge University Press.

Pogatscher, Alois. 1888. *Zur Lautlehre der Griechischen, Lateinischen und Romanischen Lehnworte im Altenglischen.* Strassburg: K. J. Trübner.

Pollington, Stephen. 2000. *Leechcraft: Early English Charms, Plant Lore and Healing.* Frithgarth: Anglo-Saxon Books [repr. 2003].

Porter, David W. 2011b. "The Antwerp-London Glossaries and the First School Text" in *Rethinking and Recontextualizing Glosses. New Perspectives in the Study of Late Anglo-Saxon Glossography.* Eds. Patrizia Lendinara, Loredana Lazzari and Claudia Di Sciaccia. Textes et Études du Moyen Age, 54. Porto: Fédération Internationale des Instituts d'Études Médiévales.

Poussa, Patricia. 1982. "The Evolution of Early Standard English: The Creolization Hypothesis" in *Studia Anglica Posnaniensia* 14: 69-85.

Quirk, Randolph, Sidney Greenbaum, Geoffrey Leech and Jan Svartvik. 1972. *A Grammar of Contemporary English.* London: Longman [8th impr. 1979].

Quirk, Randolph, Sidney Greenbaum, Geoffrey N. Leech and Jan Svartvik. 1985. *A Comprehensive Grammar of the English Language.* London: Longman.

Raible, Wolfgang. 1983. "Zur Einleitung" in *Zur Semantik des Französischen.* Eds. Helmut Stimm and Wolfgang Raible. Wiesbaden: Steiner, 1-24.

Rausch, Andrea and Brigitte Lotz. 2003. *Dumonts Kleines Kräuterlexikon.* 4th ed. Köln: DuMontmonte Verlag.

Reibel, David A. 1963. "A Grammatical Index to the Compound Nouns of Old English Verse". Indiana University, unpublished dissertation.

Riddle, John. 1974. "Theory and Practice in Medieval Medicine" in *Viator: Medieval and Renaissance Studies* V: 157-84.

Riehl, Claudia. 2004. *Sprachkontaktforschung. Eine Einführung*. Tübingen: Gunter Narr.

Rigg, A.G. and Gernot R. Wieland. 1975. "A Canterbury Classbook of the Mid-Eleventh Century (the 'Cambridge Songs' Manuscript)" in *Anglo-Saxon England*, 4: 120-9.

Roudet, Léonce. 1921. "Sur la classification psychologique des changements sémantiques" in *Journal de Psychologie* 18: 676-92.

Rubin, Stanley. 1974. *Medieval English Medicine*. London: David & Charles Limited.

Rusche, Philip G., ed. 1996. "The Cleopatra Glossaries: An Edition with Commentary on the Glosses and their Sources". Yale University, unpublished dissertation.

Rusche, Philip G. 2003. "Dioscorides' *De materia medica* and Late Old English Herbal Glossaries" in Biggam (2003a, ed.), 181-94.

Rusche, Philip G. 2008. "The Sources for Plant Names in Anglo-Saxon England and the Laud Herbal Glossary" in Dendle and Touwaide (eds.), 128-44.

Rusche, Philip. forthc. *The Cleopatra Glossaries*.

Rydén, Mats. 1981. "English Plant Name Research" in *Papers from the First Nordic Conference for English Studies*. Eds. Stig Johannson and Björn Tysdal, Universitetet i Oslo, 374-84.

Saenger, Paul. 1997. *Space between Words: The Origins of Silent Reading*. Stanford: Stanford University Press.

Sand, Andrea. 2003. "The Definite Article in Irish English and Other Contact Varieties of English" in Tristram (2003, ed.), 413-30.

Sandmann, Gert. 1975. "Studien zu altenglischen Zaubersprüchen". Münster, unpublished dissertation.

Sauer, Hans. 1992a. *Nominalkomposita im Frühmittelenglischen: Mit Ausblicken auf die Geschichte der englischen Nominalkomposition*. Tübingen: Max Niemeyer.

Sauer, Hans. 1992b. "A Linguistic Description of the Old English Plant Names" in *Words, Texts and Manuscripts: Studies in Anglo-Saxon Culture Presented to Helmut Gneuss on the Occasion of his Sixty-Fifth Birthday*. Ed. Michael Korhammer. Cambridge: D.S. Brewer, 381-408.

Sauer, Hans. 1995. "On the Analysis and Structure of Old and Middle English Plant Names" in *Papers from the VII International Conference of the Spanish Society for Medieval English Language & Literature*. Eds. Bernardo Santano Moreno, Adrian R. Birtwistle and Luis G. Girón Echevarría. Cáceres: Universidad de Extremadura, 299-325.

Sauer, Hans. 1996. "English Plant Names in the Thirteenth Century: The Trilingual Harley Vocabulary" in *Middle English Miscellany. From Vocabulary to Linguistic Variation*. Ed. Jacek Fisiak. Poznán: Motivex, 135-55.

Sauer, Hans. 2000. "The Earliest Layer of English Word-Formation: A Sketch of Word-Formation Patterns in the *Épinal-Erfurt Glossary*" in *Proceedings of the*

Conference of the German Association of University Teachers of English. Eds. Bernhard Reitz and Sigrid Rieuwerts. Trier: Wissenschaftlicher Verlag Trier, 77-90.

Sauer, Hans. 2001. "The Old English Suffix *–el/-il/-ol/-ul/-l* (> ModE *–le*, cf. *beetle, girdle, thistle*) as Attested in the *Épinal-Erfurt Glossary*" in *Innovation and Continuity in English Studies*. Ed. H. Grabes. Frankfurt am Main: Peter Lang, 289-313.

Sauer, Hans. 2003. "The Morphology of the Old English Plant-Names" in Biggam (2003a, ed.), 161-79.

Sauer, Hans. 2008. "Archbishops, Lords, and Concubines: Words for People and their Word-Formation Patterns in Early English (*Épinal-Erfurt Glossary* and Ælfric's *Glossary*) - A Sketch" in *Intertexts: Studies in Anglo-Saxon Culture Presented to Paul E. Szarmach*. Eds. Virginia Blanton and Helene Scheck. Medieval and Renaissance Texts and Studies, 334. Tempe: Arizona Center for Medieval and Renaissance Studies, 381-409.

Sauer, Hans. 2009. "Glosses, Glossaries and Dictionaries in the Medieval Period" in *The Oxford History of English Lexicography*. 2 vols. Ed. Anthony Cowie. Oxford: Clarendon Press, 1: 17-40.

Sauer, Hans. 2011. "Patterns of Loan-Influence on the Medieval English Plant Names, with Special Reference to the Influence of Greek" in *Foreign Influences on Medieval English*. Eds. Jacek Fisiak and Magdalena Bator. Frankfurt am Main: Peter Lang, 55-76.

Sauer, Hans. 2012. "Old English Plant Names with Suffixes, especially the Suffix *–el*" in *Explorations in the English Language: Middle Ages and Beyond. Festschrift for Professor Jerzy Welna on the Occasion of his 70th Birthday*. Eds. Joanna Esquibel and Wojtys Anna. Studies in English Medieval Language and Literature, 35. Frankfurt am Main: Peter Lang, 213-28.

Sauer, Hans and Ulrike Krischke. 2004. "Die altenglischen Pflanzennamen aus linguistischer und lexikographischer Sicht" in *Sudhoffs Archiv* 88: 175-208.

Sauer, Hans and Ulrike Krischke. 2009. "The Dictionary of Old English Plant Names (*DOEPN*), or: The Graz - Munich Dictionary Project" in Bierbaumer and Klug (eds.), 145-80.

Sauerhoff, Friedhelm. 2001. *Pflanzennamen im Vergleich. Studien zur Benennungstheorie und Etymologie*. Stuttgart: Steiner.

de Saussure, Ferdinand. 1916. *Cours de linguistique générale*. Eds. Charles Bally and Albert Sechehaye. 1983. *Course in General Linguistics*. Transl. and ed. by Roy Harris. London: G. Duckworth.

Scheler, Manfred. 1977. *Der englische Wortschatz*. Grundlagen der Anglistik und Amerikanistik, 9. Berlin: Erich Schmidt Verlag.

Schmid, Hans-Jörg. 2005. *Englische Morphologie und Wortbildung*. Berlin: Erich Schmidt.

Schneider, Edgar W. 2007. *Postcolonial English. Varieties Around the World*. Cambridge: Cambridge University Press.

Schrijver, Peter. 1995. *Studies in British Celtic Historical Phonology*. Amsterdam: Rodopi.

Schrijver, Peter. 2009. "Celtic Influence on Old English: Phonological and Phonetic Evidence" in *English Language and Linguistics* 13: 193-211.

Schuchardt, Hugo. 1976. *Hugo-Schuchardt-Brevier: ein Vademecum der allgemeinen Sprachwissenschaft*. 2nd ed. Darmstadt: Wissenschaftliche Buchgesellschaft [1st ed. 1890].

Schulz, Monika. 2000. *Magie oder Die Wiederherstellung der Ordnung*. Frankfurt: Peter Lang.

Schulz, Monika. 2003. *Beschwörungen im Mittelalter: Einführung und Überblick*. Heidelberg: Carl Winter.

Schurz, Gerhard. 2011. *Einführung in die Wissenschaftstheorie*. 3rd ed. Darmstadt: Wissenschaftliche Buchgesellschaft [1st ed. 2006].

Serjeantson, Mary S. 1935. *A History of Foreign Words in English*. London: Routledge and Kegan Paul [3rd impr. 1962].

Seuren, Pieter. 1988. "Lexical Meaning and Presupposition" in Hüllen and Schulze (eds.), 190-203.

Singer, Charles. 1927. "The Herbal in Antiquity" in *The Journal of Hellenic Studies* 47: 1-52.

Singer, Charles. 1928. *From Magic to Science: Essays on the Scientific Twilight*. New York: Dover Publications [reprint 1958].

Siraisi, Nancy G. 1990. *Medieval & Early Renaissance Medicine: An Introduction to Knowledge and Practice*. Chicago: The University of Chicago Press.

Skeat, Walter W. 1896. "Anglo-Saxon Plant Names", in *Notes and Queries* 9: 163-4.

Smith, A.H. 1956. *English Place-Name Elements*. 2 vols. Publications of the English Place-Name Society, 25-6. Cambridge: Cambridge University Press.

Štekauer, Pavol. 2000. *English Word-Formation: A History of Research (1960-1995)*. Tübingen: Gunter Narr.

Storms, Godfrid. 1948. *Anglo-Saxon Magic*. The Hague: Martinus Nijhoff [reprint 1974].

Strang, Barbara. 1970. *A History of English*. London: Methuen.

Talbot, Charles H. 1967. *Medicine in Medieval England*. London: Oldbourne.

Thomason, Sarah Grey. 2001. *Language Contact. An Introduction*. Edinburgh: Edinburgh University Press.

Thomason, Sarah Grey and Terrence Kaufman. 1988. *Language Contact, Creolization and Genetic Linguistics*. Berkeley: University of California Press.

Tournier, Jean. 1988. *Précis de Lexicologie Anglaise*. Paris: Éditions Nathan.

Townend, Matthew. 2008. "Contacts and Conflicts: Latin, Norse, and French" in Mugglestone (ed.), 60-85.

Trier, Jost. 1931. *Der deutsche Wortschatz im Sinnbezirk des Verstandes. Die Geschichte eines sprachlichen Feldes*. Heidelberg: Winter.

Tristram, Hildegard L.C., ed. 1997. *The Celtic Englishes I*, Heidelberg: Carl Winter.

Tristram, Hildegard L.C., ed. 2000. *The Celtic Englishes II*, Heidelberg: Carl Winter.

Tristram, Hildegard L.C., ed. 2003. *The Celtic Englishes III*, Heidelberg: Winter Universitätsverlag.

Tristram, Hildegard L.C., ed. 2006. *The Celtic Englishes IV*, Potsdam: Potsdam University Press.

Ullmann, Stephen. 1962. *Semantics: An Introduction to the Science of Meaning.* Oxford: Basil Blackwell [many reprints].

Ungerer, Friedrich and Hans-Jörg Schmid. 2006. *Introduction to Cognitive Linguistics.* 2nd ed. Harlow: Longman.

Ungerer, Friedrich, Gerhard E.H. Meier, Klaus Schäfer and Shirley B. Lechner. 1984. *Grammatik des heutigen Englisch.* Stuttgart: Ernst Klett Schulbuchverlag.

Valkhoff, Marius. 1932. *Latijn, Romaans, Roemeens. Openbare les gegeven bij de opening van zijn kolleges als privaat-docent in de roemense taal en letterkunde aan de Universiteit van Amsterdam op I November 1932 door Amersfoort.* Amersfoort.

Vennemann, Theo. 2004. "Sprachgeburt und Sprachkontakt: Die Entstehung des Englischen" in *Sprachtod und Sprachgeburt.* Ed. Peter Shrijver. Bremen: Heupen, 21-55.

Viereck, Wolfgang. 2000. "Celtic and English - An Intricate Interrelationship" in Tristram (ed.), 337-98.

Voigts, Linda Ehrsam. 1979. "Anglo-Saxon Plant Remedies and the Anglo-Saxons" in *Isis* 70: 250-68.

Waldorf, Norman O. 1953. "The Hapax Legomena in the Old English Vocabulary: A Study Based Upon the Bosworth-Toller Dictionary". Ann Arbor: University Microfilms.

Wallace, M. J. 1979. "What is an Idiom? An Applied Linguistic Approach" in *Dictionaries and Their Users: Papers from the 1978 B.A.A.L. Seminar on Lexicography.* Ed. R. R.K. Hartmann. Exeter: University of Exeter, 63-70.

Ward, Bobby J. 1999. *A Contemplation Upon Flowers: Garden Plants in Myth and Literature.* Portland: Timber Press.

von Wartburg, Walter. 1936. "Die Ausgliederung der romanischen Sprachräume" in *Zeitschrift für romanische Philologie* 56: 1-48.

Waxenberger, Gabriele. 1996. *Die Zuordnung der altenglischen Substantive zu den Flexionstypen untersucht am Buchstaben D.* Europäische Hochschulschriften, 14, Angelsächsische Sprache und Literatur, 315. Frankfurt: Lang.

Weinstock, Horst. 2006. "Romanismen im Englischen / Romanismes en anglais" in Ernst, Glessgen, Schmitt and Schweickard (2006, eds.), 1834-42.

White, David L. 2003. "Brittonic Influence in the Reductions of Middle English Nominal Morphology" in Tristram (2003, ed.), 29-45.

Winford, Donald. 2003. *An Introduction to Contact Linguistics*, Oxford: Blackwell.

Wollmann, Alfred. 1990. *Untersuchungen zu den frühen Lateinischen Lehnwörtern im Altenglischen: Phonologie und Datierung.* Münchner Universitäts-Schriften. Texte und Untersuchungen zur englischen Philologie, 15. München: W. Fink.

Wunderli, Peter. 1989. *Französische Lexikologie. Einführung in die Theorie und Geschichte des französischen Wortschatzes.* Tübingen: Niemeyer.

Yule, George. 2006. *The Study of Language.* 3rd ed. Cambridge: Cambridge University Press.

INDEXES

INDEX OF WORD FORMS

This index lists the words and word forms necessary to explain the etymological, morphological or semantic history of the Old English plant names. It does not include the German and Modern English plant names that provide the denotational meaning of the Old English plant names.
Superscript fn refers to the footnoote(s) on the respective page.

MIDDLE HIGH GERMAN

OLD HIGH GERMAN

LOW GERMAN

MIDDLE LOW GERMAN

OLD NORSE, OLD WEST NORSE

INDEX OF BOTANICAL NAMES

Superscript fn refers to the footnoote(s) on the respective page.

INDEX OF SUBJECTS

The index of subjects relates to part A of the present work. Superscript [fn] refers to the footnoote(s) on the respective page.

MÜNCHENER UNIVERSITÄTSSCHRIFTEN
Texte und Untersuchungen zur Englischen Philologie

Herausgegeben von Christoph Bode, Helmut Gneuss und Hans Sauer

Die Bände 1-20 sind im Wilhelm Fink Verlag München erschienen und sind leider alle vergriffen.

Band 1 Karl Reichl: Religiöse Dichtung im englischen Hochmittelalter.

Band 2 Mechthild Gretsch: Die *Regula Sancti Benedicti* in England und ihre altenglische Übersetzung.

Band 3 Manfred Pfister: Studien zum Wandel der Perspektivenstruktur in elisabethanischen und jakobäischen Komödien.

Band 4 Gisela Guddat-Figge: Catalogue of Manuscripts Containing Middle English Romances.

Band 5 Ingrid Hantsch: Semiotik des Erzählens.

Band 6 Michael Korhammer: Die monastischen Cantica im Mittelalter und ihre altenglischen Interlinearversionen.

Band 7 Roland Torkar: Eine altenglische Übersetzung von Alcuins *De Virtutibus et Vitiis*, Kap. 20 (Liebermanns Judex).

Band 8 Hans Sauer: Theodulfi Capitula in England.

Band 9 Ina Schabert, Werner Habicht (Hrsg.): Sympathielenkung in den Dramen Shakespeares.

Band 10 Evelyn Hentschel: Die dramatische Funktion der Requisiten bei Shakespeare.

Band 11 Angelika Lutz: Die Version G der Angelsächsischen Chronik.

Band 12 Manfred Pfister (Hrsg.): Alternative Welten.

Band 13 Walter Kluge: Der Stil der dramatischen Sprache in den Stücken George Bernard Shaws.

Band 14 Walter Hofstetter: Winchester und der spätaltenglische Sprachgebrauch.

Band 15 Alfred Wollmann: Untersuchungen zu den frühen lateinischen Lehnwörtern im Altenglischen.

Band 16 Andreas Mahler: Moderne Satireforschung und elisabethanische Verssatire.

Band 17 Lucia Kornexl: Die *Regularis Concordia* und ihre altenglische Interlinearversion.

Band 18 Sonja Fielitz: Jakob Gretser, *Timon. Comoedia Imitata* (1584). Erstausgabe von Gretsers *Timon* – Drama mit Übersetzung und einer Erörterung von dessen Stellung zu Shakespeares *Timon of Athens*.

Band 19 Martin Richter: Die altenglischen Glossen zu Aldhelms *De laudibus virginitatis* in der Handschrift BL, Royal 6 B. VII.

Band 20 Ursula Lenker: Die westsächsische Evangelienversion und die Perikopenordnungen im angelsächsischen England.

Ab Band 21 wird die Reihe im Peter Lang Verlag fortgesetzt.

Band 21 Anna Helene Feulner: Die griechischen Lehnwörter im Altenglischen. 2000.

Band 22 Sonja Fielitz: Wit, Passion and Tenderness. Ovids *Metamorphosen* im Wandel der Diskurse in England zwischen 1660 und 1800. 2000.

Band 23 Sabine Gieszinger: The History of Advertising Language. The Advertisements in *The Times* from 1788 to 1996. 2001.

Band 24 Svenja Weidinger: Revision der Empfindsamkeit. Samuel Richardsons Fortschreibungen von *Pamela* (1739-1761). 2002.

Band 25 Carolin Schreiber: King Alfred's Old English Translation of Pope Gregory the Great's *Regula pastoralis* and its Cultural Context. A Study and Partial Edition According to All Surviving Manuscripts Based on Cambridge, Corpus Christi College 12. 2003.

Band 26 Brigitte Langefeld: The Old English Version of the Enlarged Rule of Chrodegang. Edited together with the Latin Text and an English Translation. 2003.

Band 27 Berit Bettina Schubert: [*Enter* SHAKESPEARE.]. Der Dramatiker als Figur im modernen Drama. 2003.

Band 28 Ursula Kalbhen: Kentische Glossen und kentischer Dialekt im Altenglischen. Mit einer kommentierten Edition der altenglischen Glossen in der Handschrift London, British Library, Cotton Vespasian D.vi. 2003.

Band 29 Renate Bauer: *Adversus Judaeos*. Juden und Judentum im Spiegel alt- und mittelenglischer Texte. 2003.

Band 30 Lucia Kornexl / Ursula Lenker (eds.): Bookmarks from the Past. Studies in Early English Language and Literature in Honour of Helmut Gneuss. 2003.

Band 31 Katja Kraushaar: Englische Elegien. 2004.

Band 32 Oliver M. Traxel: Language Change, Writing and Textual Interference in Post-Conquest Old English Manuscripts.The Evidence of Cambridge, University Library, Ii. 1.33. 2004.

Band 33 Claudia Küster: Give me some music.... Shakespeare an der Bayerischen Staatsoper München. Auftragswerke und Erstaufführungen im Vergleich mit ihren literarischen Vorlagen. 2005.

Band 34 Matthias Galler: *O Death, thou comest when I had thee least in mind!* Der Umgang mit dem Tod in der mittelenglischen Literatur. 2007.

Band 35 Christine Elsweiler: Laȝamon's *Brut* between Old English Heroic Poetry and Middle English Romance. A Study of the Lexical Fields 'Hero', 'Warrior' and 'Knight'. 2011.

Band 36 Renate Bauer/Ulrike Krischke (eds.): More than Words. English Lexicography and Lexicology Past and Present. Essays Presented to Hans Sauer on the Occasion of his 65th Birthday – Part I. 2011.

Band 37 Renate Bauer/Ulrike Krischke (eds.): Fact and Fiction. From the Middle Ages to Modern Times. Essays Presented to Hans Sauer on the Occasion of his 65th Birthday – Part II. 2011.

Band 38 Ulrike Schenk: Die mittelenglische Romanze zwischen Imitation und Innovation. Zur Dynamik einer Diskurstradition. 2013.

Band 39 Ulrike Krischke: The Old English Complex Plant Names: A Linguistic Survey and a Catalogue. 2013.